PLANT BIOCHEMISTRY

Jones & Bartlett Learning Titles in Biological Science

Plant Structure: A Colour Guide
Brian G. Bowes, James D. Mauseth

Lewin's CELLS, Second Edition
Lynne Cassimeris
George Plopper
Vishwanath R. Lingappa

Environmental Science, Eighth Edition
Daniel D. Chiras

Human Biology, Seventh Edition
Daniel D. Chiras

Plants, Genes, and Crop Biotechnology
Maarten J. Chrispeels
David E. Sadava

Restoration Ecology
Sigurdur Greipsson

Evolution: Principles and Processes
Brian K. Hall

Strickberger's Evolution, Fourth Edition
Brian K. Hall
Benedikt Hallgrímsson

Essential Genetics: A Genomics Perspective, Fifth Edition
Daniel L. Hartl

Genetics: Analysis of Genes and Genomes, Seventh Edition
Daniel L. Hartl
Elizabeth W. Jones

Genetics of Populations, Fourth Edition
Philip W. Hedrick

Lewin's Essential GENES, Second Edition
Jocelyn E. Krebs
Elliott S. Goldstein
Stephen T. Kilpatrick

Lewin's GENES X
Jocelyn E. Krebs
Elliott S. Goldstein
Stephen T. Kilpatrick

Tropical Forests
Bernard A. Marcus

Botany: An Introduction to Plant Biology, Fourth Edition
James D. Mauseth

Environmental Science, Fourth Edition
Michael L. McKinney
Robert M. Schoch
Logan Yonavjak

RNA Interference and Model Organisms
Joanna A. Miller

Introduction to the Biology of Marine Life, Tenth Edition
John F. Morrissey
James L. Sumich

Exploring Bioinformatics: A Project-Based Approach
Caroline St. Clair
Jonathan E. Visick

Molecular Biology: Genes to Proteins, Fourth Edition
Burton E. Tropp

The Ecology of Agroecosystems
John H. Vandermeer

Mammalogy, Fifth Edition
Terry A. Vaughan
James M. Ryan
Nicholas J. Czaplewski

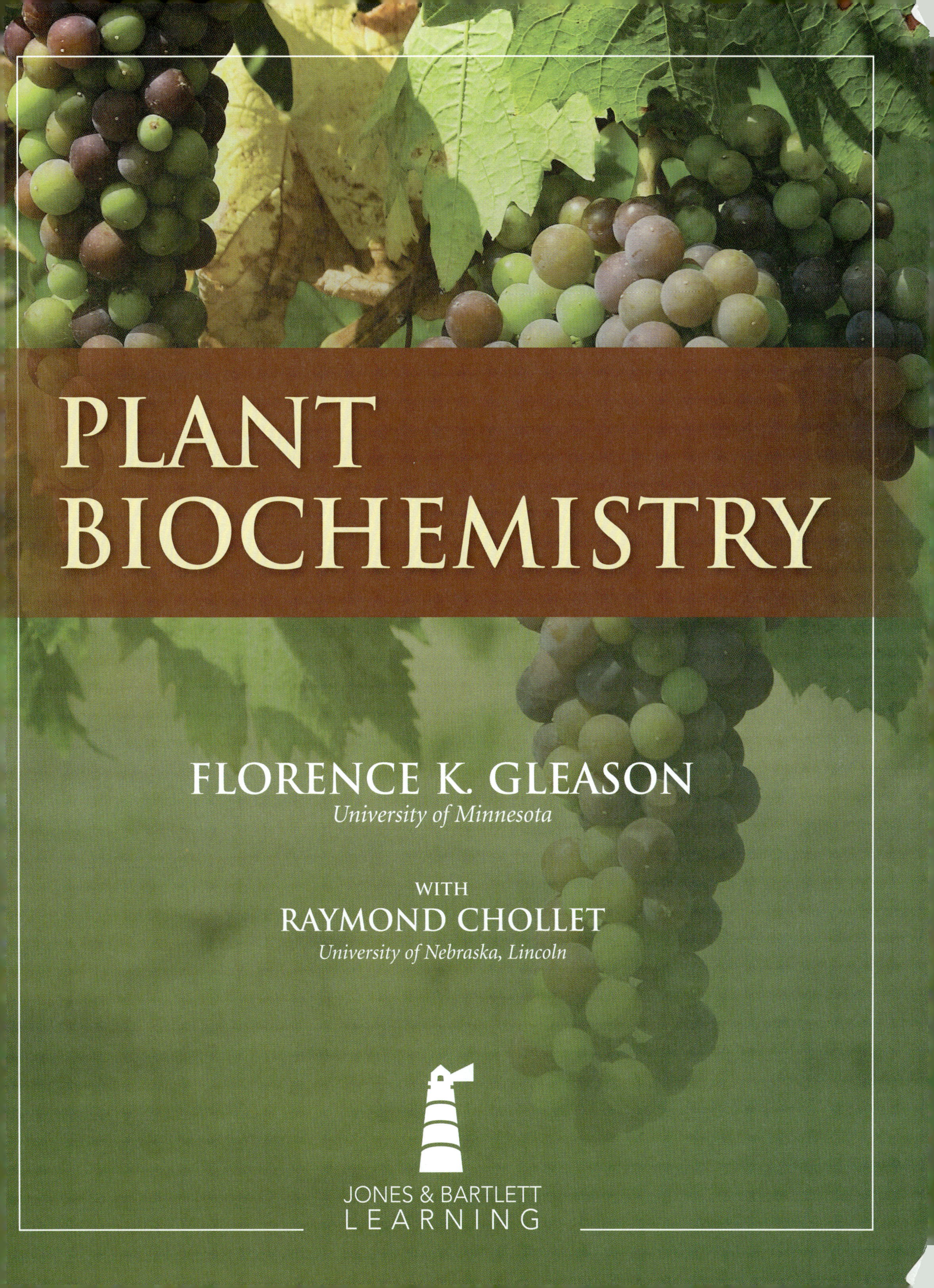

PLANT BIOCHEMISTRY

FLORENCE K. GLEASON
University of Minnesota

WITH
RAYMOND CHOLLET
University of Nebraska, Lincoln

JONES & BARTLETT
LEARNING

World Headquarters
Jones & Bartlett Learning
40 Tall Pine Drive
Sudbury, MA 01776
978-443-5000
info@jblearning.com
www.jblearning.com

Jones & Bartlett Learning Canada
6339 Ormindale Way
Mississauga, Ontario L5V 1J2
Canada

Jones & Bartlett Learning International
Barb House, Barb Mews
London W6 7PA
United Kingdom

Jones & Bartlett Learning books and products are available through most bookstores and online booksellers. To contact Jones & Bartlett Learning directly, call 800-832-0034, fax 978-443-8000, or visit our website, www.jblearning.com.

Substantial discounts on bulk quantities of Jones & Bartlett Learning publications are available to corporations, professional associations, and other qualified organizations. For details and specific discount information, contact the special sales department at Jones & Bartlett Learning via the above contact information or send an email to specialsales@jblearning.com.

Production Credits
Chief Executive Officer: Ty Field
President: James Homer
SVP, Chief Operating Officer: Don Jones, Jr.
SVP, Chief Technology Officer: Dean Fossella
SVP, Chief Marketing Officer: Alison M. Pendergast
SVP, Chief Financial Officer: Ruth Siporin
Publisher, Higher Education: Cathleen Sether
Acquisitions Editor: Molly Steinbach
Senior Associate Editor: Megan R. Turner
Editorial Assistant: Rachel Isaacs
Production Manager: Louis C. Bruno, Jr.
Senior Marketing Manager: Andrea DeFronzo
V.P., Manufacturing and Inventory Control: Therese Connell
Illustrations: CAE Solutions Corporation
Composition: CAE Solutions Corporation
Cover Design: Kristin E. Parker
Associate Artist and Photo Researcher: Carolyn Arcabascio
Cover Image: © Nataliya Ostapenko/ShutterStock, Inc.
Printing and Binding: Courier Kendallville
Cover Printing: Courier Kendallville

Library of Congress Cataloging-in-Publication Data
Gleason, Florence K.
Plant biochemistry / Florence K. Gleason and Raymond Chollet. — 1st ed.
p. ; cm.
Includes bibliographical referencs and index.
ISBN 978-0-7637-6401-2 (alk. paper)
1. Botanical chemistry. I. Chollet, Raymond, 1946- II. Title.
[DNLM: 1. Plant Physiological Phenomena. 2. Biochemistry. 3. Molecular Biology. 4. Plants—chemistry. 5. Plants—metabolism. QK 861]
QK861.G64 2012
572'.2—dc22

2010041570

6048

Printed in the United States of America
15 14 13 12 11 10 9 8 7 6 5 4 3 2 1

BRIEF CONTENTS

CONTENTS

PREFACE

The seeds for this book were planted approximately 20 years ago when I taught my first course in plant biochemistry. Although there were several excellent texts on general biochemistry, I soon realized that none of them placed an emphasis on plant metabolism. Initially, I selected a group of reviews from the literature to use in class, but these usually assumed that the readers already had a good grasp of the subject, which many did not. The resolution was to generate my own series of lecture notes for the students, the initial outlines of which quickly grew into chapters and eventually into this book.

This text assumes that the students have a background in basic organic chemistry and cell biology. The fundamental concepts of biochemistry, such as energy transfer, are introduced, but from a plant perspective. Many of the same topics are covered here as in a general biochemistry text, but I also stress how these processes differ in photoautotrophs, including photosynthetic bacteria, cyanobacteria, and plants. This book also delves into specific plant-related topics such as cell wall and natural product synthesis. Each chapter can stand alone and some sections, and even entire chapters, can be omitted based on the plant biochemistry instructor's preferences without causing a major break in the development of a particular topic. When it is appropriate, the reader is referred to chapters and appendices where more details on a given subject can be found. Literature references are kept to a minimum and include recent reviews on the chapter topics. To ensure that these references remain relevant, they will be included and updated on the book's Student Companion Web site http://biology.jbpub.com/gleason/plant biochemistry/.

When I generated a table of contents for the proposed textbook, I realized that the topics were a rather select group that faculty and students at the University of Minnesota felt were missing in their other courses. Some topics that could be included will not be found, such as light quality and perception by plants or techniques for genetic manipulation, which have been very influential in the development of modern biochemistry but are widely available in general biochemistry texts. A great deal of information is included on biosyntheses and enzyme structure and regulation. My goal is to illustrate how plants take sunlight and convert it into all the biochemical compounds needed to sustain life, reproduce, and interact with the other organisms and abiotic elements in their environment.

Chapters 1 and **2** cover photosynthesis, the light reactions, and carbon dioxide fixation. These processes involve the transformation of light energy into chemical bond energy and form the core of plant function. Photosynthesis is arguably the most important biochemical process on this planet because it is the major pathway for taking carbon dioxide from the atmosphere and putting it into the biosphere. Research on photosynthesis has had a long and inspiring history. Some of the key historical discoveries are described, including the Hill reaction and the use of 14C to trace the path of carbon in CO_2 fixation. Major breakthroughs in photosynthesis research include the use of molecular techniques to isolate genes, to overproduce chloroplast proteins, and to determine the three-dimensional structure of the membrane-embedded proteins that facilitate the light reaction. Research in photosynthesis is still an active and vibrant field as we strive to find more ways of increasing plant productivity for food and fuel.

Chapter 3 covers the catabolism of carbohydrates. The processes of glycolysis, the citric acid cycle, and electron transport in plants are similar to those found in nonphotosynthetic organisms. Some features, however, are unique to plant systems. In green tissues, the chloroplast is the ultimate source of fixed carbon compounds. In a leaf's cytoplasm, these compounds can be used to generate ATP or alternatively used to synthesize sucrose and other sugars for export to the nongreen tissues. In addition, the photosynthetic cell must maintain the pools of sugar-phosphates in the chloroplast at levels that are sufficient to sustain the C-3 cycle. The regulatory mechanisms for all these pathways in plants are governed by small molecules. Some of the regulatory mechanisms are similar to those found in animal cells, but, in many cases, light is the ultimate regulatory agent.

The catabolic pathways are generally thought of as transforming the chemical bond energy in small organic molecules to bond energy in the phosphoester bonds of ATP. However, these pathways also supply carbon intermediates for the biosyntheses of other metabolites, especially in autotrophs. These biosynthetic products are used for cell growth and reproduction, but, in plants, a large sink for sugars and other carbon compounds is cell wall synthesis. **Chapter 4** describes the synthesis of primary cell walls consisting mainly of cellulose fibrils embedded in a complex carbohydrate matrix. A wall may seem to be a rather static structure, but primary cell walls are surprisingly dynamic. They contain several types of proteins that are important in maintaining and restructuring the walls. The wall proteins facilitate wall loosening and rebuilding to accommodate cell expansion. Recent reports have described many additional cell wall proteins that function in defense and intercellular signal transduction.

Carbon dioxide, water, and sunlight are the major plant "nutrients," but other inorganic compounds are also required. A source of fixed nitrogen and sulfur is essential and must be available in the environment. **Chapter 5** deals with the uptake and metabolism of the fixed forms of nitrogen, nitrate, nitrite, and ammonia.

Molecular nitrogen (N_2) is abundant in the atmosphere, but only certain prokaryotes, including some photosynthetic microorganisms, are capable of catalyzing its reduction into these fixed forms. Many plants have taken advantage of these microorganisms and their ability to fix nitrogen by forming symbiotic relationships. These relations are especially intimate in the case of legumes and rhizobia. The process of nitrogen fixation presents plants with an especially complex challenge: how to maintain oxygen-evolving photosynthesis in the presence of the nitrogen-fixing enzymes, which are irreversibly inactivated by oxygen. Plants must provide their symbionts with carbon nutrients and maintain them in a compartment that excludes oxygen. Free-living organisms face a similar problem and have several methods to cope with what seem to be the incompatible processes of nitrogen fixation, photosynthesis, and aerobic respiration.

Nitrogen assimilation and fixation generally result in the synthesis of the amino acid, glutamate. This amino acid is the major amino donor for the synthesis of most of the other major amino acids. These pathways are discussed in Chapter 8 for aromatic amino acids and in Appendix 6. The two exceptions are the sulfur-containing amino acids, cysteine and methionine, which are covered in the second half of Chapter 5. The role of cysteine in plant metabolism is further explored by outlining the synthesis of glutathione. Its role in maintaining the redox state of the cytoplasm and binding and detoxification of heavy metals and xenobiotics are vital to cell survival. Chapter 5 ends with a description of the synthesis and action of the sulfur-containing natural products, in particular, the glucosinolates. These natural products are found mainly in the Brassica species, including vegetables that are a staple of the human diet such as cabbage and broccoli. The glucosinolates are produced by the plant as protection from herbivores. In some instances, insect herbivores have adapted and are attracted to glucosinolates. Some insects can even sequester the plant glucosinolates in their own cells to protect themselves from predators.

Chapter 6 continues with the theme of carbon metabolism by outlining the synthesis of lipids. Lipid production in plants starts with fatty acid synthesis in plastids. Fatty acids are esterified to glycerol by enzymes associated with the endoplasmic reticulum and are often modified by addition of double bonds. The resulting lipids are then transferred back to plastids or to other organelles and extensively modified. Plants produce over 300 different types of lipids, such as the unique galacto- and sulfo-lipids, which maintain the fluidity of thylakoid membranes. The greatest lipid variety occurs in seeds. These lipids include oils such as the ricinoleic acid–containing triglyceride in castor beans. This unusual hydroxylated oil is used by humans to produce lubricants and paints and is only one of many seed oils in commercial use. Plants are also a possible source of the very long chain polyunsaturated fatty acids that are required in the human diet. Most of these compounds are found in fish oils, which are, in turn, derived from algae and fungi. Genetic engineering of plants to express the genes for very long-chain polyunsaturated fatty acids (VLC-PUFA) synthesis may one day provide an alternate source of these nutrients.

Chapter 7 describes isoprenoid or terpene synthesis in plants. The mevalonic acid pathway that produces sterols in animals is found in the plant cytoplasm. In addition, plants have a second synthetic pathway for isoprenoids located in plastids. This methyl-erythritol phosphate pathway is apparently a legacy of their cyanobacterial ancestors. The total isoprenoid synthesis is divided between these two pathways. The traditional mevalonic acid pathway produces sterols and inducible defense compounds called phytoalexins. The plastid pathway generates the carotenoids essential in photosynthesis, and the volatile monoterpenes that attract insect pollinators or deter herbivores. The monoterpenes are well known to most humans. They are some of the common flavoring agents in foods such as mint and menthol and the scents in perfumes. In addition, the plastid pathway generates the precursors for a wealth of plant hormones, the cytokinins, gibberellins, brassinosteroids, abscisic acid, and strigolactones. This division of labor between the two plant pathways seems to be absolute. Although they both produce similar precursors, mutations in genes for either pathway are usually lethal.

Phenolic compounds and their derivatives are the subject of **Chapter 8.** The chapter starts with the shikimic acid pathway and the synthesis of the aromatic amino acids. Phenylalanine is especially important; removal of the amino group yields cinnamic acid and leads to the vast number and complexity of the phenylpropanoids. There are two main branches; one produces monolignols, which are essential for lignin synthesis, and a second that leads to a variety of natural products. Lignin, a random polymer of monolignols, is the structural component of secondary cell walls. It provides protection, physical support for terrestrial plants, and a rigid vascular system for water transport. Its chemical intransigence presents a major problem for humans who want to use plant products for food for themselves and animals and for commercial processes such as plant-derived biofuels. Government agencies and private interests are actively supporting research to manipulate lignin synthesis and degradation.

The monolignols are also a source of natural products. Modified monolignols often serve the same purposes as monoterpenes. Volatiles such as eugenol and myristicin may deter herbivores but generally attract humans.

The second major branch of the phenylpropanoids leads to the polyketides. The best known of theses compounds are the flavonoids and derivatives. The anthocyanins enable plants to cope with abiotic stressors such as intense sunlight and UV radiation. They may also serve to lure pollinators by imparting attractive colors to flower petals or to lure seed dispersal agents by providing brightly colored fruits. Resveratrol, the polyketide found in grapes, seems to provide major health benefits to humans such as maintaining a healthy heart and preventing diabetes.

While phenylpropanoids provide the plant with both essential components and natural products, the alkaloids described in **Chapter 9** can all be considered secondary metabolites. They are produced as defensive compounds, but their occurrence is often limited to a small number of species. The human use of alkaloids has a long and colorful history and continues today with the discovery of new compounds and uses. Many alkaloids have chemical structures similar to mammalian neuroreceptors or neurotransmitters. These alkaloids, such as opiates and tropanes, are well known for their neurological effects. While multivolume series have been written on alkaloids, this chapter explores only a few of the compounds that are of major use to humans. This includes synthesis and action of the opiates and vinca alkaloids used in modern medicine. Because of their complex structures and intricate biosynthesis, these compounds are still obtained from the producing plants rather than chemical synthesis. Research using plant tissue cultures and semisynthesis, where the precursor made by the plant is further modified by a chemist, has provided additional sources of alkaloids and analogs.

The final chapter covers proteins and peptides that are unique to plants. These include the well-characterized seed storage proteins important in human nutrition. Recent research on gene expression has shown that these proteins are often synthesized in tissues other than seeds and have functions in plant development and defense. The germins are a good example; they were once considered simply a food source for germinating embryos. Many germins and similar proteins also have enzymatic activity and most likely act to remodel primary cell walls in both embryos and developing organs. The seed proteases and protease inhibitors have well-defined defensive functions. The ribosome-inactivating proteins (RIPs) also seem to be designed to kill herbivores. Their ability to enter cells, however, is often incompatible with this role. They may actually be a defense against viral invasion.

Genomic analyses of several plants have produced a long list of peptides labeled defensins. These peptides are synthesized as large precursor proteins that are subsequently cleaved and modified usually by the formation of disulfide bonds. Their small size and stability have facilitated determination of their tertiary structures. Defensins have a variety of tertiary structures, and some of them are rather unusual, such as the cyclic, knotted cyclotides. These peptides can disrupt bacterial cell membranes and, thus, protect against microbial invaders. Recent reports, however, show that not all defensins are for defense. Some of these peptides are actually plant hormones or other signaling molecules. At this point, we have reached a frontier in plant biology.

Ancillaries

For the Student

A student companion website, hosting study quiz questions, an interactive glossary, crossword puzzles, interactive flashcards, and research and reference links, has been developed to accompany Plant Biochemistry and is available at http://biology.jbpub.com/gleason/plantBiochemistry/.

For Instructors

An Instructor's Media CD-Rom containing PowerPoint® Lecture Outlines and an Image Bank are available to instructors using Plant Biochemistry in their course instruction.

Acknowledgments

I would like to thank my students over the years and my colleagues in plant biology at the University of Minnesota for stimulating me to write about the biochemistry of plants. I would also like to thank all the outside reviewers who corrected my errors and made many useful suggestions:

Brian Ayre, University of North Texas
Donald Briskin, University of Illinois, Urbana-Champaign
David G. Clark, University of Florida
Peter Facchini, University of Calgary
Ahmed Faik, Ohio University
Bob Houtz, University of Kentucky
Hisashi Koiwa, Texas A&M University
Jia Li, University of Oklahoma
David Lightfoot, Southern Illinois University, Carbondale
Davin Malasarn, University of California, Los Angeles
Johannes Stratmann, University of South Carolina
Andrew Wood, Southern Illinois University, Carbondale

CHAPTER 1

Photosynthesis: The Light Reaction

CHAPTER OUTLINE

1.1 Overview

Life on Earth runs on solar energy. The biological solar age started 3.5 to 3.0 billion years before the present when ancient microorganisms first evolved the ability to convert light energy from the sun into chemical bond energy of organic molecules. These ancient microorganisms lived in an anoxic world and used sunlight to produce adenosine triphosphate (ATP). Using this photochemically derived energy, they were thus able to take advantage of the various reduced compounds, such as H_2S, on the primitive Earth and subsequently reduce atmospheric CO_2 to sugars. About 0.5 billion years later, cyanobacteria developed a slightly different process, replacing reduced sulfur and organic compounds with H_2O as their reducing agent and releasing oxygen into the primitive atmosphere. The oxygen released from this oxygenic photosynthesis slowly built up in the Earth's atmosphere, reaching a partial pressure of 1 kPa approximately 1.5 billion years ago. At this time the oxygen concentration was sufficient for formation of ozone (O_3) in the stratosphere that screened out the more energetic solar radiation and made it possible for the first organisms to colonize the land. Today, oxygenic photosynthesis is the most common biochemical reaction on our planet, both in aquatic and terrestrial systems, and sustains almost all heterotrophic life forms. This complex series of reactions, often referred to as green plant photosynthesis, has been conserved throughout Earth's history and is the mechanism for energy transduction in cyanobacteria, algae, and nonvascular and vascular plants.

The overall chemical process of green plant photosynthesis was first described in the nineteenth century by the following equation:

$$6CO_2 + 6H_2O \xrightarrow{\text{Light}} C_6H_{12}O_6 + 6O_2$$

A large amount of energy is required to drive this reaction; the standard free energy change (ΔG^o) is calculated to be 2,880 kJ mol^{-1}. This energy is provided by visible light with wavelengths of 400 to 700 nm. The equation, as written, describes an oxidation-reduction (redox) reaction. Historically, the reaction was not recognized as a redox reaction because water does not readily release electrons and thus is not an effective reducing agent. However, in the 1930s Cornelis van Niel and his coworkers demonstrated that anaerobic photosynthetic bacteria perform the following reaction:

$$nCO_2 + nH_2A \xrightarrow{\text{Light}} (CH_2O)_n + nA$$

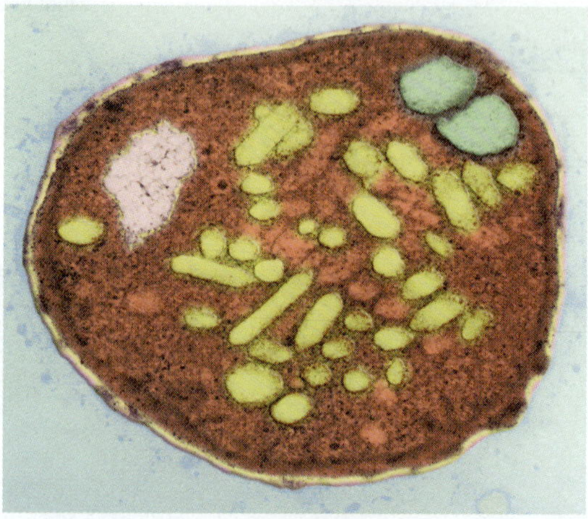

FIGURE 1.1 Light micrograph of a purple sulfur bacterium. This example of a Proteobacterium photosynthesizes only under anoxic conditions. Bacteriochlorophylls, other pigments, and proteins needed to harvest light energy and produce ATP are located in its cell membrane. Hydrogen sulfide (H_2S) is the reducing agent giving rise to the grains of elemental sulfur.

In anaerobic sulfur bacteria such as *Rhodospirillum rubrum*, H_2A is H_2S, which is a good reducing agent, and elemental sulfur is excreted to the outside of the cell (**FIGURE 1.1**) or oxidized to other compounds. By analogy, H_2A in green plant photosynthesis is H_2O and oxygen is released. Subsequent experiments by Ruben and Kamen in 1941 using ^{18}O-labeled water clearly showed that the oxygen generated in green plant photosynthesis is derived from the oxidation of water. Unlike anaerobic bacteria, plants absorb sufficient light energy to remove electrons from water. Subsequent research in the past 67 years has sought to elucidate the mechanisms behind this deceptively simple redox reaction.

1.2 The Hill Reaction

Photosynthesis occurs in the chloroplasts of eukaryotic algae and plants and in the cytoplasm of cyanobacteria. The chloroplast is an organelle with two or more outer membranes and a set of internal membranes called thylakoids. The soluble portion of the organelle is called the stroma (**FIGURE 1.2**).

Although the chloroplast is clearly the site of photosynthesis in plants, experiments with isolated chloroplasts in the early twentieth century were disappointing. When removed from the plant cell, these organelles were inactive. However, in 1937 Robin Hill showed that isolated chloroplasts evolve oxygen when supplied with an artificial electron acceptor such as ferricyanide:

$$4Fe^{3+} + 2H_2O \xrightarrow{\text{Light}} 4Fe^{2+} + 4H^+ + O_2$$

FIGURE 1.2 Electron micrograph (a) and diagram (b) of a chloroplast from maize. Chloroplasts in eukaryotic algae and plants have two or more outer membranes. In addition, the organelle has an extensive array of interior membranes called thylakoids. In angiosperms (flowering plants), these are arranged in stacks called grana (singular, granum). The thylakoid membranes are the site of the light reaction. These membranes are small sacs with an internal space or lumen. The soluble portion of the chloroplast is the stroma. It contains enzymes and cofactors needed for CO_2 fixation.

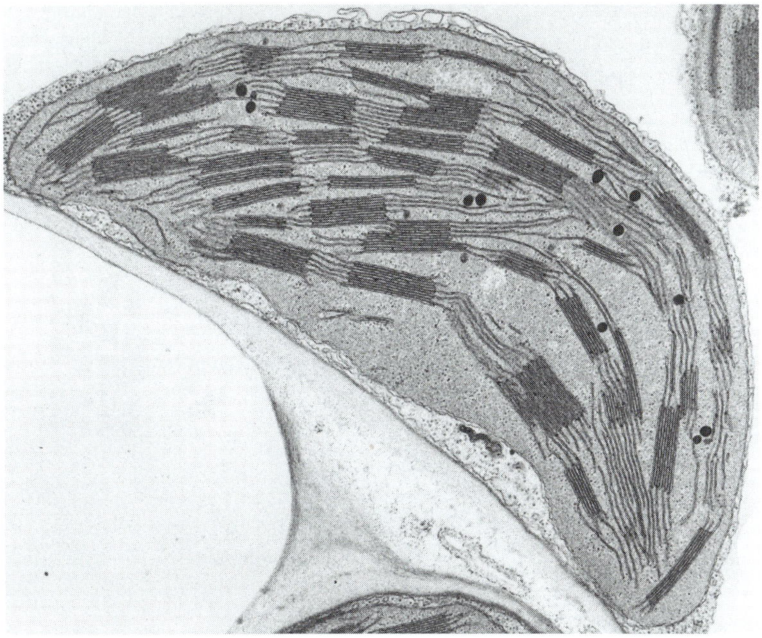

(a)

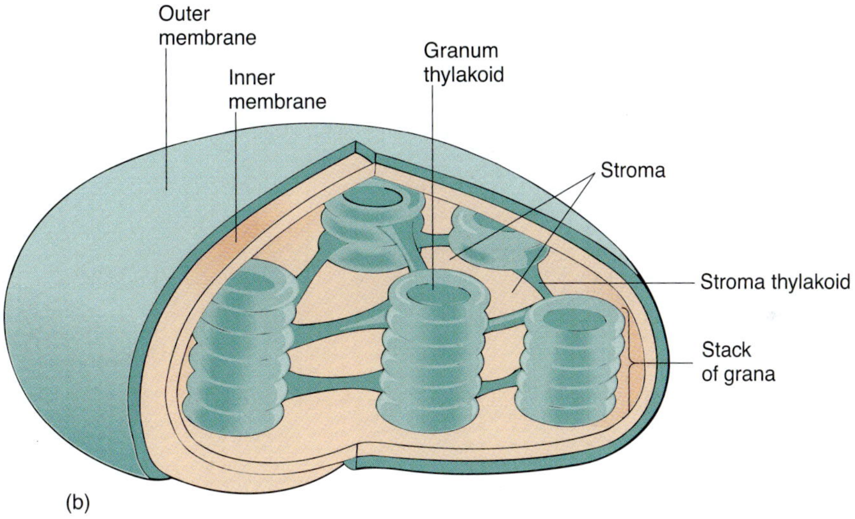

(b)

He concluded that the biological electron acceptor must be a water-soluble compound that is leached out of the organelle during the isolation procedure. In the most commonly used methods for chloroplast isolation, the outer membranes are damaged and soluble components of the stroma can be lost. The membranes, including thylakoid membranes, remain associated with each other and can be separated by centrifugation into a pellet. The components of the stroma are present in the supernatant fraction, usually in a highly dilute form. Isolated chloroplasts evolve O_2 (oxidize H_2O) but cannot produce carbohydrates (reduce CO_2). Gentle procedures for chloroplast isolation that preserved intact organelles capable of the complete photosynthetic process were not devised until the 1960s.

However, Hill's work was a major breakthrough in the study of photosynthesis. He clearly showed that photosynthesis is basically two oxidation-reduction reactions. The light reaction involves the oxidation of water, release of oxygen, and reduction of a soluble component. This occurs in thylakoid membranes. The second redox reaction, the reduction of carbon dioxide and oxidation of the soluble component, occurs in the water-soluble portion (i.e., the stroma) of the chloroplast. The Hill reaction can be written as follows:

$$2H_2O + 2A \xrightarrow{\text{Light}} 2AH_2 + O_2$$

where A is a redox dye such as dichloroindophenol, methyl viologen, or ferricyanide. Hill's

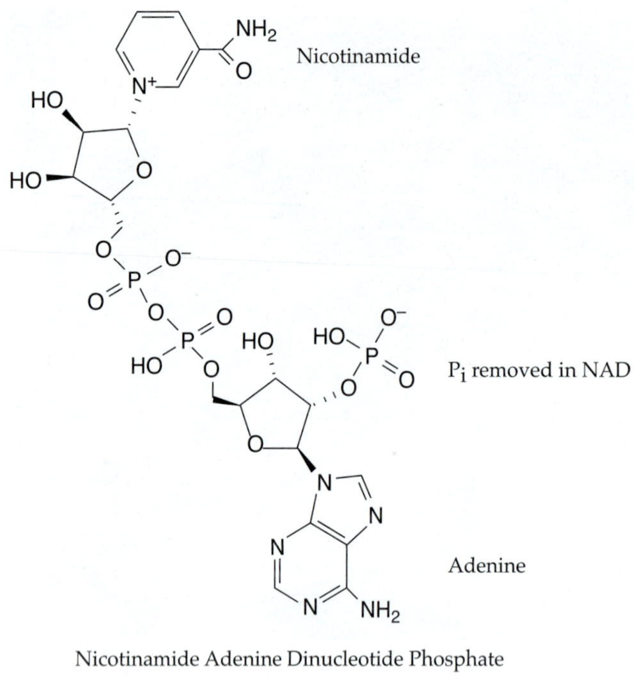

(a)

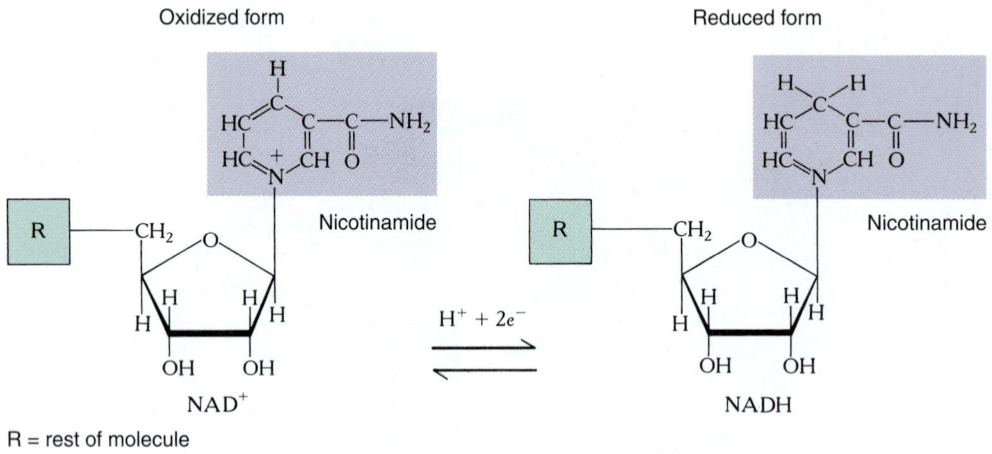

(b)

FIGURE 1.3 Nicotinamide adenine dinucleotide phosphate (NADP). (a) NADP consists of two nucleotides connected by a phosphodiester bond. A similar redox cofactor, NAD, lacks the second phosphate group on the adenine ribose. (b) The nicotinamide ring accepts two electrons and one proton (hydride, H^- ion) to produce the reduced cofactor, NADPH.

contribution made it possible to study the light reaction in isolated organelles either spectrophotometrically or polarographically by monitoring the evolution of oxygen. In the 1950s Daniel Arnon and coworkers demonstrated that the soluble cofactor, NADP (nicotinamide adenine dinucleotide phosphate; **FIGURE 1.3**), is the physiological electron acceptor. This research group also showed that isolated chloroplasts produce ATP when adenosine diphosphate (ADP) and PO_4^{3-} (inorganic phosphate, Pi) are added to the incubation mixture. The light reaction could now be written in its complete form:

$$2\text{NADP} + \text{Pi} + \text{ADP} + 2\text{H}_2\text{O} \xrightarrow{\text{Light}} 2\text{NADPH} + 2\text{H}^+ + \text{ATP} + \text{O}_2$$

1.3 Photosynthetic Pigments

Chlorophyll *a*

This basic energy conversion, light energy to chemical bond energy, occurs in the thylakoid membranes. Visible light supplies the energy for the oxidation of water and the transfer of electrons to NADP. Electron transfer also provides the energy for the synthesis of ATP. Light energy is captured by photosynthetic pigments. All organisms that perform oxygenic photosynthesis use a pigment called chlorophyll *a* (**FIGURE 1.4**). The basic structural element of all chlorophylls is a tetrapyrrole ring that is also found in related molecules containing heme groups such as cytochromes. In the chlorophyll ring the four nitrogens are coordinated to a Mg ion, as opposed

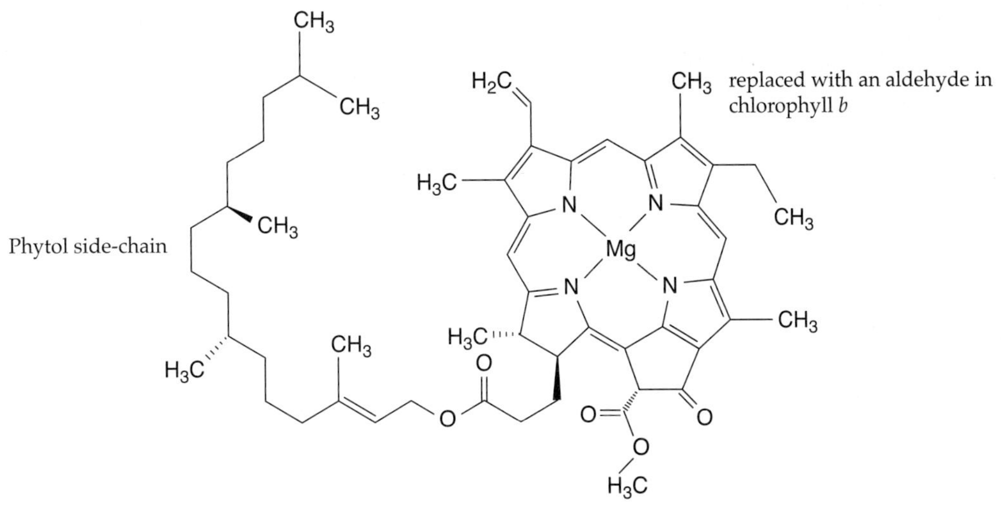

Chlorophyll *a*

FIGURE 1.4 Structure of chlorophyll *a*. The tetrapyrrole ring is a conjugated system, making this portion of the molecule rigid and planar. The four nitrogens of the tetrapyrrole ring are conjugated to a Mg ion. Various substitutions can be made to the groups on the ring to produce other types of chlorophylls. For example, in chlorophyll *b* a methyl group is substituted with an aldehyde. The 20-carbon long phytol side chain makes the entire molecule very hydrophobic.

to an Fe in heme groups. The ring is a conjugated system (i.e., alternating double and single bonds), which means that it has a rigid, planar structure because of the overlap of the hybrid molecular orbitals. In addition, chlorophylls have a 20-carbon hydrocarbon side chain, called a phytol unit, making these molecules very hydrophobic. Chlorophylls are associated with integral membrane proteins and can be extracted using organic solvents such as acetone or ethanol.

In addition to chlorophyll *a*, most plants also contain other types of chlorophylls with different chemical groups on the side chains of the tetrapyrrole ring. Chlorophylls absorb visible light in the blue and red regions of the visible spectrum (**FIGURE 1.5**; **see Appendix 7 for discussion of spectrophotometry**). The absorption spectrum is influenced by the chemical environment (e.g., the solvent in which the molecules are dissolved). Figure 1.5 shows the spectrum of isolated chlorophyll *a* and associated carotenoid pigments dissolved in ethanol. The major absorption peak at approximately 430 nm is characteristic of all tetrapyrrole rings and is called the Soret band. The absorption bands in the red region of the spectrum (approximately 630 and 660 nm) are characteristic of chlorophyll *a* and are called the Q bands. Most photosynthetic organisms also synthesize some accessory pigments that absorb light in other regions of the visible spectrum. In chlorophyll *b*, a common accessory pigment in angiosperms, an aldehyde replaces a methyl group in the tetrapyrrole ring (Figure 1.4). This substitution shifts the Soret band toward the red end of the spectrum and the Q bands toward the blue end of the spectrum compared with chlorophyll *a*, thus providing a wider range of light absorption (**FIGURE 1.6**). "Green" light, at approximately 500 to 600 nm, is not absorbed but is

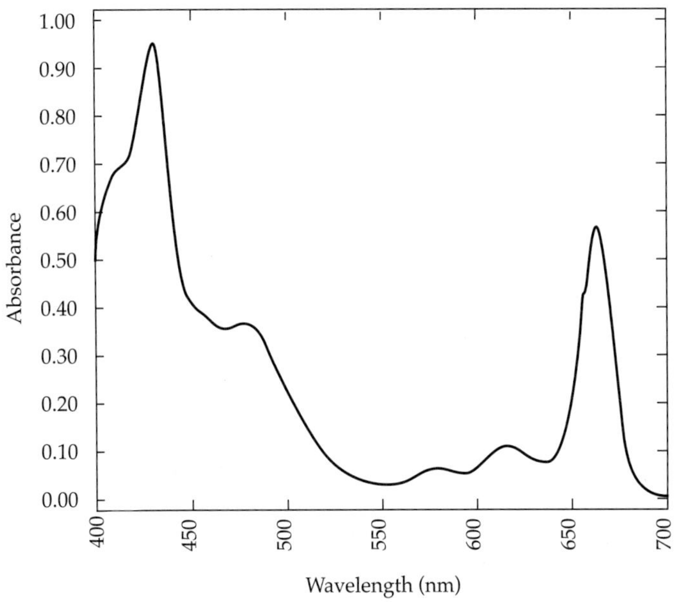

FIGURE 1.5 The visible absorption spectrum of photosynthetic pigments from a cyanobacterium, *Gloeocapsa alpicola*. The photosynthetic pigments were extracted from the cells with ethanol. The major peak at 430 nm is called the Soret band and is characteristic of most cyclic tetrapyrroles. The bands at 630 and 660 nm, Q bands, are characteristic of chlorophyll *a*. The shoulder at approximately 470 nm is due to absorption of light by accessory carotenoid pigments.

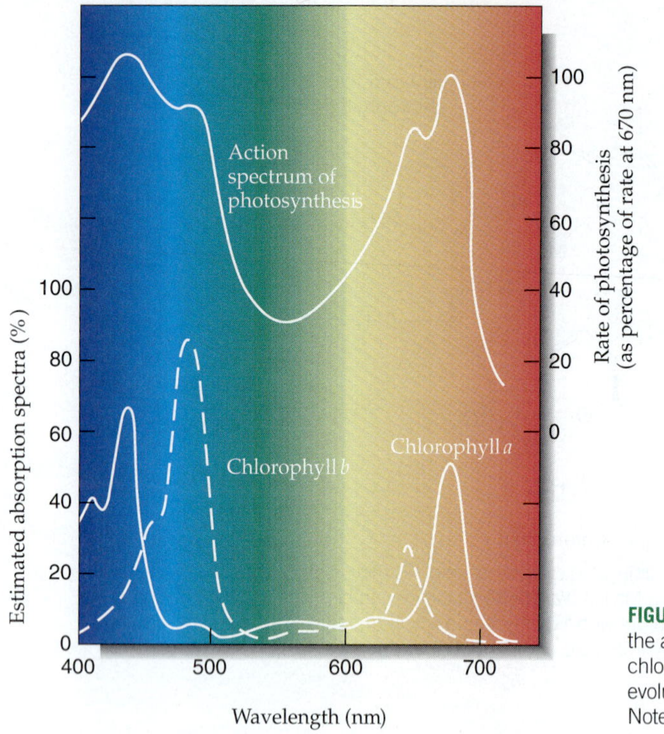

transmitted or reflected, and the solution appears green to human eyes.

■ Accessory Pigments

Other types of chlorophylls and pigments are found in photosynthetic organisms. Bacteriochlorophylls are characteristic of the purple and green photosynthetic bacteria. These absorb light at longer wavelengths (800–890 nm) than chlorophyll *a*. Chlorophylls *c* and *d* are accessory pigments in algae such as diatoms. In addition to various chlorophylls, most plants also contain accessory pigments called carotenoids. These are 40-carbon, conjugated hydrocarbons that absorb light in the violet and blue regions of the visible spectrum (**FIGURE 1.7**). They function in

FIGURE 1.6 Action spectrum of the light reaction. In this diagram the action spectrum is superimposed above the absorption spectra of chlorophylls *a* and *b*. The rate of photosynthesis, measured as oxygen evolution, roughly parallels the absorption spectra of the chlorophylls. Note the abrupt "red drop" in the rate of photosynthesis above 680 nm.

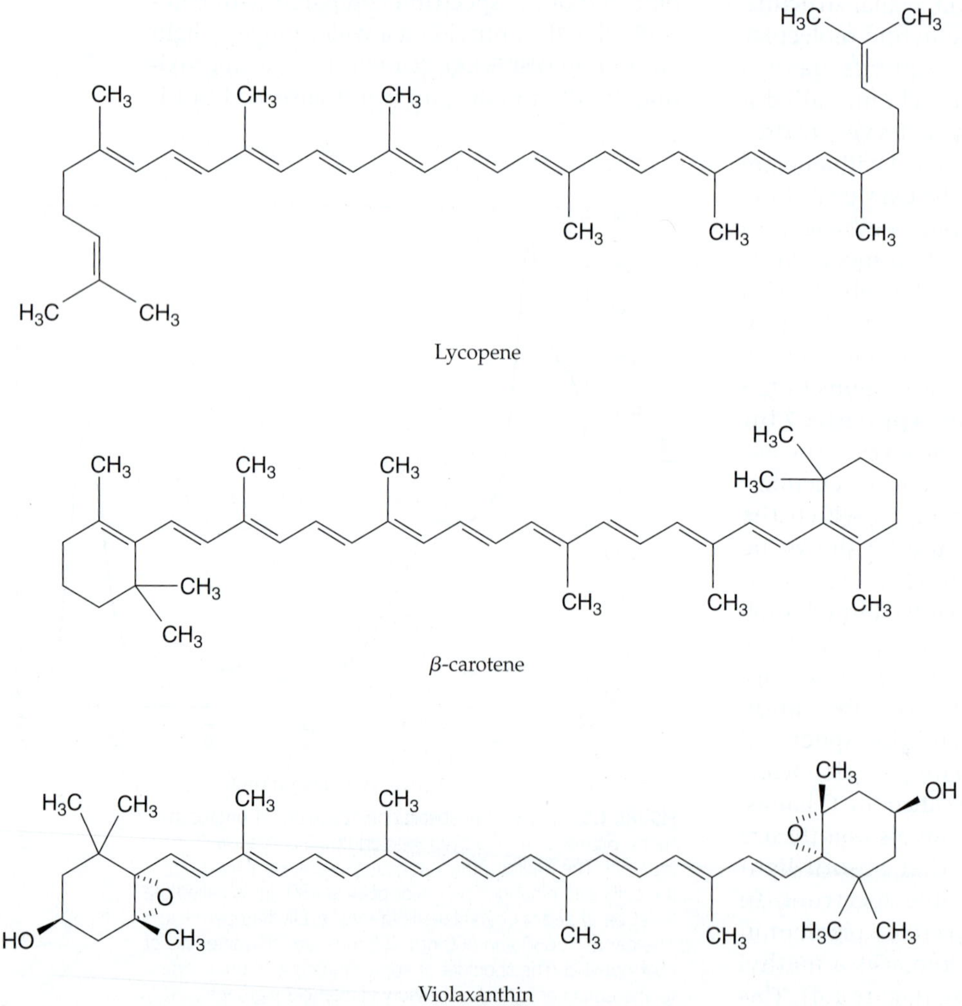

FIGURE 1.7 Examples of some common carotenoids. These 40-carbon-long hydrocarbons are conjugated systems that absorb visible light in the blue region of the spectrum. They act as accessory pigments in the light reaction and also protect the proteins in the photosynthetic reaction centers from excessive light and free radical damage.

harvesting light energy and in protecting the photosynthetic apparatus from photo-oxidative damage by excess illumination. Phycobilins are linear tetrapyrroles that absorb light at various wavelengths, including the green region of the spectrum. These are water-soluble pigments found in cyanobacteria and some eukaryotic algae. Most photosynthetic pigments, however, are not soluble in water and are an integral part of a cellular membrane. Chromatophores in photosynthetic bacteria are invaginations of the cell membrane called chlorosomes. Thylakoid membranes are found in the cytoplasm of cyanobacteria. In eukaryotic algae and plants, these membranes are in the chloroplast and are not connected to the organelle's outer membranes. They can often have elaborate stacked arrangements, as seen in Figure 1.2.

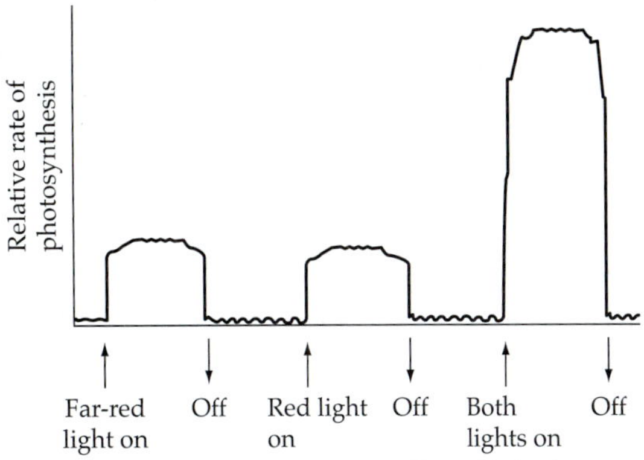

FIGURE 1.8 Emerson enhancement effect. Algal cultures are briefly exposed to far red light (700 nm) or/and red light (680 nm). The rate of photosynthesis with both types of light is more than the sum of activity with either single light source. These data suggest that the light reaction requires two RCs working in series for maximum efficiency. (*Adapted from:* Taiz, L. and Zeiger, E., *Plant Physiology*, 2nd ed. Sunderland, MA: Sinauer Associates, Inc., 1998. Pg. 165, Fig. 7.10.)

1.4 Origin of the Z-Scheme

Emerson Enhancement

Although Robin Hill's discovery of the oxygen-evolving capacity of isolated chloroplasts made it possible to study the light reaction in isolated organelles, unicellular algae were often a better experimental system to study the whole process of photosynthesis. The eukaryotic green alga, *Chlorella* sp., was commonly used. In the 1950s, R. Emerson and his colleagues measured photosynthetic efficiency in *Chlorella* cultures with very short light flashes. They concluded that the light reaction was much faster than CO_2 fixation. They also found that approximately 2,500 chlorophyll molecules were required to evolve one O_2 molecule. Most of these pigment molecules served to harvest light energy (i.e., they functioned as an antenna system). Sunlight is a diffuse source of energy that must be concentrated to be effective. Light energy is also easily dissipated before it can be used. Most of these 2,500 chlorophylls are associated with integral membrane proteins called light-harvesting complexes (LHCs). LHCs consist of several polypeptides that generally have alpha helices traversing the thylakoid membranes. The pigments are noncovalently attached to the membrane-embedded part of these proteins. The LHCs trap light energy and transfer it to a photosynthetic reaction center (RC) that includes only a small fraction of the total pigment molecules in the membrane. Emerson's group determined the action spectrum for photosynthesis by measuring the rate of oxygen evolution from *Chlorella* cultures at different wavelengths (Figure 1.7). Oxygen evolution parallels the absorption spectrum of chlorophyll and accessory pigments except for the abrupt decline in efficiency at the red end of the spectrum. Chlorophyll molecules that absorb light energy at wavelengths longer than 680 nm do not appear to be as efficient as those that absorb at shorter wavelengths. This anomaly was referred to as the "red drop." Using brief flashes of light at different wavelengths, these researchers also found that exposure to far-red light (greater than 680 nm), when supplemented with a brief flash of red light (less than 680 nm), yielded a rate greater than the sum of exposure to each wavelength alone. This phenomenon was called the Emerson enhancement effect (**FIGURE 1.8**).

Photosystems I and II

The data of the Emerson group were interpreted as a requirement for several LHCs to gather and transfer light energy to a reaction center (RC) and for two RCs, working in series, for maximum efficiency. This led to the proposed "Z-scheme" of photosynthesis that was first published in 1960 (**FIGURE 1.9**). Two RCs, photosystem I (PSI), and photosystem II (PSII), which constitute only 1% of the total pigment-protein complexes in thylakoid membranes, act together to transfer electrons from H_2O to NADP. PSI is a supramolecular complex of proteins and pigments with maximal absorption at 700 nm (P_{700}). PSI performs the light-induced reduction of a soluble 2Fe-2S protein, ferredoxin, which then transfers electrons to NADP via a flavoprotein, ferredoxin-NADP reductase (FNR).

$$\text{Light} \rightarrow P_{700} \rightarrow P_{700}^{+} + e^{-} \rightarrow \text{Acceptor-}e^{-} \rightarrow \text{Ferredoxin} \rightarrow \text{NADP} \rightarrow \text{NADPH}$$

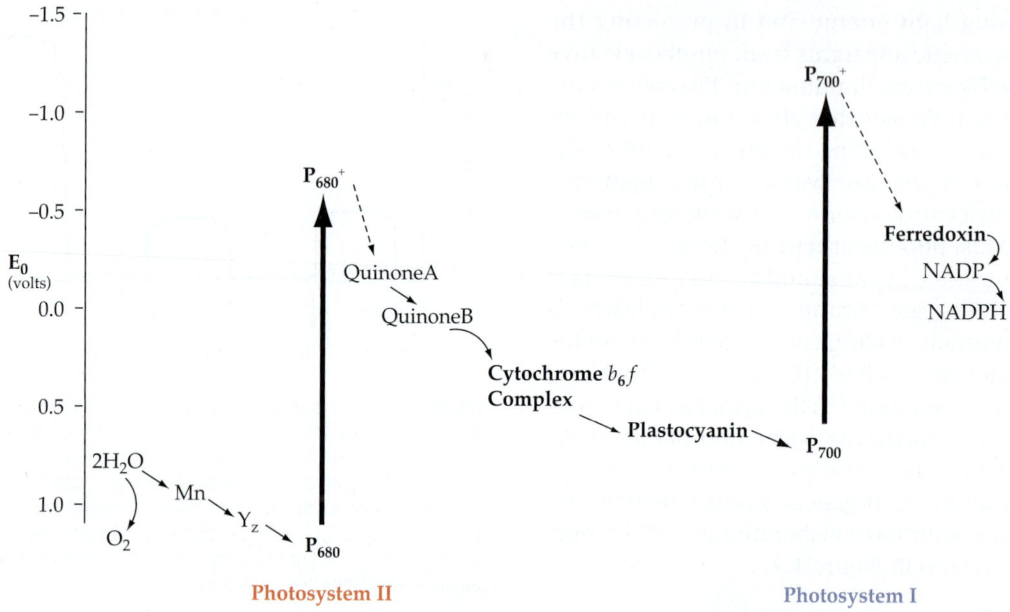

FIGURE 1.9 Z-scheme. After being published in the 1960s, this basic outline became the paradigm for further experimental work on the light reaction. PSI (P_{700}) absorbs far red light and transfers an electron to ferredoxin and ultimately to NADP. This electron is replaced in the RC by transfer of electrons from PSII via a cytochrome b_6f complex. The shorter wavelength of light absorbed by PSII (P_{680}) also provides sufficient energy to oxidize H_2O. PSII uses H_2O as the ultimate source of electrons.

P_{700}^+ is reduced by plastocyanin, a membrane-associated copper-containing redox protein located in the thylakoid space (lumen). PSII is a supramolecular complex of proteins and pigments with maximal absorption at 680 nm (P_{680}). It carries out the light-induced oxidation of H_2O and the reduction of a small molecule called a quinone.

$$\text{Light} \rightarrow P_{680} \rightarrow P_{680}^+ + e^- \rightarrow \text{Quinone B reduced}$$

After accepting two electrons and acquiring two protons, the reduced quinone leaves the PSII reaction center and diffuses within the lipid bilayer of the thylakoid membrane. It is reoxidized by a cytochrome b_6f complex that transfers electrons to plastocyanin. The PSII RC simultaneously catalyzes the oxidation of water:

$$2H_2O \rightarrow 4H^+ + O_2 + 4e^- \rightarrow P_{680}^+ \rightarrow P_{680}$$

The four electrons are channeled, one at a time, into the PSII reaction center to replace those transferred to quinone B. The oxygen escapes from the cell, but the H^+s remain in the thylakoid lumen, contributing to a proton gradient.

The two photosystems are connected by a cytochrome b_6f complex. Cytochromes are proteins that contain heme groups (i.e., tetrapyrrole rings coordinated to iron). The b_6f complex contains both *b*-type and *f*-type cytochromes in addition to a Rieske 2 Fe-2S protein consisting of Fe coordinated to sulfur groups in the protein. This complex oxidizes the quinones reduced in PSII and reduces plastocyanin to replace electrons transferred out of PSI (Figure 1.9). It also transfers protons from the stroma to the thylakoid lumen. The proton gradient in the thylakoid lumen drives the synthesis of ATP via a fourth membrane-bound component of the light reaction, the chloroplast F0F1 ATP synthase.

1.5 Light Reaction: Details

The Reaction Center of Purple Bacteria: A Model for PSII

The purple bacteria, members of the Proteobacteria phylum, photosynthesize under anaerobic conditions, using sulfide, thiosulfate, or hydrogen as their reducing agent. They have a single RC containing proteins and bacteriochlorophylls associated with the cell membrane. Purple bacteria use light energy to produce a transmembrane proton gradient that drives the synthesis of ATP. Mechanistically, their RCs have much in common with PSII, and these organisms have been used as a model system to study green plant photosynthesis. A major breakthrough in research on the mechanism of photosynthesis occurred in the 1980s when a group led by J. Deisenhofer, H. Michel, and R. Huber reported the x-ray crystal structure of the RC from *Rhodopseudomonas viridis* (renamed *Blastochloris viridis*). In addition to its direct relevance to photosynthesis, their work also described procedures for the crystallization and structure determination of membrane-bound proteins using octyl glucoside detergents to isolate and maintain the proteins in a native state. Previously, integral membrane proteins were almost impossible to isolate and characterize using crystallographic techniques.

Structure of the Bacterial Reaction Center

The transmembrane core of the *R. viridis* RC consists of three proteins, designated H, M, and L. These letters refer to their relative mobility during sodium dodecyl sulfate-polyacrylamide gel electrophoresis, where H has the largest relative molecular weight, M is intermediate in size, and L is the smallest subunit. Subsequent amino acid analyses showed that in reality H is the smallest protein, L is intermediate in size, and M is the heaviest; this illustrates the anomalous behavior of membrane proteins in denaturing gels designed to separate water-soluble proteins. These RC preparations often contained a *c*-type cytochrome that was easily lost during purification. In addition to these four proteins, the RC also contained four bacteriochlorophylls, two bacteriopheophytins (chlorophyll molecules lacking the coordinated Mg), two quinones, an Fe^{2+}, and a carotenoid. The L and M proteins are anchored in the membrane. As shown in **FIGURE 1.10**, each has five membrane-spanning helices. All RC cofactors are associated with these two core proteins. The H polypeptide has one helix embedded in the membrane with the bulk of the protein in the cytoplasm. The cytochrome is positioned on the edge of the cell membrane on the periplasmic side.

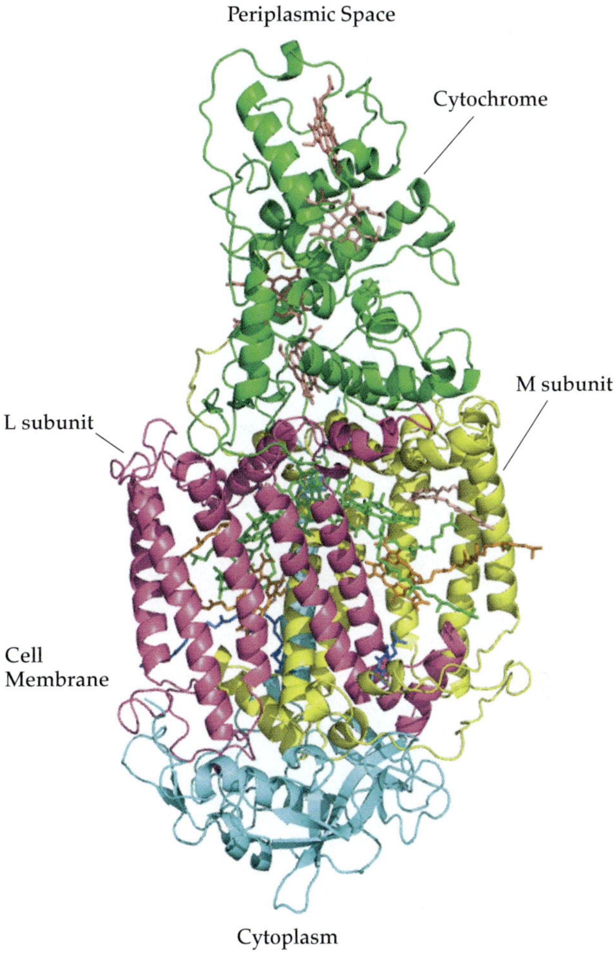

FIGURE 1.10 Reaction center of the Proteobacterium, *Rhodopseudomonas viridis.* The L subunit (magenta) and the M subunit (yellow) each contain five helices that span the cell membrane. The H subunit (blue) is located mainly in the cytoplasm with one helix embedded in the membrane. The cytochrome (green) sits atop the L-M dimer in the periplasm. Bacteriochlorophylls in L and M are shown in green, bacteriopheophytins in orange, and the quinones in deep blue. The heme groups in the cytochrome are shown in pink. (Structure from Protein Data Bank, 1prc. J. Deisenhofer, O. Epp, K. Miki, R. Huber, and H. Michel, *J. Mol. Biol.* 246 [1995]: 429–457.)

Light Energy Is Used To Reduce a Quinone

How does this simple RC work? Light energy is absorbed and transferred by the antenna pigments into the "special pair" of bacteriochlorophyll molecules in the RC (P_{870}). An excited electron is freed from this primary electron donor and captured by the primary acceptor, bacteriopheophytin, in the L subunit. This transfer takes approximately 3 picoseconds (ps). The M subunit, with its corresponding cofactors, is placed in a symmetrical position, but electron transfer does not seem to proceed through this pathway under normal physiological conditions. From the reduced bacteriopheophytin, the electron is transferred to quinone A in approximately 200 ps. Electron transfer to this quinone results in a one electron-reduced quinone (i.e., a semiquinone or quinone radical anion). Fully reduced quinones (quinols) carry two electrons and two protons (**FIGURE 1.11**). From quinone A the electron is transferred to quinone B in 100 μs. The Fe^{2+} positioned between quinones A and B does not directly participate in electron transfer but acts to stabilize the semiquinones (**FIGURE 1.12**). The transfer proceeds if this Fe is removed, but at a much slower rate. The cycle repeats so that quinone B receives two electrons and two protons. The protons are transferred from amino acid side chains in the RC proteins but ultimately come from water in the cytoplasm. Fully reduced quinone B no longer binds to the RC and diffuses into

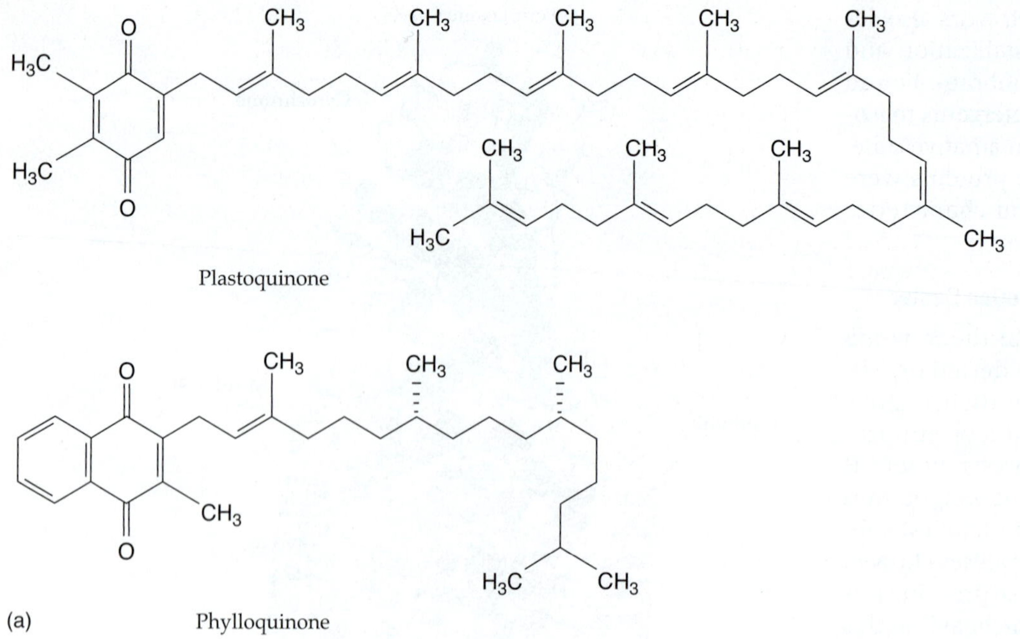

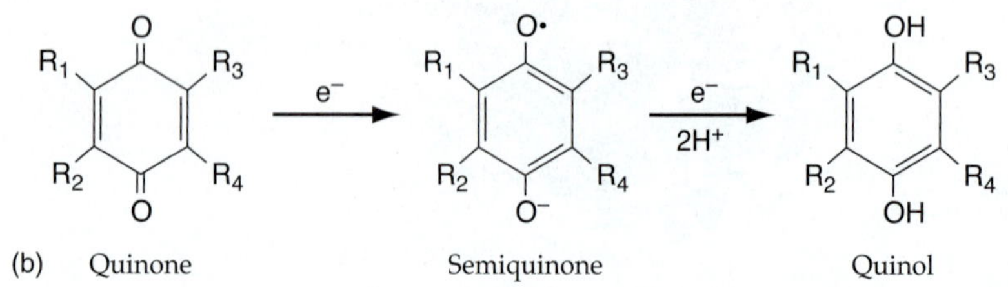

FIGURE 1.11 Structure and redox activity of quinones. Part a shows the structure of plastoquinone associated with PSII and phylloquinone in PSI. The long hydrocarbon side chains make these molecules very hydrophobic. Part b shows the redox activity of a quinone. A quinone can accept one electron and become a semistable radical anion. A fully reduced quinone (quinol) carries two electrons and two protons.

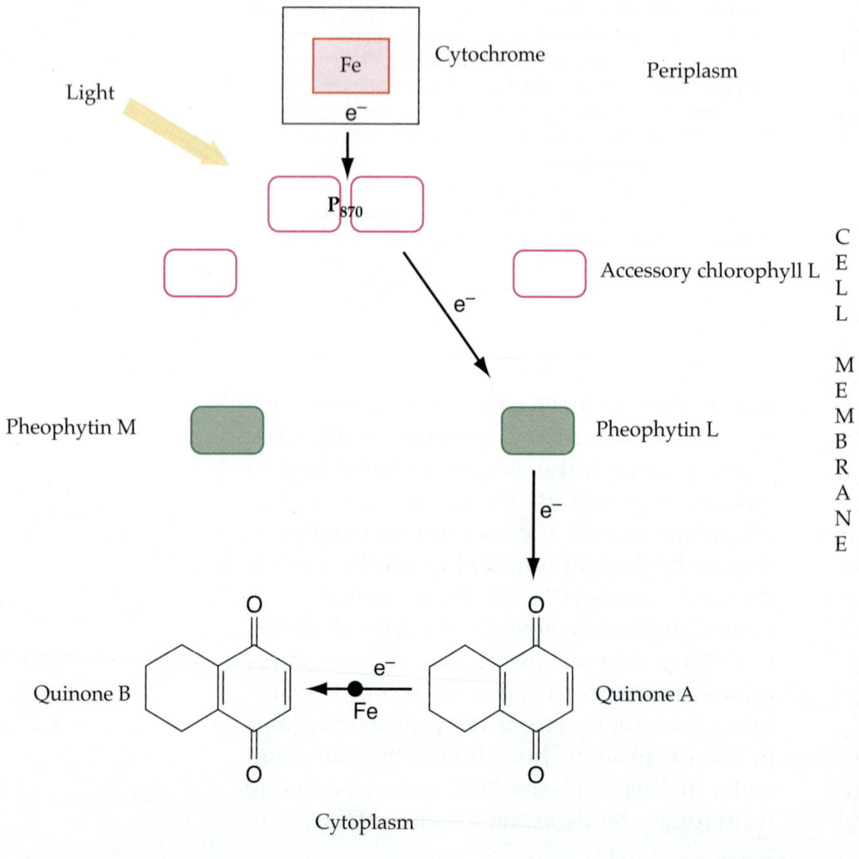

FIGURE 1.12 Electron transfer in the *R. viridis* reaction center. The special pair of bacteriochlorophylls (P_{870}) absorb light at far red wavelengths. An electron is rapidly transferred to the bacteriopheophytin in the L subunit and subsequently to quinone A and finally to quinone B. The electron lost by P_{870} is replaced by transfer from the cytochrome in the periplasm. Compare the diagram with the crystal structure of the RC shown in Figure 1.10. After two photo events and proton transfer from the cytoplasm, quinone B is fully reduced and diffuses out of the RC into the lipid bilayer. It is reoxidized by a cytochrome complex in the cell membrane.

the membrane. The rapid electron transfer in the RC ensures that the back reaction or radical production does not occur.

The electron lost from the oxidized special pair of bacteriochlorophylls in the P_{870}^+ RC is replaced by an electron transferred from the cytochrome positioned in the periplasm. This cytochrome, in turn, is reduced by a cytochrome bc_1 complex located in the cell membrane in the vicinity of the RC. The cytochrome bc_1 complex oxidizes the reduced quinone B and then reduces cytochrome *c*. In the electron transfer process, it also transfers protons from the cytoplasm into the periplasmic space. The bacterial RC and associated cytochromes perform a cyclic electron transfer that serves to create a proton gradient across the cell membrane. The energy stored in this H^+ gradient can then be used to generate ATP.

1.6 Photosystem II

Electron Transfer

The purple bacterial RC serves as a good model for the plant PSII reaction center. Gene and protein sequencing data show significant similarities between the L and M proteins of the bacterial RC and the core proteins of PSII, confirming that a bacterial type II center was the evolutionary precursor of PSII RC in cyanobacteria and plants. PSII consists of approximately 25 proteins that can be isolated as a complex from thylakoid membranes. The proteins are called Psb proteins with the corresponding *psb* genes, but some of the proteins are still referred to by their original, pre-genomic designations. The RC core of PSII consists of the D1 (PsbA) and D2 (PsbD) proteins. Like the L and M proteins of the bacterial RC, the D1–D2 core consists mainly of five transmembrane helices. Cofactors associated with the D1–D2 core include six chlorophyll molecules, two pheophytin molecules, two quinones, an Fe^{2+}, and a four Mn-one Ca cluster that is unique to oxygenic photosynthesis (**FIGURE 1.13**).

The mechanism of action is similar to that in the bacterial RC. Light energy is funneled into a special pair of chlorophyll *a* molecules (P_{680}) located at the interface of the D1 and D2 proteins. Electron transfer proceeds in 3 to 4 ps to the pheophytin in the D1 protein. As in the bacterial center, electrons are presumably not funneled through the symmetric cofactors in the D2 protein. The electron is then transferred to a quinone A and, subsequently, to quinone B (plastoquinones). After two electron transfers and addition of two H^+, quinone B is fully reduced and the quinol diffuses into the thylakoid bilayer. It is replaced by an oxidized quinone and the cycle continues. The oxidized chlorophyll special pair is reduced by electron transfer from a transient tyrosine radical in the D1 protein referred to as Yz. This specific tyrosine residue is the electron transfer conduit from the water-oxidizing reaction, which is the ultimate source of electrons in green plant photosynthesis (**FIGURE 1.14**).

Oxidation of Water

The water-splitting/oxygen-evolving reaction takes place on the lumenal side of the D1 protein.

$$2H_2O \rightarrow 4H^+ + 4e^- + O_2$$

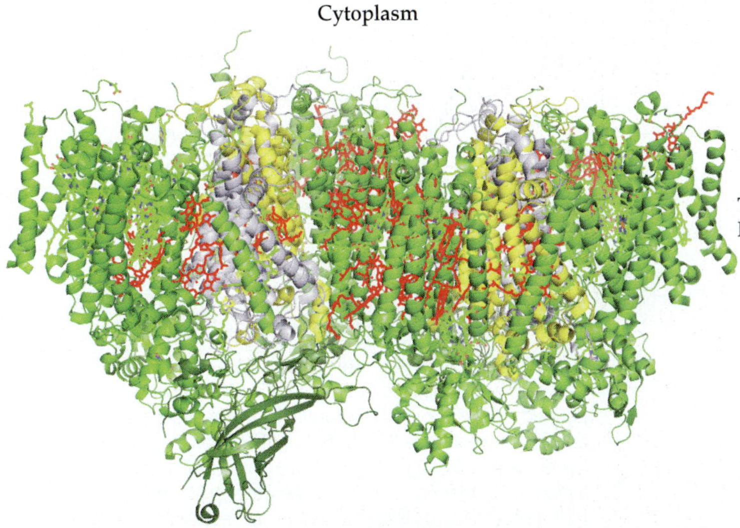

FIGURE 1.13 Structure of PSII from the cyanobacterium *Thermosynechococcus elongatus*. This view of PSII is parallel to the thylakoid lipid bilayer with the cytoplasmic side at the top and the thylakoid lumen at the bottom. The PSII complex crystallizes as a dimer with an axis in the center. Each monomer consists of 19 protein subunits. The D1 protein is in light blue and D2 in yellow. These two proteins, which correspond to the L and M subunits of the bacterial RC in Figure 1.10, consist of five transmembrane helices oriented in the membrane. In addition, each subunit contains 36 chlorophyll *a*, 7 carotenoids, 4 Mn, 1 Ca, 2 plastoquinones, 2 pheophytins, and 1 Fe. Some of the chlorophylls are depicted in red. The majority of the chlorophylls are found in light-harvesting proteins that flank the D1–D2 core. The PsbO protein that stabilizes the oxygen-evolving complex with its Mn–Ca cluster is located on the lumenal side and is shown in dark green. (Structure from Protein Data Bank, 1s51. K. N. Ferreira, T. M. Iverson, K. Maghlaoui, J. Barber, and S. Iwata, *Science* 303 [2004]: 1831–1838.)

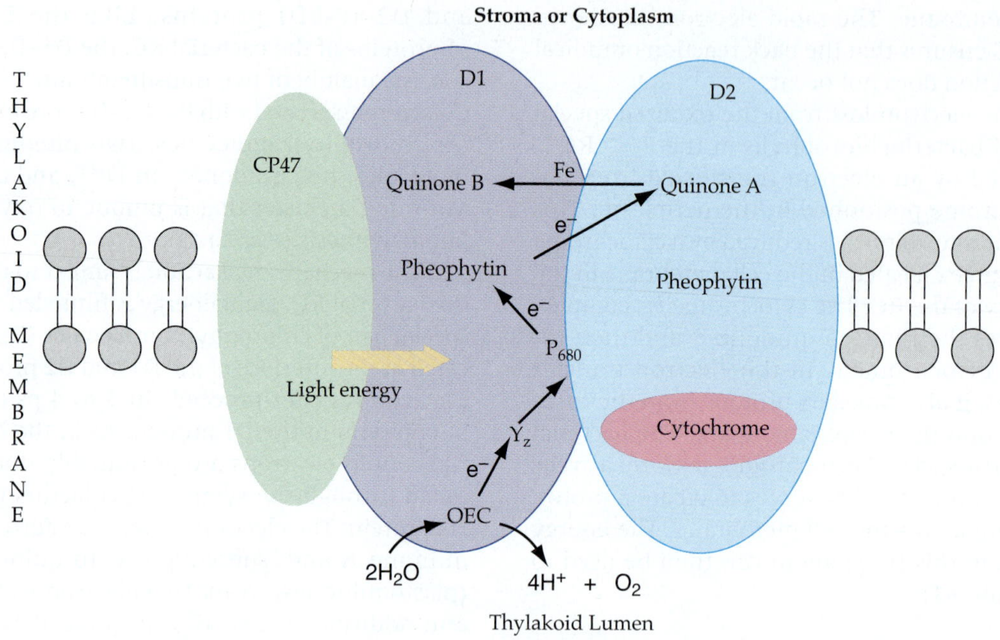

FIGURE 1.14 Electron transfer in PSII. The D1–D2 core proteins are flanked by light harvesting complexes. One of these antenna proteins, chlorophyll protein (CP) 47, is shown transferring light excitation energy into P_{680}, which consists of two chlorophyll *a* molecules centered between the D proteins. An electron is freed from P_{680} and captured by a pheophytin in the D1 protein. The electron is then transferred to quinone A and subsequently to quinone B. An Fe^{2+} in the D1–D2 core facilitates this transfer but does not undergo any redox changes. The electron lost by P_{680} is replaced by a tyrosine, Yz in the D1 protein. This amino acid residue channels electrons one at a time from the oxygen-evolving complex, which consists of 4 Mn and 1 Ca ion and uses light energy to split H_2O and channel electrons to Yz. The core complex of PSII also includes one or more cytochromes. These cytochromes may serve as alternate electron pathways to protect the core complex from radical damage and/or to provide for cyclic electron transport in PSII.

The function of the water-oxidizing complex is to transfer the four electrons, one at a time, through Yz to P_{680}^+ without releasing free radical intermediates. In 1967, P. Joliot and B. Kok set up an experimental system with isolated thylakoid membranes that they subjected to very brief, intense flashes of light. They found that oxygen evolution exhibited a periodicity of approximately one O_2 evolved per three to four flashes (**FIGURE 1.15**). Based on their experimental data, these researchers proposed the S state-cycle or "water-oxidizing clock." Five different states are possible, S_0 to S_4, corresponding to different oxidation states of the Mn ions associated with D1. Electrons are channeled one at a time to the tyrosine residue, Y_Z. O_2 is released only in the transition from S_4 to S_0 when the clock is "reset" (**FIGURE 1.16**). The oxygen-evolving complex has a cofactor containing four Mn, one Ca, and one Cl ions. A crystal structure of PSII from cyanobacteria was published in 2001, but the oxygen-evolving part of the D1 protein was not resolved until very recently. Although the oxygen-evolving complex is part of D_1, another protein, PsbO (also called MSP [manganese-stabilizing protein]), is required to maintain the stability of the complex. Because of the instability of the isolated complex, it is difficult to crystallize; in addition, the x-ray beam reduces all the Mn ions to the same redox state. Biophysical techniques have led to several models of the complex that

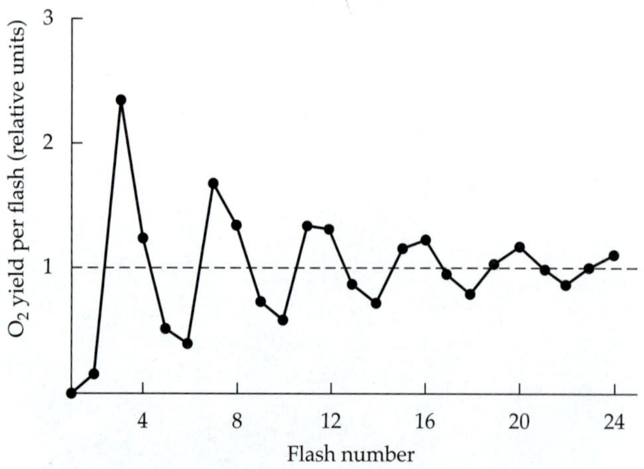

FIGURE 1.15 Joliot-Kok experiment. When removed from darkness and subjected to brief, intense flashes of light, isolated thylakoid membranes release O_2 in discrete bursts of one O_2 for every three to four flashes. This type of experiment led to the proposed "water clock" model that has served as a guide to further investigation of the water-splitting reaction in PSII. (*Adapted from*: Blankenship, R., *Molecular Mechanisms of Photosynthesis*, Wiley-Blackwell, 2002. Pg. 114, Fig. 6.10A.)

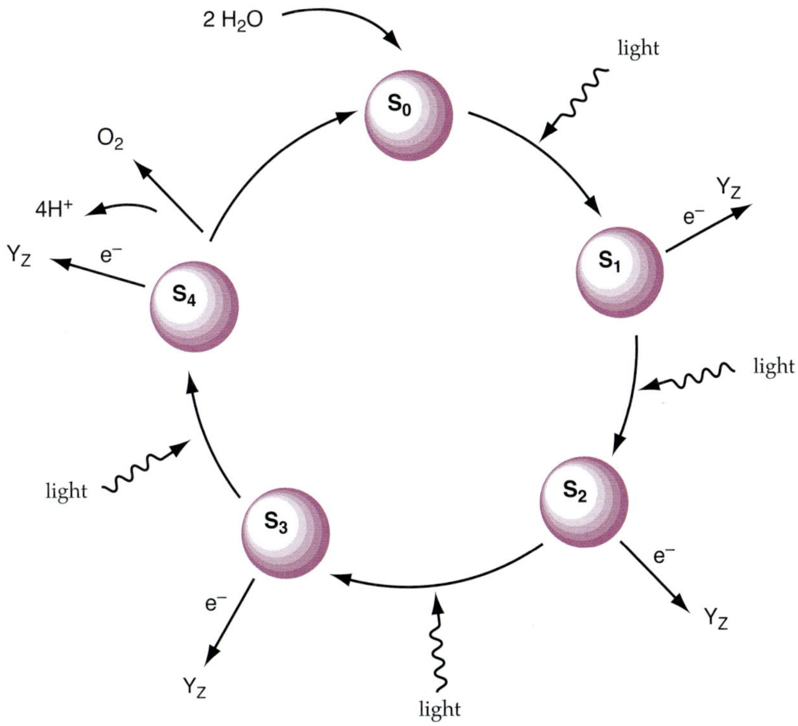

FIGURE 1.16 The "water clock" in PSII. Based on the Joliot-Kok experiment (Figure 1.15), five states of the 4Mn–1Ca ions in the oxygen-evolving complex were proposed. Two H_2O molecules are bound in a "resting" state, S_0. Light absorbed by P_{680} serves to change the redox state of the Mn ions, leading to S_1, and results in the transfer of one electron from the complex to a tyrosine residue, Yz. Subsequent light absorption by P_{680} continues the process through three additional states. O_2, in various oxidation states, remains bound until it is released as a stable molecule at S_4. Protons are also released from the complex at various states but remain in the thylakoid lumen.

involve binding and sequential breakdown of two water molecules coordinated with redox changes in the Mn ions. The precise details whereby $2H_2O$ molecules are oxidized without release of intermediate radicals are still unknown.

1.7 Photosystem I

The bacterial counterpart of PSI is the type of RC found in green sulfur bacteria. In these microorganisms the RC catalyzes cyclic electron transfer as in the purple bacteria or can in some cases reduce NAD. H_2S serves as the reducing agent and replaces the electrons transferred out of the RC. X-ray crystal structures of PSI have been solved for complexes isolated from cyanobacteria and angiosperms. The overall structure resembles that of PSII, suggesting the core protein complexes evolved from the same ancestral RC. The PSI RC in angiosperms consists of 12 core proteins, designated Psa proteins with *psa* corresponding genes. The core of the complex consists of two integral membrane proteins, PsaA and PsaB, six chlorophyll *a* molecules, two phylloquinones, and an 4Fe-4S cluster. The PsaA and PsaB proteins correspond to the D proteins in PSII. They are both helical, integral membrane proteins with a mass of approximately 84 kDa each. These proteins bind the RC pigments, and the primary events in electron transfer occur here.

Light energy is channeled into a pair of chlorophyll *a* molecules (P_{700}) where an electron is freed and rapidly transferred to a phylloquinone in PsaA protein and then to the Fe-S cluster located between the PsaA and B proteins (**FIGURE 1.17**). A membrane-associated protein, PsaC, on the stromal side of the thylakoid accepts the electron from the PsaA cluster. PsaC has a mass of approximately 9 kDa and contains two 4Fe-4S clusters. The PsaC protein binds and reduces the soluble electron carrier, ferredoxin (Fdx). As in other RCs, there are two parallel pathways of electron transfer to the iron-sulfur acceptors. Biophysical studies suggest that, unlike PSII, in PSI both branches are used. Ferredoxin is a soluble protein with a transient association with the PsaD protein on the stromal side of the thylakoid membrane. Fdx is a 2Fe-2S protein in most plants and carries one electron. It associates with the Psa proteins of the RC by ionic interactions. In reduced Fdx these interactions are weakened,

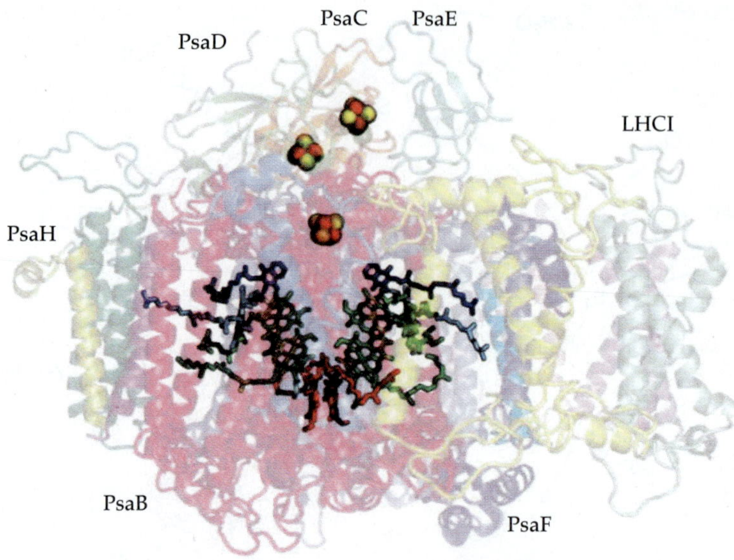

FIGURE 1.17 Structure of PSI from *Pisum sativum.* The crystal structure is shown in the background, viewed parallel to the thylakoid membrane. The stromal side is at the top and the lumenal side is at the bottom. PSI from pea is monomeric with 16 protein subunits and 127 cofactors, mostly chlorophyll a molecules associated with LHCs. Subunits PsaA and PsaB form the core of PSI. The cofactors that function in electron transfer are superimposed in the foreground. The special pair of chlorophyll molecules, P_{700}, are shown in red and the accessory pigments in green. The phylloquinones are colored blue and the iron (red)–sulfur (yellow) cluster is depicted by balls in the center of the core. Electron transfer from this cluster proceeds through two additional clusters located in the PsaC protein. The final electron acceptor, ferredoxin, binds on the stromal side to sites on the PsaC, D, and E proteins. On the lumenal side, the PsaF protein binds reduced plastocyanin that replaces the electron lost by P_{700}. (*Reproduced from*: Nelson, N. and Yocum, C.F., "Structure and Function of Photosystems I and II," *Annu. Rev. Plant Biol.* 57 [2006]: 521–565, Fig. 5.)

and Fdx diffuses away from PSI and reduces NADP on the stromal side of the thylakoid membrane.

$$2Fdx_{reduced} \rightarrow NADP + H^+ \rightarrow NADPH + 2Fdx_{oxidized}$$

The above reaction is catalyzed by a flavoprotein, ferredoxin-NADP reductase (FNR).

The electron lost from the PSI chlorophyll pair is replaced by transfer from plastocyanin. PsaF is a 19-kDa protein found on the lumenal side of the thylakoid membrane. It docks reduced plastocyanin, a Cu-containing redox protein, and facilitates reduction of the oxidized primary electron donor, $P_{700}{}^+$.

1.8 Cytochrome b_6f Complex

The integral membrane proteins that form an electron transfer pathway between PSII and PSI consist of a cytochrome-Fe complex. This complex in thylakoids resembles the cytochrome bc_1 complex in purple bacteria and a similar electron-transfer complex found in the mitochondrial inner membrane. The chloroplast complex consists of several proteins, including two cytochrome b_6 molecules, cytochrome *f*, and a Rieske 2Fe-2S protein. Its function is to oxidize quinols to quinones and reduce plastocyanin. In addition, it transports protons from the stroma into the thylakoid lumen.

$$2QH_2 + 2PC_{ox} + 2H^+{}_{stroma} + Q \rightarrow 2Q + 2PC_{red} + 4H^+{}_{lumen} + QH_2$$

The proton-pumping activity of this complex is referred to as a "Q cycle" and results from the properties of the redox components. Reduced quinones carry two electrons and two protons. Cytochromes, nonheme iron proteins, and plastocyanin each carry one electron and no protons. In step 1 of the Q cycle, a reduced quinone (QH_2) from PSII binds to the b_6f complex (**FIGURE 1.18**). One of its electrons is transferred via the Rieske iron-sulfur protein and cytochrome *f* to plastocyanin. The second electron from the bound quinone is transferred to the cytochrome *b* complex. The two protons carried by the reduced quinone are left in the thylakoid lumen as the oxidized quinone is released from the cytochrome complex.

In step 2 oxidized quinone binds to a second quinone binding site on the complex. The reduced cytochrome *b* transfers its electron to this oxidized quinone producing a semiquinone at the second binding site. Now the cytochrome complex is reset to its initial state and can bind another reduced quinone from PSII.

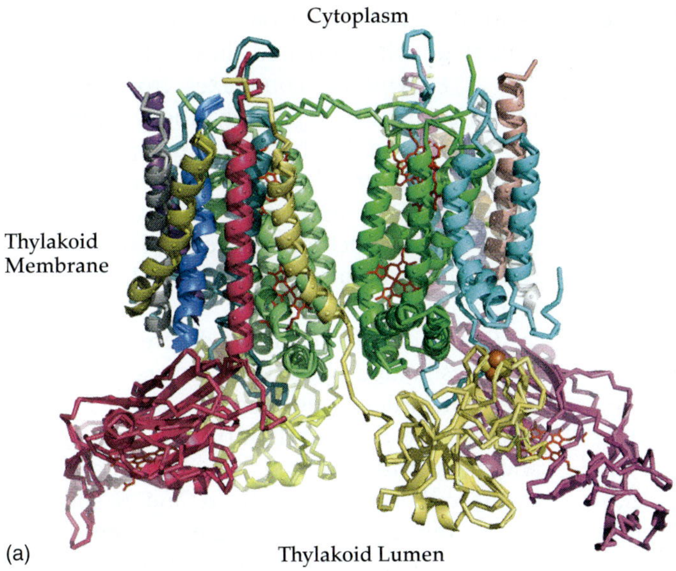

FIGURE 1.18 Structure of the cytochrome b_6f complex from the cyanobacterium, *Mastigocladus laminosus*. (a) The complex is shown parallel to the thylakoid membrane with the cytoplasmic side at the top and the lumenal side at the bottom. The cytochrome complex is a dimer with a twofold axis running perpendicular through the center of the complex. The monomer consists of four large and four small protein subunits and cofactors. The helical portion of the complex is embedded in the thylakoid membrane. Four heme groups are shown in red. A Rieske 2Fe-2S center is shown as red and yellow spheres. The cytochrome *f* protein is on the lumenal side of the complex and is shown as a magenta stick diagram. It provides a binding site for oxidized plastocyanin (not shown). (b) Q cycle in the cytochrome b_6f complex. Reduced plastoquinone binds at the left side of the figure to the Q_0 side. One electron is transferred to the Rieske 2Fe-2S cluster and then to the heme group in cytochrome *f*. Plastocyanin docks near this part of the complex on the lumenal side of the thylakoid membrane and accepts an electron from the cytochrome. The second electron from the reduced quinone is transferred to a heme group in cytochrome *b*. An oxidized quinone then binds at the Q_1 site on the right side of the figure. It accepts one electron from cytochrome *b* and remains bound to the complex. The cycle then repeats. Protons are transferred to the thylakoid lumen when the quinones are oxidized. (Structure from Protein Data Bank, 1vf5. G. Kurisu, H. Zhang, J. L. Smith, and W. A. Cramer, *Science* 302 [2003]: 1009–1014.)

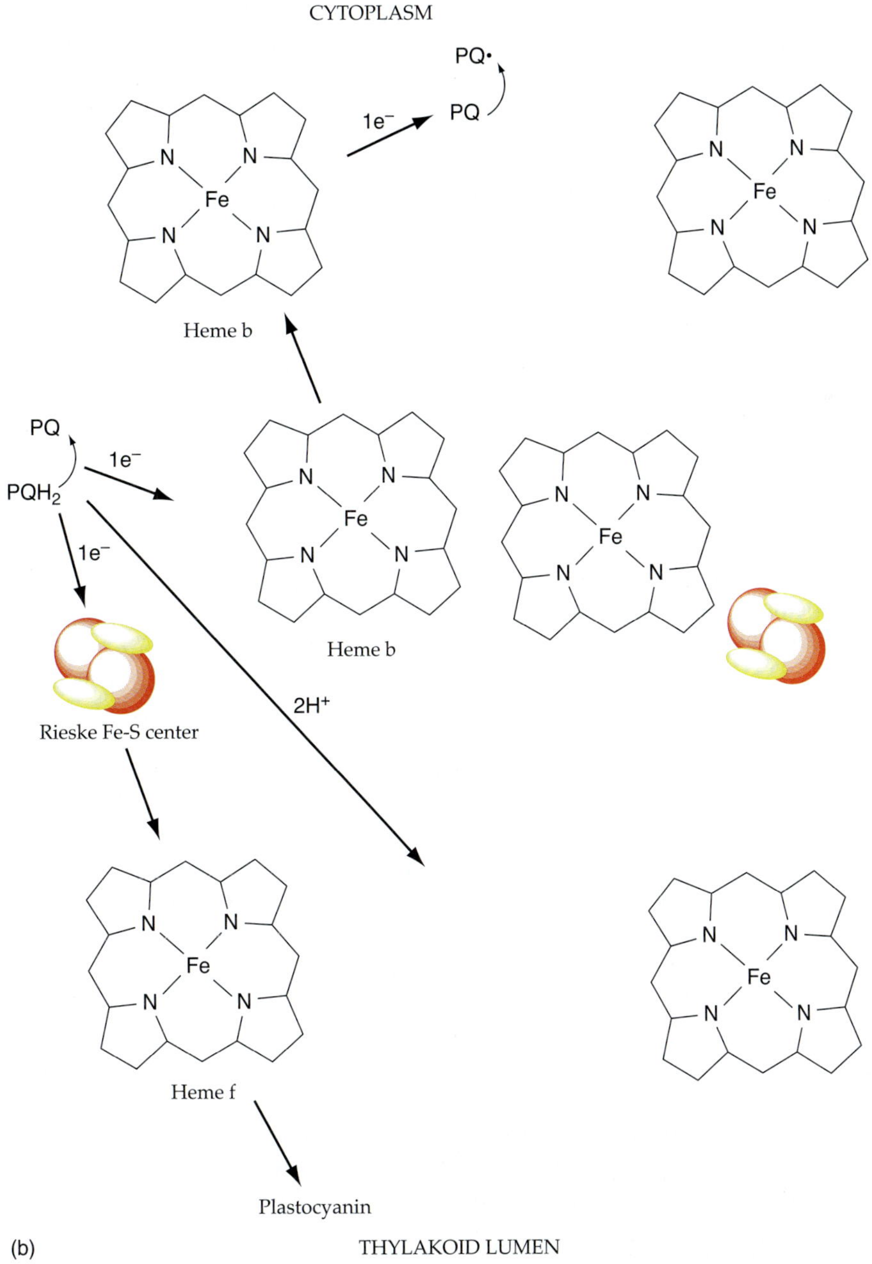

During this second turnover, the semiquinone that is still bound to the complex is fully reduced, which requires two protons that are obtained from the stroma and transferred through the complex. The fully reduced quinol is released from its binding site and can be reoxidized by binding to the other site on the complex. The net result is that two protons are transferred from the stroma to the thylakoid lumen for every one electron that is transferred through the complex to plastocyanin. This enhances the production of an electrochemical potential (proton gradient) across the thylakoid membrane that can be used to drive the synthesis of ATP.

1.9 Photophosphorylation (ATP Synthesis)

The hydrolysis of ATP yields the energy needed to drive most cellular reactions.

$$ATP + H_2O \rightarrow ADP + Pi + H^+$$
$$\Delta G^o = -30 \text{ kJ mol}^{-1}$$

Under conditions in the chloroplast, $\Delta G \approx -52 \text{ kJ mol}^{-1}$.

The energy needed to drive the reaction in the direction of ATP synthesis is obtained from the proton gradient that builds up in the thylakoid lumen during the light reaction. As a result of water oxidation and proton pumping in the Q cycle, the lumen becomes more acidic and the stroma, more alkaline.

Various methods have been used to demonstrate that the stroma becomes more alkaline by 1 to 1.5 pH units during the light reaction. A corresponding decrease in the lumen pH has been calculated so that a proton motive force can be described as

$$\Delta p = \Delta\Psi - 59\Delta pH$$

where p is the proton motive force and Ψ is the electrical potential. In chloroplasts, Mg^{2+} ions are released from the lumen in the light. This approximately balances the charge on the H^+ ions accumulated in the lumen, and, therefore, $\Delta\Psi$ is negligible. This is not true in mitochondria and in some photosynthetic bacteria.

The synthesis of ATP requires a membrane protein complex called the CF0CF1 ATP synthase (**FIGURE 1.19**). CF0 is a membrane-spanning protein complex that provides a "passage" for protons from the lumen to the stroma. CF1 is a protein complex that is connected to CF0 and protrudes into the stroma. It is the catalytic unit that, when isolated, has ATPase activity. The ATP synthase complexes from bacteria, mitochondria, and photosynthetic organisms are highly similar both at the primary and tertiary structural levels. Crystal structures of these enzyme complexes from *Escherichia coli* and bovine mitochondria are available at high resolution, and the spinach chloroplast enzyme is partially resolved.

CF1 is composed of five different polypeptides designated with Greek letters: α, β, γ, δ, and ε, in order of decreasing mass. The three α and three β subunits form the bulk of CF1 and have three binding sites for nucleotides at the α/β interfaces. The α and β subunits are arranged around a single γ protein that, together with the ε protein, form a connection to the CF0 complex.

CF0 is embedded in the thylakoid membrane. It is composed of a single *a* subunit, a single *b* subunit, and a variable number of *c* subunits. The *c* subunits provide a passage for protons into the stroma through the otherwise impermeable thylakoid membrane. A specific aspartic acid residue (asp 61 in the *E. coli* enzyme) on *c* is protonated in the lumen and the uncharged protein then rotates into the hydrophobic bilayer. Deprotonation occurs at the *a/c* interface on the stromal side aided by an arginine residue in the *c* subunit (arg 210). The deprotonated aspartate must have an unusually high pK_a because it would be easier to move an uncharged residue through the bilayer back to a position facing the lumen. The physical rotation of the *c* subunits together with the connected γ/ε proteins provides the energy for a series of conformational changes in the three α/β subunit pairs. In the open conformation, an α/β unit binds ADP and Pi. In the subsequent closed conformation, ATP is produced in an anhydrous compartment where thermodynamics favors ATP synthesis.

A final, energy-requiring conformational change is required to return to the open conformation and release the newly synthesized ATP. In the above model the *b* and δ subunits form a stator that ensures that the *a* subunit and the α/β units do not rotate along with the γ/ε and *c* subunits and dissipate the kinetic energy generated by proton movement into the stroma. This elaborate ATP synthase complex converts the kinetic energy of proton movement into chemical bond energy of ATP.

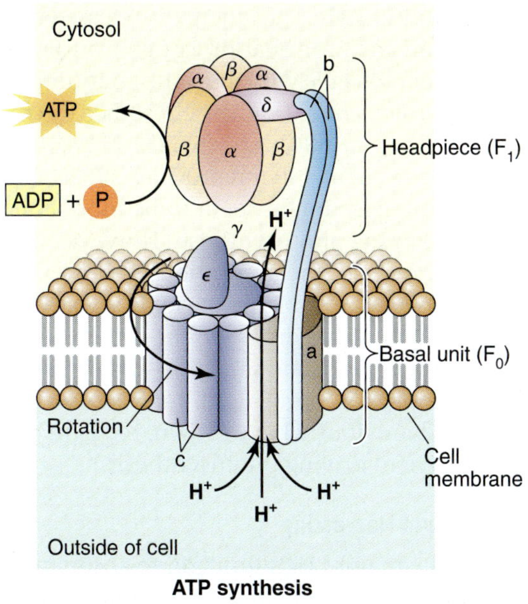

ATP synthesis

(a)

FIGURE 1.19 Structure and function of ATP synthase. The model shown here is based on the crystal structure of the *E. coli* complex. Part a illustrates the basic architecture of the F_0 subunit composed of a, b, and several c polypeptides. This part of the complex is embedded in the cell membrane in the bacterium but would occupy a similar position in thylakoids of photosynthetic organisms. The γ peptide of F_1 protrudes into the cytoplasm or the chloroplast stroma, and is embedded in the center of the $\alpha_3\beta_3$ cluster of the complete F_0F_1 synthase. Protons in the periplasm or lumen bind to an aspartic acid residue in a C subunit that then rotates into the membrane. The proton is released on the cytoplasmic or stromal side of the membrane. This movement provides energy for conformational changes in the α/β subunits. Part b illustrates the nucleotide binding sites at the α/β interfaces. At A, ADP and inorganic phosphate bind to the loose (L) conformation. Proton transport in F_0 causes a conformation change to B where the binding site is closed or tight (T). The phosphodiester bond of ATP is formed in this state. Another conformation change, shown at D, is required to produce the open (O) state and release the ATP.

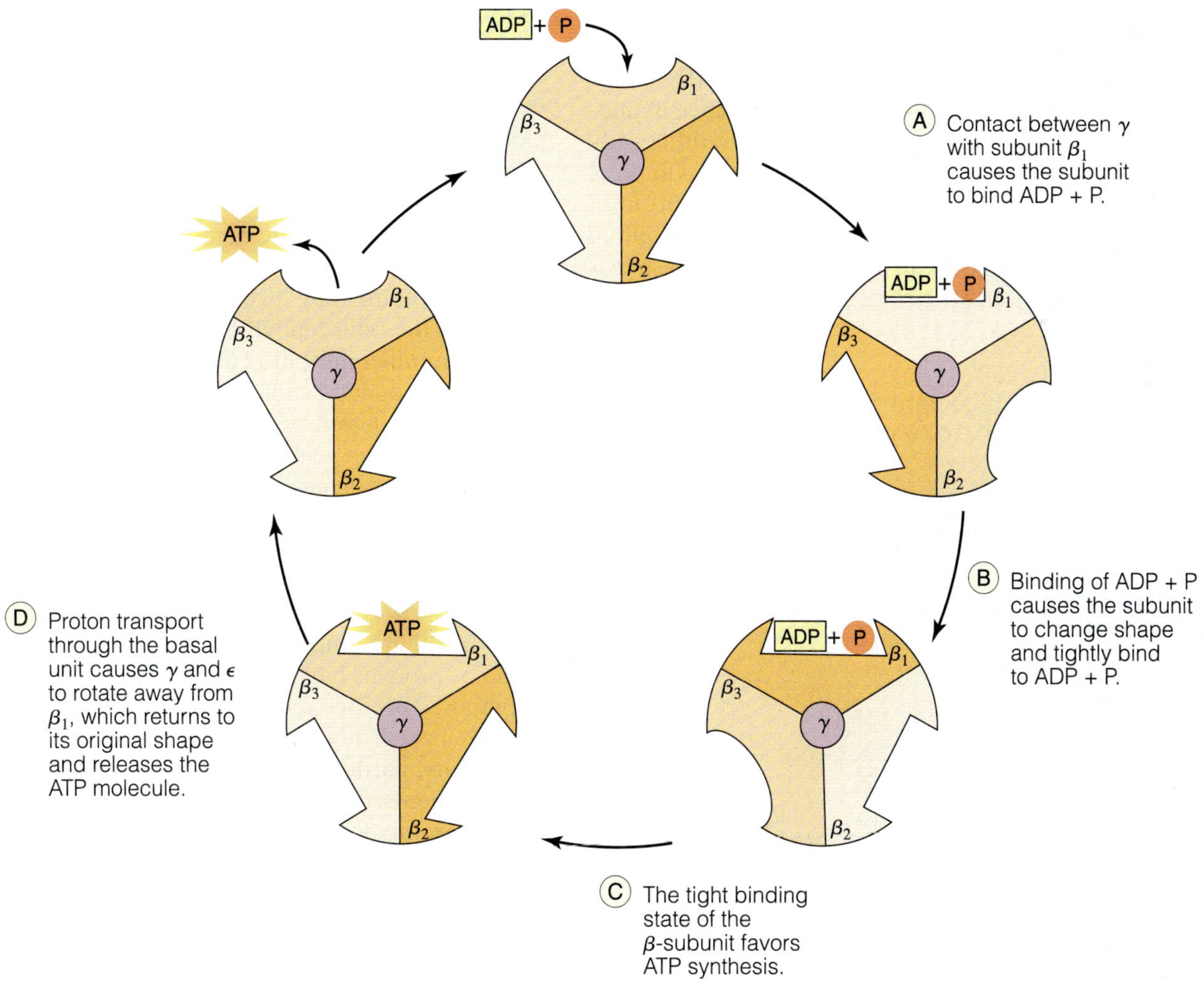

(b)

1.10 Light-Harvesting Complexes

Light Energy

The photosynthetic apparatus in the thylakoid membranes can be divided into two components, the electron transport chain described above and the antenna system, which consists of the proteins of the light-harvesting complexes (LHCs). The light energy that initiates electron transfer in PSI and PSII is rarely the result of direct excitation of the RCs from sunlight. Instead, solar energy is captured by the antenna system and funneled into the RCs. The light used in photosynthesis, 400 to 700 nm, has energy that is inversely proportional to the wavelength: $E = hc/\lambda$, where h is Planck's constant and c is the velocity of light in a vacuum (see Appendix 7 for definitions and values of constants). Light, however, can also be envisioned as massless particles called photons. The energy of a photon is a unit called a quantum. A quantum must be absorbed totally by the material with which it interacts. In the case of chlorophyll, this includes photons in the blue and red regions of the visible spectrum (Figure 1.7 and **FIGURE 1.20**). The light that is effective in photosynthesis is referred to as photosynthetically active radiation (PAR, usually expressed in µmol photons $m^{-2}s^{-1}$). Absorption of a quantum of energy boosts an electron in the chlorophyll molecule to a higher orbital. This excited state is unstable, and the electron will quickly return to the ground state. The energy is released as light (fluorescence) or heat or can be transferred to another molecule with the same quantum states. When calculating the energy absorbed by chlorophyll, light at 680 nm provides 175 kJ mol^{-1}, whereas light at 400 nm provides 296 kJ mol^{-1}.

Blue light might seem to be more efficient at promoting photosynthesis. However, as seen in Figure 1.20, the absorption of shorter wavelengths leads to a highly unstable second singlet state that quickly decays to the somewhat more stable energetic state similar to that produced by the absorption of red light. As experiments described previously have shown, red light is most effective in driving photosynthesis.

Light Harvesting

The major LHCs function to absorb most of the photosynthetically active radiation from light and transfer the excitation energy to the RCs. LHCs consist of several protein subunits, designated Lhca for those associated with LHCI and Lhcb for peptides associated with LHCII. These proteins bind various chlorophylls and carotenoids. LHCI is the complex associated with PSI, and LHCII is most often associated with PSII. The antenna peptides are sometimes referred to as cabs, chlorophyll *a*/*b* binding proteins, although in some organisms chlorophyll *b* is replaced by other types of chlorophylls. These proteins may also be referred to as CP proteins (i.e., chlorophyll-binding proteins) and have a number indicating their relative mobility on sodium dodecyl sulfate-polyacrylamide gels. Because membrane proteins often have anomalous mobility on gels, the numbers may not reflect an accurate mass of the protein.

The proteins of the LHCs are encoded by a family of nuclear genes referred to as *lhca* genes for proteins associated with LHCI and PSI and *lhcb* genes for proteins associated with LHCII. LHCI and LHCII are multimers of these Lhc proteins. All Lhc proteins have a similar tertiary structure composed of three membrane-spanning helices (**FIGURE 1.21**). They have significant amino acid similarity, particularly in the regions that bind the pigments; these include, in most cases, 12 to 15 chlorophylls and 2 to 4 carotenoids.

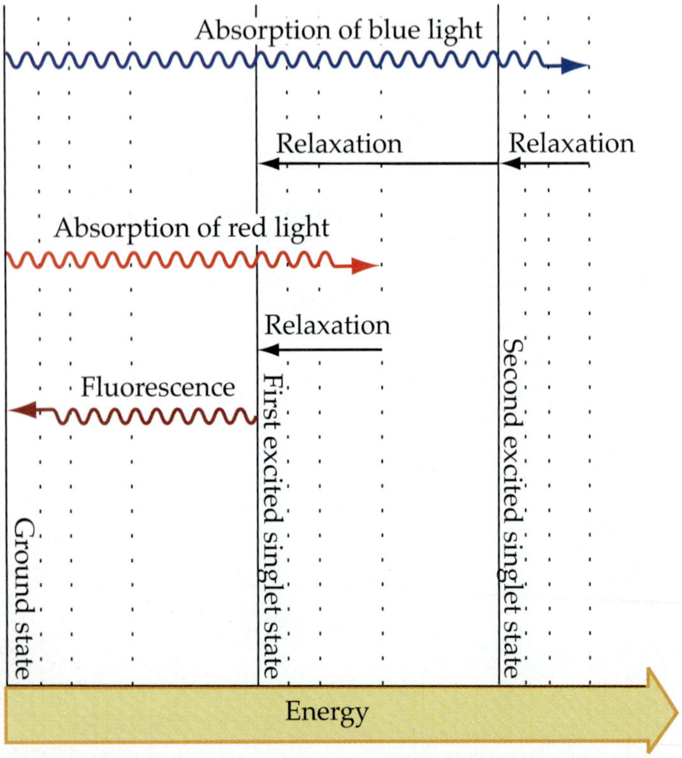

FIGURE 1.20 Molecular orbitals of chlorophyll. The diagram shows that excitation with blue light sends an electron to the second excited singlet state that quickly relaxes to the first excited singlet state. The energy is lost as heat. The first excited singlet state corresponds to the excitation due to red light. Basically, only energy from this state, corresponding to red light absorption, is available to drive photosynthesis. (*Adapted from*: Buchanan, et al., *Biochemistry and Molecular Biology of Plants*, Wiley-Blackwell, 2002. Pg. 573, Fig. 12.3.)

■ State Transitions

The number of LHCIs associated with PSI appears to be constant, whereas the LHCIIs associated with PSII depend on environmental conditions. Plants grown in high light (PAR > 400) usually have fewer LHCIIs than those grown under low light conditions. In general, PSI is more efficient at electron transfer than PSII. The D1 protein in PSII is also vulnerable to light-induced damage and must be constantly synthesized and replaced. The amount and positioning of LHCII is therefore controlled at several levels. The regulation of the amount of Lhcb polypeptides has been shown to require both transcriptional and translational controls. A more rapid regulation, however, occurs by the relocation of the LHCII within the thylakoid membrane, a process referred to as "state transitions."

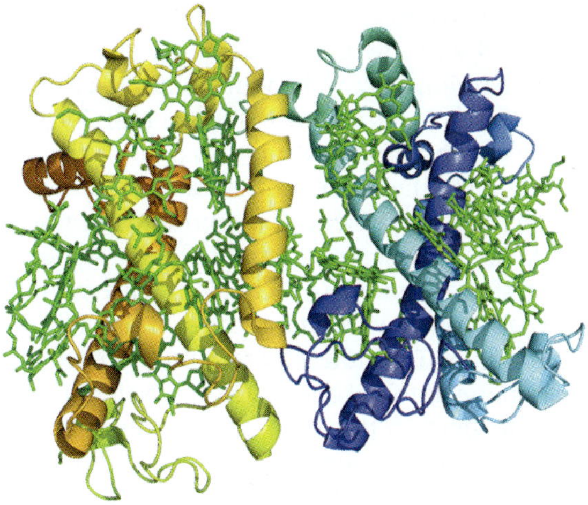

FIGURE 1.21 Structure of the LHCII from spinach. The crystal structure shows a helical molecule with 14 chlorophylls and 4 carotenoids, all colored green. The pigments are not covalently bound to the protein. The LHCs are integral membrane proteins; the helices and attached pigments span the lipid bilayer of the thylakoid membranes. (Structure from Protein Data Bank, 1rwt. Z. Liu, H. Yan, K. Wang, T. Kuang, J. Zhang, L. Gui, X. An, and W. Chang, *Nature* 428 [2004]: 287–292.)

In the normal functioning of the Z-scheme, there is an equal distribution of light to both photosystems; this is referred to as state 1. Under conditions when more light energy is funneled into PSII than can be used for electron transfer, the plastoquinone pool will be highly reduced. This triggers movement of the LHCIIs away from PSII, and energy is then mainly directed into PSI, referred to as state 2. This lateral movement of LHCII refers to its redistribution in the thylakoid membranes of vascular plants with their elaborate architecture of granal and stromal thylakoids.

As shown in **FIGURE 1.22**, the PSI–LHCI complexes are mainly located in the stromal

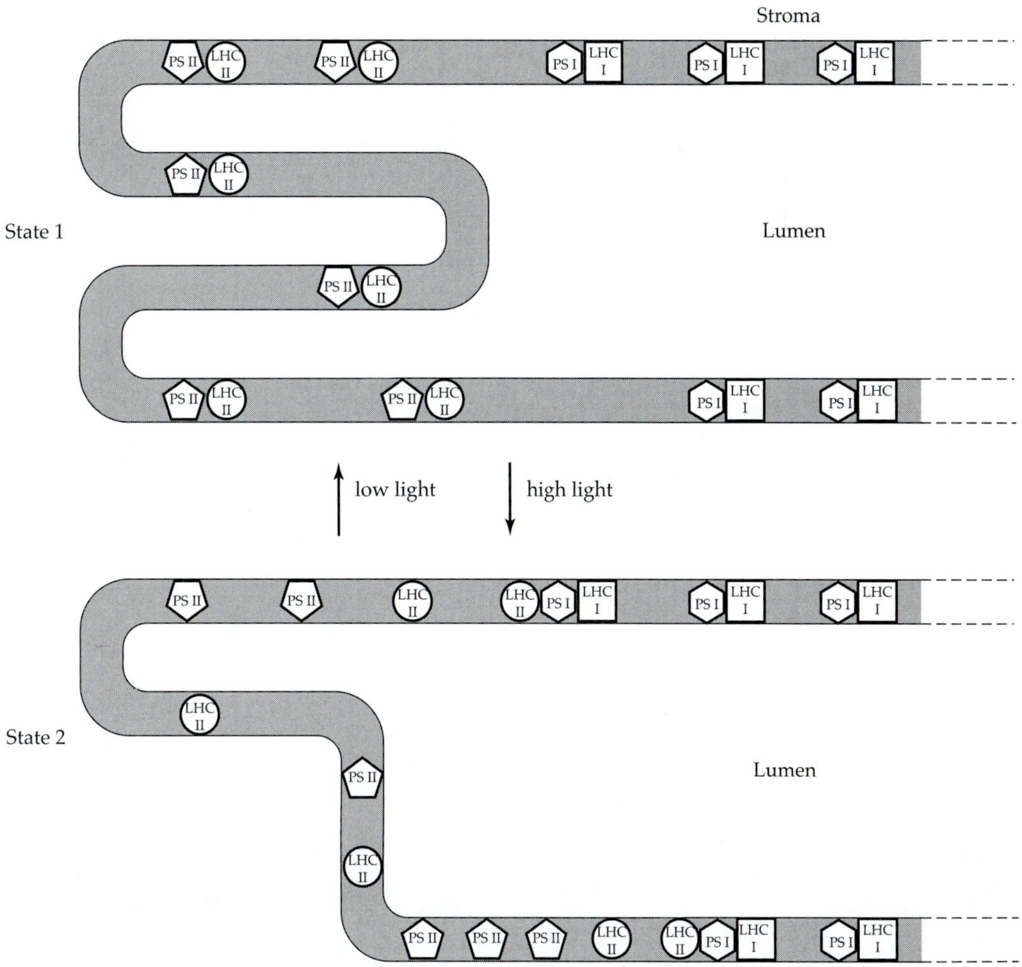

FIGURE 1.22 State transitions. The four integral membrane components of the light reaction are not distributed randomly in the thylakoids. In angiosperms, PSII is mainly located in the granal stacks, whereas PSI is found in the single, stromal portion of the thylakoids. The amount of LHCI associated with PSI is constant, but the association of LHCII with PSII is variable. In low to moderate light conditions, LHCII is associated with PSII in the grana; this is referred to as state 1. Under high light conditions, LHCII is phosphorylated and moves away from PSII and can associate with PSI. More light energy is then channeled into PSI as shown in state 2. State transitions may also be accompanied by changes in the stacking of the thylakoids.

thylakoids, whereas PSII–LHCII complexes are associated with the granal stacks. In state 2 a reorganization occurs in which LHCII is no longer closely associated with PSII and can form a complex with PSI. This may also involve a reorientation of the membrane stacks in addition to a physical movement of the proteins in the thylakoids. A loosening of the thylakoid structure in state 2 could also facilitate replacement of the D1 protein without total disruption of the PSII complex. The state 2 arrangement funnels more energy into PSI and promotes cyclic electron transport. In this state, ferredoxin that is reduced by PSI donates electrons to the cytochrome b_6f complex by a reversal of the ferredoxin-NADP reductase reaction. NADPH can also serve as a cyclic electron donor via a membrane-bound NADPH dehydrogenase. Cyclic electron transport ensures continued production of a proton gradient in the lumen and ATP synthesis without NADP reduction.

What promotes this movement of LHCII and/or membrane reorganization? It has been shown that one of the proteins of the LHCII can be phosphorylated on a threonine residue near its N-terminus by a thylakoid-associated kinase. This protein kinase is activated by redox conditions in the chloroplast, such as a highly reduced plastoquinone pool and elevated pH in the stroma. A phosphorylated LHCII is electrostatically repelled by PSII and moves out of its immediate vicinity, constituting a state 2 configuration with more energy directed into PSI. A change in the chloroplast redox state, as in darkness, inactivates the kinase. A phosphatase removes the Pi from LHCII, allowing it to reassociate with PSII and return to the state 1 configuration.

1.11 Nonphotochemical Quenching

Dissipation of excess light energy can also occur by a process called nonphotochemical quenching (NPQ) during which light energy is converted to radiation that is not active in driving photochemistry. Nonphotochemical quenching is facilitated by specific carotenoids organized into a pathway called the xanthophyll cycle (**FIGURE 1.23**). This cycle is triggered by excess acidification of the

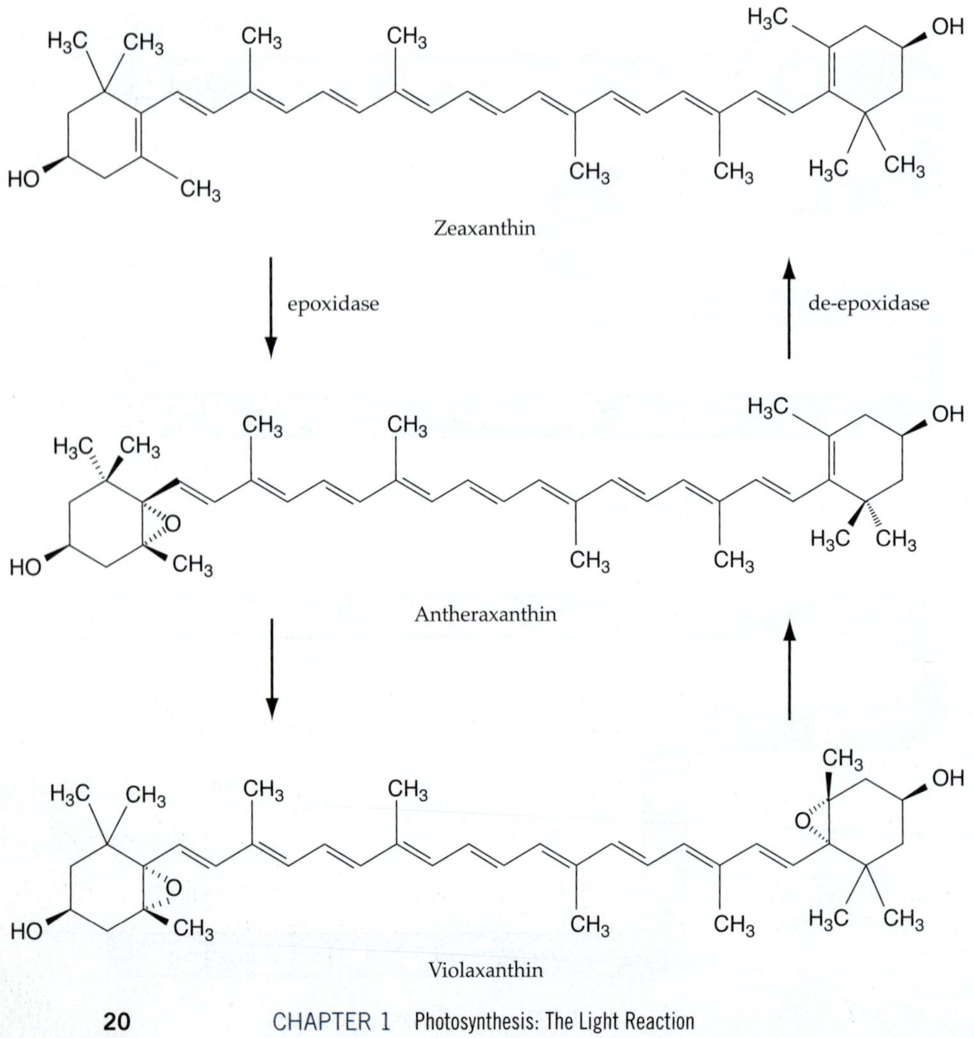

FIGURE 1.23 Xanthophyll cycle. Excess light energy can be dissipated as heat via processes called nonphotochemical quenching (NPQ). The best-characterized mechanism is the de-epoxidation of violaxanthin. The de-epoxidase enzyme is located in the thylakoid lumen and is activated at low pH; it catalyzes a transition from violaxanthin to zeaxanthin. The additional conjugation of the zeaxanthin molecule puts it at a lower energy level than the epoxide carotenoids, which accounts for the heat lost in these reactions.

thylakoid lumen. The low pH activates violaxanthin de-epoxidase, a water-soluble enzyme found in the lumen, which catalyzes the de-epoxidation of violaxanthin to zeaxanthin as shown in Figure 1.23. Zeaxanthin has a higher degree of conjugation and is therefore at a lower energy state. Epoxidation conversely requires energy, and thus the cycle serves to absorb excess light energy and uses it to alter chemical bonds and produce heat.

1.12 Photosynthesis Inhibitors: Herbicides

It is a common practice in biochemical research to use inhibitors that are specific for the reaction or pathway being investigated. Several types of inhibitors have been used to study the light reaction, some of which are also used commercially as herbicides. The substituted ureas and triazines (**FIGURE 1.24**) block electron transport in PSII. They compete for the quinone B binding site on the D1 protein. Some plants, such as maize, are resistant to triazines because they have an effective system to detoxify the herbicide. Both compounds are presumably specific for PSII and thus should pose no toxicity to animals or other nontarget organisms. However, atrazine, at high concentrations, inhibits mitochondrial electron transport. In addition, because of their chlorine substituents, the triazines are particularly resistant to microbial decay. In temperate environments, atrazine may accumulate to toxic levels in drinking water supplies.

Paraquat (Figure 1.24) inhibits electron transport from PSI. It can accept an electron from PSI and inhibit reduction of ferredoxin and NADP. In addition, it can transfer an electron to oxygen, causing the formation of superoxide and hydroxyl radicals that damage the photosynthetic membranes (see Appendix 5). This dual effect makes paraquat an extremely effective herbicide. However, it also causes the formation of oxygen radicals in many other organisms. Use of paraquat as a herbicide is banned in the United States because of its potential for causing DNA damage and cancer in humans and other animals.

FIGURE 1.24 Herbicides that inhibit the light reaction. DCMU and atrazine compete with oxidized quinone B at its binding site on the D1 protein in PSII. Paraquat will accept an electron from PSI and inhibit reduction of ferredoxin. The paraquat radical will also initiate chemical reactions that can cause membrane damage.

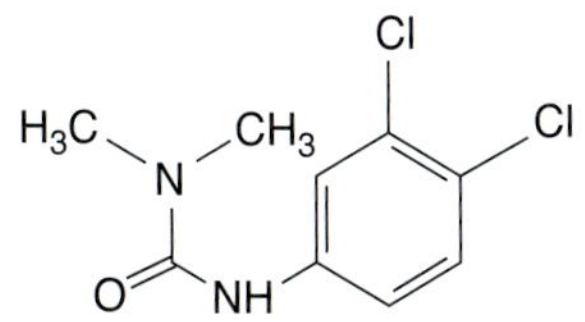

Dichlorophenyl dimethyl urea (DCMU, Diuron)

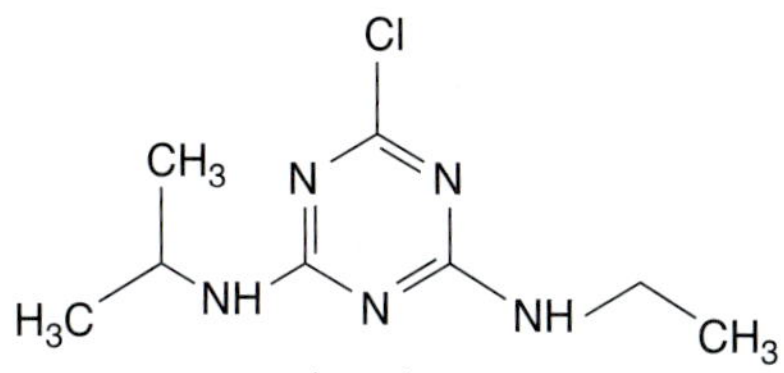

Atrazine

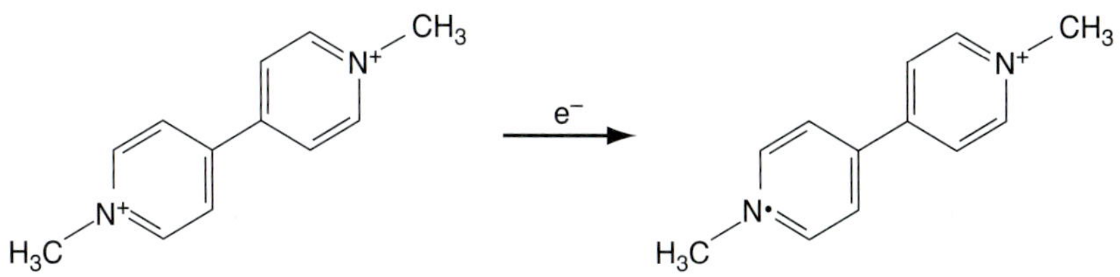

Paraquat

CHAPTER SUMMARY

Photosynthesis is the essential process that supports most life on our planet. Both bacterial and green plant photosynthesis are characterized by similar processes:

a. Capture of light energy by antenna pigments
b. Transfer of the light energy to RCs
c. Using the light energy to free an electron from the RC and transfer it to cofactors that are then used to reduce CO_2
d. Using light energy to generate a proton gradient that can be used to synthesize ATP.

$$2NADP + Pi + ADP + 2H_2O \xrightarrow{\text{Light}} 2NADPH + 2H^+ + ATP + O_2$$

The overall light reaction shown above is common to organisms that perform oxygen-evolving photosynthesis. This reaction requires two RCs that work in series and

CHAPTER SUMMARY, CONT.

are embedded in thylakoid membranes. A precise orientation of the components in the membrane is essential for their function.

PSI consists of several antenna pigments that transfer excitation energy to the RC that contains two chlorophyll *a* molecules, P_{700}. P_{700} is oxidized, and the electron is transferred through a series of pigments to an Fe-S subunit in the RC. This subunit then reduces a peripheral membrane protein, ferredoxin, located on the stromal side of the thylakoid membrane. Ferredoxin, in turn, donates electrons to NADP. P_{700}^+ is reduced by plastocyanin, another peripheral membrane protein located on the lumenal side of the thylakoid membrane. Plastocyanin is reduced by a series of electron transfer events that originate in PSII. PSII also consists of several antenna pigments that transfer excitation energy to a pair of chlorophyll molecules (P_{680}) in the RC core. P_{680} is oxidized, and an electron is transferred to quinone B. Two photo events are required to reduce the quinone to the quinol, which then leaves the RC and diffuses into to the thylakoid lipid bilayer where it is reoxidized by a cytochrome b_6f complex. This complex reduces plastocyanin and also pumps protons into the thylakoid lumen. P_{680}^+ is reduced by a 4Mn-1Ca cluster in the RC that oxidizes H_2O and transfers electrons sequentially to P_{680}. The protons generated by the oxidation of water remain in the lumen. The proton gradient that builds up in the lumen supplies the energy for ATP synthesis via an integral membrane complex, the chloroplast ATP synthase. The NADPH and ATP generated in the light reaction are used to drive CO_2 fixation in the chloroplast stroma, as described in Chapter 2. The oxygen generated in the light reaction is released from the cell but is essential for all organisms, including plants, which also catabolize carbohydrates by aerobic pathways described in Chapter 3.

REFERENCES FOR FURTHER INFORMATION

Allen, J.P., Seng, C., & Larson, C. (2009). Structures of proteins and cofactors: X-ray crystallography. *Photosyn. Res.* **102:**231–240.

Amunts, A. & Nelson, N. (2008). Functional organization of a plant photosystem I: Evolution of a highly efficient photochemical machine. *Plant Physiol. & Biochem.* **46:**228–237.

Blankenship, R.E. (2002). History and early developments in photosynthesis, In: *Molecular Mechanisms of Photosynthesis*, Oxford, UK: Blackwell Science, pp. 26–41.

Boyer, P.D. (2002). A research journey with ATP synthase. *J. Biol. Chem.* **277:**39045–39061.

Kargul, J. & Barber, J. (2008). Photosynthetic acclimation: Structural reorganization of light harvesting antenna—role of redox-dependent phosphorylation of major and minor chlorophyll a/b binding proteins. *FEBS J.* **275:**1056–1068.

Nelson, N. & Yocum, C.E. (2006). Structure and function of photosystems I and II. *Annu. Rev. Plant Biol.* **57:**521–565.

PLANT BIOCHEMISTRY *on the web*

Visit the *Plant Biochemistry* student companion Web site at http://biology.jbpub.com/gleason/plantbiochemistry/ for additional online study resources on these topics.

CHAPTER 2

Carbon Dioxide Fixation

CHAPTER OUTLINE

2.1 Overview

The process of photosynthesis consists of two overall redox reactions. In the light reaction, described in Chapter 1, light energy is converted into chemical energy in the form of reduced cofactors (NADPH) and ATP. These reactions take place in the thylakoid membranes. In the second overall reaction in photosynthesis, CO_2 is reduced to C-H bonds in carbohydrates using the energy in the reduced NADP and ATP. These reactions occur in an aqueous environment, the stroma of the chloroplast, or the cytoplasm of a photosynthetic prokaryotic cell. Unlike the light reaction, CO_2 fixation was more difficult to investigate in isolated organelles because most of the soluble components were lost if the outer membranes were damaged. The experimental systems used initially by most researchers were unicellular green algae such as *Chlorella* sp. Using cultures of *C. pyrenoidosa*, M. Calvin, A. Benson, J. Bassham, and many coworkers at the University of California in the 1950s described the pathway for CO_2 fixation. They took advantage of the fact that ^{14}C, the radioactive isotope of ^{12}C, could at that time be obtained from the nearby Lawrence-Livermore National Laboratory near Berkeley. By feeding $^{14}CO_2$ to cultures of algae and isolating and identifying the ^{14}C-labeled products, they found that the first product of fixation was a three-carbon compound, glyceric acid-3-phosphate. By continuing to label cells for various time periods and isolating metabolites by paper chromatography, they were able to describe the complete pathway, which is now known as the C-3 pathway or Calvin cycle.

Because there is no direct input of light, the C-3 pathway has often been referred to as the "dark reactions." In the absence of the light reaction, however, insufficient NADPH or ATP is available to drive CO_2 fixation. In addition, many of the enzymes in the C-3 pathway are light activated and function slowly, if at all, in the dark. The term "dark reactions" is therefore clearly misleading.

The C-3 cycle (**FIGURE 2.1**) can be divided into three major processes:

1. Carboxylation of ribulose-bis-phosphate (RuBP) to yield two equivalents of glyceric acid-3-phosphate
2. Reduction of glyceric acid-3-phosphate to glyceraldehyde-3-phosphate (GAP)
3. Further metabolism of GAP to generate sugar intermediates and regenerate the initial substrate, RuBP

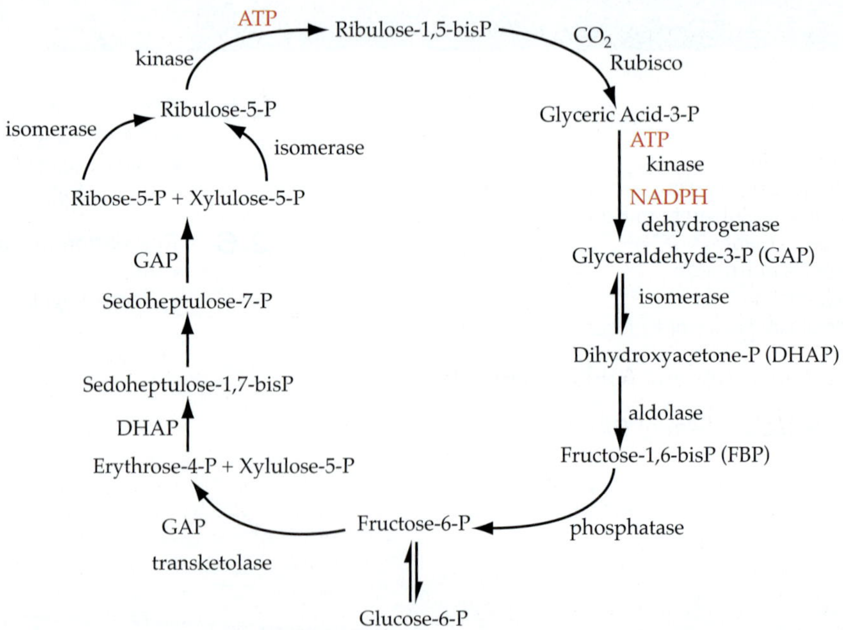

FIGURE 2.1 Outline of the C-3 cycle. The carboxylation reaction starts with the 5-C sugar, ribulose-1,5-bisphosphate. CO_2 is added and two molecules of the 3-C compound, glyceric acid-3-P, are produced. In the second phase of the cycle, reduction, the acids are reduced to glyceraldehyde-3-P (GAP), which is isomerized to dihydroxyacetone phosphate (DHAP). The final part of the cycle consists of several reactions involving GAP and DHAP. The products are sugars, which provide intermediates for further biosyntheses. The final step is catalyzed by ribulose-5-P kinase, which regenerates the substrate, ribulose-1,5-bisP, and completes the cycle.

2.2 Carboxylation

The first step of the cycle involves the addition of CO_2 to the pentose-phosphate, ribulose-1, 5-bisphosphate. This step is energetically favorable with a $\Delta G = -35$ kJ/mol. It is catalyzed by the enzyme, ribulose bisphosphate carboxylase (Rubisco). This enzyme is the major catalyst for production of fixed carbon in the biosphere and the world's most abundant protein. It constitutes approximately 50% of the soluble protein in plant leaves and is often seen under the electron microscope as large crystalline inclusions in plant and algal cells. The enzyme's structure is similar in all oxygenic photosynthetic organisms. It is composed of eight large (L) subunits with a mass of 51 to 58 kDa and eight small (S) subunits with a mass of 12 to 18 kDa. In eukaryotic organisms, the gene for the large subunit is found in the chloroplast genome, and its transcription is light regulated. The genes for the small subunit form a gene family and are found in the nuclear genome. S subunits are synthesized in the cytoplasm with an N-terminal transit or signal sequence for transport into the chloroplast. Folding and assembly of the subunits occurs in the chloroplast stroma and requires a specific protein called a chaperonin that binds the various subunits and facilitates formation of the active complex.

The genes for both Rubisco subunits have been isolated from several plant species and cloned into *Escherichia coli* to provide large quantities of enzyme and to facilitate site-directed mutagenesis. Unfortunately, the expression of the cloned proteins often yields aggregates called inclusion bodies, and relatively little functional L8S8 Rubisco is obtained. In cyanobacteria, the L and S subunits constitute an operon. Assembly does not seem to require a specific chaperonin. An active cyanobacterial L8S8 Rubisco can be expressed in *E. coli* if the bacterial chaperones, GroES and GroEL, are present. Because it could be cloned and produced in *E. coli,* initial biochemical analyses were done with recombinant cyanobacterial Rubisco.

Structure of Rubisco

The purple nonsulfur bacteria, such as *Rhodospirillum rubrum*, have a simpler Rubisco structure than that found in cyanobacteria and plants. In these organisms the enzyme is a dimer of L subunits. The bacterial enzyme has approximately 28% amino acid identity to the L subunit from spinach, suggesting descent from a common ancestor and facilitating extrapolation of data on the bacterial enzyme to the plant Rubisco. The structure of the *R. rubrum* Rubisco was the first of this enzyme family to be solved by x-ray crystallography. The structure was determined to a high resolution (1.7 Angstrom), which means that amino acid side chains can be clearly located. (See Appendix 4 for discussion of structure analysis.) There are two active sites per L2 consisting of amino acid residues from the C-terminal domain of one subunit and the N-terminal domain of the other. The C-terminal domain forms an eight-stranded α/β barrel. This type of structure is one of the most common protein tertiary structures described to date. The active site is located in the center of the barrel, and amino acid residues that function in catalysis are brought together by loops that connect the eight β-strands to the eight α-helices (**FIGURE 2.2**).

X-ray structure analysis of spinach Rubisco shows a similar barrel structure in each L subunit. The plant L8S8 enzyme is a tetramer of L2 dimers with two active sites per dimer as in the *R. rubrum* carboxylase. Although there is minimal amino acid similarity overall, the two types of L subunits have a similar barrel structure with the active site in the center of the barrel. Conserved residues in the C-domain of one L subunit and a few residues from the N-terminal domain of the neighboring L subunit constitute the sugar-binding site.

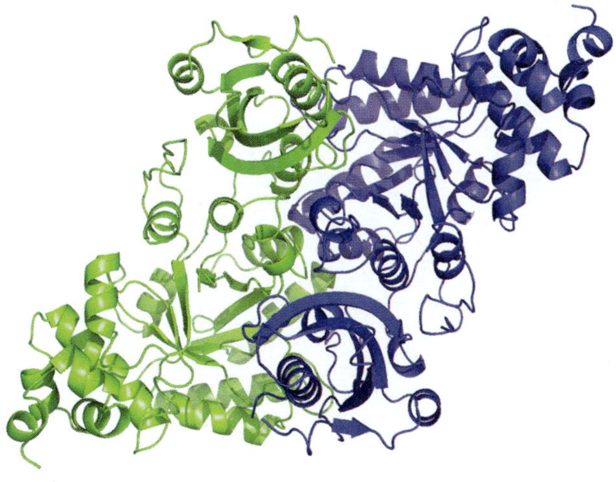

FIGURE 2.2 Crystal structure of the L2 Rubisco from *Rhodospirillum rubrum.* The image shows one L subunit in green and one in blue. The α/β barrel conformation in the C-terminal domain of the protein is seen in the green subunit viewed from the top. The "staves" of the barrel are beta strands shown as arrows. The active site consists of loops connecting the eight beta strands to the eight alpha helices of the barrel. Part of the active site is contributed by amino acid side chains from the N-terminal domain of the blue subunit. (Structure from Protein Data Bank, 5rub. G. Schneider, Y. Lindqvist, and T. Lundqvist, *J. Mol. Biol.* 211 [1990]: 989–1008.)

Binding of Substrate to Rubisco

Rubisco must be activated before it is capable of catalysis by binding a molecule of CO_2. This activator CO_2 forms a carbamate by binding to the epsilon amino group of a conserved lysine residue at the active site (K_{201} in the spinach L subunit). The "activator CO_2" remains bound; it is not a substrate but forms part of the catalytic mechanism. A Mg^{2+} ion is subsequently bound near the carbamate. The Mg^{2+} ion is coordinated by oxygen of the carbamate and by adjacent acidic side chains of aspartic acid 203 and glutamic acid 204. Nonactivated enzyme binds ribulose bisphosphate but in an unnatural configuration, and no catalysis occurs. A combination of structural studies, site-directed mutagenesis, and kinetic analyses has provided a clear picture of the amino acid residues important in sugar binding and catalysis. Some of these are shown in **FIGURE 2.3** with numbering from the spinach enzyme. The sugar-phosphate binds in an extended configuration rather than as a ring. The phosphate on C-1 is located near three glycine residues. There are relatively few ionic interactions with the enzyme in this region, and the substrate binding is relatively flexible. The phosphate on C-5 is held in position by stronger ionic interactions with histidine and arginine residues. The Mg^{2+} ion interacts with OH groups on carbons 2 and 3 of the sugar. Binding of the substrate also causes large conformational changes in the enzyme. Loop 6 of the barrel moves to an extended position, closing the top of the barrel like a lid.

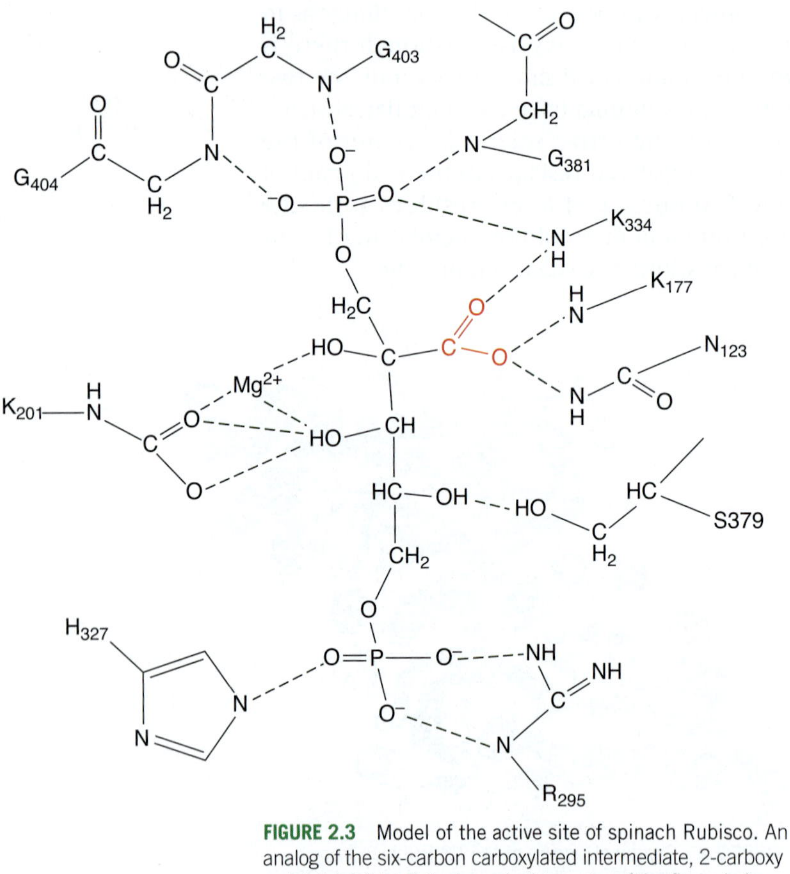

FIGURE 2.3 Model of the active site of spinach Rubisco. An analog of the six-carbon carboxylated intermediate, 2-carboxy arabinitol-1,5-bisP, is shown in the center of the figure in its extended configuration. The phosphate group on C-1 is H-bonded to three glycine residues in the L subunit. Conversely, the phosphate group on C-5 is more tightly bound by a His and an Arg residue. Lysine-201 is carbamylated, and the coordinated Mg^{2+} is positioned nearby. Both serve to bind to the OH groups on C-2 and C-3 and facilitate enolization. The carboxyl group on C-2 is held in position by H bonds from two lysines and an asparagine residue. Several water molecules are also present in the active site that bind to the sugar but are not shown in this figure.

Rubisco Catalysis

After enzyme activation and binding of the sugar substrate, the next step in catalysis is enolization between C-2 and C-3 (**FIGURE 2.4**), which involves removal of a proton from C-3 of ribulose bisphosphate. The carbamate on lysine-201 is in resonance equilibrium with the epsilon-N of this lysine, forming a base to facilitate H^+ removal and enolization of the sugar. Asp 203 and Glu 204, two nearby acidic residues, also influence this process. The carbon of the substrate CO_2, which is slightly positively charged, attacks C-2 of the enol substrate. Although several residues in the active site region may influence CO_2 binding, there is no specific amino acid that binds CO_2 before its attack on C-2 of the sugar. A six-carbon intermediate is formed, 2-carboxy-3-ketoarabinitol-1,5-bisphosphate (3-keto-CABP). Acid treatment of the enzyme *in vitro* quenches the reaction and traps this intermediate. It can be isolated and is relatively stable in solution; when added to activated enzyme, product is formed.

The 3-keto-CABP is then hydrated to a gem-diol, followed by deprotonization of the OH group on C-3. The C-C bond between C-2 and C-3 is cleaved, forming one molecule of glyceric acid-3-phosphate and a C-2 carbanion of glyceric acid-3-P. A stereo-specific protonation of the carbanion is followed by opening of the barrel and release of two molecules of glyceric acid-3-phosphate.

What is the function of the S subunits in the L8S8 Rubisco? Plant Rubiscos are inactive if cloned and produced without the S subunits; these are required to stabilize the quaternary structure and prevent aggregation of the L subunits. However, the simpler *R. rubrum* L2 enzyme functions without S subunits. Cyanobacterial mutant Rubisco without S subunits is assembled in vivo and has about 1% of the carboxylation activity of the wild-type enzyme. It has a lower maximal velocity and produces unnatural products due to improper protonation of the intermediate. Site-directed mutagenesis of the S subunits indicates they influence the kinetic parameters of the holoenzyme but they do not appear to be directly involved in the chemistry of catalysis shown in Figure 2.4.

Kinetics of Rubisco

Rubisco is not an especially efficient catalyst. The Michaelis constant (K_m; see Appendix 3 for review of enzyme kinetics) for substrate CO_2 is approximately 12 μM. The K_m is the concentration of substrate at which the reaction is proceeding at half the maximum velocity (V_{max}). Because the concentration of CO_2 in water under standard conditions is approximately 11 μM, Rubisco is only working at half the rate that is theoretically possible. In addition, its kcat is 3 to 4 s^{-1}. The kcat, or turnover number (V_{max}/[E]), is a measure of the number of substrate molecules converted to product per unit time. For most enzymes in primary metabolism, this number is usually 500 to 1,000 s^{-1}. Compared with other biological catalysts, Rubisco is a very slow enzyme. This may partially account for the fact that plant cells synthesize large quantities of this protein.

Oxygenation

Rubisco binds substrate CO_2 directly to the 2,3-enol of ribulose-bisP and will also bind and fix O_2, producing one molecule of glyceric acid-3-P and one molecule of 2-phosphoglycolic acid (**FIGURE 2.5**). Phosphoglycolic acid is basically a waste product and in many algae and aquatic plants is dephosphorylated by a specific stromal phosphatase to glycolic acid and excreted into the water. Terrestrial plants salvage approximately 75% of the carbon in the glycolic acid via the glycolate cycle, also called the photorespiratory carbon oxidation pathway (**FIGURE 2.6**). This process, called photorespiration, involves the uptake of oxygen and release of carbon dioxide. It also involves transport of intermediates through three organelles, chloroplasts, peroxisomes, mitochondria, and back to the chloroplast. The expenditure of energy for transport and phosphorylation of the salvaged glyceric acid, which can be returned to the C-3 cycle, lowers the overall efficiency of photosynthesis. Basically, O_2 is a competitive inhibitor for CO_2 in the Rubisco reaction. The K_m for O_2 (inhibition constant, K_i, for CO_2) is 200 to 500 μM, and the concentration of O_2 in

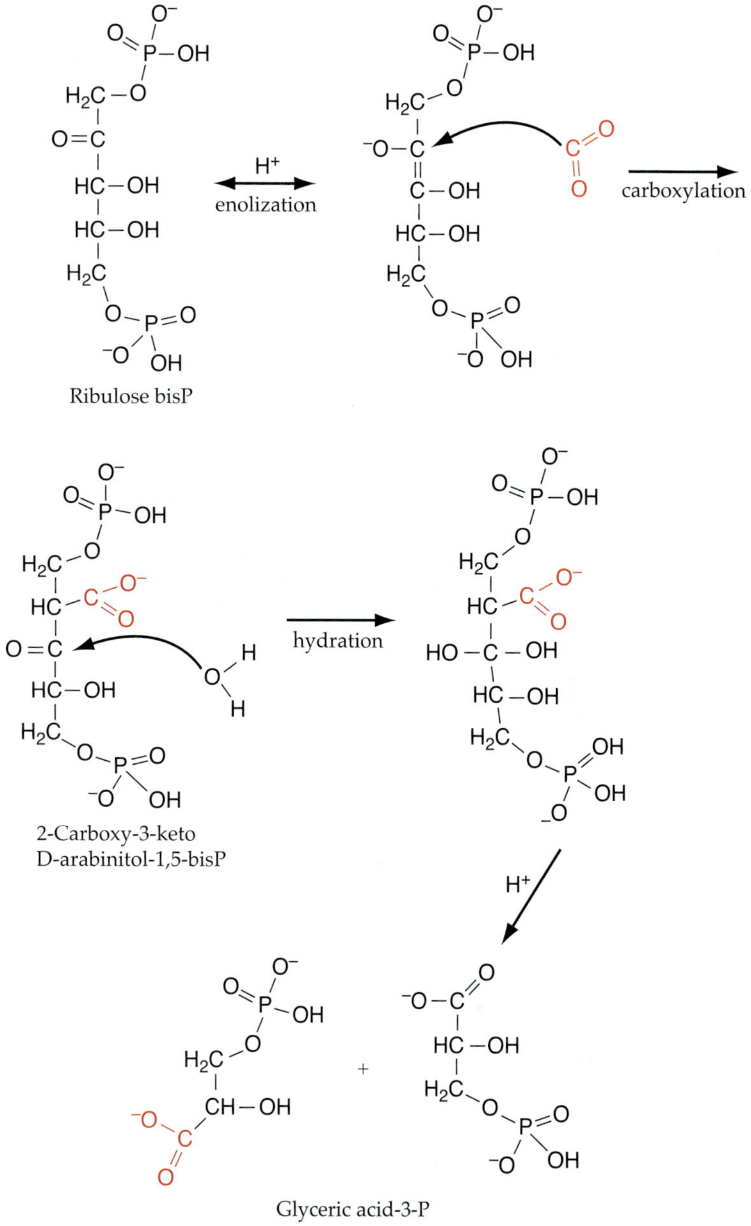

FIGURE 2.4 Carboxylation of ribulose-1,5-bisphosphate by Rubisco. After binding, ribulose bisphosphate undergoes enolization assisted by the carbamylated lysine and Mg^{2+} ion in the activated enzyme (see Figure 2.3). The carbon of the substrate CO_2, which is slightly positively charged, attacks the electron-rich C-2 of the substrate leading to a 6-C intermediate. Hydration of C-3 is followed by cleavage of the C-2–C-3 bond and protonation of a carbanion, resulting in two molecules of glyceric acid-3-P. C–C bond cleavage also results in a conformational change that opens the enzyme active site to release these products.

water under standard conditions is approximately 265 μM. Using these values it is estimated that 20% to 30% of the turnovers of Rubisco result in the fixation of O_2 rather than CO_2. The oxygen-fixing reaction may be even higher under certain physiological or environmental conditions.

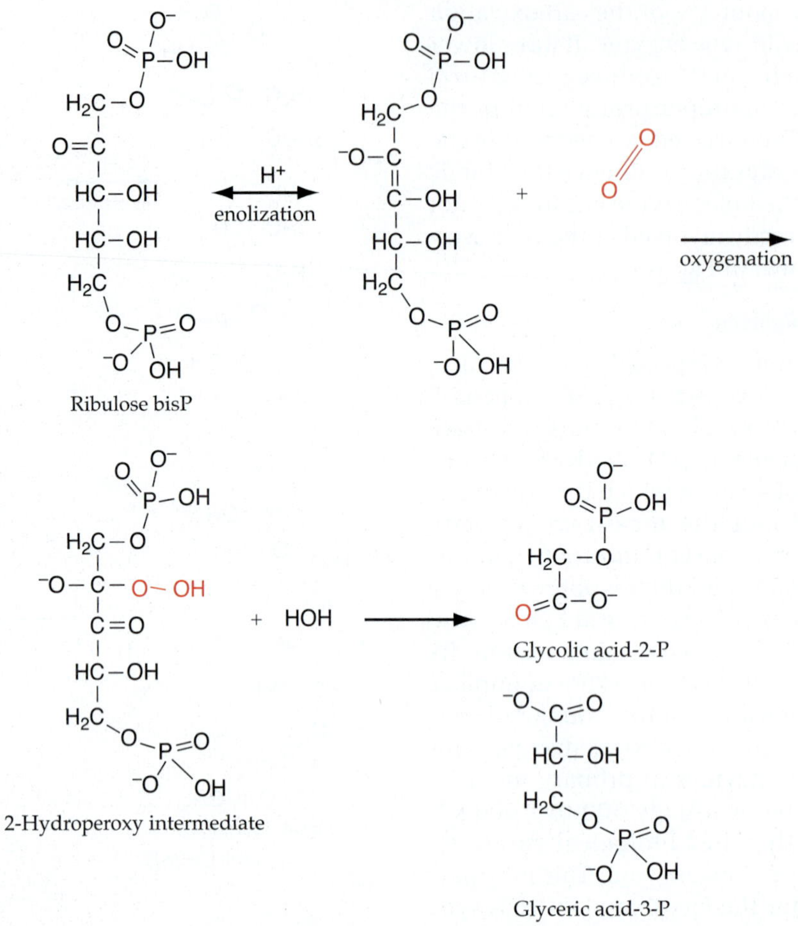

FIGURE 2.5 Oxygenation of ribulose-1,5-bisphosphate by Rubisco. As in Figure 2.4, ribulose-1,5-bisphosphate binds and undergoes enolization. An O_2 molecule can then attack C-2, leading to an unstable hydroperoxy intermediate. Subsequent hydration and C-2–C-3 bond cleavage results in one molecule of 3-phosphoglyceric acid and one molecule of a two-carbon compound, 2-phosphoglycolic acid.

Specificity Factor of Rubisco

The catalytic efficiency of an enzyme is determined by dividing the kcat by the K_m. The ratio of the catalytic efficiency of Rubisco as a carboxylase to its efficiency as an oxygenase is defined as the relative specificity factor. If this factor is then multiplied by the ratio of $[CO_2]/[O_2]$ in the environment of Rubisco, it is a measure of how well the enzyme fixes CO_2 compared with the fixation of O_2. Increasing the specificity factor of Rubisco would minimize the wasteful process of photorespiration and increase plant productivity. The specificity factor for cyanobacterial Rubisco is approximately 50 to 70, whereas that for angiosperms is 75 to 110 (**TABLE 2.1**). This suggests that the process of natural selection has resulted in an enzyme with greater efficiency as a carboxylase as the ratio of atmospheric oxygen to carbon dioxide has increased over geological time.

Can modern techniques in genetic engineering produce an enzyme with a still higher specificity? There are a minimum of 20 critical residues in the active site region of the enzyme, and these are almost completely conserved among the L8S8 Rubiscos. Site-directed mutagenesis has shown that these residues are required for maximum kcat. Very few mutations result in enzymes with increased specificity, and these inevitably show a reduced rate of reaction. To date, no genetic engineering of Rubisco has resulted in a more efficient carboxylase. Structural studies show that many of the active site residues are in van der Waals contact with the S subunits so that engineering of these subunits can also cause changes in the specificity factor. However, most experimental data indicate that, at the protein level, Rubisco is optimized for CO_2 fixation.

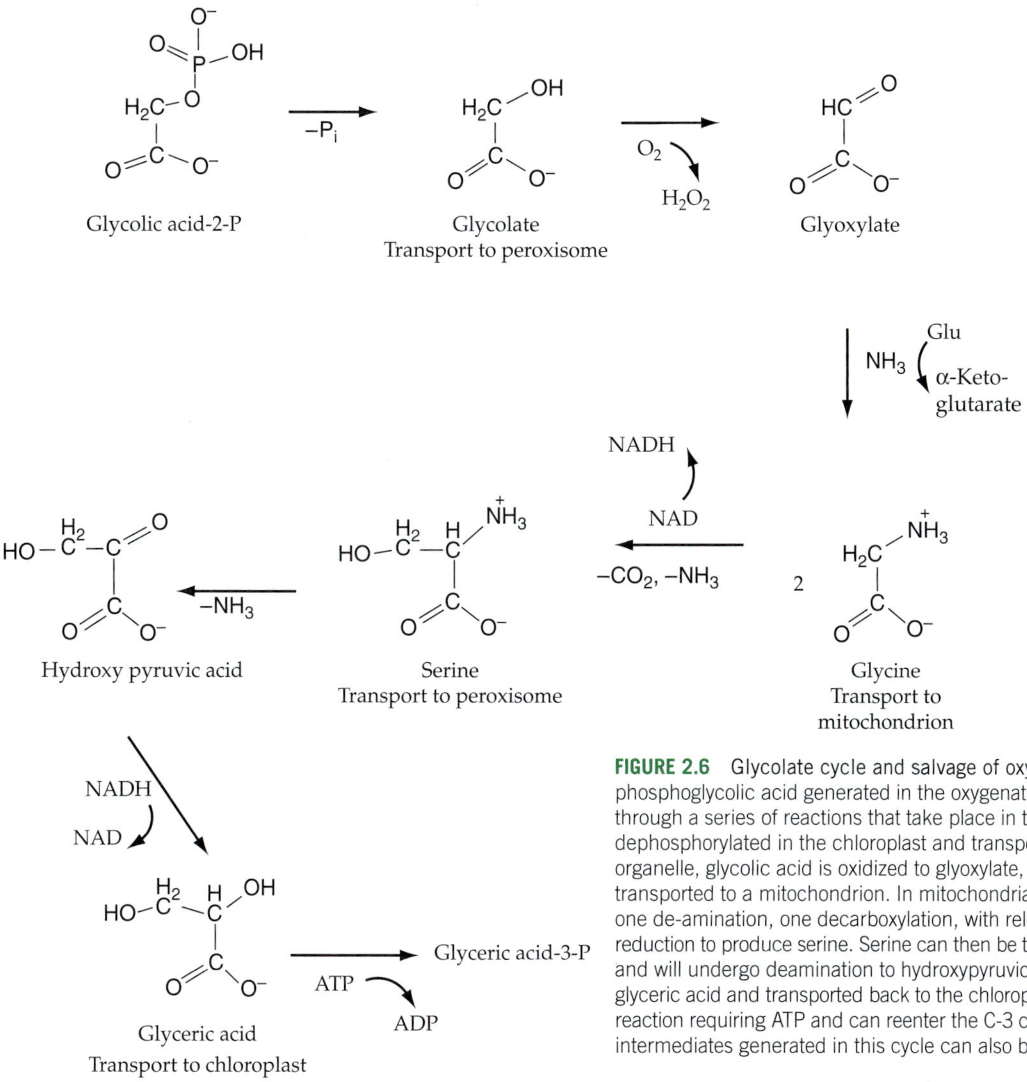

FIGURE 2.6 Glycolate cycle and salvage of oxygen fixation. In terrestrial plants the phosphoglycolic acid generated in the oxygenation reaction of Rubisco is recovered through a series of reactions that take place in three organelles. Glycolic acid-2-P is dephosphorylated in the chloroplast and transported to a peroxisome. In this organelle, glycolic acid is oxidized to glyoxylate, transminated to glycine, and transported to a mitochondrion. In mitochondria, two glycine molecules undergo one de-amination, one decarboxylation, with release of CO_2, and condensation and reduction to produce serine. Serine can then be transported back to the peroxisome and will undergo deamination to hydroxypyruvic acid. This acid can be reduced to glyceric acid and transported back to the chloroplast. It is then phosphorylated in a reaction requiring ATP and can reenter the C-3 cycle. In addition to salvage, the intermediates generated in this cycle can also be used in other metabolic reactions.

Table 2.1 Specificity factors for rubisco from different species

Type	S_{rel}
Bacterial L2	10–20
Bacterial L8S8	50–80
Cyanobacteria	40–60
Eukaryotic algae: green	60–70
Eukaryotic algae: non-green	100–170
C-3 plants	80–100
C-4 plants	70–100

S_{rel} is defined as the catalytic efficiency of Rubisco as a carboxylase divided by its catalytic efficiency as an oxygenase. When multiplied by the ratio of CO_2 to O_2 in the cell, the result provides an estimate of how much CO_2 is fixed vs. O_2 fixed.

Data were obtained from several sources. (See recent summary in Figure 2 in Savir, Y., Noor, E., Milo, R., and Tlusty, T. (2010). Cross-species analysis traces adaptation of Rubisco toward optimality in a low-dimensional landscape, *Proc. Natl. Acad. Sci.* **107**:3475–3480.)

2.3 Reduction of Glyceric Acid-3-Phosphate

The second reaction in the C-3 cycle is the reduction of glyceric acid-3-phosphate to the corresponding aldehyde (GAP). The reaction has a large positive ΔG and requires both ATP and reduced NADP, the products of the light reaction. The steps are shown in **FIGURE 2.7**. The first step is phosphorylation to 1,3-bis-phosphoglyceric acid, which requires ATP and is catalyzed by a kinase. The bisphosphate then binds to a cysteine-SH group in the active site of a dehydrogenase and is reduced to the aldehyde. This reduction is one of the major energy requiring steps of the C-3 cycle. GAP is readily isomerized to the corresponding ketone, dihydroxyacetone phosphate (DHAP), catalyzed by an isomerase. GAP and DHAP are the major intermediates involved in reactions in the remainder of the C-3 cycle.

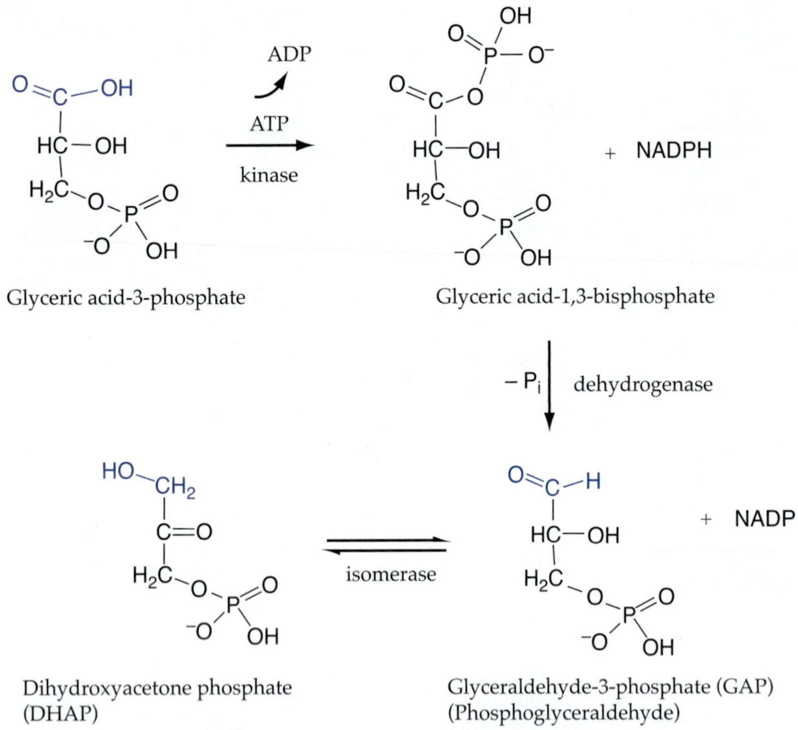

FIGURE 2.7 Reduction of glyceric acid-3-P. This reaction requires a large input of energy. It is initiated by a kinase reaction in which the acid is phosphorylated to glyceric acid-1,3-bis-P. The bisphosphate binds to the chloroplast glyceraldehyde-3-phosphate dehydrogenase, followed by hydrolysis of the phosphate group on C-1 and reduction to the aldehyde. The product, GAP, readily isomerizes to DHAP in a reaction catalyzed by an isomerase. These three reactions provide the two major intermediates for the rest of the C-3 cycle.

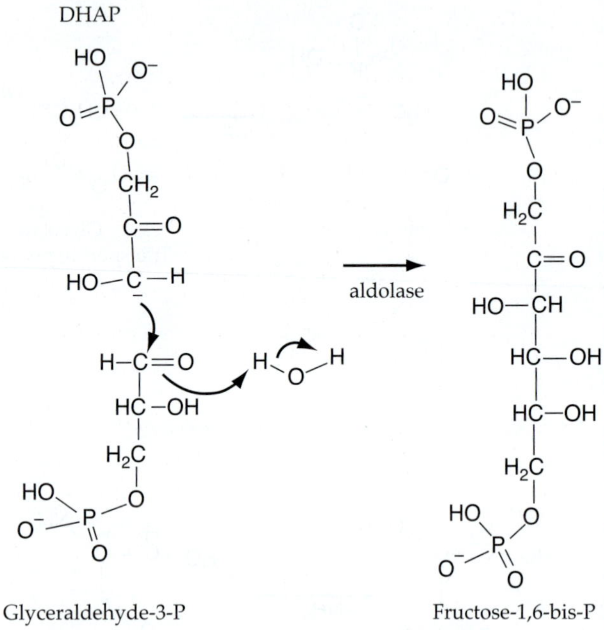

FIGURE 2.8 Aldol condensation reaction. This step in the C-3 cycle is an example of a common reaction in metabolism. The ketone, DHAP, binds to the active site of an aldolase. Removal of a proton from C-3 generates a carbanion that attacks the aldehyde carbon of GAP. Formation of a C–C bond yields a six carbon sugar, fructose-1,6-bis-P. A reverse aldol reaction can be catalyzed by the same enzyme.

2.4 Further Metabolism of GAP/DHAP

The remainder of the C-3 cycle involves various intermediates formed from GAP and its isomer, DHAP (Figure 2.1). These two triose phosphates are combined by an aldol condensation reaction to form a six-carbon sugar. The reaction, shown in **FIGURE 2.8**, is catalyzed by an aldolase. The product is fructose-1,6-bisphosphate. In the next step a phosphatase removes the phosphate from C-1 and yields fructose-6-P, a key intermediate. Fructose-6-P can be removed from the cycle and isomerized to glucose for starch synthesis. In most plants starch is synthesized and stored in plastids. Alternatively, fructose-6-P can undergo a series of reactions with GAP or DHAP to produce 4-, 5-, and 7-C sugar-phosphate intermediates that are used in a variety of other biosyntheses that occur in the chloroplast. To complete the C-3 cycle and maintain a constant supply of substrate for Rubisco, ribulose-5-P is phosphorylated by a kinase, which is the second energy-requiring step of the C-3 cycle. The overall reaction for the C-3 cycle to produce a hexose can be written as follows:

$$6CO_2 + 18ATP + 12NADPH \rightarrow C_6H_{12}O_6 + 18ADP + 18Pi + 12NADP$$

This indicates a minimum stoichiometry of three ATP and two NADPH required per one CO_2 fixed.

2.5 CO_2 Concentrating Mechanisms

Many photosynthetic organisms have evolved biochemical strategies to improve the efficiency of the C-3 cycle by eliminating the oxygenation of ribulose bisphosphate by Rubisco. These generally involve some mechanism to concentrate CO_2 intracellularly.

C-4 Pathway

In the 1960s H. Hatch and R. Slack attempted to repeat the $^{14}CO_2$-labeling experiments of Calvin and coworkers using sugar cane as their experimental organism. They found that the first products labeled with ^{14}C were four-carbon dicarboxylic acids rather than glyceric acid. Subsequent surveys showed that this was the case with many tropical and subtropical plants, principally monocots (grasses). These are now designated C-4 plants. C-4 plants add CO_2 to phosphoenolpyruvic acid (PEP) to generate the 4 C intermediate, oxaloacetic acid. The enzyme, PEP carboxylase (PEPC), uses bicarbonate ion (HCO_3^-) as a substrate, unlike Rubisco, which uses gaseous CO_2. The crystal structure of the enzyme from *Zea mays* has been solved to

3 Å (angstrom = 10^{-10} m) resolution but shows remarkable similarity to a PEPC in *E. coli*, which has been refined to a higher resolution. The carboxylase subunit has a molecular mass of approximately 110 kDa, and the four subunits of the active enzyme constitute a dimer of dimers. The active domain is an eight-stranded beta barrel structure with the rest of the enzyme consisting of alpha helices. A Mg^{2+} ion is bound at the active site followed by binding of bicarbonate to arginine and histidine residues. O_2 does not compete with bicarbonate, and there is no oxygenation reaction (**FIGURE 2.9**).

When CO_2 is dissolved in water, it slowly forms carbonic acid. This reaction can be catalyzed by carbonic anhydrase, another enzyme found in high concentrations in the cytoplasm of C-4 leaf mesophyll cells.

$$CO_2 + H_2O \xrightarrow[\text{carbonic anhydrase}]{} H_2CO_3 \rightarrow HCO_3^- + H^+ \quad (pK_a = 3.5)$$

In many C-4 plants, such as sugarcane and maize, the product of PEPC, oxaloacetic acid, is reduced to malic acid by a stromal NADP-dependent dehydrogenase. In other plants such as millet, it can be converted by a transamination reaction to aspartic acid. In one of three common versions of the C-4 pathway, malate is transported out of the mesophyll cells where it is produced to cells that surround the vascular bundles of the plant, called bundle sheath cells; this wreath-like arrangement of cells is called Kranz anatomy (**FIGURE 2.10**). Unlike C-3 plants, the mesophyll cells of the C-4 leaf do not contain Rubisco in their chloroplasts. Instead,

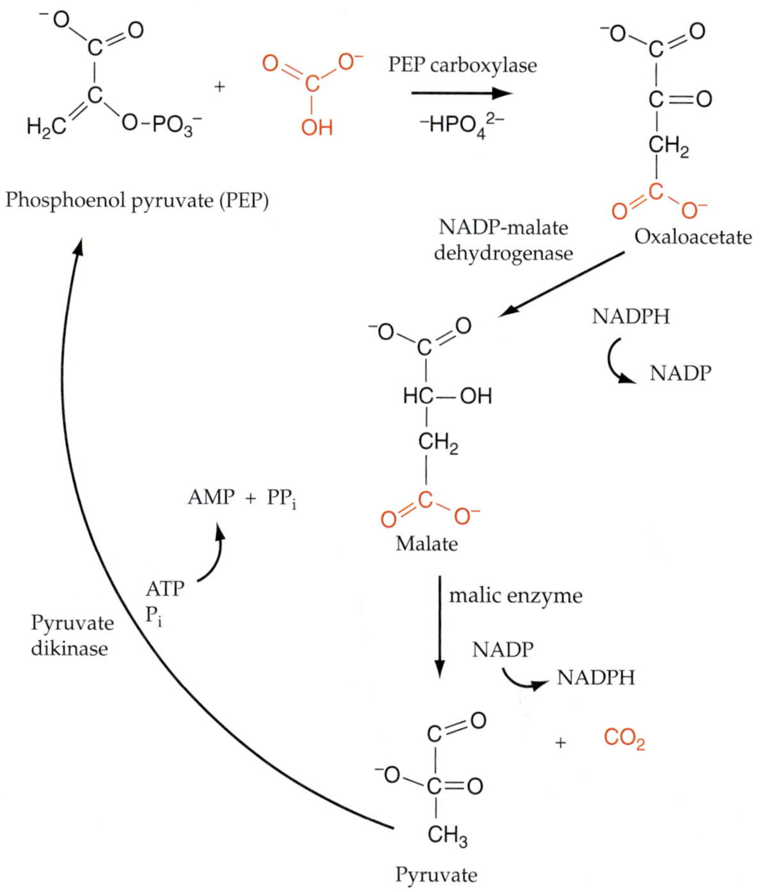

FIGURE 2.9 C-4 cycle in sugarcane and maize. The initial step in the cycle, carboxylation of PEP, is a reaction similar to that catalyzed by Rubisco in that CO_2 attacks a substrate enol intermediate. The important difference here is that PEPC binds bicarbonate anion rather than gaseous CO_2. Removal of a phosphate group yields the product, oxaloacetic acid. In many C-4 plants this is reduced to malate, which can be transported to bundle sheath cells or stored intracellularly. Malate is subsequently oxidatively decarboxylated by malic enzyme yielding pyruvate and CO_2. The resulting CO_2 is fixed by Rubisco. The pyruvate is recycled in a kinase type reaction that requires the hydrolysis of two phosphoester bonds in ATP. A phosphate group is then added to yield the starting compound, PEP.

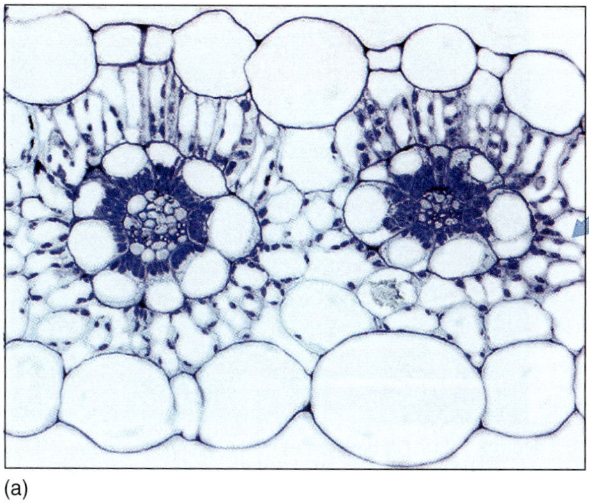

(a)

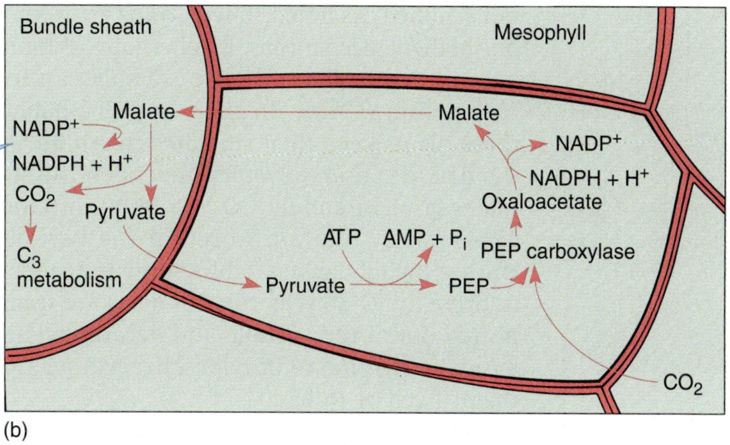

(b)

FIGURE 2.10 Kranz tissue anatomy. In many C-4 plants the photosynthetic cells in the leaf are arranged to provide an internal "compartment" with elevated CO_2 concentration. The leaf mesophyll cells (a), which are the site of the C-3 cycle in C-3 plants, contain relatively few chloroplasts in a C-4 plant. Instead, the cytoplasm has elevated levels of carbonic anhydrase and PEP carboxylase. The product of C-4 fixation, oxaloacetate (b), is reduced by NADP-malate dehydrogenase and transported to cells that surround the vascular tissue in the leaf, the bundle sheath cells. These cells have an abundance of chloroplasts containing Rubisco and the other enzymes of the C-3 cycle. Malic enzyme catalyzes the oxidative decarboxylation of malate and release of CO_2, which is 10 or more times higher in concentration in these cells than in the atmosphere. Fixation in these cells greatly reduces the oxygenation reaction of Rubisco.

Rubisco and the other enzymes of the C-3 pathway are in chloroplasts of the bundle sheath cells. Malate, transported from the mesophyll cells, undergoes oxidative decarboxylation to pyruvate catalyzed by malic enzyme. The CO_2 that is released is fixed by Rubisco.

$$\text{Malate} + \text{NADP} \rightarrow \text{Pyruvate} + CO_2 + \text{NADPH}$$

The actual net fixation of carbon into sugars then proceeds as in C-3 plants. The C-4 pathway is a modification where the initial fixation into four-carbon acids is an appendage to the C-3 cycle. Although this may seem like a costly mechanism to get around photorespiration, C-4 plants concentrate CO_2 in the bundle sheath cells to approximately 70 μM as opposed to the 5 μM concentration that is common in the mesophyll cells of C-3 plants. Thus, maize and other C-4 plants lose almost no carbon or energy in photorespiration. They are particularly efficient at high temperatures under either dry or humid conditions when C-3 plants keep their stomates closed to conserve water and, therefore, perform little CO_2 fixation. In contrast, in a cool, moist environment C-3 plants generally out-compete C-4 plants due to the extra ATP and transport costs of C-4 metabolism.

Surveys found that C-4 metabolism has evolved independently in approximately 45 different plant families, providing an excellent example of convergent evolution. The enzymes found in the C-4 pathway are present in almost all plants where they have various metabolic functions. In C-4 plants, isoenzyme forms have been co-opted and modified to serve in CO_2 fixation. A major evolutionary innovation is a mechanism to sequester malate after it is formed so that the cytoplasm of leaf mesophyll cells is not acidified. As noted above, Kranz anatomy is one of these mechanisms. An alternative is found in some desert plants that use Crassulacean acid metabolism (CAM). In their hot, dry habitat these plants keep their stomates closed during the day to conserve water. They open the stomates at night and fix CO_2 via PEPC into four carbon acids that are stored in the vacuole. During the day, when light energy is available to drive the C-3 cycle, the C-4 acids are transported out of the vacuole and decarboxylated by malic enzyme to provide intracellular CO_2 that is fixed by Rubisco.

The ATP required to regenerate PEP is provided by respiration, and, in addition, CAM uses a membrane-bound ATPase to pump acids across the vacuolar membrane. Despite the high cost of CAM, it allows plants to survive in environments that are otherwise not conducive to plant growth. As expected, CAM plants grow more slowly than C-3 plants but some CAM species are facultative (i.e., they can switch between C-3 and CAM depending on environmental conditions). Unexpectedly, CAM is also common in aquatic plants. This enables them to take up and fix CO_2 into malate and other four carbon acids at night when their algal competitors are not photosynthetically active.

Carboxysomes and Other Carbon Concentrating Mechanisms

Cyanobacteria and some eukaryotic algae have evolved various other mechanisms to increase intracellular CO_2 concentrations. In many instances the cells take up dissolved CO_2 and convert it to bicarbonate via carbonic anhydrase. In cyanobacteria, the bicarbonate can be stored intracellularly in large inclusion bodies called carboxysomes (**FIGURE 2.11**). This organelle-like body consists of carbonate salts, carbonic anhydrase, and Rubisco. A dynamic equilibrium of CO_2 uptake, conversion to HCO_3^-, and back to CO_2 provides a high concentration of substrate

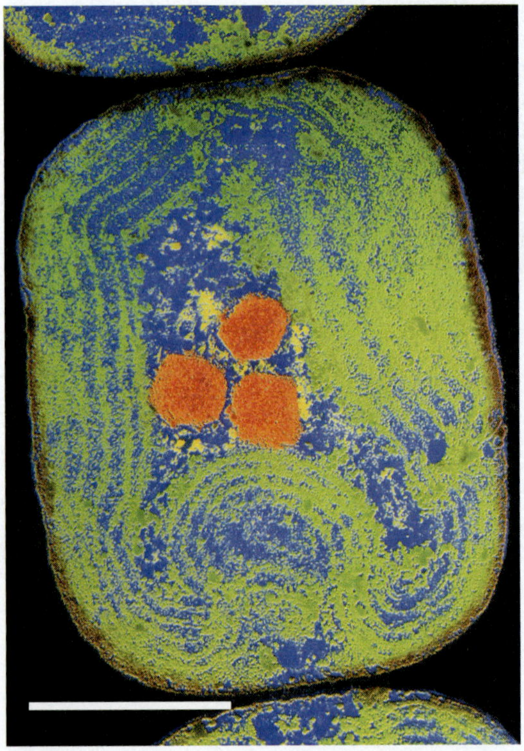

FIGURE 2.11 Carboxysomes in the cyanobacterium, *Pseudanabaena* sp. This is a false color transmission electron micrograph of a cyanobacterial cell. The thylakoid membranes are shown in green. The orange polyhedral bodies in the center of the cytoplasm are carboxysomes. These contain high concentrations of carbonic anhydrase and Rubisco in a semicrystalline state. Intracellular concentrations of CO_2 are several fold higher than in the water environment, thus minimizing the oxygenation reaction of Rubisco. Bar = 1 μm.

for Rubisco with little competition from O_2. Despite the low specificity factor of cyanobacterial Rubisco, these organisms are very efficient at fixing carbon. Eukaryotic algae have also evolved mechanisms to store HCO_3^- in cellular compartments such as vacuoles or the thylakoid lumen. It has been proposed that the L8S8 Rubiscos found in modern plants have the highest specificity possible under prevailing atmospheric conditions. This proposal is supported by the fact that plants and algae have evolved various mechanisms to increase intracellular concentrations of CO_2, thereby increasing substrate rather than producing new genetic alterations to the enzyme. Further study of the steps in these adaptative processes may provide a better route to genetically engineering plants with reduced photorespiration.

2.6 Regulation of Carbon Dioxide Fixation

As mentioned previously, the C-3 cycle does not operate in the dark. In addition to decreased levels of ATP and NADPH in the stroma when the light reaction is not operating, several enzymes involved in CO_2 fixation and sugar conversion are regulated by light. These include Rubisco, NADP-glyceraldehyde-3-P dehydrogenase, fructose-bisphosphatase, sedoheptulose-bisphosphate, and phosphoribulose kinase in the C-3 cycle; PEPC and pyruvate dikinase in the C-4 cycle; and enzymes associated with these cycles, such as NADP-malate dehydrogenase, glucose-6-P dehydrogenase, and ADP-glucose pyrophosphorylase. These enzymes have no chromophores that directly absorb light but are influenced by physiological changes in the stroma that are a result of the light reaction. Because of proton pumping into the thylakoid lumen, the pH of the stroma rises and free Mg^{2+} concentrations increase in the light to maintain a constant electrochemical potential. There is also a rise in the ATP/ADP ratio in the light and an overall more negative redox potential due to increases in reduced ferredoxin and NADPH. All these parameters can affect enzyme activity in the stroma.

Rubisco Activation

The carbamylation of an active site lysine residue (K201, see Figure 2.3) in Rubisco is facilitated at higher pH due to the high pK_a (10.5) of the lysine side chain. It is an unprotonated lysine that binds the activator CO_2. The increase in stromal Mg^{2+} concentration also promotes ion binding to the active site. In addition, many plants have a specific enzyme, Rubisco activase, that is required for enzyme activation. Phosphorylated sugars can bind to un-carbamylated Rubisco and cause the conformational changes in the barrel structure that lead to closing of the active site. Rubisco activase removes bound sugars in a reaction that requires hydrolysis of ATP by the activase. In some plants the activase itself is further subject to redox regulation.

Redox Regulation

Some chloroplast enzymes are regulated by reducing agents. Fructose-bis-phosphatase, sedoheptulose-bisphosphatase, phosphoribulokinase, and the NADP-dependent malate dehydrogenase are activated by reduction, whereas glucose-6-P dehydrogenase in the pentose phosphate pathway is down-regulated. When isolated from chloroplasts, the first three enzymes have very low activity *in vitro*. Activity can be enhanced by adding artificial reducing agents such as dithiothreitol, which reduces protein disulfides (cystine S-S bonds) to sulfhydryls (cys-SH). Reduction often leads to large conformational changes in the proteins that, in turn, modulate activity. Several physiological reducing agents can function like dithiothreitol *in vivo*. A general intracellular reducing agent is the tripeptide glutathione (GSH) (**FIGURE 2.12**). Reduced GSH is a general reducing agent found in most cells at millimolar concentrations. It forms a disulfide-linked hexapeptide when oxidized. The oxidized form (GSSG) is reduced by NADPH and a flavoprotein, glutathione reductase. GSH is the predominant form of the peptide in plant cells in the light and serves to maintain a reducing environment.

Activation by Thioredoxin Another redox agent that specifically reduces other protein disulfides is thioredoxin. Thioredoxins are a family of relatively small proteins (MW = 10−12 kDa) that act as protein disulfide reductases. Several thioredoxin isoforms are found in the cytoplasm, mitochondria, and chloroplasts. These proteins contain a conserved active site sequence, Trp-Cys-Gly-Pro-Cys, that is located on the surface of the protein. Variable regions outside the active site determine the target proteins a particular thioredoxin will bind to and reduce. Thioredoxins, in turn, are reduced by NADPH and a flavoprotein thioredoxin reductase or, in chloroplasts, by a ferredoxin-dependent thioredoxin reductase.

Biochemical and structural studies have demonstrated the site of action of thioredoxin on some of the light-activated chloroplast enzymes. For example, phosphoribulokinase is

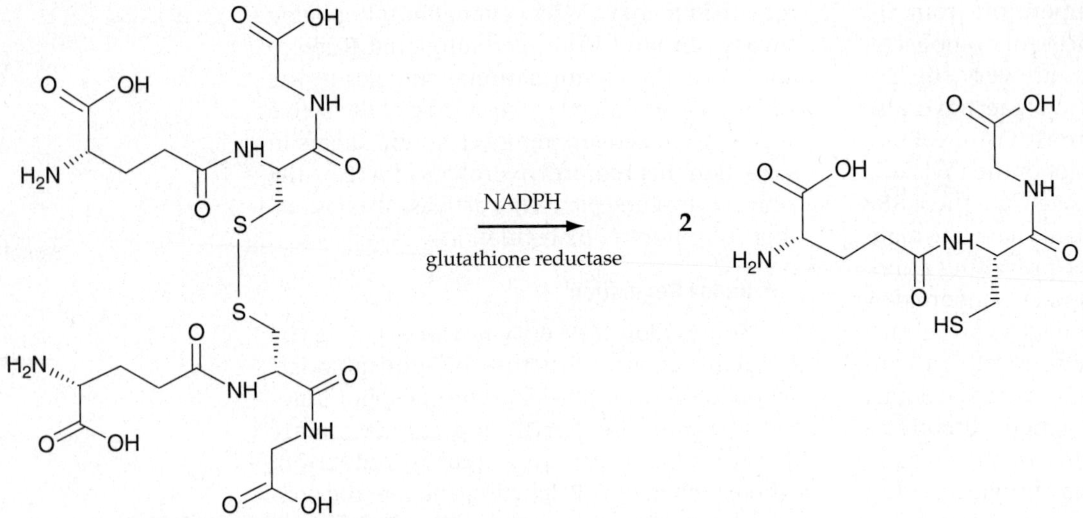

FIGURE 2.12 Glutathione: an intracellular redox buffer. GSH is a tripeptide, γ-glutamyl-cysteinyl-glycine. In the oxidized form, it is a disulfide-bonded dimer (GSSG). The hexapeptide is reduced by NADPH in a reaction catalyzed by a flavoprotein, glutathione reductase. Most cells contain a high GSH/GSSG ratio that maintains proteins and other metabolites in the cytoplasm and chloroplast stroma in a reduced state.

an enzyme that is unique to the C-3 cycle. It is composed of two identical subunits each with a mass of approximately 40 kDa. In the oxidized, inactive spinach enzyme, cys16 forms a disulfide bond with cys55. Cys55 is located in the sugar binding site, and the oxidized enzyme is unable to bind substrate. A specific chloroplast thioredoxin, thioredoxin f, is especially efficient at reducing this disulfide bond and restoring kinase activity (**FIGURE 2.13**).

Thioredoxin f also reduces and activates NADP-dependent malate dehydrogenase. This dehydrogenase catalyzes the reduction of oxaloacetate in the C-4 pathway. A similar malate dehydrogenase is found in C-3 chloroplasts where it serves to shuttle reducing equivalents across the chloroplast membranes. The C-4 enzyme in sorghum has been investigated extensively; it is composed of two identical subunits with eight cysteine residues per subunit.

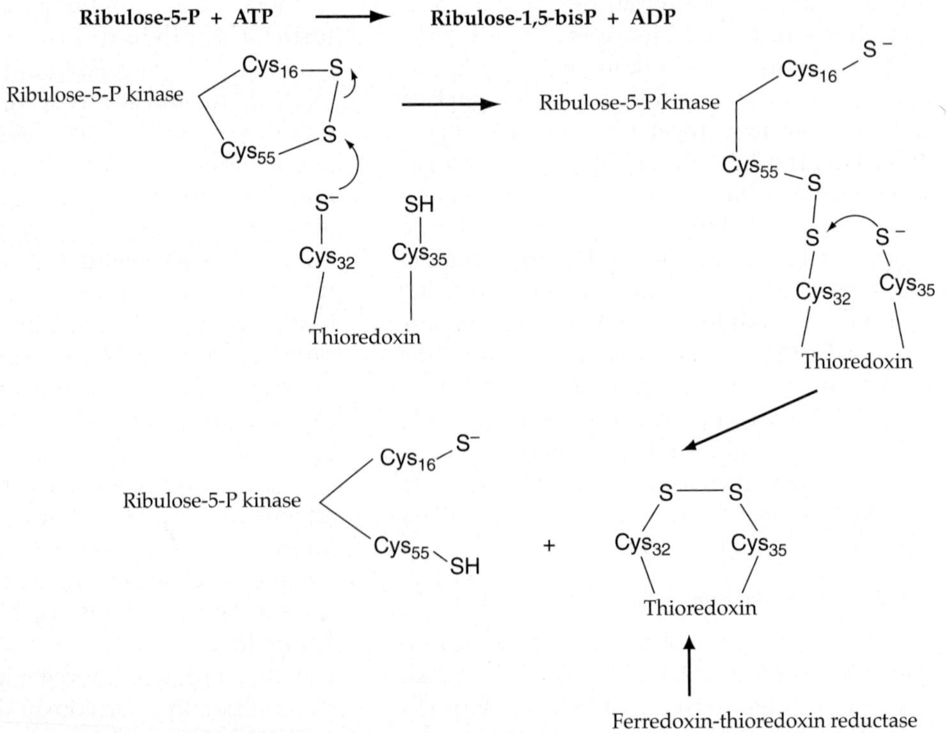
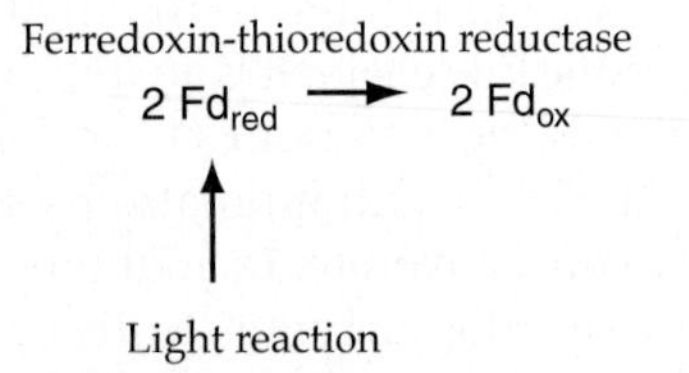

FIGURE 2.13 Activation of ribulose-5-P kinase by thioredoxin. In the spinach kinase, a disulfide bond between Cys 16 and Cys 55 blocks access to the active site. The thiolate anion of Cys 32 in thioredoxin attacks the Cys 55 in the kinase, and a transient interprotein disulfide is formed. Attack of the anion of Cys35 of thioredoxin forms a disulfide at the active site of thioredoxin and releases the reduced, activated kinase. Oxidized thioredoxin in chloroplasts is reduced by two equivalents of ferredoxin and a specific reductase.

Unlike the phosphoribulokinase, none of the cysteines in malate dehydrogenase are in the active site. In the oxidized enzyme, two disulfide bridges are formed between cysteine residues in both the N-terminal and C-terminal domains of the enzyme. X-ray crystal structure analysis shows that the disulfide near the C-terminal blocks the active site. Reduction of this disulfide by thioredoxin restores some activity by opening access to the active site. Full activity is regained by large conformational changes in the protein that occur when the N-terminal disulfide is also reduced. Initial *in vitro* experiments showed that a specific thioredoxin, designated thioredoxin m, was the reducing agent in this activation. However, subsequent biochemical and mutant analyses have demonstrated that thioredoxin f may actually be more efficient. Plants have more thioredoxin types than nonphotosynthetic organisms. This suggests that there are many light/redox-activated proteins in plants but the parameters that govern the specificity of thioredoxins for a particular target protein are still unresolved.

Regulation by Phosphorylation The PEPC in C-4 plants is regulated by reversible phosphorylation. The enzyme itself is subject to allosteric inhibition by malate and activation by glucose-6-phosphate. A PEPC kinase phosphorylates a conserved serine residue located near the N-terminal of the carboxylase. Phosphorylation leads to a decrease in binding of malate and thus greater activity under cellular conditions. The phosphate is removed in the dark by a protein phosphatase. The PEPC kinase is one of the smallest protein kinases known (30–35 kDa). It is present in vascular plant tissues in very low abundance presumably due to its rapid turnover. The kinase phosphorylates PEPC in the light and may itself be subject to redox regulation by thioredoxin. The PEPC in CAM plants is also regulated by a kinase that is, in turn, regulated by a circadian signal.

CHAPTER SUMMARY

As described by Robin Hill and other researchers in the first part of the twentieth century, photosynthesis consists of two redox reactions. The light reaction, covered in Chapter 1, outlines the conversion of light energy into chemical bond energy in the thylakoid membranes. H_2O is oxidized, NADP is reduced, and a proton gradient is generated in the thylakoid lumen to produce ATP. In the C-3 cycle, CO_2 is reduced to sugars using the NADPH and ATP generated in the light reaction. Although discussed separately, the two reactions are absolutely dependent on each other for energy input and recycling of ferredoxin, NADP, and ADP. The CO_2 that is reduced in the C-3 cycle provides the precursors for the metabolic intermediates required for catabolism and ATP production in the dark and the biosynthesis of all other metabolites required for plant growth and reproduction. Chapter 3 outlines the transport of fixed carbon out of the chloroplast and utilization in the basic catabolic pathways of glycolysis and aerobic metabolism. Subsequent chapters return to the chloroplast and the C-3 pathway to obtain intermediates for biosyntheses of both primary and secondary metabolites.

REFERENCES FOR FURTHER INFORMATION

Andersson, I. (2008). Catalysis and regulation in Rubisco. *J. Exp. Bot.* **59:**1555–1568.

Blankenship, R.E. (2002). Carbon metabolism. In: *Molecular Mechanisms of Photosynthesis*. Oxford, UK: Blackwell Science, pp. 171–203.

Izui, K., Matusumura H., Furumoto, T., & Kai, Y. (2004). Phosphoenol pyruvate carboxylase: A new era of structural biology. *Annu. Rev. Plant Biol.* **55:**69–84.

Portis, A.R., Li, C., Wang, D., & Salvucci, M.E. (2008). Regulation of Rubisco activase and its interaction with Rubisco. *J. Exp. Bot.* **59:**1597–1604.

Schürmann, P. & Jacquot J.P. (2000). Plant thioredoxin systems revisited. *Annu. Rev. Plant Physiol. & Plant Mol. Biol.* **51:**371–400.

PLANT BIOCHEMISTRY ***on the web***

Visit the *Plant Biochemistry* student companion Web site at http://biology.jbpub.com/gleason/plantbiochemistry/ for additional online study resources on these topics.

CHAPTER 3

Storage and Utilization of Fixed Carbon

CHAPTER OUTLINE

3.1 Overview

The light reaction of photosynthesis is the essential process in plants that captures light energy and changes it into chemical bond energy in ATP and reduced cofactors. The C-3 cycle uses this chemical bond energy to fix atmospheric CO_2 into phosphorylated sugars. These sugars subsequently serve as intermediates in various biosyntheses that occur in the chloroplast stroma. Excess fixed carbon is polymerized into starch and stored in plastids. In chloroplasts, starch is remobilized in the dark to provide reduced cofactors, phosphorylated sugars, and other metabolites. Intermediates of the C-3 cycle are also transported out of the chloroplast where they are metabolized in the cytoplasm to produce sucrose, which is translocated to nonphotosynthetic "sink" tissues such as flowers and storage organs or catabolized to yield ATP. The major sink for fixed carbon in plants, however, is the synthesis of primary cell wall polymers, principally cellulose. All these processes involve sugar-phosphate precursors.

3.2 Introduction to Sugars

Sugars are defined as polyhydroxy-aldehydes or ketones. The simplest molecules that fit this description are the trioses, glyceraldehyde, and dihydroxyacetone (**FIGURE 3.1**). Note that glyceraldehyde has a chiral center at C-2 that results in two isomeric forms. These have the D (R) or L (S) configuration as shown. Addition of another C–OH to D-glyceraldehyde generates two, four-carbon sugars with the D configuration, D-erythrose and D-threose. Likewise, adding another C–OH to L-glyceraldehyde generates the corresponding L sugars. The sugars in Figure 3.1 are illustrated in an extended form called a Fischer projection that shows the stereochemistry at each chiral center. A sugar is D or L depending on the configuration at the highest numbered chiral center. The four-carbon sugars shown are both D sugars because the OH group at C-3 is placed on the right side. D-erythrose and D-threose differ in configuration at C-2. These two tetroses are not mirror image isomers; they are referred to as epimers. Most sugars in biological systems are D isomers; unless otherwise noted, all sugars discussed in this chapter have the D configuration.

In aqueous solution, sugars with five or more carbon atoms form ring structures. The reaction of a hydroxyl group with an aldehyde produces a hemiacetal. C-1 in the ring is now an asymmetrical carbon, referred to as an anomeric carbon, and two configurations (anomers) are possible, alpha and beta (**FIGURE 3.2**). The ring structures in

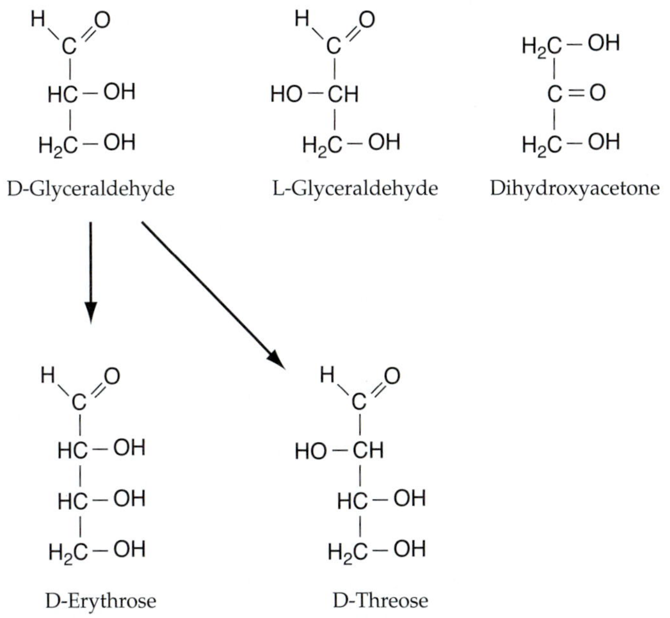

FIGURE 3.1 Structures of sugars. The simplest sugars are the trioses, glyceraldehyde and dihydroxyacetone. C-2 in glyceraldehyde is a chiral center; therefore two configurations, D (R) and L (S), are possible. Most sugars in biological systems have the D configuration. Addition of another carbon to D-glyceraldehyde generates tetroses with the D configuration. Erythrose and threose are shown as Fischer projections. The hydroxyl group on the highest numbered chiral center is placed on the right side of the drawing. These tetroses differ at C-2 and are isomers called epimers.

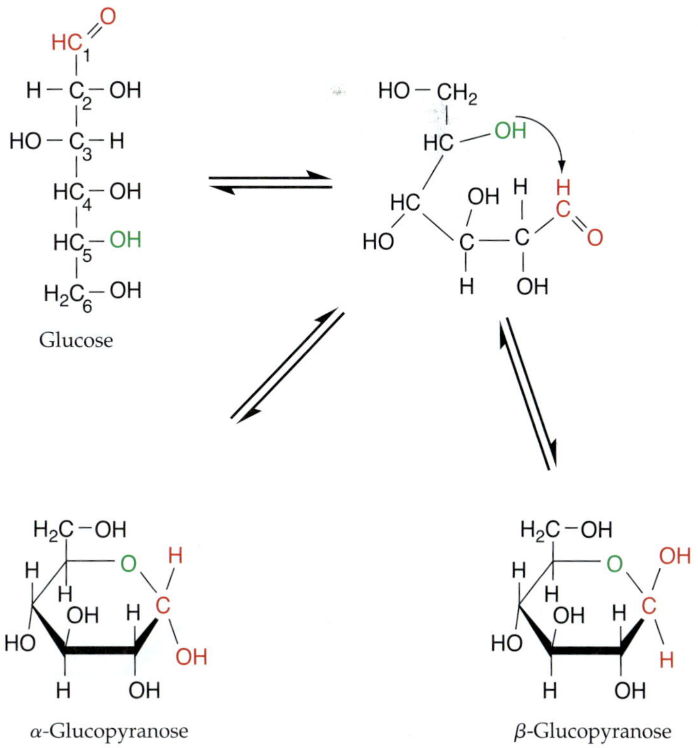

FIGURE 3.2 Ring formation of glucose. In solution glucose forms a six-membered ring. The hydroxyl group at C-5 attacks the carbonyl carbon to form a hemiacetal. Sugars in the six-membered ring conformation resemble the six-sided heterocycle, pyran, and can be referred to as pyranoses. In the ring form, the carbonyl carbon is now a chiral center and two configurations, alpha and beta, are possible. These are interconvertible via the straight chain.

Figure 3.2 are Haworth representations showing the bottom part of the ring protruding from the plane of the paper and the top part of the ring in the plane. The six-carbon aldoses usually form six-membered rings resembling pyran and are referred to as pyranoses. Formation of five-membered rings is also possible.

The actual structures of sugars in solution are less regular and are referred to as conformational models (**FIGURE 3.3**). The shape of the glucose molecules in this figure is called the chair conformation because of its resemblance to a folding lawn chair. The chair is the most stable conformation, but others are possible, such as the boat and the envelope.

Returning to dihydroxyacetone, there is one less chiral center in the ketoses and, correspondingly, fewer isomers. In general, the ketoses are named in a similar manner to the aldoses with the "ul" inserted, for example, erythrose, the aldose; erythrulose, the corresponding four-carbon ketose. Common ketoses with the "D" configuration are shown in **FIGURE 3.4**. In solution ketoses also form ring structures. An OH group reacts with the ketone to form a hemiketal (**FIGURE 3.5**). For six-carbon sugars, these are generally five-membered rings similar to furan. The corresponding sugars are referred to as furanoses.

Sugars can be polymerized by removing water to form a glycosidic bond. If the bond includes the anomeric carbon, an alpha or beta linkage is possible. For example, the disaccharide maltose is formed by an alpha linkage between C-1 of the first glucose and C-4 of the second. Cellobiose is a related disaccharide with a β-1,4 linkage (**FIGURE 3.6**). Disaccharides such as maltose have a free anomeric carbon and readily react with additional sugars or other compounds. Such sugars are called reducing sugars. Note that the disaccharide sucrose does not have a free anomeric carbon and is not a reducing sugar (Figure 3.6). Because it is chemically inert, sucrose is the most common sugar that is translocated in vascular plants. Polymers composed of two to five sugar monomers are referred to as oligosaccharides. Those with more than five monomers are called polysaccharides.

Sugars are often used to modify other molecules in the cell such as terpenes and flavonoids. Modification generally makes molecules more water soluble and less reactive. Glycosylated natural products are often inert storage forms. Sugars can also be O-linked to serine and threonine residues in proteins and N-linked to asparagine

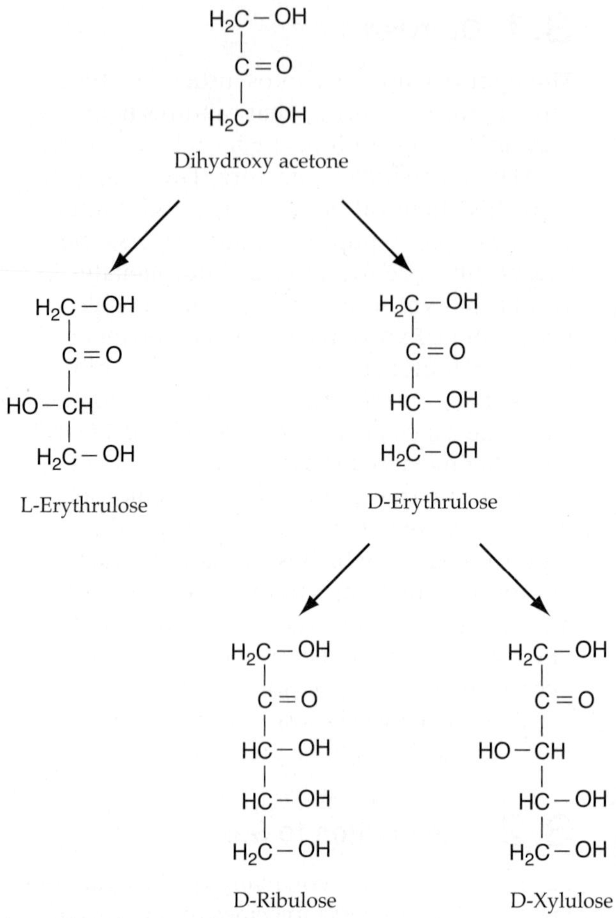

FIGURE 3.4 Structures of ketoses. Adding one carbon to dihydroxyacetone gives rise to two ketoses, D and L erythrulose. As in the aldoses, the D and L assignment is based on the position of the hydroxyl group on the highest numbered chiral center. The five carbon ketoses with the D configuration can have different arrangements at C-3 giving rise to two epimers, ribulose and xylulose.

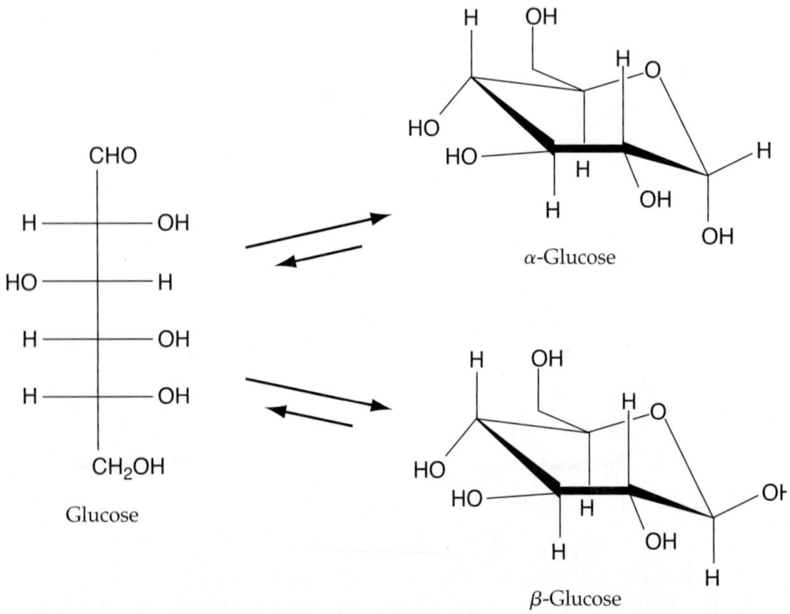

FIGURE 3.3 Conformational model of glucose. Ring formation of glucose results in alpha and beta configurations with the chair conformation. Other forms are possible, but the chair shown here is the most stable conformation due to favorable orientation of substituents on the ring.

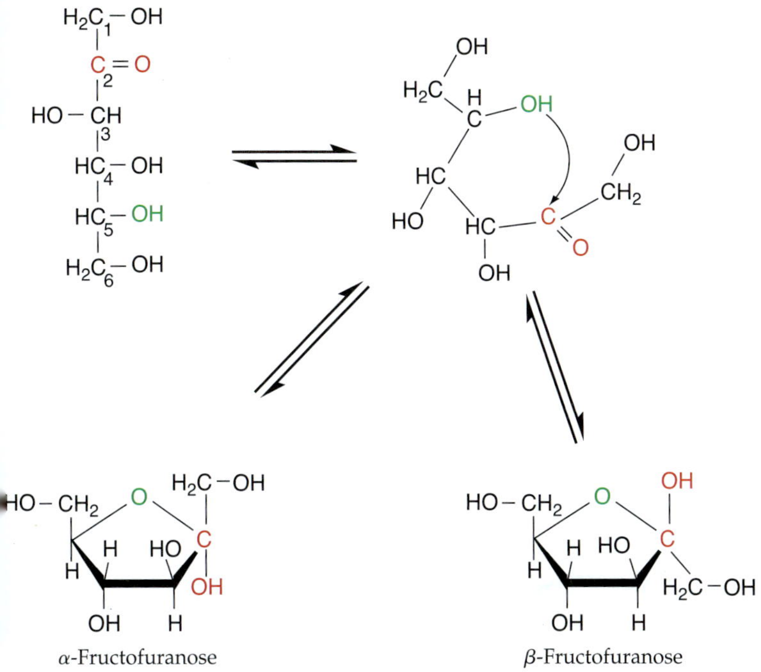

FIGURE 3.5 Ring formation in fructose. As in aldoses, ketoses form ring structures in solution. The hydroxyl group at C-5 attacks the carbonyl carbon (C-2) forming a hemiketal. These resemble the heterocycle, furan, and can be referred to as furanoses. Two arrangements are possible around C-2 (the anomeric C), giving rise to alpha and beta isomers.

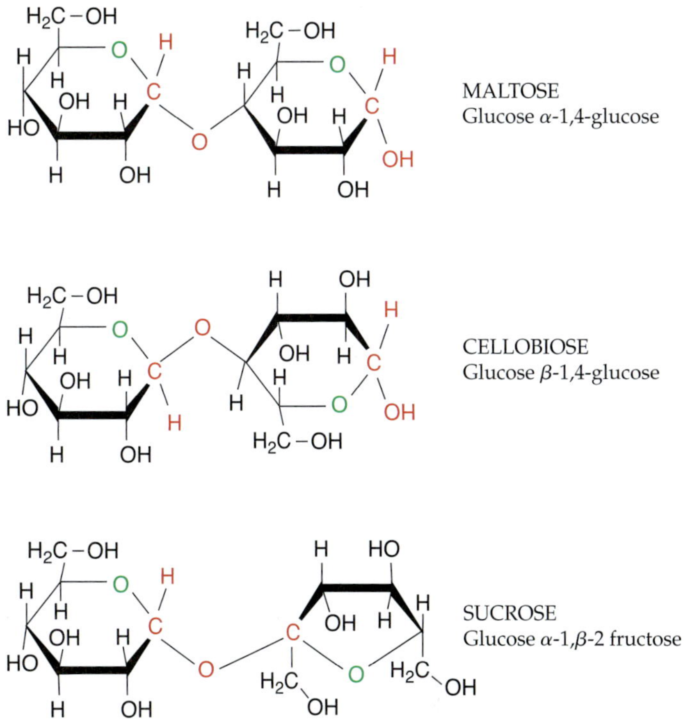

FIGURE 3.6 Structures of three common disaccharides. The anomeric carbons are shown in red.

residues. Glycoproteins are common in eukaryotes but rare in prokaryotes. The glycosylation reactions occur in the Golgi apparatus or the endoplasmic reticulum. Lipids can also be glycosylated. Galactolipids are a common constituent of thylakoid membranes in plants and help to maintain membrane fluidity during temperature fluctuations.

3.3 Starch Metabolism

Starch is a glucose polymer that is synthesized, stored, and metabolized in plastids of most plants (**FIGURE 3.7**). Plant starches are composed of two types of glucose polymers, amylose and amylopectin. Amylose is a linear polymer of α-1,4-linked glucose residues ranging from 100 to 2,000 residues. It constitutes about 30% of a typical starch molecule. Amylopectin is also a polymer of α-1,4-linked glucose residues with the addition of α-1,6 branches occurring about every 20 residues. Amylopectin is the major polysaccharide in starch.

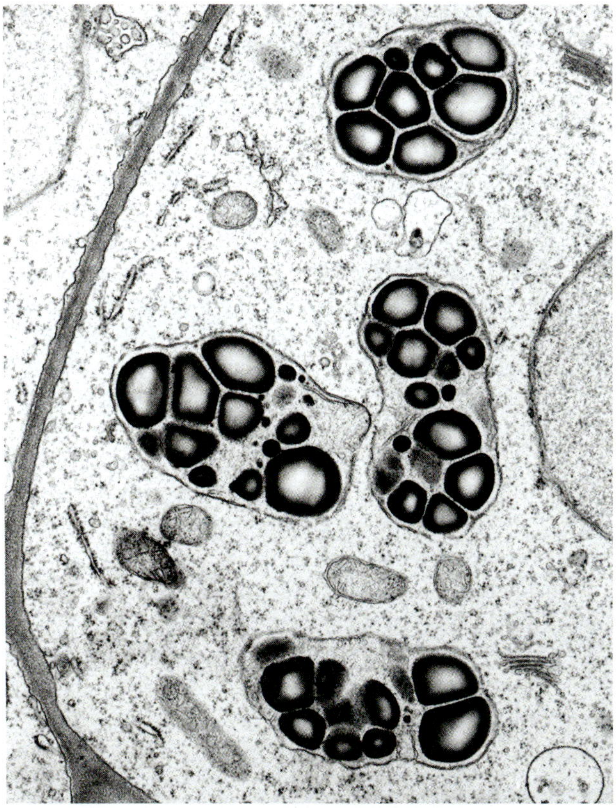

FIGURE 3.7 Storage plastids (amyloplasts) with starch granules. This micrograph of a bean root tip shows several plastids containing starch grains. Magnification = 77,000×.

Biosynthesis

Starch synthesis in leaves starts in the chloroplast stroma with intermediates of the C-3 cycle, mainly, fructose-6-phosphate (F6P). A hexose-phosphate isomerase catalyzes its

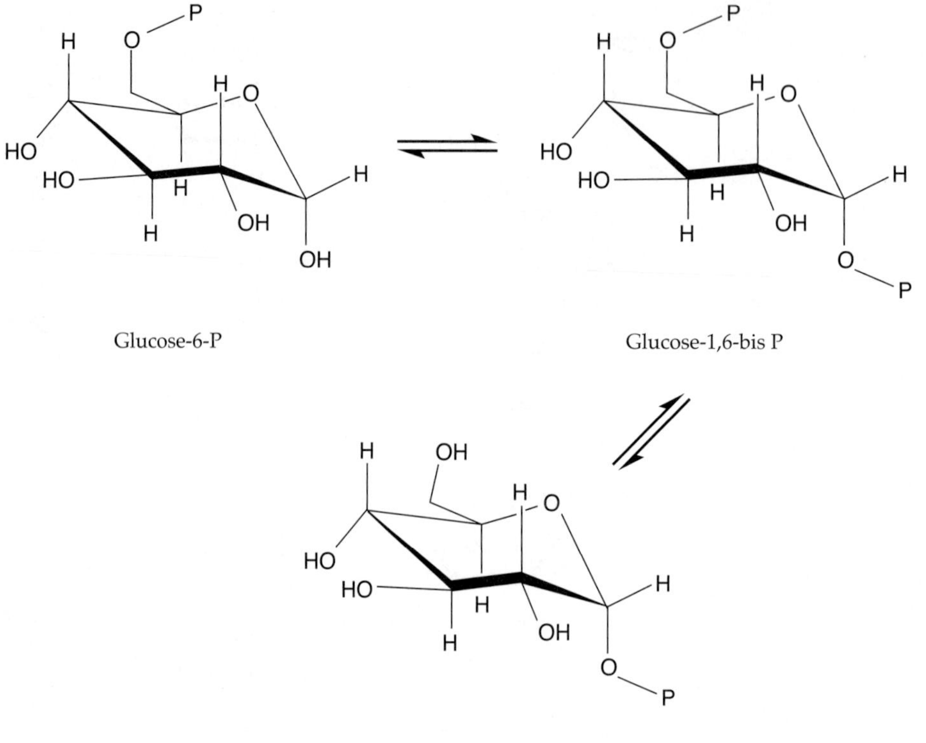

FIGURE 3.8 Phosphoglucomutase reaction. This rearrangement proceeds through a bisphosphate intermediate. A phosphate ion is bound to a serine residue in the active site of the enzyme and is transferred to C-1 of the substrate, G6P. The phosphate ion at C-6 is then removed by this same serine residue and the product, glucose-1-phosphate, is released from the enzyme.

rearrangement to the corresponding aldose, glucose-6-phosphate (G6P). The phosphate is moved from C-6 to C-1 by the enzyme, phosphoglucomutase, which has a phosphate bound to its active site. The rearrangement involves a 1,6-bisphosphate intermediate (**FIGURE 3.8**). Glucose-1-phosphate (G1P) and ATP are then used to produce the activated nucleotide sugar donor, ADP-glucose, needed for starch synthesis (**FIGURE 3.9**). ADP-glucose is the substrate for the enzyme, starch synthase. This enzyme catalyzes the formation of α-1,4 bonds between glucose residues with the release of ADP and inorganic phosphate (Pi) and elongates both amylose and amylopectin. Formation of amylopectin requires a starch branching enzyme that hydrolyzes an α-1,4-linked polysaccharide and forms an α-1,6 glucosidic bond between the end of the cut chain and C-6 of a glucose in another amylose molecule. Debranching enzymes cleave branch points and catalyze reattachment, resulting in further modification of amylopectin.

Starch grains grow from several points, forming a complex polymer. Pairs of amylopectin molecules pack together in organized helical arrays. These semicrystalline layers alternate with more amorphous layers composed of the unbranched portions of amylopectin and amylose. Starch granules show alternating rings of amorphous and semicrystalline polymers (**FIGURE 3.10**).

In storage organs, such as roots, tubers, and seeds, starch is synthesized in nonphotosynthetic plastids called amyloplasts. Sucrose is transported from the leaves by way of the phloem and is hydrolyzed to its constituent sugars in the cytoplasm of "sink" tissues. Several different enzymes may catalyze this hydrolysis, such as invertase or a reversible sucrose synthase. G6P or ADP-glucose are transported through the amyloplast membranes, and starch is synthesized as in green tissues.

Starch in chloroplasts is often phosphorylated. The enzymes, alpha-glucan water dikinases, use ATP to add phosphate groups to C-3 or C-6 hydroxyl groups in a random manner. Phosphorylation facilitates degradation and rapid starch turnover, which is necessary in leaf chloroplasts. Starch in storage organs such as seeds and tubers is more stable. Starch is degraded and provides glucose residues for energy generation. A variety of hydrolytic enzymes cuts the macromolecule into smaller pieces. Alpha-amylases are endoglucosidases that hydrolyze

FIGURE 3.9 Synthesis of ADP-glucose. The reaction is catalyzed by ADP-glucose pyrophosphorylase. The pyrophosphate (diphosphate group, PPi) is a good leaving group and the resulting AMP is esterified to glucose-1-phosphate forming ADP-glucose.

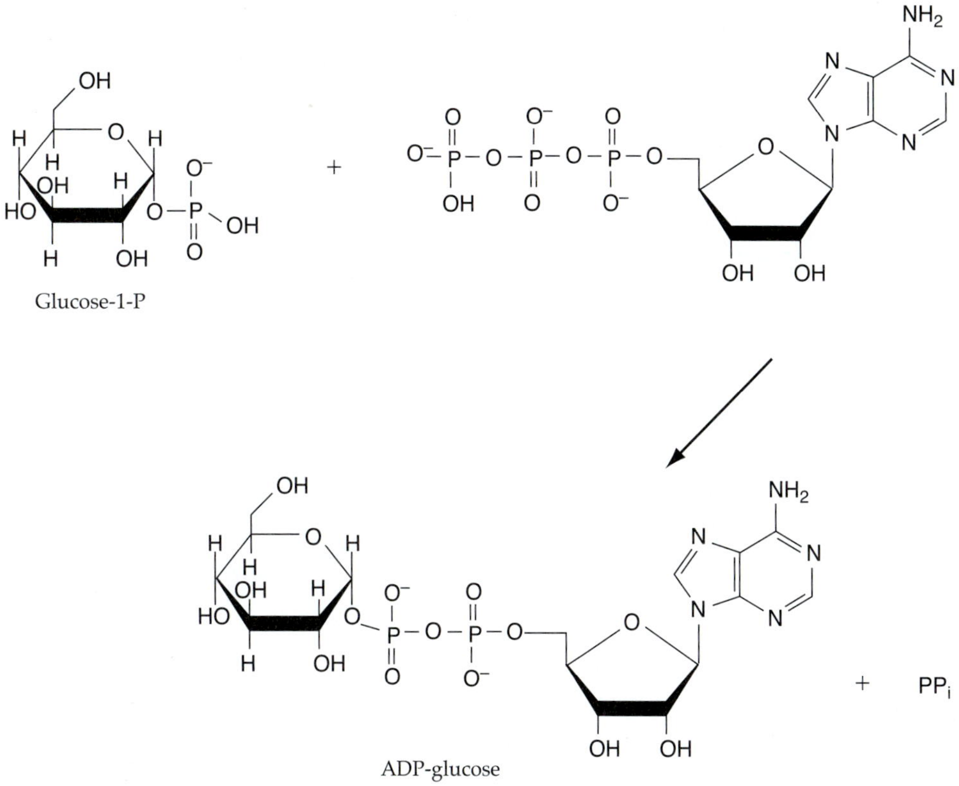

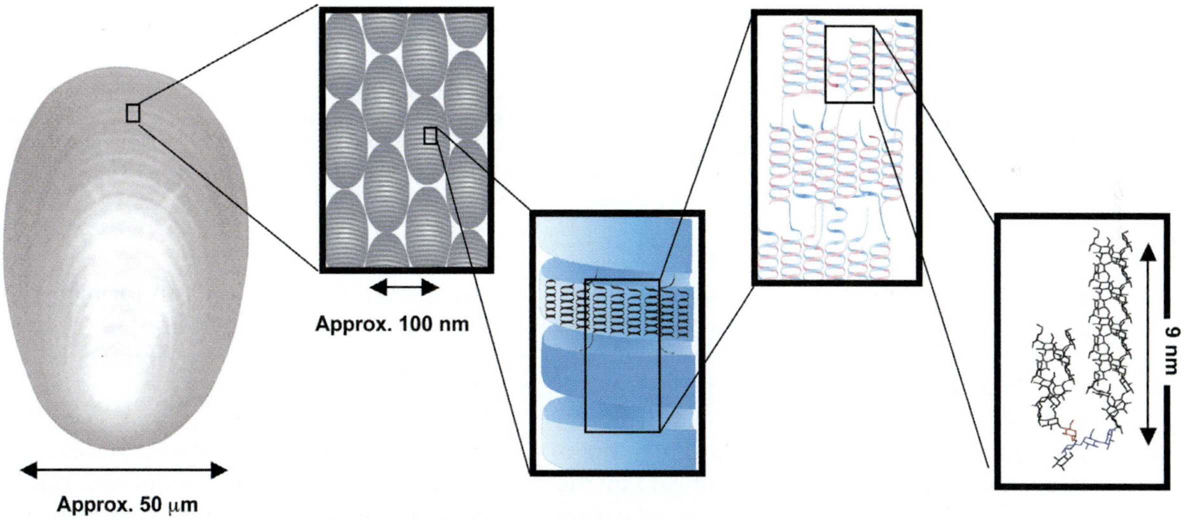

FIGURE 3.10 Structure of starch granules. Although they appear amorphous, starch granules have a semicrystalline structure. The highly branched amylopectin molecules form helical structures as shown on the right. These helices are packed between the less structured regions of amylopectin and straight-chain amylose yielding a layered structure as shown on the left. (*Reproduced from*: A. Blennow, T. H. Nielsena, L. Baunsgaarda, R. Mikkelsena, and S. B. Engelsenb, *Trends in Plant Science* 7, No. 10 [2002]: 445–450. Reprinted with permission from Elsevier.)

α-1,4-bonds in the middle of the starch molecule and give rise to shorter oligosaccharides. Beta-amylases catalyze the hydrolysis of maltose units from the reducing end of a starch molecule. The debranching enzyme is required to catalyze the hydrolysis of α-1,6 bonds. Most plants have integral membrane proteins that transport glucose and/or maltose out of the chloroplast for further metabolism in the cytoplasm.

3.4 Oxidative Pentose Phosphate Pathway

In the dark, chloroplast starch phosphorylase catalyzes the addition of a phosphate group to the end of a starch polymer, generating glucose-1-phosphate (G1P). This is isomerized to G6P by phosphoglucomutase and catabolized in the stroma by the oxidative pentose phosphate

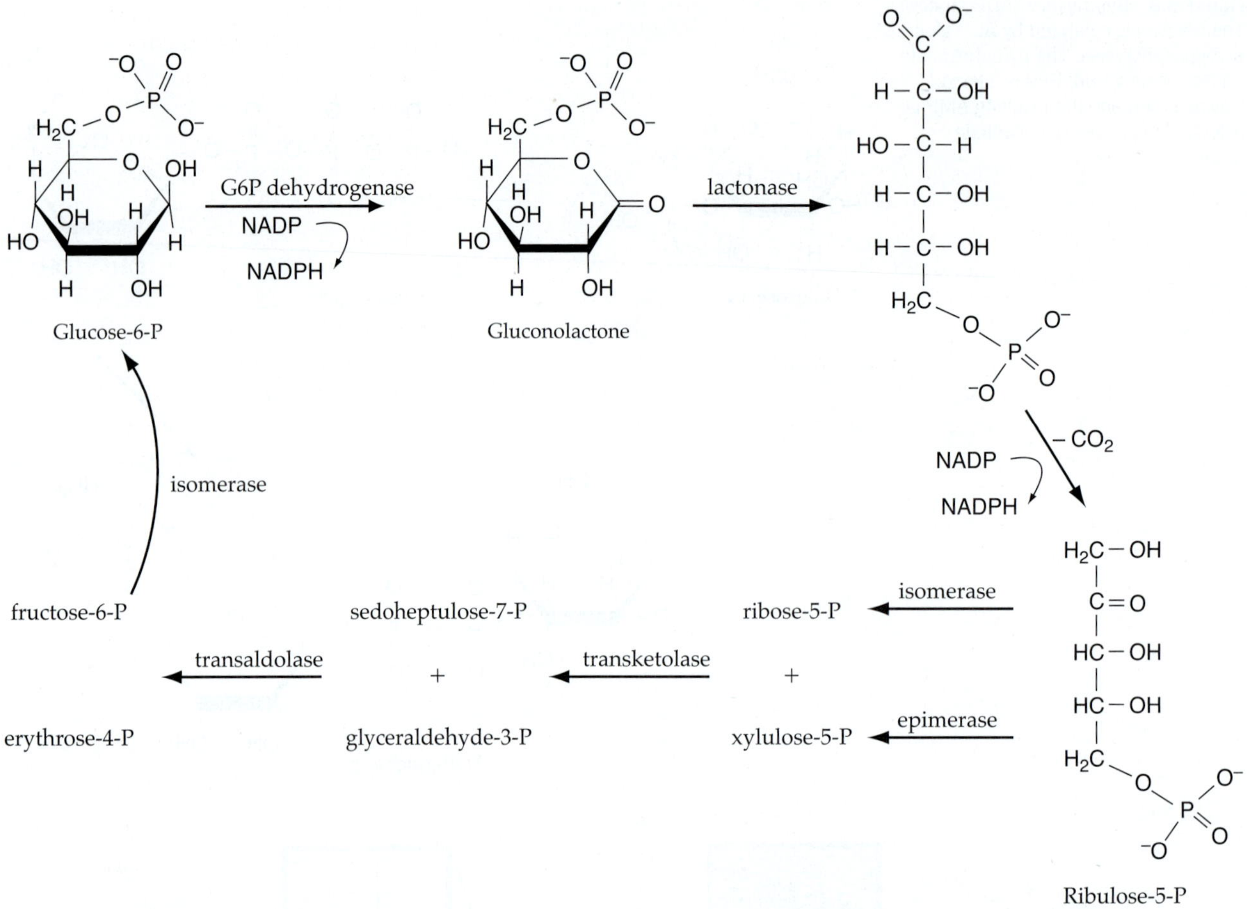

FIGURE 3.11 Major reactions in the oxidative pentose phosphate pathway. This figure shows some of the reactions that occur in the stromal pathway. The first three reactions, catalyzed by G6P dehydrogenase, lactonase, and gluconic acid-6-P dehydrogenase, are unique to this pathway. These reactions generate reduced NADP in the dark. The subsequent reactions, including those not shown here, regenerate sugar-P intermediates to replenish the C-3 cycle and provide substrates for biosyntheses that occur in the chloroplast.

pathway, also called the hexose monophosphate shunt (**FIGURE 3.11**). The first reaction is catalyzed by glucose-6-phosphate dehydrogenase. This oxidation of glucose supplies NADPH in the dark. The reaction is essentially irreversible, and the dehydrogenase is tightly regulated. Its activity is inhibited in the light by the higher pH in the stroma and the reducing agents, glutathione and thioredoxin. This light-mediated inhibition prevents a futile cycle of sugar synthesis in the C-3 cycle and degradation of sugars by the pentose pathway from occurring simultaneously. The second enzyme, gluconate-6-phosphate lactonase, cleaves the cyclic ester ring to yield a straight chain conformation. Gluconate-6-phosphate is oxidized and decarboxylated to yield ribulose-5-phosphate. In addition to generating NADPH in the dark, this pathway also yields sugar-P intermediates required by chloroplast biosynthetic pathways such as the C-3 cycle. Some of the intermediates, such as G6P, can be dephosphorylated and transported to the cytoplasm for biosynthetic reactions or energy generation.

3.5 Sucrose Synthesis

The disaccharide, sucrose, is synthesized in the cytoplasm and is the most common sugar translocated in vascular plants. Carbon that is fixed in the chloroplast can be transported to the mesophyll cell cytoplasm by a number of different transporters located in the chloroplast membranes. The best characterized is the triose-P-inorganic phosphate translocator, an anti-porter protein system that transports dihydroxyacetone phosphate (DHAP) out of the stroma in exchange for an influx of phosphate ion.

In a series of biosynthetic reactions in the cytoplasm, DHAP can be converted to F6P. DHAP is isomerized to glyceraldehyde-3-phosphate, and the two sugars undergo an aldol condensation to yield fructose-1,6-bisphosphate (FBP). A phosphatase catalyzes the removal of the phosphate on C-1 yielding F6P. This cytoplasmic form of fructose bisphosphatase is different from the enzyme found in the chloroplast. It is regulated by an unusual metabolite, fructose-2,6-bisphosphate (F-2,6-BP). F-2,6-BP is a regulatory metabolite produced from F6P and ATP by a specific kinase. This F6P kinase, in turn, is activated when the triose phosphate pool in the cytoplasm is depleted. The inhibitory action of F-2,6-BP on fructose bisphosphatase depletes the pool of hexoses available for sucrose synthesis and export. Fixed carbon then remains in the cell and is used to produce intermediates for the C-3 and other pathways and for starch synthesis.

Sucrose synthesis proceeds in the cytoplasm from F6P, which is isomerized to G6P and acted on by a mutase to yield G1P. G1P is a substrate for the enzyme, uridine diphosphate-glucose pyrophosphorylase, that catalyzes the production of the sugar nucleotide, UDP-glucose. Sucrose-phosphate synthase transfers this glucose to F6P to produce sucrose-phosphate. A sucrose phosphatase removes the phosphate, which makes this a unidirectional reaction (**FIGURE 3.12**).

Sucrose is a relatively unreactive sugar. Because both anomeric carbons are linked in the glycosidic bond, it has no reducing end. It can be transported in the phloem to nonphotosynthetic tissues of the plant such as roots, flowers, and seeds. When reaching its destination, sucrose can be catabolized to its constituent sugars by the enzymes, invertase, and a sucrose synthase that catalyzes the following reversible reaction:

$$\text{sucrose} + \text{UDP} \leftrightarrow \text{UDP-glucose} + \text{fructose}$$

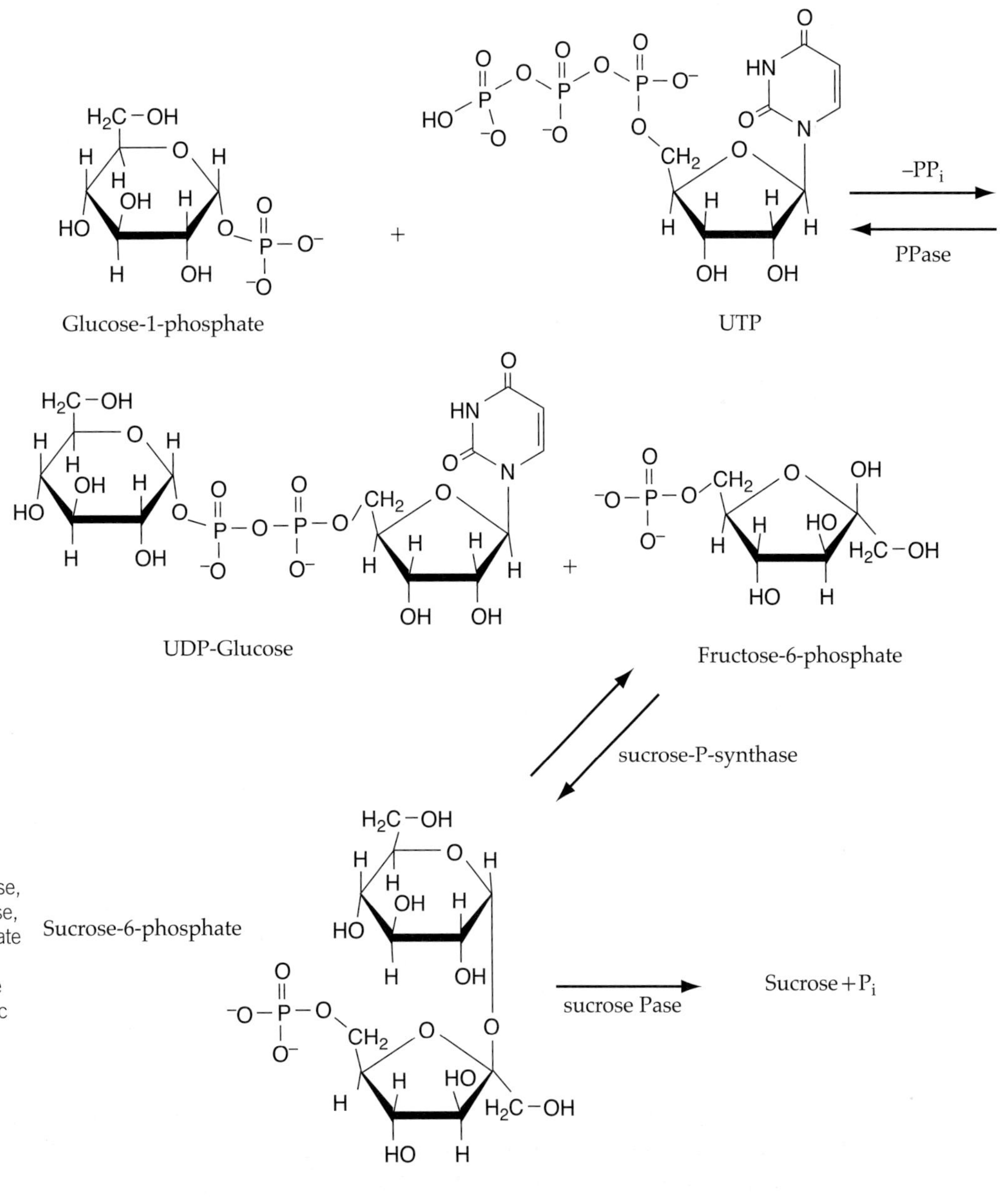

FIGURE 3.12 Synthesis of sucrose. The enzyme, UDPG pyrophosphorylase, catalyzes the synthesis of UDP-glucose, the sugar nucleotide that is the substrate for sucrose-phosphate synthase. This enzyme catalyzes the formation of the glycosidic bond between the anomeric carbons of both monosaccharides. These reactions are readily reversible depending on the concentration of intermediates in the cytoplasm. The final phosphatase reaction removes the phosphate from the disaccharide and is irreversible.

In storage organs the resulting glucose or nucleotide derivative can be transported across the plastid membranes and used to synthesize starch, which is stored in amyloplasts.

Some plants synthesize and transport sugars other than sucrose. For example, some Rosaceae, such as apple trees, transport carbon as sorbitol, a six-carbon sugar alcohol. Some deciduous trees, like the elm, transport the nonreducing trisaccharide, raffinose. This trisaccharide has the composition galactose-($1\alpha \rightarrow 6$)-glucose-($1\alpha \rightarrow \beta$)-fructose and is also found in some seeds such as cotton. Many animals, including humans, cannot digest the galactose-1α-6-glucose bond.

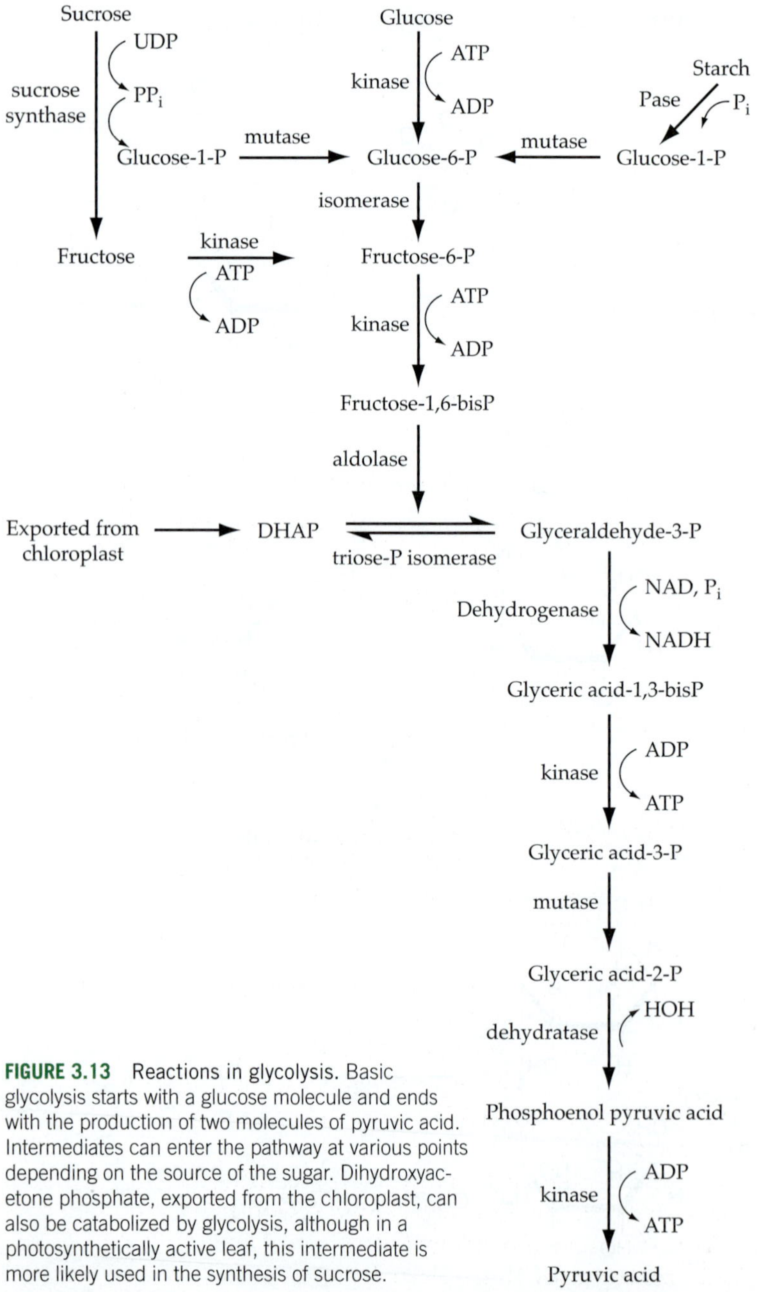

FIGURE 3.13 Reactions in glycolysis. Basic glycolysis starts with a glucose molecule and ends with the production of two molecules of pyruvic acid. Intermediates can enter the pathway at various points depending on the source of the sugar. Dihydroxyacetone phosphate, exported from the chloroplast, can also be catabolized by glycolysis, although in a photosynthetically active leaf, this intermediate is more likely used in the synthesis of sucrose.

3.6 Catabolism of Glucose

Glycolysis

Glycolysis and the citric acid cycle are the major pathways for catabolism of glucose used by aerobic organisms, including plants.

$$C_6H_{12}O_6 + 6O_2 \rightarrow 6CO_2 + 6H_2O$$
$$\Delta G^o = -2{,}800 \text{ kJ/mole}$$

The oxidation occurs in several enzyme-catalyzed steps, and the energy released is used to drive the synthesis of ATP. The processes are said to be "coupled."

$$ADP + Pi \rightarrow ATP + H_2O$$
$$\Delta G^o = +30 \text{ kJ/mole}$$

The oxidation of glucose starts in the cell cytoplasm with a series of 10 enzymatic steps that result in the production of two molecules of pyruvic acid from the input of one glucose. If sucrose is the starting material, the disaccharide can be hydrolyzed to glucose and fructose by invertase or to G1P and fructose by the reversible sucrose synthase (**FIGURE 3.13**). Starch can also be hydrolyzed to glucose or to maltose by amylases. Many plants have membrane-bound translocators that move these sugars into the cytoplasm at night.

When starting with glucose, the sugar must be activated by phosphorylation before it can be oxidized. A hexokinase uses ATP to produce G6P. This is isomerized to F6P. A second kinase uses ATP to produce FBP.

As described previously, FBPase can act on this substrate to catalyze the back reaction to F6P. The enzyme is subject to complex regulation depending on the tissue type and energetic needs of the cell. The major regulatory molecule is F-2,6-BP, produced by a kinase reaction. The F-2,6-BP kinase is an unusual bifunctional enzyme. Although many enzymes catalyze reversible reactions, this bifunctional enzyme has separate active sites for kinase and phosphatase activities. It appears to be the result of an ancient fusion of two genes involved in carbohydrate metabolism. The active kinase/phosphatase is a tetramer of 83-kDa subunits. The activities are regulated by a number of small molecules that act as allosteric effectors.

Kinase activity – stimulated by Pi, F6P
– inhibited by triose phosphates

$$F6P + ATP \rightarrow F\text{-}2,6\text{-}BP + ADP$$

F-2,6-BP affects mainly the F-1,6-BPase reaction in plants. In the presence of excess triose phosphates exported from the chloroplast during the day, less F-2,6-BP is made and FBPase activity is increased. This provides more hexose-phosphates for sucrose synthesis that occurs mainly during the day. In contrast, a large pool of F6P results in a higher concentration of F-2,6-BP and lower F-1,6-BPase activity. This inhibition slows further export of fixed carbon from the chloroplast and directs cytosolic sugars into glycolysis or production of other intracellular metabolic intermediates.

The bifunctional enzyme also degrades F-2,6-BP.

Phosphatase activity – inhibited by Pi, F6P, FBP

$$\text{F-2,6-BP} + H_2O \rightarrow \text{F6P} + \text{Pi}$$

Lower amounts of F-2,6-BP again result in greater FBPase activity and larger hexose-phosphate pools. This complex regulation of the FBPase reaction serves to maintain a steady supply of sugar phosphates for intracellular metabolism but also provides for exporting any surplus to other plant tissues.

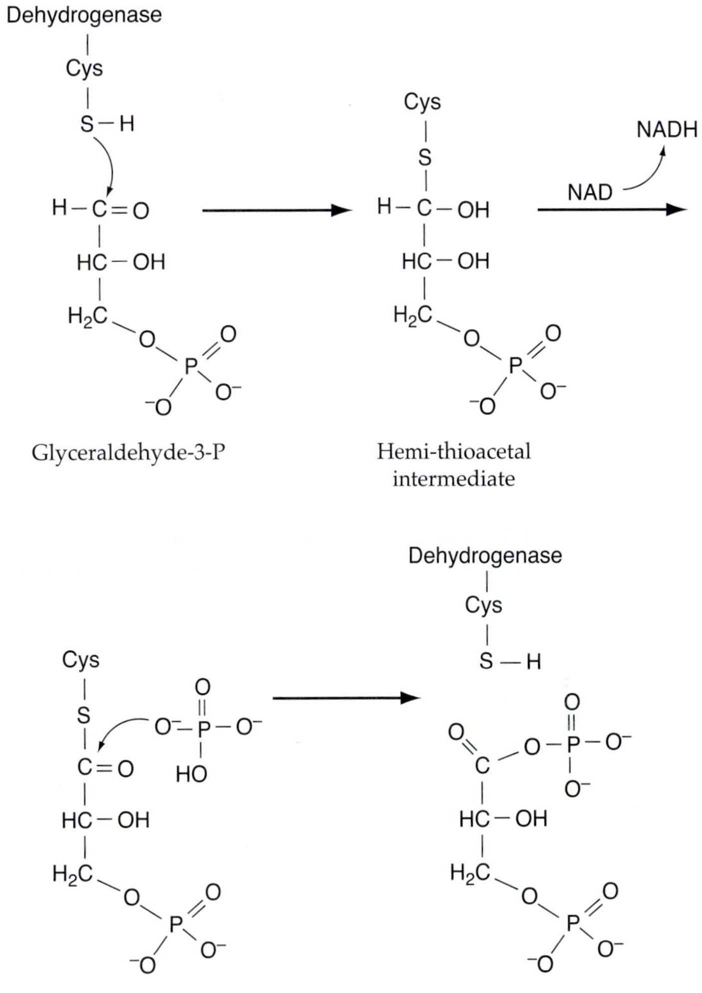

Glyceric acid-1,3-bisP

FIGURE 3.14 Glyceraldehyde-3-phosphate dehydrogenase reaction. One redox reaction occurs in glycolysis, catalyzed by the cytoplasmic form of the glyceraldehyde-3-P dehydrogenase. The substrate is bound to a cysteine residue in the active site of the enzyme, producing the hemithioacetal intermediate that is oxidized by NAD. A phosphate ion attacks the carbonyl carbon of the intermediate and releases the phosphorylated product, glyceric acid bisP. The overall reaction is exergonic and is a good example of a substrate-level phosphorylation. Glyceric acid-1,3-bisphosphate is a "high energy" intermediate that is used to generate ATP in subsequent steps.

Glycolysis continues with an aldolase that cleaves FBP into two, three-carbon sugars, glyceraldehyde-3-P and DHAP. DHAP is also exported from chloroplasts in the light and can enter the glycolytic pathway at this point. DHAP is easily isomerized to glyceraldehyde-3-P by triose phosphate isomerase. Glyceradehyde-3-P is oxidized to glyceric acid-3-P by glyceric acid-1,3-bisP dehydrogenase. This reaction is essentially the reverse of the reduction step in the C-3 cycle (see Chapter 2). The cofactor used by the cytoplasmic form of the enzyme is nicotinamide adenine dinucleotide (NAD) rather than NADP. In the C-3 cycle, the input of ATP was required to drive the reaction. In this reverse reaction the energy released by the oxidation of the aldehyde can be used to drive the synthesis of ATP. The reaction can be visualized as occurring in two steps. Glyceradehyde-3-P is bound to the dehydrogenase by a cysteine residue at the active site, producing a hemi-thioacetal intermediate, followed by oxidation and cleavage of the S-C bond by a Pi ion (**FIGURE 3.14**). The product is glyceric acid-1,3-bisP. This is an excellent example of a coupled reaction in which the energy released from one step is used to drive a second, energy-requiring reaction. The overall ΔG is therefore a negative number. In this subsequent step, often referred to as a substrate-level phosphorylation, phosphate is transferred from glyceric acid-bisP by a kinase and results in the synthesis of ATP. The phosphate on C-3 of the product, glyceric acid-3-P, is moved to C-2 by a mutase. A water molecule is removed from glyceric acid-2-P by an enolase, yielding a "high energy" intermediate, phosphoenolpyruvate. Another kinase transfers the phosphate from phosphoenolpyruvate to ADP to yield ATP and pyruvic acid.

Under anaerobic conditions, plant cells can generate sufficient ATP to maintain basic life processes using glycolysis alone. This requires an additional redox reaction to regenerate the oxidized cofactor, NAD. In most plants the process is similar to that found in many fungi. Pyruvate decarboxylase removes CO_2, producing acetaldehyde. Alcohol dehydrogenase uses NADH to reduce the aldehyde to ethanol and thus regenerates the oxidized cofactor for

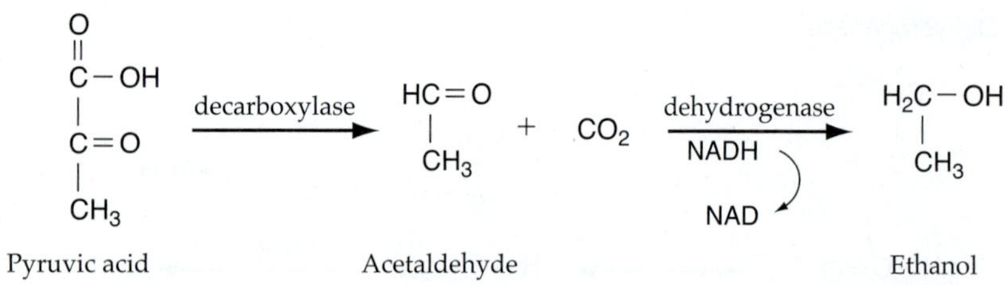

FIGURE 3.15 Fermentation reactions in plants. Under anaerobic conditions, plant cells can continue to generate ATP by glycolysis without the subsequent oxygen-requiring processes that occur in the mitochondria. This fermentation requires regeneration of the oxidized cofactor required in the glyceraldehyde-3-P dehydrogenase reaction. In plants, a decarboxylase removes one carbon and the resulting acetaldehyde is reduced by NADH catalyzed by alcohol dehydrogenase. The ethanol generated is not metabolized further and is excreted from the cell if possible.

glycolysis (**FIGURE 3.15**). The ethanol is often not metabolized further but can be excreted. This anaerobic pathway is especially important in plant roots under flooded conditions or for certain seeds that normally germinate under water, such as rice.

Citric Acid Cycle

Under aerobic conditions, the subsequent catabolism of pyruvate occurs in the mitochondrion. Pyruvate is transported across the double membrane into the soluble matrix (**FIGURE 3.16**). The decarboxylation of pyruvate to acetate is catalyzed by the pyruvate dehydrogenase complex, which then esterifies acetate to the cofactor, coenzyme A. This enzyme complex consists of three proteins. The first enzyme, pyruvate dehydrogenase, binds pyruvic acid and catalyzes a decarboxylation. The resulting two-carbon fragment remains bound to the enzyme's cofactor, thiamine pyrophosphate. The second enzyme in the complex, a transacetylase, catalyzes the transfer of the two-carbon fragment from thiamine pyrophosphate to its cofactor, lipoic acid. During this transfer, the lipoic acid is reduced and the two-carbon fragment is oxidized to acetate. The acetate is then transferred to a third cofactor, coenzyme A. The third enzyme in the complex, lipoyl dehydrogenase, catalyzes the oxidation of lipoic acid that is bound to the transacetylase using NAD as its oxidizing agent (**FIGURE 3.17**).

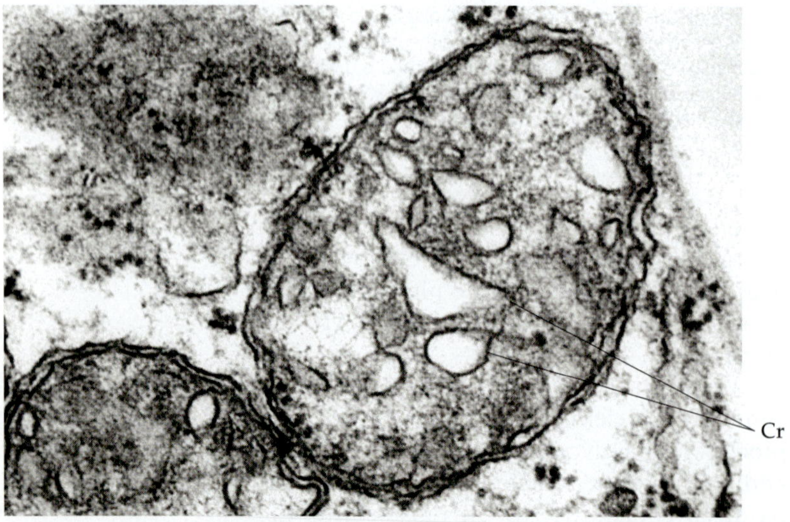

FIGURE 3.16 Structure of plant mitochondria. Two mitochondria in a cotton cell are shown. In the micrograph, the outer membrane and several tubular folds of the inner membrane, the cristae (Cr) are visible. In general, plant mitochondria contain fewer cristae than those of animal cells, reflecting plant cells lower rate of aerobic metabolism. Magnification is approximately 74,000x.

The pyruvate dehydrogenase complex is often associated with a specific kinase and a phosphatase. The kinase can add phosphates to serine residues in the first enzyme in the complex and down-regulate dehydrogenase activity. The kinase is, in turn, inhibited by ADP and pyruvate. The phosphatase activity, in contrast, is due to a generic type of phosphatase and does not apparently respond to metabolic regulators. The result is that the pyruvate dehydrogenase complex is most active when levels of ADP and pyruvate are high in the mitochondrion but is lower in the presence of excess ATP. How phosphorylation actually changes enzymatic activity at the chemical level is unknown.

The citric acid cycle begins with an aldol condensation of acetyl-CoA to the four-carbon oxaloacetic acid to yield the six-carbon citric acid and coenzyme A. Hydrolysis of the CoA ester is an exergonic reaction that drives this step in the direction of citric acid synthesis. A series of eight enzymatic reactions occurs in the cycle and results in the oxidation of two of the carbons in the intermediates to CO_2 and the regeneration of oxaloacetic acid (**FIGURE 3.18**). No oxygen is used during the cycle. The oxidizing agents are the cofactors NAD and flavin adenine dinucleotide (FAD). The α-ketoglutarate (oxoglutarate) dehydrogenase is a three-enzyme complex with activity and cofactors similar to the pyruvate dehydrogenase complex, resulting in the production of CO_2, NADH, and succinyl-CoA. Hydrolysis of the CoA thioester by succinyl-CoA synthetase is an exergonic reaction and yields one molecule of ATP for every turn of the cycle.

Other compounds can also be metabolized by the citric acid cycle. Acetyl-CoA can be produced directly from the oxidation of fatty acids. Malate is produced in the cytoplasm of most

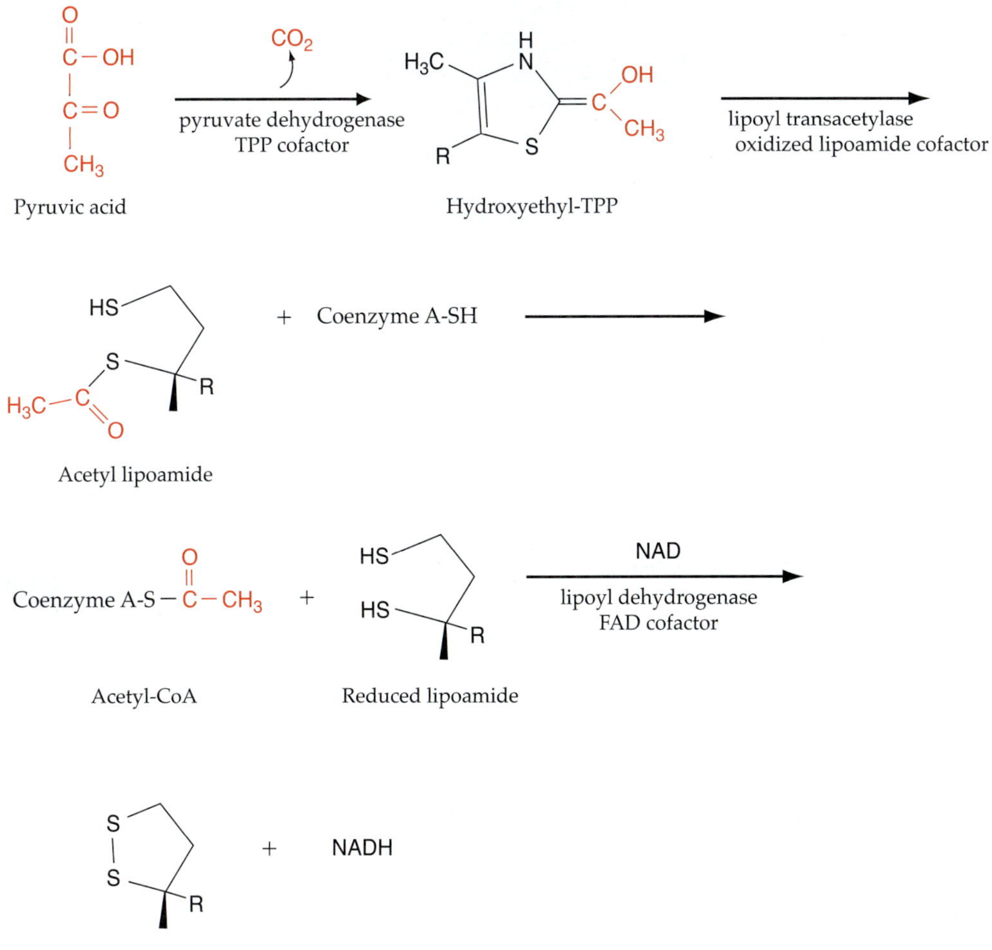

FIGURE 3.17 Reactions in the pyruvate dehydrogenase complex. The first enzyme of the complex, pyruvate dehydrogenase, catalyzes a decarboxylation of pyruvate, leaving the remaining acetyl fragment bound to its cofactor, thiamine pyrophosphate (TPP). The acetyl fragment is then transferred to the lipoic acid cofactor of the second enzyme of the complex, lipoyl transacetylase. The transacetylase transfers the acetyl group to coenzyme A which then diffuses out of the complex. The third enzyme in the complex, lipoyl dehydrogenase, has a flavin adenine dinucleotide (FAD) cofactor that reoxidizes the lipoic acid cofactor of the transacetylase and reduces NAD. The complex is now in its initial state and ready for another turnover. The overall reaction catalyzed by the complex is

$$\text{pyruvate} + \text{CoA} + \text{NAD} \rightarrow \text{acetyl-CoA} + CO_2 + \text{NADH}$$

plants by a phosphoenolpyruvate carboxylase reaction and is transported across the mitochondrial membranes and catabolized in the cycle. However, most of the malate produced in both compartments is used primarily in biosynthetic pathways rather than for energy generation. In the heterotrophic cells of plant roots, organic acids such as citrate and malate are often excreted into the soil. These serve to chelate metals such as iron and facilitate their uptake into the cell.

Electron Transport System

The electron transport system (ETS) consists of a series of integral membrane proteins and cofactors in the inner membrane of the mitochondrion. The function of the ETS is to oxidize the cofactors and transfer the electrons to oxygen; this allows the citric acid cycle to continue to operate. The cristae (Figure 3.16) are folded extensions of the inner membrane that bring the ETS in close proximity to the site of cofactor reduction in the matrix. In the ETS, NADH and $FADH_2$ are oxidized, O_2 is reduced to H_2O, and H^+s are transported into the intermembrane space.

The ETS consists of four protein complexes (**FIGURE 3.19**). Complex I, the NADH dehydrogenase, oxidizes NADH and reduces ubiquinone. In plants, complex I consists of approximately 20 protein subunits. Reduction of the quinone requires two electrons and two protons. This

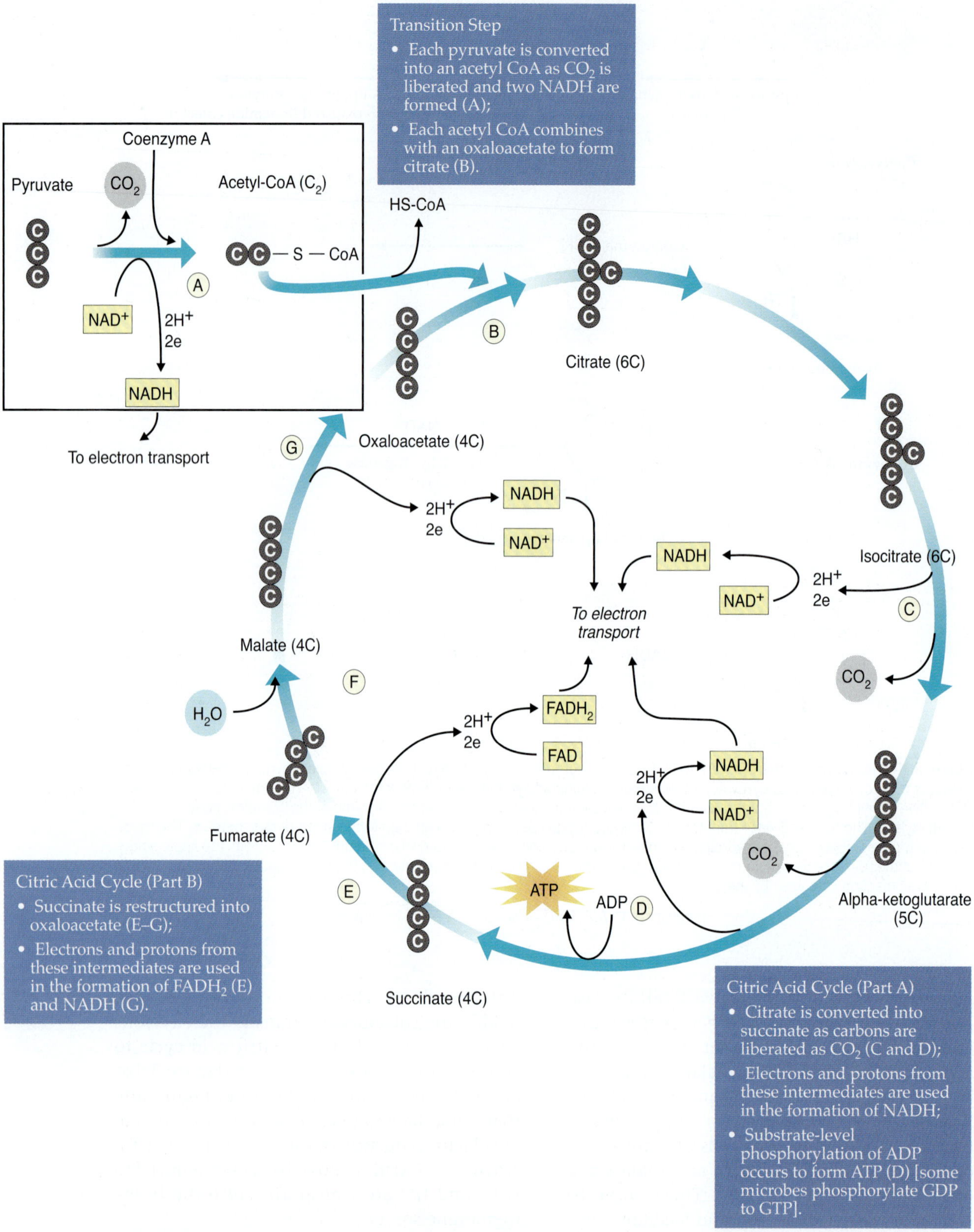

FIGURE 3.18 Citric acid cycle. Pyruvate is oxidatively decarboxylated to acetyl-CoA by the pyruvate dehydrogenase complex at A. Acetyl-CoA and oxaloacetate undergo an aldol condensation catalyzed by citrate synthase to produce citric acid at B. Isomerization to isocitrate is catalyzed by aconitase, and oxidative decarboxylation, by isocitrate dehydrogenase at C, yielding alpha-ketoglutarate (oxoglutarate and NADH). The ketoglutarate dehydrogenase complex at D resembles the pyruvate dehydrogenase complex described in Figure 3.17. The product, succinyl-CoA, is hydrolyzed to succinate and yields one equivalent of ATP. The remaining steps reconstitute oxaloacetate. Succinate is oxidized to fumarate by an FAD-dependent dehydrogenase (complex II) at E. Fumarate is hydrated to malate by fumarase at F. Finally, malate is oxidized to oxaloacetate by a dehydrogenase at G and the cycle continues. Reduced cofactors are reoxidized by the electron transport system.

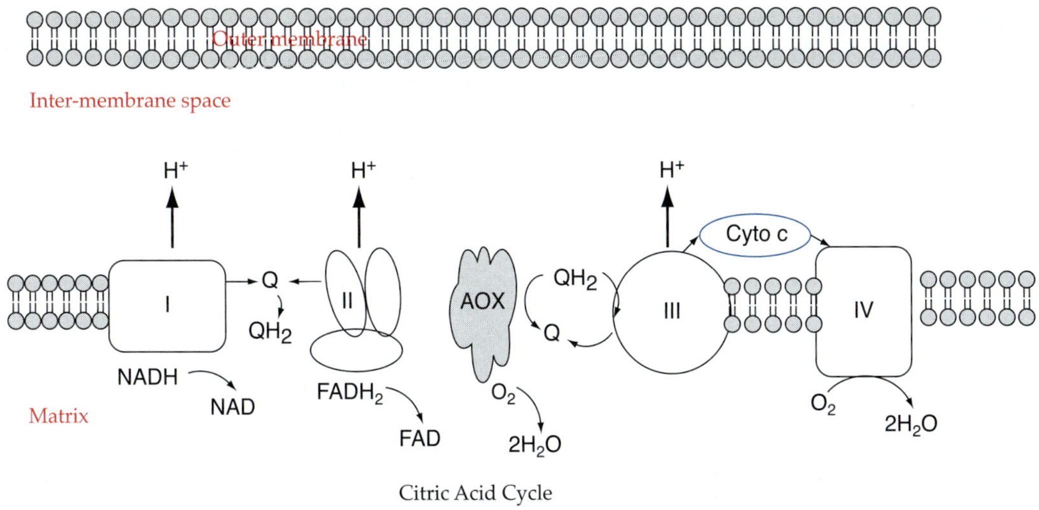

FIGURE 3.19 Mitochondrial ETS. The components of the ETS are located in the inner membrane (cristae) of the mitochondrion. Protein complex I oxidizes NADH and reduces ubiquinone (Q). Complex II oxidizes $FADH_2$ and reduces Q. Electrons are transported by QH_2 to complex III. Complex III reduces cytochrome c (Cyto c) and reoxidizes the quinones by a Q cycle comparable to that which occurs in the cytochrome b_6f complex in the thylakoid membranes of the chloroplast (Figure 1.18). Reduced cytochrome c transfers electrons to complex IV, which reduces O_2 to water. During electron transport, the ETS "pumps" protons into the intermembrane space. Plant mitochondria also contain an alternate oxidase complex that oxidizes QH_2 and transfers the electrons directly to O_2 but does not transfer protons into the intermembrane space.

reaction facilitates the transfer of H^+s from the matrix to the intermembrane space. Complex II, succinic dehydrogenase, is actually a part of the citric acid cycle. It oxidizes $FADH_2$ and reduces ubiquinone. As in complex I, protons are required for this reduction and are derived from water in the matrix.

Complex III is a cytochrome bc_1 protein complex similar to the cytochrome b_6f complex in the thylakoid membranes (see Chapter 1). It consists of approximately 10 subunits, including a *c*-type cytochrome, two *b*-type cytochromes, and a Reiske 2Fe-2S protein. It oxidizes ubiquinone and reduces cytochrome *c* in a Q cycle that results in the "pumping" of protons into the intermembrane space. Complex IV, the cytochrome *c* oxidase, transfers electrons to oxygen. It consists of approximately 12 subunits in plants, including two heme proteins and two copper-containing subunits.

Electron transport through complex IV can be inhibited by cyanide. Many plants are capable of cyanide-insensitive respiration. They have an alternate oxidase associated with the inner mitochondrial membrane. This is an Fe-containing protein complex that oxidizes ubiquinol and reduces oxygen but does not transport protons from quinol oxidation across the inner mitochondrial membrane. The function of this alternate oxidase is uncertain. It acts as an uncoupler (i.e., separating the processes of electron transport from phosphorylation of ADP). It may provide a route for cofactor oxidation and continued operation of the citric acid cycle when complex IV is blocked or when the supply of other factors, such as Pi, limit ATP synthesis. The energy lost in this alternate oxidase is released as heat. It has been proposed that this slight increase in temperature may facilitate the release of volatile compounds that can attract insects for pollination or defense. This is true for the voodoo lily (*Sauromatum guttatum*), which generates heat in its flowers that releases fly-attracting amines, but the effects of heat production in other plants, particularly leaves, have not been demonstrated.

The mitochondrial membranes also contain NADH dehydrogenases, one in the inner membrane facing the matrix and a second dehydrogenase in the outer membrane, facing the cytoplasm. Both enzymes oxidize NADH without contributing to the proton gradient in the intermembrane space. The dehydrogenase facing the cytoplasm oxidizes NADH generated in glycolysis so that no translocator is required to move reduced cofactor across the mitochondrial membranes to be oxidized. The dehydrogenase in the inner membrane oxidizes reduced cofactor that is generated in the matrix. Recall that in photorespiration, glycine is metabolized to serine in the mitochondrion with the generation of reduced NAD (see Chapter 2). The inner membrane dehydrogenase may act as a "safety valve," ensuring a steady supply of oxidized cofactor to operate the pyruvate dehydrogenase complex and the citric acid cycle when electron transport is inadequate.

Chemiosmotic Coupling

Chemiosmotic coupling is an ingenious method by which the cell uses membranes to couple the energy released by the oxidation of cofactors to the production of a chemical bond, yielding ATP. The ETS "pumps" protons from the matrix into the intermembrane space in the mitochondrion. Activity of the ETS results in a proton gradient in the space (Figure 3.19) similar to the proton gradient produced in the thylakoid lumen during the light reaction of photosynthesis. Movement of protons into the intermembrane space results in a large difference in pH and electrical charge between the space and the matrix. An ATP synthase complex facilitates movement of protons back into the matrix. The synthase uses the energy of this proton flow to synthesize ATP from ADP and phosphate ions. As in the chloroplast, the mitochondrial ATP synthase consists of two major complexes. The F0 subunit consists of a, b, and a variable number of c subunits and is embedded in the inner mitochondrial membrane. The F1 subunit consists of α, β, γ, δ, and ε peptides that protrude into the matrix (see Figure 1.19). ADP and Pi are bound in an active site formed from the α and β subunits. As in the chloroplast enzyme, closing of the active site creates an anhydrous cavity in which the phosphoester bond forms spontaneously. Energy is required to open the site and release ATP. This energy is supplied by rotation around the γ subunits, which in turn causes a conformational change in the active site. This rotation is caused by the movement of protons to the matrix *via* the c subunits (see Figure 1.19b). In the mitochondrion, pumping of protons results in both a pH gradient (ΔpH) and an electrochemical gradient ($\Delta\Psi$). The pH difference is smaller in mitochondria than in chloroplasts. An increase of 0.2 to 0.5 pH units has been measured in the matrix of active mitochondria. However, the change in the matrix potential may be as much as -150 mV. This $\Delta\Psi$ is the primary factor that drives mitochondrial ATP synthesis.

When adding up the net ATP that is synthesized in glycolysis, the citric acid cycle, and chemiosmotic coupling, the complete oxidation of glucose yields 36 to 38 moles of ATP per mole of glucose. The chemical equations indicate that 36 moles of ATP could be synthesized by these reactions. In the cell, however, the energy yield will be decreased by other factors, such as the cost of transporting compounds such as pyruvate, ADP, and cofactors, into the mitochondrion. Overall, catabolism of glucose is approximately 35% efficient in converting the energy in C-H bonds of glucose to phosphoester bonds in ATP. The remainder of the energy is lost as heat.

Plant mitochondria oxidize organic acids and produce ATP in the same way as animal mitochondria. However, these organelles play a more prominent role in plants in the synthesis of intermediates for biosynthetic pathways such as synthesis of nucleotides, amino acids, and vitamins. Many of the organic acids produced in the citric acid cycle are also exported for biosyntheses that occur outside the organelle.

CHAPTER SUMMARY

Sugar-phosphates produced in the C-3 cycle provide both intermediates for biosyntheses and energy for all the metabolic reactions in plants. Hexose phosphates are stored in the chloroplasts of leaves as starch. Starch synthesis provides free phosphate for additional reactions, and starch is an osmotically neutral storage molecule. Leaf starch can be catabolized in the dark and provides both intermediates and reduced cofactors for chloroplast metabolism. DHAP and sugars are exported to the cytoplasm for sucrose synthesis and transport to nonphotosynthetic tissues or used to generate ATP by glycolysis and the citric acid cycle and electron transport that occur in the mitochondria. In plants, however, the chloroplast is the real powerhouse of the cell and the catabolic pathways mainly produce intermediates used in the biosynthesis of all other organic compounds required for plant life.

One of the major uses of fixed carbon in plants is the synthesis of cell walls. Cellulose is probably the most abundant polymer on Earth. Cellulose fibrils form the backbone of the primary plant cell wall, whereas a variety of additional carbohydrates fill in the matrix. Chapter 4 covers the synthesis and assembly of primary cell walls. In secondary walls, the basic carbohydrate building blocks are complexed with lignin polymers to provide additional support and protection for plant tissues. Lignin synthesis is covered in Chapter 8 on synthesis of phenylpropanoid compounds.

REFERENCES FOR FURTHER INFORMATION

Fernie, A.R., Carrari, F., & Sweetlove, L.J. (2004). Respiratory metabolism: Glycolysis, the TCA cycle and mitochondrial electron transport. *Curr. Opin. Plant Biol.* **7**:254–261.

Plaxton, W.C. & Podesta, F.E. (2006). The functional organization and control of plant respiration. *Crit. Rev. Plant Sci.* **25**:159–198.

Smith, A.M. (2008). Prospects for increasing starch and sucrose yields for bioethanol production. *Plant J.* **54**:546–558.

Zeeman, A.C., Kossmann, J., & Smith, A.M. (2010). Starch: Its metabolism, evolution, and biotechnological modification in plants. *Annu. Rev. Plant Biol.* **61**:209–234.

PLANT BIOCHEMISTRY ***on the web***

Visit the *Plant Biochemistry* student companion Web site at http://biology.jbpub.com/gleason/plantbiochemistry/ for additional online study resources on these topics.

CHAPTER 4

Primary Cell Walls

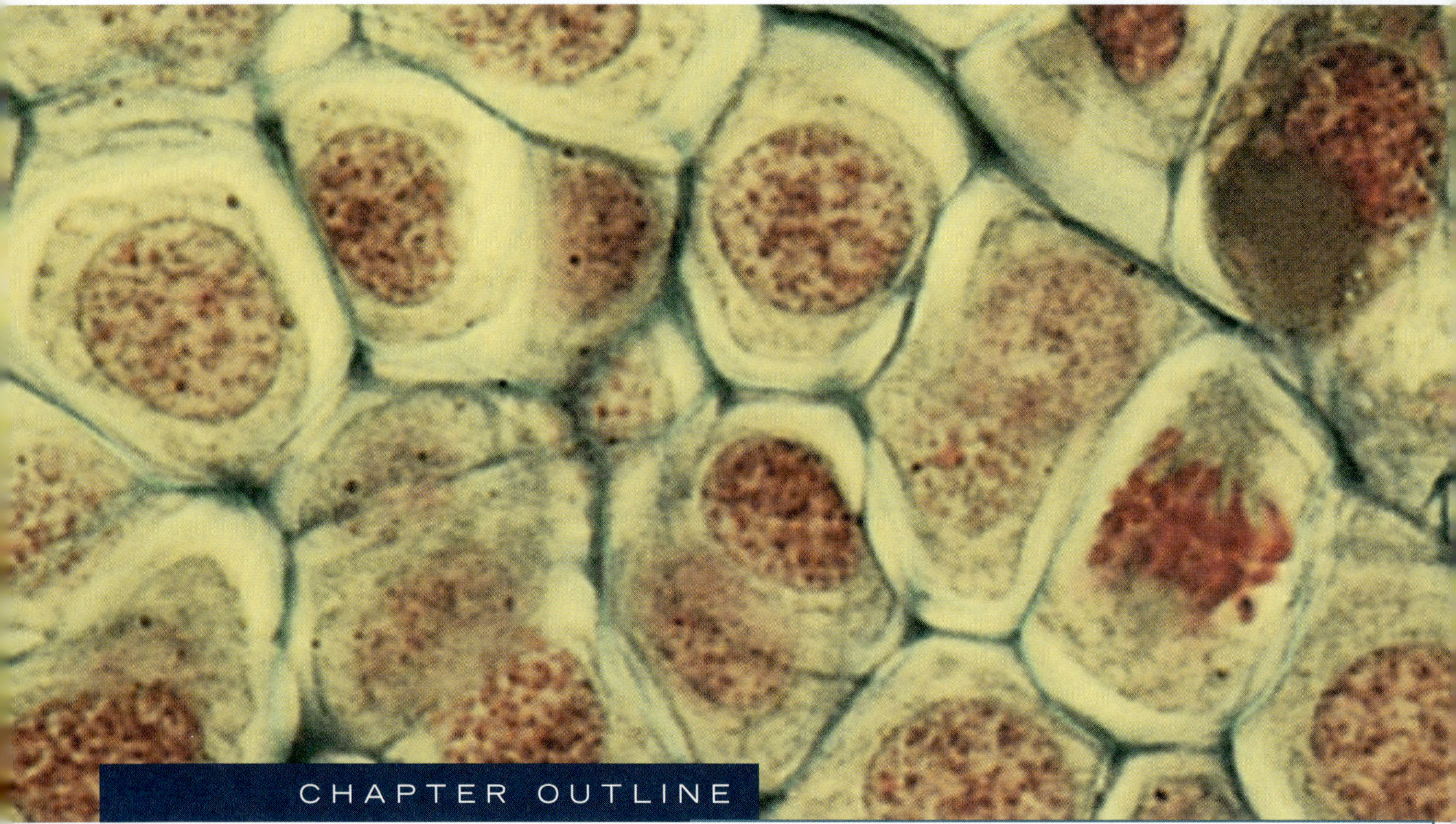

CHAPTER OUTLINE

4.1 Overview

Rigid cell walls are one of the defining characteristics of plants. These walls provide support for multicellular tissues, particularly the xylem, which is under positive water pressure. Plant walls are also the first line of defense against pathogens and herbivores.

Most of the carbon fixed in photosynthesis is ultimately exported to the exterior to form cell walls. Primary cell walls are composed of cellulose, a polymer of β-1,4-linked glucose. Several of these linear polymers associate to form rigid fibrils. The fibrils are embedded in an amorphous matrix of hemicelluloses and pectin that are composed of diverse and complex carbohydrates. The components of the primary walls are held together by hydrogen and ionic bonds. These relatively weak bonds provide mechanical strength but are easily made and broken to allow sufficient elasticity for growth and morphological changes. Primary cell walls are dynamic structures. Proteins embedded in the matrix have various enzymatic activities that modify the carbohydrate cross-linking and elasticity of the walls. Intercellular signals required to initiate defense mechanisms after pathogen invasion are also provided by wall components. Mature plant cells add aromatic compounds to the primary wall components and polymerize these to a rigid matrix called lignin. Synthesis of secondary walls and lignin are covered in Chapter 8. In this chapter the structure and synthesis of the primary wall of vascular plants is described.

4.2 Cellulose

Cellulose is a linear polymer of β-1,4-D-glucose as shown in **FIGURE 4.1**. Note that each glucose unit is rotated by 180 degrees relative to its neighbors. The repeating unit is the disaccharide, cellobiose. Unlike starch, cellulose chains have no branches. Because of the rotation of the pyranose rings, hydrogen bonding is possible between the C-3–OH group of one glucose residue and the ring oxygen of the adjacent glucose. This impedes rotation of the residues, leading to formation of a stiff, band-like molecule. Each cellulose molecule is 8,000 to 15,000 glucose units long, but these single molecules can associate to form microfibrils that are several micrometers in length. The fibrils are composed of individual cellulose molecules and are further organized into paracrystalline arrays of 36 or more cellulose molecules linked to each other by hydrogen bonds. The microfibrils are 5 to 12 nm wide and are large enough to be seen in the electron microscope (**FIGURE 4.2**).

Cellulose Synthesis

The enzyme, cellulose synthase, uses UDP-glucose as a substrate. UDP-glucose is found in millimolar concentrations in most leaf tissue. There are two major reactions that supply the nucleotide sugar for cellulose synthesis. Sucrose synthase catalyzes the following reversible reaction:

$$\text{sucrose} + \text{UDP} \rightarrow \text{fructose} + \text{UDP-glucose}$$

An alternative source is the UDP-glucose pyrophosphorylase reaction:

$$\text{glucose-1-P} + \text{UTP} \rightarrow \text{UDP-glucose} + \text{PPi}$$

The enzyme, invertase, which cleaves sucrose to glucose and fructose, may also supply glucose for cellulose synthesis.

Cellulose synthase is encoded by a gene family designated *cesA*. There are 10 *cesA* genes in *Arabidopsis* and 18 in poplar. The *cesA* genes belong to a superfamily of enzymes, the glycosyltransferases. There are approximately 415 genes for these nucleotide-sugar transferases in plants, based on their derived amino acid sequence comparisons. Glycosyltransferases use nucleotide sugars as substrates and transfer the sugar moiety to an acceptor molecule. All these enzymes have a recognizable transferase motif. In the cellulose synthases and cellulose synthase-like enzymes, this motif consists of three

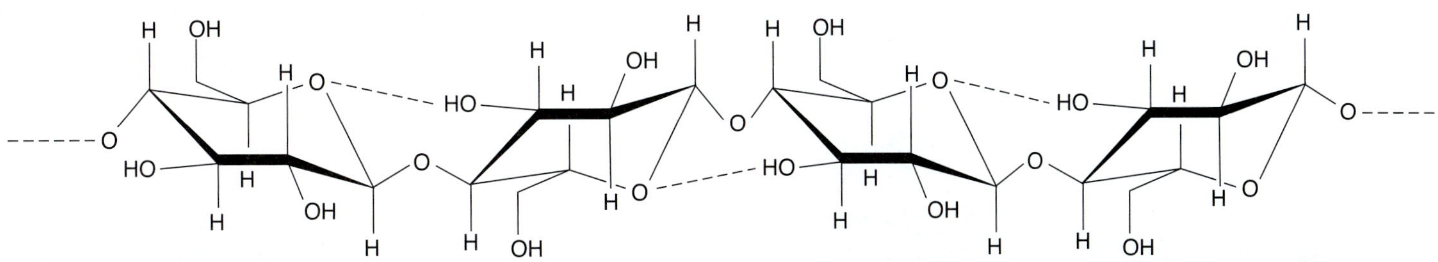

FIGURE 4.1 Structure of cellulose. Cellulose is a polymer of β-1,4-linked glucose residues. The repeating unit is the disaccharide, cellobiose. Hydrogen bonds can form between adjacent residues to yield a rigid, linear macromolecule.

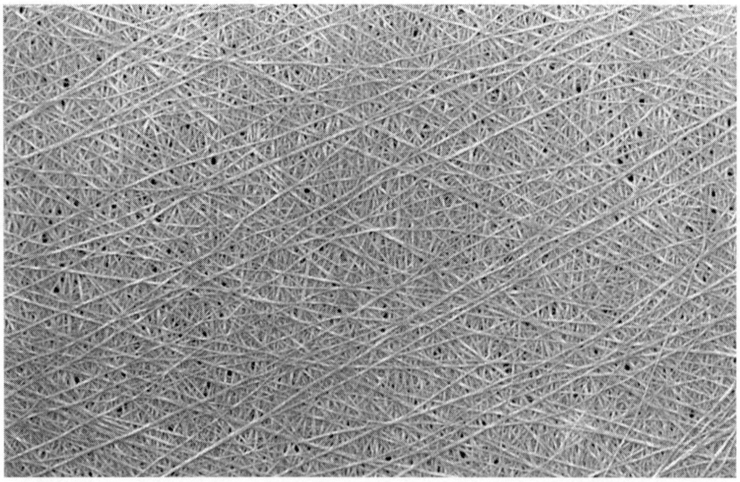

FIGURE 4.2 Cellulose fibrils in a primary cell wall of *Populus sieboldii* (Japanese aspen). Several layers of cellulose microfibrils are visible; other layers are present deeper in the wall. Each layer provides strength in the direction parallel to the microfibrils. Other wall components have been dissolved away to reveal the cellulose. Magnification = ×7,000.

aspartic acid residues and a QxxRW active site sequence that binds the nucleotide sugar. Transferases are often highly specific and bind a different nucleotide sugar and catalyze transfer to various other molecules, but genetic information is not sufficient to predict substrates and products. Biochemical confirmation of enzyme function of these enzymes in plants is generally lacking.

Cellulose synthase (CesA) transfers glucose to the end of an existing cellulose molecule. The enzyme is embedded in the cytoplasmic membrane and is difficult to purify because of its relative insolubility in water and large size. Much of the early research on this enzyme has been done using bacteria that synthesize cellulose, such as some species of *Bacillus, Agrobacterium,* and *Acetobacter*. Unlike plants, these bacteria do not make fibrils but build an amorphous extracellular matrix of cellulose. Cotton fibers are almost pure cellulose and have also been used as a source of synthase. In most plant preparations, the enzymes are not active when solubilized with detergents. The alternative procedure is to isolate plasma membrane preparations that will yield active synthase but also contain a number of proteins and other components.

The cellulose synthase monomer is a large protein of approximately 1,100 amino acids. It is active as a hexamer consisting of three different CesA isoforms. The hexamers further associate to form larger complexes of 36 subunits, called rosettes (a hexamer of hexamers). These are large enough to be visible by electron microscopy in freeze-fracture preparations of cell membranes (FIGURE 4.3). The rosettes are often associated with proteins on the cytoplasmic side of the membrane that have sucrose synthase activity. From a combination of biochemical and morphological analyses and sequence comparisons, a model for the general structure of CesA proteins has been proposed (FIGURE 4.4). The N-terminal domain of the protein is in the cytoplasm and contains a putative Zn-binding site. The Zn motif (sequence of amino acids) resembles the classical Zn finger domain found in some transcription factors. This finger consists of two histidine residues and two cysteine residues forming a short loop. Binding of a Zn ion stabilizes the helix-strand-loop into a finger-like structure. Zn fingers are best characterized as

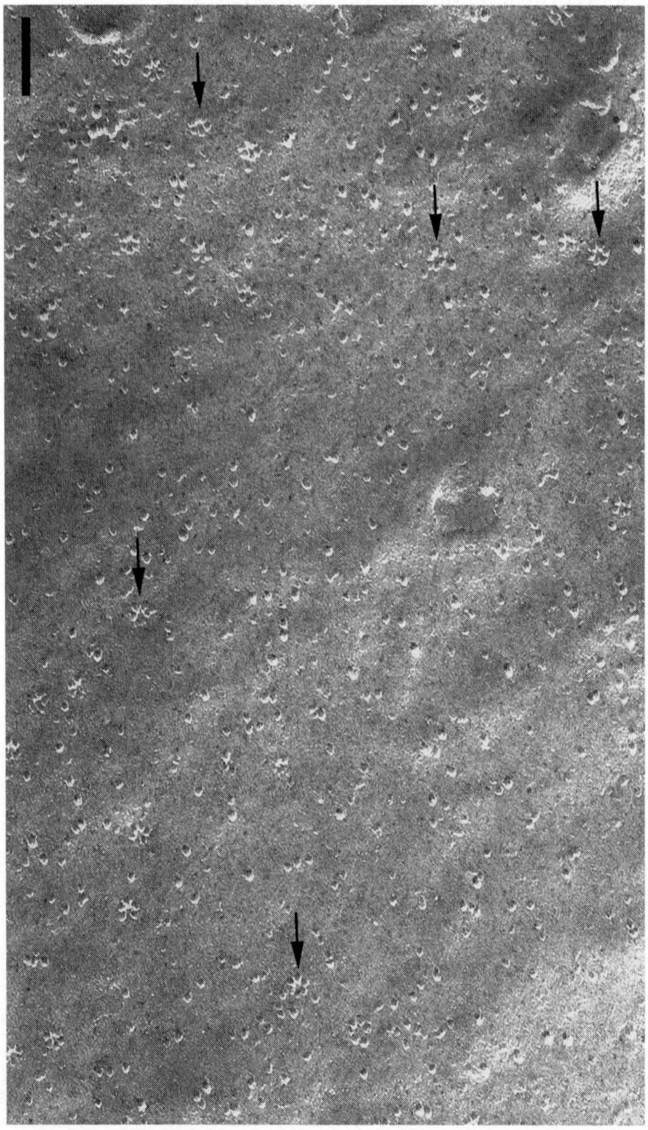

FIGURE 4.3 Freeze fracture replica of a cell membrane from the alga, *Micrasterias denticulata.* The arrows point to rosettes of cellulose synthase subunits embedded in the cell membrane. Bar = 0.1 μm.

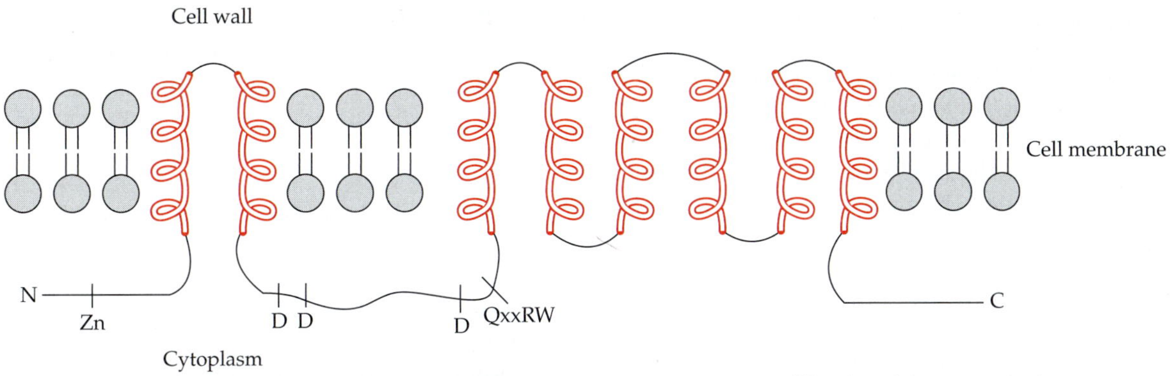

FIGURE 4.4 Proposed intracellular structure of cellulose synthase. The amino terminal (N) region of the enzyme is shown at the left side of the diagram. A Zn-binding domain (Zn) facilitates interaction with other synthase subunits to form a hexamer. This interaction domain is followed by two transmembrane helices predicted from the hydrophobic properties of the residues. The active site is on the cytoplasmic side of the protein and consists of two aspartic acid residues (D,D) separated by two to four variable amino acids, a third aspartic acid, and the conserved transferase motif: glutamine, two variable residues (xx), arginine, and tryptophan. The C-terminal domain consists of six predicted transmembrane helices that provide an exit pore for the nascent cellulose molecule.

DNA binding domains. Similar structures, however, also stabilize protein–protein interactions. It is assumed that the Zn finger-like region of the cellulose synthase monomer stabilizes the quaternary structure of the complex. This domain is followed by two transmembrane helices. Looping back into the cytoplasm, the protein has two hypervariable regions that are not highly conserved among the different Ces isoforms in any single species. Presumably these residues play no role in catalysis. The C-terminal domain of the protein consists of six transmembrane helices that may form a pore through which the newly synthesized cellulose molecule exits the cell.

A sequence motif that is characteristic of many glycosyltransferases, asp, asp, asp, gln, X, X, arg, trp, is found in the cytoplasmic portion of the protein, as indicated in Figure 4.4. This is the presumed binding site for UDP-glucose. Fluorescent labeling of Ces complexes *in vivo* shows that the entire rosette moves through the plane of the membrane at approximately 300 nm per minute. This corresponds to the addition of a glucose to a preexisting cellulose molecule at a rate of 300 to 1,000 min^{-1} (k_{cat}).

The progressive synthesis of the cellulose chains outside the plasma membrane seems to be guided by microtubules in the cytoplasm that are oriented parallel to the direction of fiber synthesis. There is some controversy about the precise role of the microtubules. Herbicides that are reported to inhibit cellulose synthesis, such as isoxaben and 2,6-dichlorobenzonitrile, do not inhibit catalytic activity. Instead, formation of a defective cell wall results from abnormal orientations of microtubules that occur after herbicide treatment. These findings indicate that microtubules are essential for guiding rosette movement and fibril orientation.

Several additional proteins are associated with the cellulose synthase rosettes. Sucrose synthase was mentioned previously. An endoglucanase called Korrigan (Kor) is required for synthesis of a primer to initiate cellulose fibril formation. Kor may also function in loosening the wall components during cell expansion. A membrane-associated thioredoxin fraction has been reported to regulate synthase activity in response to changes in cellular redox potential.

Callose

In addition to cellulose, most plants also produce a glucose polymer called callose composed of β-1,3-linked glucose units with variable sugar branches. Callose is made in response to stress and wounding and also forms part of the cell plate during cytokinesis. Cellulose synthase catalyzes the formation of β-1,3-glucose linkages *in vitro* but apparently does not do this *in vivo*. A specific enzyme is required that is a product of the *calS* gene. This gene encodes a large glucosyltransferase with approximately 2,000 residues. It is embedded in the cell membrane but does not form rosette structures. Like cellulose synthase, the enzyme uses UDP-glucose as a substrate to yield polymers, but these β-1,3-linked chains do not produce fibrils. The product is a rather irregular, globular polymer. In wounded tissues, callose is synthesized and forms a large, clot-like structure that is a barrier to invasion by microorganisms.

4.3 Wall Matrix

The matrix of the cell wall is composed of a variable number of sugar polymers with complex structures. Most of these are synthesized by glycosyltransferases associated with the membranes of the endoplasmic reticulum or the Golgi apparatus. The products are transported to the cell membrane in vesicles (**FIGURE 4.5**). After being expelled from the cell, these polymers are organized into a gel-like matrix that cross-links the cellulose fibrils. Matrix polymers may be further modified by various enzymes found in the wall.

The matrix polysaccharides can be divided into two major groups. The hemicelluloses are generally linear polymers of β-1,4-linked glucose residues with α-1,6 branches to other sugars, most often xylose. The pectins are variable polymers rich in galacturonic acid and are further subdivided depending on their sugar composition. Typical dicot cell walls are composed of approximately 30% cellulose, 30% hemicellulose, 35% pectin, and 1% to 5% protein. In contrast, monocot primary walls are more variable in composition, but in noncereal grasses they have approximately 25% cellulose, 55% hemicellulose, 10% pectin, and 10% protein.

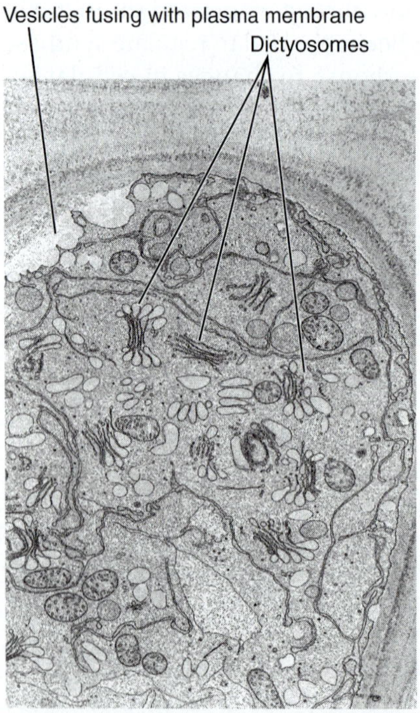

FIGURE 4.5 Golgi vesicles exporting wall polysaccharides. Several stacks of Golgi membranes (dictyosomes) are shown. The ER and Golgi in these cells are actively producing cell wall polysaccharides. These products are transported in the vesicles that fuse with the cell membrane and release their contents to the exterior.

Hemicellulose

Hemicelluloses were originally defined as that fraction of the cell wall matrix that could be extracted with strong base. These include a variety of different polysaccharides characteristic of a particular species of plant. Hemicelluloses in general, however, consist of two major types of polymers, xyloglucan and xylans. Xyloglucan is the most commonly occurring component and consists of a polymer of β-1,4-linked glucose units with α-1,6-D-xylose branches on three consecutive glucose units. The xylose branches may be further extended with other sugars such as arabinose and fucose. Xylans are more complex but are basically polymers of β-1,4-linked xylose units with branches consisting arabinose (arabinoxylans) and other sugars or derivatives such as modified glucuronic acids. Less common polymers include the mannans and glucomannans that consist of a backbone of β-1,4-linked mannose units with different sugar branches.

Hemicellulose polymers are not as long as cellulose molecules but are long enough to form hydrogen bonds to cellulose molecules in different fibrils. They provide cross-links between the fibrils. Although cross-linking serves to stabilize the wall, the hemicellulose links actually allow greater flexibility by preventing aggregation of cellulose fibrils.

Hemicellulose polymers are synthesized by glycosyltransferases, designated cellulose synthase-like (Csl) enzymes located in the endoplasmic reticulum and the Golgi complex. There is evidence, however, that some of these polymers may be produced at the plasma membrane by these Csl enzymes. Plant genomes contain several *csl* genes that show some sequence similarity to cellulose synthase (*cesA*) genes. Most of the Csl enzymes have not been characterized biochemically, but some of these proteins have been found associated with the plasma membrane as well as embedded in membranes of the Golgi complex. Mutations in the *csl* genes lead to irregularities in cell wall structure in root hairs and embryos. These genetic analyses have been used in lieu of biochemical assays for enzyme activity. The mutant phenotypes show a lack of microfibrillar cross-linking, suggesting that Csl proteins are involved in hemicellulose synthesis.

Hemicelluloses are generally exported in vesicles and are often modified by enzymes located in the cell wall. Glycosidases (hydrolases) often remove sugars from the branches. Acetylation and methylation of sugar residues is also possible and may occur before transport as well as in the apoplast (exterior of the plant cell).

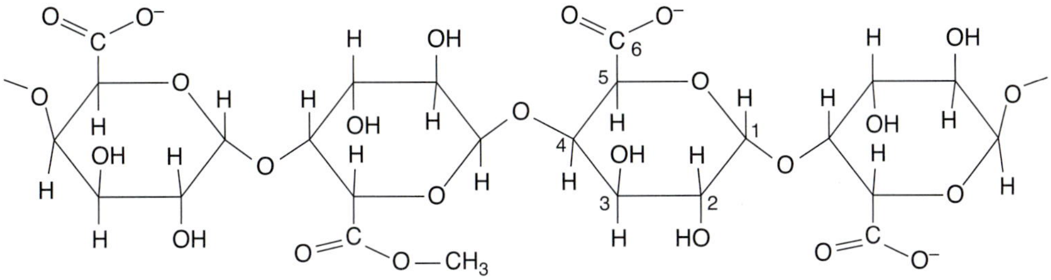

FIGURE 4.6 Galacturonic acid backbone polymer found in pectin. Galacturonic acid, from the oxidation of galactose, is the basic building block of pectin polymers. The acids are connected by α-1,4-glycosidic bonds. These acid groups can form salt bridges, usually with Ca^{2+} ions, to other polymers. The acid groups can be modified by methylation as shown, which would inhibit ionic cross-linking. The C2 and C3 OH groups are acetylated in some plant walls.

Pectin

Pectin is experimentally defined as that portion of the wall matrix that can be extracted with calcium chelators such as EDTA (ethylene diamine tetra-acetic acid). Pectins can be quite variable but most commonly consist of polymers of galacturonic acid, the acid of galactose (**FIGURE 4.6**). The linear polymer is subsequently modified with various sugar branches. Like the hemicelluloses, pectin polymers are synthesized by specific glycosyltransferases located in the endoplasmic reticulum and Golgi apparatus. The products are exported in vesicles (see Figure 4.5). The galactose precursor is oxidized to the acid in the cytoplasm. The substrate for pectin synthesis is UDP-galacturonic acid. After export, the polymers can be further modified by enzymes located in the wall matrix.

Pectin consists of four major types of polysaccharides:

1. Homogalacturonan is a linear polymer of α-1,4-galacturonic acid. Ca^{2+} ions form cross-links between homogalacturonan polymers, providing rigidity to the matrix. The carboxyl group at C6 is often modified by O-methylation or acetylation. Hydroxyl groups on C2 and C3 can also be modified by acetyl transferases. These modifications can change the ion binding properties of the polymers and disrupt cross-linking.
2. Rhamnogalacturonan I is a polymer of an α-1,4-galacturonic acid (homogalacturonan) with α-1,2-rhamnose repeating unit. The rhamnose residues can be further modified with additional sugars such

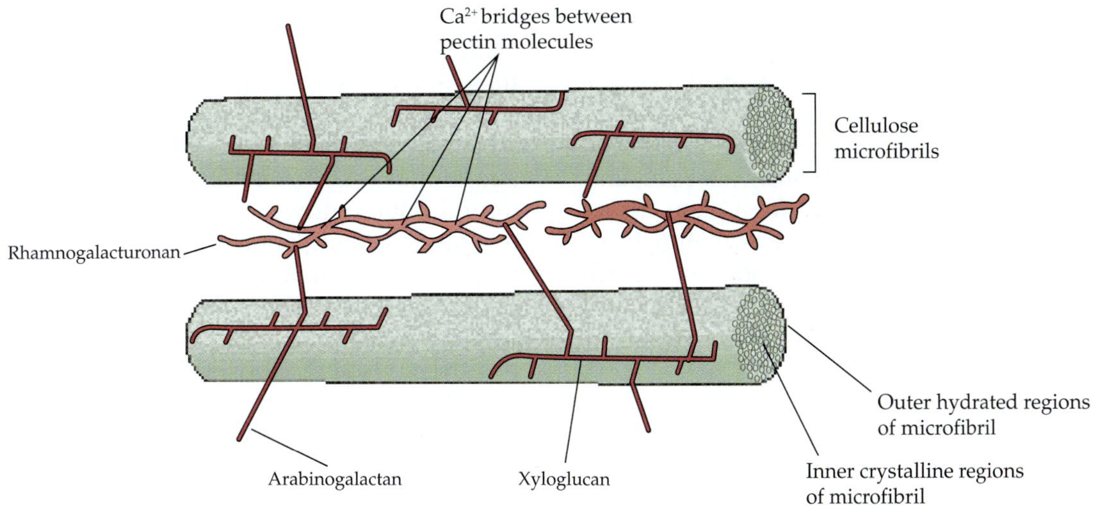

FIGURE 4.7 Simple model of hemicelluloses and pectin cross-linking to cellulose fibrils. A simple model of the various carbohydrate components in the primary cell wall was first published in 1991. It shows the cellulose microfibrils with a rigid semicrystalline structure. The fibrils are linked to matrix polymers, hemicelluloses and pectins, by hydrogen bonds. The galacturonic acid residues of pectin can bind Ca^{2+} and form ionic cross-links between different polymers in the matrix. The matrix polymers serve to bind the various carbohydrates but also keep the cellulose fibrils from associating with each other. This basic architecture provides both stability and flexibility to the primary cell wall.

as galactose or arabinose. This rhamnogalacturonan I polymer also forms Ca^{2+} cross-links to other pectin or hemicellulose polymers in the matrix (**FIGURE 4.7**).

3. Rhamnogalacturonan II has the same basic backbone structure as rhamnogalacturonan I but has much more variability in the sugar branching. At least 12 different sugars have been reported in rhamnogalacturonan II fractions with 20 different linkages. Borate ions (BO_3^{3-}) form borate complexes between adjacent sugar hydroxyl groups and cross-link the rhamnogalacturonan IIs. It has also been reported that borate diester links can form spontaneously and cross-link these polysaccharides.
4. Xylogalacturonans and apiogalacturonans are found in a small number of plants. Xylogalacturonans have been characterized in *Arabidopsis* primary walls; they are typical polymers of galacturonic acid with β-1,3-linked xylose branches (xylogalacturonans), which can be extended with other sugars such as galactose or fucose. The apiogalacturonans are found in aquatic monocots where the galacturonic acid polymers have branches with O-2 or O-3-linked apiose units.

Pectins provide strength to walls by cross-linking the various polymers with ionic bridges as well as hydrogen bonds. This variable cross-linking also controls porosity of the wall, regulating passage of water and small molecules, including ions. Pectins are important in linking cells to each other in plant tissues. It has also been proposed that methylated pectins are a source of methane that seems to be released from some plants into the atmosphere. Most methane in the atmosphere is the product of microbial fermentations or geological processes. It has been reported, however, that extremely small amounts of methane can be released from plant tissues under stress conditions, such as high light or excessive heat. Although the source of plant methane is still controversial, one plausible initial step is the demethylation of pectin esters.

Proteins: Enzymes

A number of enzymes are found in the matrix of the cell wall. The methyl esterases and acetyl transferases that modify glucuronic acid residues in pectin were mentioned previously. These control the degree of cross-linking of the polysaccharides in the wall and allow for changes in extensibility. Other enzymes are glucanases, such as Kor. This endo-1,4-β-glucanase is associated with primer formation of cellulose fibrils but can also digest the bonds in cellulose and hemicelluloses, thereby loosening the wall structure. In contrast, peroxidases catalyze the formation of cross-links between the carbohydrates and various structural proteins and contribute to wall rigidity. Wall-associated kinases have been described. These are attached to the plasma membrane by sugar-phosphatidyl inositol linkages to lipids in the bilayer. Many of these kinases have lectin domains. Lectins are proteins that bind to specific carbohydrates. Wall kinases bind polymer fragments and initiate signal transduction to the cytoplasm for cell defense after bacterial or fungal invasion of the wall.

Proteins: Structural

The wall matrix also contains several types of structural proteins. Extensins are proteins with leucine or proline-rich repeated regions. The prolines are usually hydroxylated and peroxidases catalyze linkages between these proteins and tyrosine residues in other wall-associated proteins. The additional cross-linking provides a barrier to pathogen invasion. Expansins are small proteins that are exported in Golgi vesicles. They bind to cellulose fibrils and disrupt the noncovalent linkages to matrix polysaccharides. This facilitates cell expansion by loosening the structure of the wall. Arabinogalactan proteins are extensively modified with carbohydrates. Their function is unknown, but they are believed to act as signaling molecules to regulate cell expansion and differentiation.

A diagram showing the complex structure of the primary wall is shown in **FIGURE 4.8**. Several technical constraints have made it difficult to provide a more precise description of this vital plant organelle. Most enzymes involved in synthesis of wall components are integrated into the membranes of the Golgi apparatus or the plasma membrane. Others are embedded in the matrix and cannot easily be extracted in their native form. Until recently, most chemical procedures to analyze carbohydrates have involved harsh reagents and relatively imprecise characterizations. Newer chromatographic techniques and readily available mass spectrometers are starting to provide a better picture of the polymers in the wall. The variable nature of the wall carbohydrates among species and even among the various tissues of a single species, however, limits any general description of wall components and their functions. Despite the wealth of data provided by genome sequencing, mechanisms to describe how plant cell wall assembly, expansion, and repair are coordinated with the processes of cell division and differentiation are still lacking.

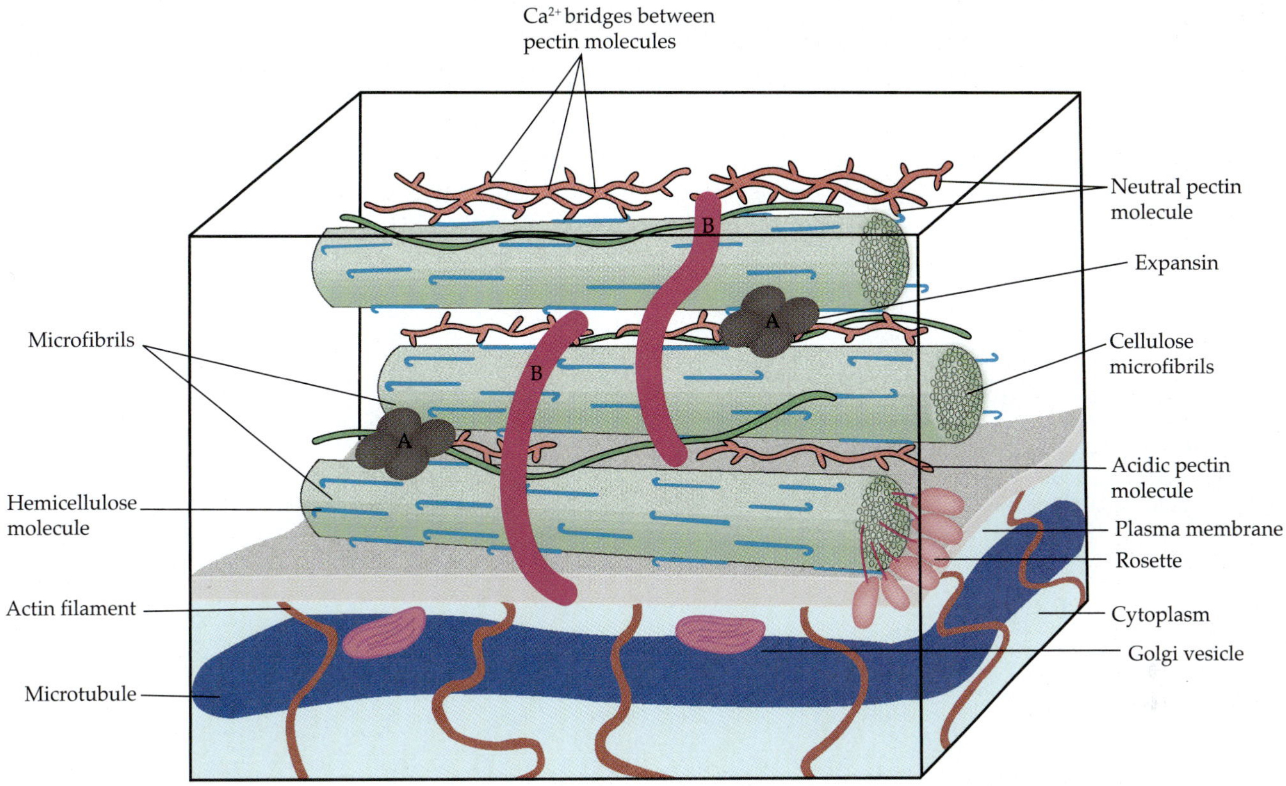

FIGURE 4.8 Current model of primary cell wall structure. Cellulose microfibrils are generated by cellulose synthase rosettes shown embedded in the plasma (cell) membrane. Microtubules in the cytoplasm serve to guide the movement of the rosettes and fibril synthesis. Hemicelluloses are shown as relatively short blue lines, whereas the pectins are depicted as more irregular branched polymers. Expansin proteins at A break cross-links between fibrils and matrix polymers. Extensin proteins at B provide additional cross-links between fibrils.

CHAPTER SUMMARY

A great deal of the carbon fixed by plants is used in the synthesis of cell walls. Primary cell walls consist of cellulose fibrils embedded in an amorphous carbohydrate matrix. Cellulose is a polymer of β-1,4-linked glucose residues; individual molecules associate to form long fibrils. The matrix consists of hemicellulose, a polymer resembling a short chain of cellulose with carbohydrate branches, and pectin, which is composed of several different sugars attached to a polygalacturonic acid backbone. Matrix polysaccharides, together with Ca^{2+} and borate ions, provide cross-links between the cellulose fibrils. They also function in wall permeability and extensibility. Secondary walls include these components plus lignin, which is synthesized from phenylalanine derivatives. The general chemical composition of primary walls has been determined for many plant species. The enzymes involved in synthesis of wall polymers, however, have not been characterized extensively due to their large size and membrane association. Genetic analyses have identified many enzymes and isoforms that are associated with cell wall synthesis, but without further biochemical investigation, the activities of these enzymes remain speculative.

Cell walls are dynamic structures in plants. Permeability changes can regulate the extracellular movement of ions and other small molecules through the apoplast. Changes in cross-linking provide protection from invading pathogens, and signaling molecules initiate defenses that allow the plant to resist further invasion. Cell wall material and composition are also of vital interest to humans. Cell walls constitute a large portion of food for both humans and livestock, provide material for clothing, have provided fuel in the past, and will probably provide a major supply of biofuels in the future. Newer methods of determining wall enzyme activities and regulation are just starting to unravel the complex synthesis of the deceptively simple structure.

REFERENCES FOR FURTHER INFORMATION

Cosgrove, D.J. (2005). Growth of the plant cell wall. *Nat. Rev. Mol. Cell Biol.* **6:**850–861.

Mohnen, D. (2008). Pectin structure and biosynthesis. *Curr. Opin. Plant Biol.* **11:**266–277.

Mutwil, M., Debolt, S., & Persson, S. (2008). Cellulose synthesis: A complex complex. *Curr. Opin. Plant Biol.* **11:**252–257.

Somerville, C. (2006). Cellulose synthesis in higher plants. *Annu. Rev. Cell Dev. Biol.* **22:**53–78.

Somerville, C., Bauer, S., Brininstool, G., Facette, M., Hamann, T., Milne, J., Osborne, E., Paradez, A., Persson, S., Raab, T., Vorwerk, S., & Youngs, H. (2004). Toward a systems approach to understanding plant cell walls. *Science* **306:**2206–2211.

PLANT BIOCHEMISTRY *on the web*

Visit the *Plant Biochemistry* student companion Web site at http://biology.jbpub.com/gleason/plantbiochemistry/ for additional online study resources on these topics.

CHAPTER 5

Nitrogen and Sulfur Metabolism

CHAPTER OUTLINE

5.1 Overview

The major nutrients for photosynthetic organisms are CO_2, H_2O, and sunlight. Using light energy, they are able to produce all the organic compounds needed for growth and reproduction. Additional inorganic nutrients are also required, however, such as an abundant source of phosphorus required for ATP and sugar-phosphate synthesis and lesser amounts of other inorganic nutrients such as iron, manganese, and copper, which all have roles as cofactors in the proteins that convert light energy into chemical bond energy (see Chapter 1). Inorganic nutrients are generally divided into two major groups: macronutrients, such as phosphorus, nitrogen, sulfur, and potassium, are required in relatively large amounts, and micronutrients, such as chloride ion, boron, and iron, are essential elements needed in much smaller quantities.

Two of these essential macronutrients, N and S, are the subject of this chapter. Primary metabolites, such as amino acids and nucleic acids, and many cofactors, such as NADP and biotin, contain nitrogen. Nitrogen constitutes 1% to 3% of the dry weight of most terrestrial plants and is the element that most often limits plant growth. Sulfur is present in smaller quantities but is essential for the synthesis of the amino acids cysteine and methionine. It is also required for production of sulfolipids and cofactors such as S-adenosyl methionine (SAM) and 3′-phosphoadenosine-5′-phosphosulfate (PAPS). Unlike N, sources of sulfur are generally abundant and seldom limit plant growth.

5.2 Nitrogen Metabolism

Nitrogen gas (N_2) constitutes approximately 78% of the Earth's atmosphere. However, the N≡N triple bond is extraordinarily stable, and nitrogen gas is basically inert. Utilization of atmospheric nitrogen as a nutrient is limited to certain prokaryotes that can "fix" N_2 into ammonia (NH_3). Eukaryotes require the fixed forms of nitrogen. For plants and algae these include nitrate (NO_3^-), nitrite (NO_2^-), and ammonia (NH_3). Most plants can also utilize organic forms of fixed nitrogen such as urea and amino acids, but these are not generally available in their environment under natural conditions. Commercial fertilizers, designated N:P:K, often contain N as ammonium nitrate, which are the two major forms of nitrogen readily imported and assimilated by plants.

Nitrate Assimilation

The nitrate anion (NO_3^-) is soluble in water and is fairly mobile in most alkaline soils. Ammonia is generally present as ammonium cation (NH_4^+) and often bound to negatively charged soil particles, making it less mobile than nitrate. Nitrate is taken up by plant roots in an active transport process. Two types of nitrate transporters are found in the cell membranes of root epidermal cells. High affinity transporters (HATs) have binding constants for nitrate in the 1 μM to 1 mM range and are present in both constitutive and inducible isoforms. Low affinity transporters (LATs) take up nitrate from the soil when it is present at greater than 1 mM concentration and are produced constitutively.

The HAT proteins belong to a superfamily of major facilitator proteins involved in active transport of ions. The *hat* genes code for proteins with 500 to 600 amino acids. Homology modeling suggests that the HAT proteins contain two sets of six transmembrane helices connected by a cytoplasmic loop. Three types of HATs have been described in the unicellular green alga, *Chlamydomonas*. HAT 2.1 binds both nitrate and nitrite, whereas HAT 2.2 is specific for nitrate and HAT 2.3 seems to transport mostly nitrite. Genes for similar transporters have been found in *Arabidopsis thaliana* and other vascular plants. LATs have limited amino acid identity to the HATs but are predicted to have a similar three-dimensional, membrane-spanning structure. LATs can also take up organic forms of nitrogen when these are present in the environment.

The transport activities of the HATs and LATs are dependent on the function of an integral membrane ATPase that pumps H^+ ions out of the cell. It has been estimated that two to three ATP are consumed per nitrate or nitrite ion transported into the cell. The activity of the transporters is subject to a complex regulation at the protein level. Nitrate transport follows a diurnal cycle with peak uptake occurring during the day to coincide with the maximal rate of photosynthesis. After uptake into root epidermal cells, nitrate must be further transported into the inner cortical cells and into the xylem vessels for movement into photosynthetic tissues for assimilation. Nitrate can also be taken up by transporters in the vacuole membrane for transient storage. Nitrite is more chemically reactive than nitrate and must be metabolized into more stable forms. Although some processing of nitrate and nitrite may occur in roots, reduction of nitrate and final conversion into amino acids requires a large amount of cellular energy in the form of ATP and reduced cofactors and therefore primarily occurs in leaves.

Ammonia Transport

Gaseous ammonia is readily soluble in water and is generally present as ammonium ion:

$$NH_3 + H^+ \rightarrow NH_4^+ \quad pK_a = 9.25$$

Although less important than nitrate as an N source, it is the major form of fixed nitrogen found in rice paddies and the acidic soils of conifer forests. As in the case of nitrate, two types of transporters for ammonium ion are found in root cell membranes. A HAT-type binds ammonium ion at micromolar concentrations, whereas a LAT takes up ammonium ion at greater than 1 mM concentration. These proteins are homologous to other ion transporters, and modeling predicts 10 transmembrane helices. In addition, water transporters may also take up NH_3 because it is similar in size to H_2O molecules. In some cases potassium transporters also move NH_4^+ because the hydrated forms of both the K^+ and NH_4^+ ions are equivalent.

Ammonium transport, like nitrate assimilation, is an active process and requires the presence of a membrane-bound H^+-ATPase. In some plants, ammonium ion can be stored in vacuoles. Unlike nitrate, the ammonium ion is not transported to leaves but assimilated directly in root cells to produce the amino acid, glutamine. Processing requires the amino acid, glutamic acid, and ATP.

$$\text{Glutamate} + NH_4^+ + \text{ATP} \rightarrow \text{Glutamine} + \text{ADP} + \text{Pi}$$

The enzyme that catalyzes this assimilatory reaction, Mg^{2+}-dependent glutamine synthetase (GS), is present in the cytoplasm of root cells. Glutamine is the major form of N-transport molecule in the xylem of plants that assimilate ammonia.

Nitrate Processing

Conversion of nitrate to ammonia and assimilation into amino acids generally occurs in photosynthetic tissues. The first enzyme, nitrate reductase, reduces nitrate to nitrite and is a soluble protein found in the cytoplasm of mesophyll cells. The enzyme monomer contains approximately 900 residues and functions as a dimer, although multimers may be present *in vivo*. The monomer consists of three domains, each of which binds a different redox cofactor (**FIGURE 5.1**). These include a flavin domain at the C-terminal portion of the protein containing flavin adenine dinucleotide, a cytochrome b_5 domain containing a heme group, and a third domain at the N-terminal that has a unique molybdenum cofactor (**FIGURE 5.2**). NADH is the ultimate electron donor and transfers two electrons to flavin adenine dinucleotide. Electrons are then transferred one at a time through the cytochrome to the molybdenum cofactor. Nitrate binds to the active site in the molybdenum cofactor domain and is reduced to nitrite.

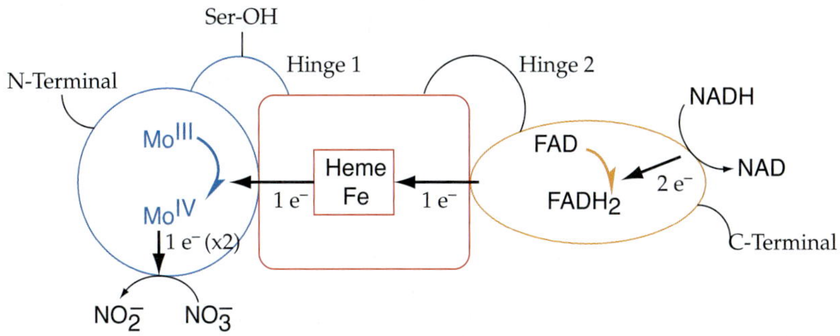

FIGURE 5.1 Diagram of the three catalytic subdomains of nitrate reductase (NR). Each NR monomer consists of three structural domains connected by loop regions designated Hinge 1 and Hinge 2. The C-terminal domain (orange) contains a flavin adenine dinucleotide (FAD) cofactor that is reduced by NADH, a two-electron carrier. The two electrons are transported one at a time to the central domain (red) where a covalently bound b-type cytochrome is located. The N-terminal domain (blue) is the binding site for NO_3^- and has a unique molybdenum cofactor (MoCo) (see Figure 5.2). This cofactor facilitates two successive electron transfers to the substrate and releases NO_2^-. The molybdenum cofactor domain also promotes protein dimerization. Hinge 1 includes an exposed serine residue that is the site of regulatory phosphorylation by NR protein kinase.

Nitrite is very reactive and binds to free amino groups such as lysine residues in proteins. It is presumably transported as the free ion into the chloroplast and reduced by a soluble enzyme in the stroma. Nitrite reductase catalyzes an unusual six electron transfer to reduce NO_2^- to ammonium ion. The enzyme contains two redox cofactors, a 4Fe-4S cluster and a unique siroheme, which is a modified form of the usual

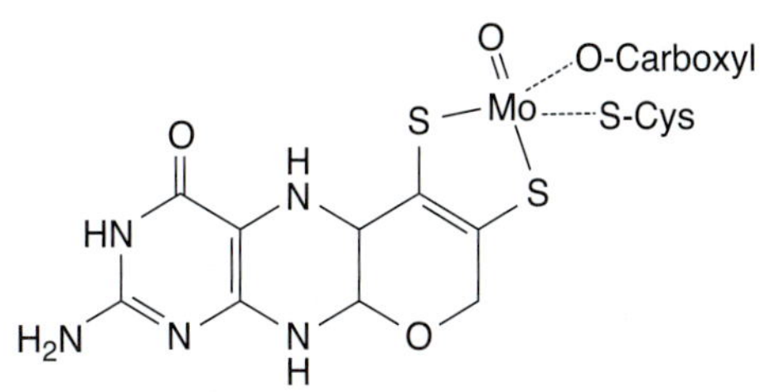

FIGURE 5.2 Structure of the molybdenum cofactor (MoCo) of nitrate reductase. This redox cofactor is composed of a molybdate ion bound to two sulfide groups in a tricyclic ring called a pterin. This type of cofactor is also found in a few other enzymes such as sulfite oxidase in peroxisomes; xanthine dehydrogenase, which is important for purine degradation; and aldehyde oxidase, the enzyme that catalyzes the last step in abscisic acid synthesis (see Chapter 7). In nitrate reductase, the cofactor is coordinated to a cysteine-SH group and an oxygen from a carboxyl group.

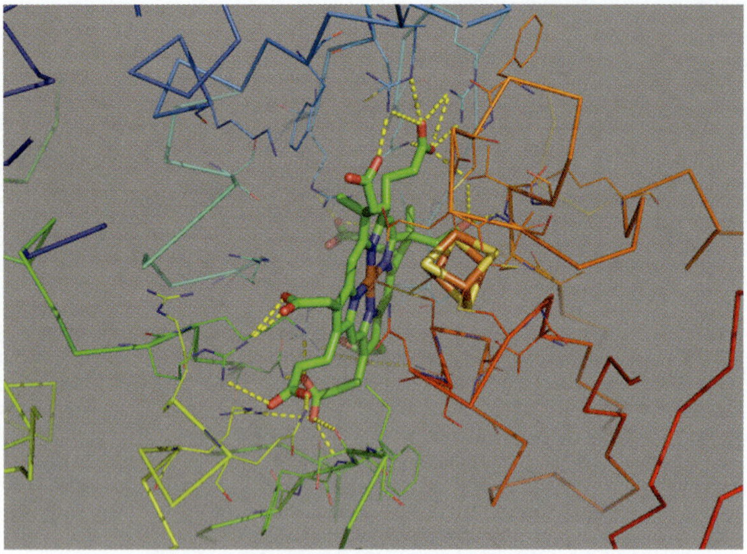

FIGURE 5.3 Tertiary structure of the active site of spinach nitrite reductase (NiR). Monomeric NiR consists of three domains partially shown as red, green, and blue C-alpha backbone traces. The 4Fe-4S cluster (orange-yellow) and the siroheme (green-blue) cofactors are located at the interface of the three domains. The C-atoms of the heme tetrapyrrole ring are colored in green with the Fe at the center in orange. The heme is liganded to cysteine residues and is held in position by hydrogen bonds to residues in the protein. One of the cysteine ligands is shown forming a bridge to one of the irons in the 4Fe-4S cluster that lies parallel to the plane of the tetrapyrrole ring. Other ligand colors are blue, for nitrogen, and red, for oxygen. (Structure from Protein Data Bank 2akj. U. Swamy, M. Wang, J. N. Tripathy, S. -K. Kim, M. Hirasawa, D. B. Knaff, and J. P. Allen, *Biochemistry* 44 [2005]: 16054–16063.)

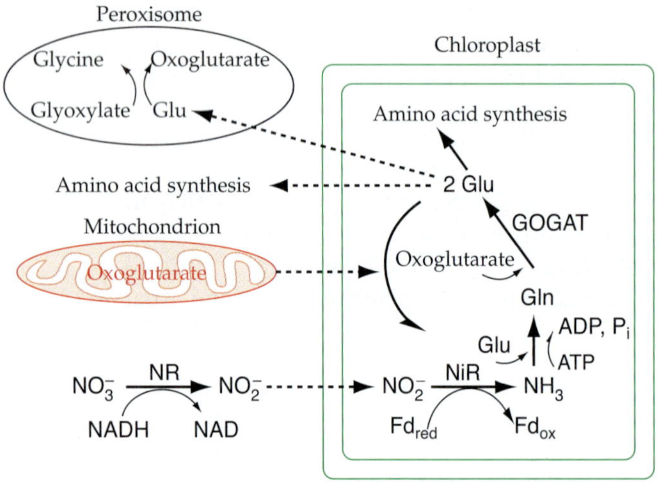

FIGURE 5.4 Summary of nitrate assimilation in photosynthetic cells. NO_3^- is reduced by nitrate reductase (NR) in the cytoplasm and the product, NO_2^-, is transported into the chloroplast stroma (green box). Nitrite reductase (NiR) in the stroma catalyzes a six-electron reduction of NO_2^- to NH_3, which is protonated to an ammonium ion. The resulting NH_4^+ is incorporated into glutamine by glutamine synthetase (GS). Glutamine is a substrate for glutamine-oxoglutarate synthase (GOGAT), which yields two equivalents of glutamate. The other substrate, oxoglutarate, is a product of the citric acid cycle in the mitochondrial matrix and is presumably transported into the chloroplast stroma. The glutamate that is produced in these reactions is used in amino acid syntheses in both plastids and cytoplasm and as an amino group donor in other amino transferase reactions. Glutamate is also a key amino donor in the formation of glycine from glyoxylate in leaf peroxisomes (glyoxysomes) in photorespiration (see Chapter 2).

heme cofactors found in cytochromes. The x-ray crystal structure of nitrite reductase from spinach has been solved to 2.8Å resolution. The enzyme consists of three alpha/beta domains with the siroheme at the domain interface (**FIGURE 5.3**). The 4Fe-4S cluster is found in domain 3 facing the siroheme and is coordinated by four cysteine residues, one of which bridges a 4.2Å gap to the siroheme. The redox cofactors are surrounded by ionizable residues that provide a hydrophilic environment for binding the anionic substrate, NO_2^-. Nitrite binds to the axial position of the heme opposite the Fe-S cluster. The reducing agent is ferredoxin, which can be generated directly by photosystem I in the light. In nonphotosynthetic tissues or at night, NADPH from carbon metabolism is used to reduce ferredoxin via a ferredoxin-NADP reductase.

How does this multielectron transfer in nitrite reductase occur? Electrons are fed into the enzyme one at a time from reduced ferredoxin until a total of six electrons have accumulated to reduce nitrite. A total of eight protons are also required.

$$NO_2^- + 6Fd_{red} + 8H^+ \rightarrow NH_4^+ + 6Fd_{ox} + 2H_2O$$

Reduced ferredoxin binds to the enzyme near the Fe-S cluster, which is reduced, and the cluster, in turn, transfers an electron to the siroheme. The reduced siroheme binds NO_2^-. Five additional electrons must be transferred into the active site until the substrate is fully reduced. Because the siroheme transfers one electron at a time, partially reduced intermediates must be produced but are not released from the enzyme and have not been characterized. No substrate or intermediates were present when the crystal structure was solved. The protons most likely arise from water and are transferred into the active site by the ionizable groups surrounding the siroheme.

The ammonium product of the nitrite reductase reaction is immediately incorporated into glutamate by the combined action of stromal glutamine synthetase (GS) and glutamate-oxoglutarate amino transferase (GOGAT), also called glutamate synthase.

1. Glu + NH_4^+ + ATP → Gln + ADP + Pi
2. Gln + oxoglutarate + $2Fd_{red}$ → 2Glu + $2Fd_{ox}$
3. Net: NH_4^+ + ATP + oxoglutarate + $2Fd_{red}$ → Glu + ADP + Pi + $2Fd_{ox}$

One of the glutamates produced in reaction 2 is recycled to the GS reaction and the other is used in biosyntheses, which require amino groups either in the chloroplast or in other organelles (**FIGURE 5.4**). Oxoglutarate (α-ketoglutarate) is an intermediate in the citric acid cycle in the mitochondria (see Chapter 3) and can be imported into chloroplasts by a specific dicarboxylate translocator.

Most plants contain two major isoforms of glutamine synthetase, GS1 and GS2. The cytoplasmic GS1 in roots that processes soil-derived ammonia was mentioned previously. The GS2 isoform in plastids is the major form required for processing ammonia. A GS1 isoform is also located in the cytoplasm where it may serve a variety of functions such as recycling the NH_3 released by the phenylalanine ammonia lyase reaction, which is the first step in phenylpropanoid synthesis (see Chapter 8). Both isoforms are octamers with molecular masses of 320 to 380 kDa and are regulated, in general, by phosphorylation and subsequent binding of 14-3-3 proteins. In contrast, the chloroplast ferredoxin-dependent glutamate-oxoglutarate amino transferase is a monomer of approximately 150 kDa, containing three redox cofactors, flavin mononucleotide, flavin adenine dinucleotide, and one 3Fe-4S cluster.

Nitrate uptake and assimilation require a large input of energy, including two to four equivalents of ATP, one NADH, and six reduced ferredoxins per NO_3^- assimilated to ammonium ion. Nitrate, however, can be transported into the cell and stored in vacuoles when photosynthesis is not actively supplying energy for further processing into amino acids. Under these conditions nitrate reductase activity is down-regulated by phosphorylation of an exposed serine residue in the first hinge region between the heme and molybdenum cofactor domains (see Figure 5.1). *In vitro* assays have demonstrated that phosphorylation alone does not alter enzyme activity. A second factor, the nitrate reductase inhibitory protein, is also required. Nitrate reductase inhibitory protein binds to the phosphorylated nitrate reductase and blocks the active site. It may also facilitate formation of inactive multimers that are sometimes observed *in vivo*. It has been shown that nitrate reductase inhibitory protein belongs to the family of regulatory proteins known as the 14-3-3 proteins. The numbers refer to protein fractions that were originally isolated from human tissue, fractionated by column chromatography, and subsequently separated by gel electrophoresis. They are key regulatory proteins found in all eukaryotes.

Several different isoforms of the 14-3-3 proteins are synthesized in eukaryotic cells and can function as homo- or heterodimers with molecular masses of approximately 60 kDa per dimer. The proteins contain nine antiparallel alpha helices that form a molecular "basket-like" structure. They bind specifically to certain phosphorylated target proteins, depending on the isoform composition of the 14-3-3 dimers. Binding to a target protein can alter enzyme activity, as in the case of nitrate reductase and GS, and/or facilitate interaction with other proteins via its basket-like structure. Activity can be quickly restored to nitrate reductase by the action of type 2A protein phosphatases that cleave the serine-phosphoester bond and release the enzyme from the 14-3-3 proteins. Phosphorylation and 14-3-3 binding provide a rapid mechanism for down-regulation of nitrate assimilation by leaves in the dark. However, the nitrate transporters in the roots are also inhibited in the dark. How the signal is transferred from leaves to roots is unknown and probably requires a "systems biology" approach to determine the details of this long-range signal transduction.

Glutamate and glutamine are the major sources of amino groups used in the synthesis of other amino acids. Amino acid synthesis requires the transamination of a precursor. In the synthesis of aspartate, a single step transfer to the keto-acid, oxaloacetate, is catalyzed by aspartate aminotransferase:

$$\text{Oxaloacetate} + \text{glutamate} \rightarrow \text{aspartate} + \alpha\text{-ketoglutarate}$$

Aspartic acid is the precursor for asparagine, also a single-step reaction requiring glutamine as the amino donor. This energy-requiring reaction is catalyzed by asparagine synthetase:

$$\text{Aspartate} + \text{glutamine} + \text{ATP} \rightarrow \text{asparagine} + \text{glutamate} + \text{ADP} + \text{Pi}$$

Aspartic acid is also the precursor for additional aliphatic amino acids, threonine, lysine, and methionine. The biosynthesis of methionine and cysteine, both sulfur-containing amino acids, is covered in this chapter. The pathways for the production of the aromatic amino acids can be found in Chapter 8. A brief outline of the pathways for synthesis of the other essential amino acids required in protein synthesis can be found in Appendix 6.

Nitrogen Fixation

As mentioned previously, only prokaryotes can reduce atmospheric nitrogen to ammonia. This includes several species of cyanobacteria, some photosynthetic bacteria such as *Rhodospirillum*, many soil bacteria belonging to the genus *Rhizobium* and *Clostridium*, and even some archaea such as *Methanosarcinia*. The nitrogenase enzyme complex is believed to have evolved in primitive organisms early in Earth's history to supply organic nitrogen. The nitrogenase complex is irreversibly inactivated by oxygen, which was not abundant in the atmosphere 3 billion years ago. However, under present-day aerobic conditions, only organisms that live in anaerobic habitats or can provide an oxygen-free environment for nitrogenase can utilize atmospheric N_2.

The nitrogenase complex consists of two enzymes that catalyze the following overall reaction:

$$N_2 + 16ATP + 8H^+ + 8\text{ electrons} \rightarrow 2NH_3 + H_2 + 16ADP + 16Pi$$

The first enzyme of the complex has ATPase activity and contains a 4Fe-4S cluster that is reduced by ferredoxin. This homodimeric protein is referred to as the iron protein (Fe-protein) or dinitrogenase reductase. The second enzyme, a heterotetramer, contains an iron-sulfur cluster and a molybdenum-iron cofactor. It binds and reduces dinitrogen with electrons transferred from the Fe-protein. This second component is referred to as the dinitrogenase or molybdenum-iron protein (MoFe protein). The x-ray crystal structures of both proteins have been solved separately and in complex.

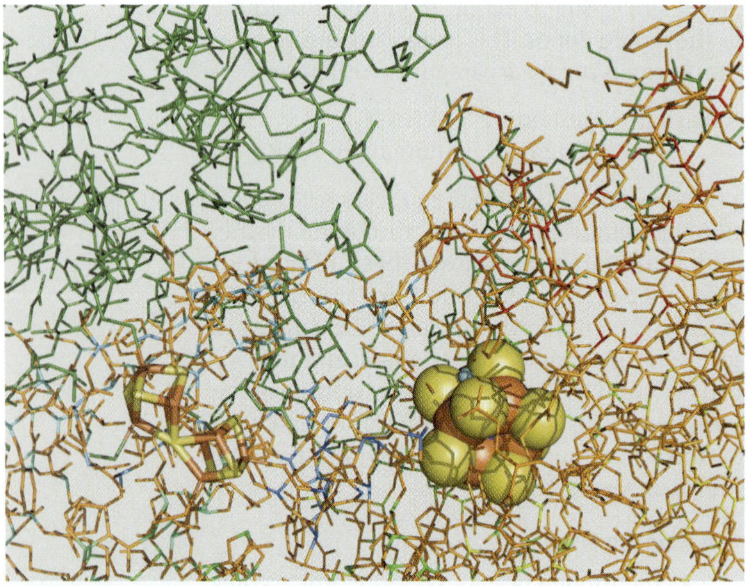

FIGURE 5.5 Active site of the MoFe protein (dinitrogenase) heterodimer. The Fe-Mo cofactor is shown as spheres with yellow for sulfurs, red for the irons, and a blue molybdenum. This cofactor is located in the α subunit of the enzyme partially shown with red and yellow sticks to represent the C-α backbone. The P cluster is shown in a stick representation with red irons and yellow sulfurs. It is located at the periphery of the β subunit shown in blue and green. There is an approximately 14 Å separation between the two cofactors. (Structure from Protein Data Bank 1m1n. O. Einsle, F. A. Tezcan, S. L. A. Andrade, B. Schmid, M. Yoshida, J. B. Howard, and D. C. Rees, *Science* 297 [2002]: 1696–1700.)

The Fe-protein is a homodimer with a molecular mass of approximately 64,000. The dimer contains one 4Fe-4S cluster that bridges the subunits. The iron-sulfur cluster is reduced by ferredoxin or, in some cases, a small flavoprotein called flavodoxin. The 4Fe-4S cluster is in a highly exposed position between the protein subunits. Although it is covalently bound to four cysteine residues in the dimer, it readily reacts with molecular oxygen, resulting in the loss of 2Fe-2S from the cluster and irreversible inactivation of the enzyme.

The Fe-protein also contains two nucleotide binding sites. On reduction of the iron-sulfur cluster and binding of Mg-ATP to both nucleotide sites, the Fe-protein now can interact with the MoFe protein. Hydrolysis of the two bound ATP molecules changes the conformation of both enzymes and facilitates the transfer of one electron to the MoFe protein. The Fe-protein is then released from the MoFe protein and the cycle repeats to transfer the full eight electrons to the MoFe protein:

$$\text{Fe-protein}_{red} + 2MgATP + \text{MoFe protein} \rightarrow \text{Complex} \rightarrow \text{Fe-protein}_{ox} + 2MgADP + 2Pi + \text{MoFe protein}_{1electron}$$

The molybdenum-iron protein functions as a tetramer ($\alpha 2\beta 2$) with a mass of approximately 250 kDa. It is functionally a dimer of α-β dimers. Each α-β protein contains two cofactors, an 8Fe-7S cluster called a P cluster and an iron-molybdenum cofactor that consists of 7 Fe-1Mo-9S attached to a homocitrate (**FIGURE 5.5**). The Fe-Mo cofactor is bound to the α subunit by a cysteine that is attached to one of the irons. The Fe-Mo cofactor is the binding site for dinitrogen. Each α-β unit also has one site for interaction with the Fe-protein. The three-dimensional structure of a fully assembled complex is shown in **FIGURE 5.6**. Electrons from the Fe-protein are transferred to the P-cluster of the MoFe protein. The P-cluster is approximately 14 Å from the Fe-Mo cofactor and bridges the α-β dimer. Presumably, bound water and the homocitrate moiety of the Fe-Mo cofactor serve as an electron, and possibly a proton, conduit for reduction of N_2. The P-cluster seems to be a one-electron transfer factor, but its geometry is unusual compared with other Fe-S clusters, and it may be able to store and transfer more than one electron at a time to the Fe-Mo cofactor.

Although the mechanistic details are still lacking, the MoFe protein must accumulate the full eight electrons before the product, NH_3, is released. Data from a variety of spectroscopic techniques combined with the three-dimensional

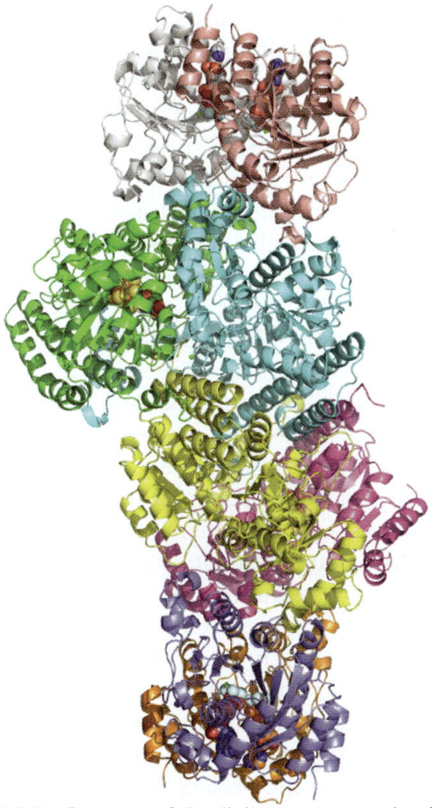

FIGURE 5.6 Structure of the dinitrogenase complex. The MoFe heterotetrameric protein is shown in the middle of the figure. One α/β dimer is green/blue and the other is yellow/magenta. The Fe-Mo cofactor can be seen in the green α-subunit as yellow (S) and red (Fe) spheres. The corresponding cofactor in the magenta α subunit faces in the opposite direction, into the plane of the paper in this diagram. Each α/β unit of the MoFe protein binds one homodimeric Fe protein, shown at the top and bottom of the MoFe proteins. The Fe protein dimer at the top of the figure is shown in pink and gray and has two bound ATP analogs shown as spheres. (Structure from Protein Data Bank 1n2c. H. Schindelin, C. Kisker, and D. C. Rees, *Nature* 387 [1997]: 370–376.)

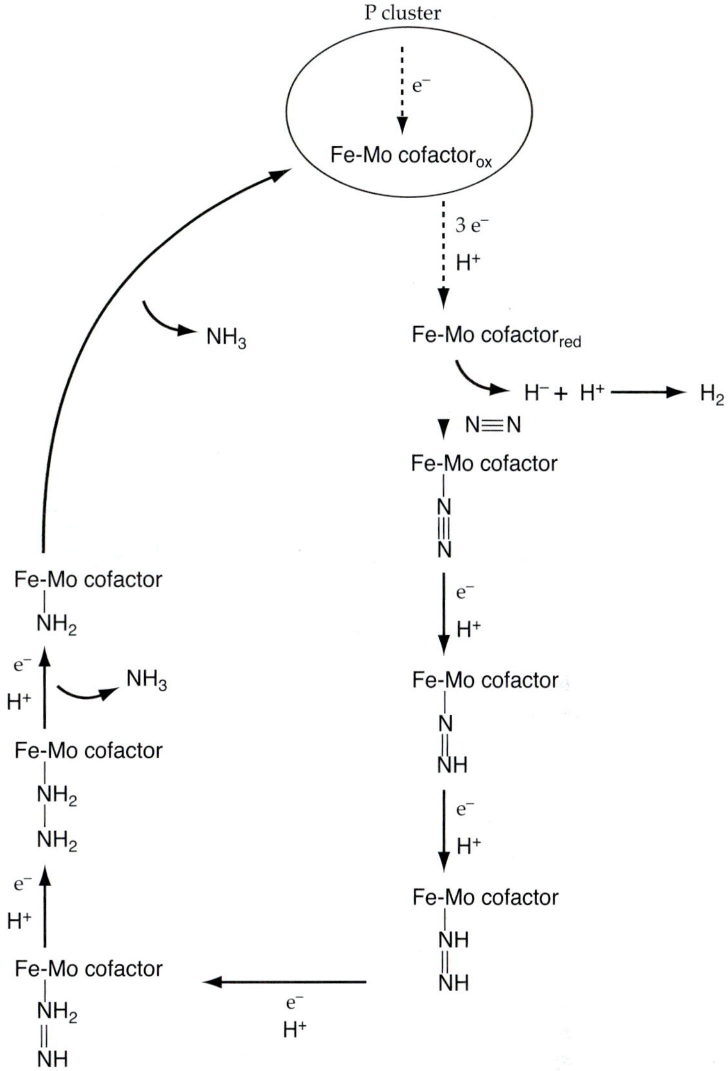

FIGURE 5.7 Proposed mechanism for dinitrogen reduction. The oxidized Fe-Mo cofactor is reduced by electron transfer from the 8Fe-7S-P cluster (see Figure 5.5). After accumulation of three electrons and protons, the substrate, N_2, is bound, displacing a hydride ion that forms H_2 and is released from the enzyme. Further electron and proton transfers give rise to a series of reduced nitrogen intermediates that remain bound to the active site. Ammonia is released in the last two steps. This figure implies that N_2 and other reaction intermediates are bound to the Mo in the Fe-Mo cofactor, but the exact binding site is unknown.

structural information have led to a proposed eight-redox state scheme for dinitrogen reduction shown in **FIGURE 5.7**. In this mechanism, the Fe-Mo cofactor must accumulate three to four electrons before dinitrogen is bound. The substrate binds to the Mo or to a nearby Fe. Binding of N_2 is assumed to displace a hydride from the active site, which is protonated by water and results in the release of one H_2 per dinitrogen reduced. In addition to protons, the nitrogenase complex can also reduce acetylene (C_2H_2) to ethylene (C_2H_4). Acetylene reduction provides a convenient assay for nitrogenase activity because the product, ethylene, is readily detected and quantitated by gas chromatography. Although acetylene binds to the Fe-Mo cofactor, it is not a competitive inhibitor of dinitrogen. This indicates that acetylene binds to one of the intermediate redox states of the enzyme before the state that binds N_2.

The Fe-Mo cofactor is essential to dinitrogen reduction, but some forms of dinitrogenase do not contain this cofactor. Alternative nitrogenases have been described that contain vanadium in place of molybdenum or have an iron-only cofactor. Some microorganisms, such as *Azotobacter vinelandii*, have the genetic information for producing the MoFe nitrogenase and the FeV and Fe-only variants. Although the reaction scheme presented in Figure 5.7 shows substrate binding to molybdenum, the actual binding site and mechanism of electron transfer in nitrogenase is still an ongoing investigation with several unknown parameters.

FIGURE 5.8 Filamentous cyanobacterium with heterocysts. The *Nostoc* sp. shown in the micrograph is an example of cell differentiation and specialization in a prokaryote. The small, green cells in the filament are typical free-living cyanobacteria and are called vegetative cells. The larger, thick-walled cells, the heterocysts, are vegetative cells that have differentiated into specialized compartments for nitrogen fixation. The pale color of the heterocysts is due to the absence of photosystem II and associated pigments in these cells. Because they lack photosystem II and the water-splitting complex, the heterocysts are relatively anoxic and are the only cells in the filament that contain the dinitrogenase complex. Heterocysts fix nitrogen into amino acids that are transported to the vegetative cells. Magnification, 350×.

Similar to the Fe-protein, the MoFe protein is irreversibly inactivated by oxygen. Organisms that fix atmospheric nitrogen must have some mechanism to protect these enzymes. The obvious solution is to fix nitrogen only under anaerobic conditions. Many facultative anaerobic bacteria found in soil, such as various *Clostridia* and *Klebsiella,* induce the genes for the nitrogenase complex only in an anoxic environment. Similar induction occurs in photosynthetic bacteria such as *Rhodospirillum. Trichodesmium,* a filamentous cyanobacterium that is the principal nitrogen-fixer in marine ecosystems, uses temporal separation of the processes of oxygenic photosynthesis and nitrogen fixation. Filamentous cyanobacteria in the Nostocales (*Nostoc, Anabaena*) fix nitrogen in specialized cells called heterocysts. These large, thick-walled cells lack photosystem II with its oxygen-evolving complex and are dedicated to nitrogen fixation. These specialized cells supply nitrogen as glutamate to the vegetative cells in the rest of the filament (**FIGURE 5.8**). The most enigmatic nitrogen fixers are aerobic organisms such as *Azotobacter* species, which are obligate aerobes. These organisms protect their nitrogenase complex using oxygen scavengers to supply their membrane-bound respiratory chain with oxygen but otherwise maintain anoxic conditions in their cytoplasm at the site of nitrogen fixation.

Some nitrogen-fixing prokaryotes form symbiotic relations with plants and rely on the host to provide anaerobic conditions for the nitrogenase complex. The water fern *Azolla* develops a cavity on the underside of its leaves that traps the cyanobacterium, *Anabaena azolla*. Under the low oxygen conditions in this cavity, the cyanobacterium develops a greater number of heterocysts than in the free-living state and supplies amino acids to the fern. The plant, in turn, excretes sucrose into the cavity to help nourish its symbionts. The aquatic fern is an important source of fixed nitrogen in aqueous systems such as rice paddies. A similarly loose association with a microorganism is found in the rhizosphere of some tropical grasses that associate with nitrogen-fixing actinomycetes such as *Azospirillum* sp.

A more highly structured and regulated symbiosis occurs between legumes and bacteria in the Rhizobia class. The Rhizobia are soil bacteria that do not fix nitrogen in the free-living state. However, they can form specific associations with legumes; for example, *Sinorhizobium meliloti* is attracted to alfalfa (*Medicago sativa*), and *Bradyrhizobium japonicum* is found only with soybeans (*Glycine max*). When the fixed forms of nitrogen are unavailable, the plant root cells produce and release specific flavonoids into the soil. For example, *Medicago* species release the flavonoid, luteolin (**FIGURE 5.9**; also see Chapter 8). These flavonoids bind to specific receptors in the cell membranes of the *Sinorhizobia* called Nod proteins and induce *nod* genes for the production of Nod factors. Nod factors are oligosaccharides consisting of three to five *N*-acetylglucosamine residues. A fatty acid is linked to the terminal sugar, and various species-specific modifications

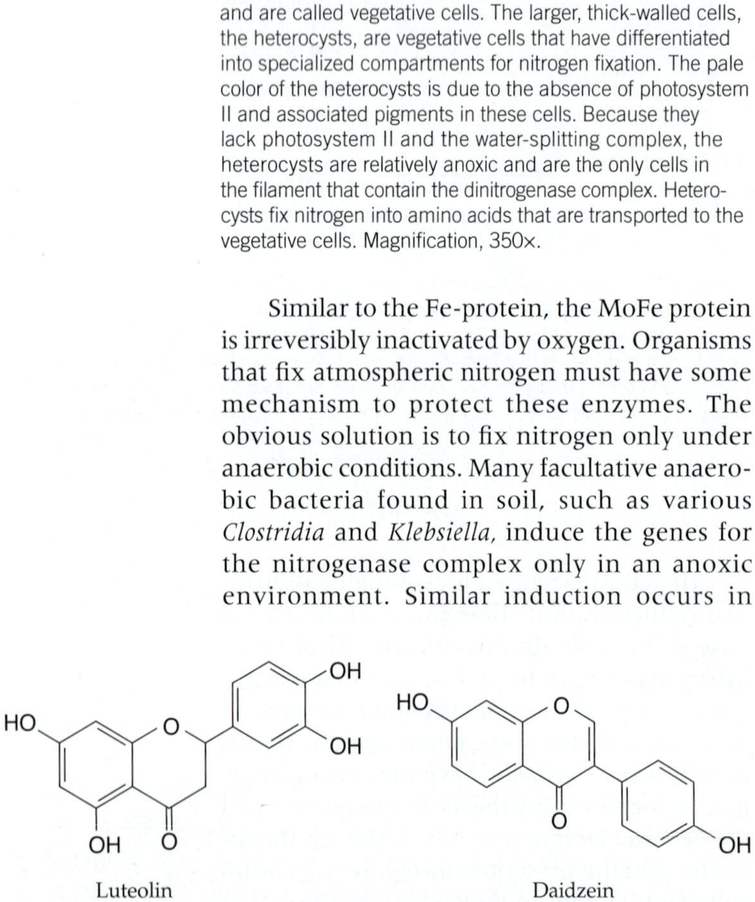

Luteolin Daidzein

FIGURE 5.9 Examples of flavonoids released by legume roots. Luteolin is an example of a flavone released into the soil by root cells of *Medicago* sp. (alfalfa). It specifically attracts *Sinorhizobium meliloti*. Daidzein is an example of an isoflavone. It is released by soybean (*Glycine max*) and attracts *Bradyrhizobium japonicum*.

are made to the other residues (**FIGURE 5.10**). Nod factors are released from the bacterial cells and bind to species specific receptor glycoproteins (lectins) in the host plant's cell walls. Nod factors act like plant hormones. Nod factor receptors are usually integral membrane kinases that are activated on binding the oligosaccharide. This initial interaction triggers a series of events in the root cells including release of Ca^{2+} ions and a signaling cascade that results in the induction of a series of genes to introduce the Rhizobia into the root cells and to facilitate a symbiotic relationship with the bacteria.

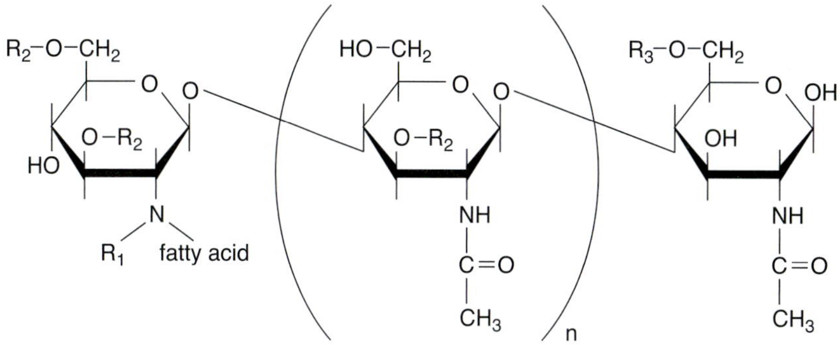

n = 1–3

R_1, R_2, R_3 = H or other modification

$R_1 = CH_3$

$R_2 = CH_3$ or $–CH_2–CH_3$

$R_3 = SO_3^-$ or sugar

FIGURE 5.10 Structures of Nod factors. Nod factors are lipo-oligosaccharides produced by Rhizobia to induce nodule formation in their host legumes. The factors consist of a backbone of three to five *N*-acetyl glucosamine units with various species-specific modifications. The nonreducing end of the Nod factor always binds a fatty acid that can be 16 to 18 carbons in length and is usually unsaturated. Nod factors interact specifically with lectin-like receptors in the root cell membranes.

The bacteria infect their host plant either by way of the root hairs or through cracks in the cortical tissues. Instead of initiating a defense response against these microorganisms, the plant surrounds them with a membrane and moves them into the cortical cells through a structure called an infection thread. Plant cells in the root cortex are stimulated to divide and form an enlarged structure called a nodule (**FIGURE 5.11**). The bacteria also divide and take up residence in the cells of the nodule. The bacteria remain surrounded by the plant cell membrane, and this structure, containing one or several bacteria, is called a symbiosome. Once inside the low-oxygen microenvironment in the nodule, the genes for nitrogenase and associated proteins are induced in the Rhizobia. Plant cells in the nodule show an increase in sucrose synthase activity. This enzyme breaks down sucrose in the roots, and the hexose-phosphates are further metabolized to produce dicarboxylic acids such as malate that are transported into the symbiosome. The plants also provide additional nutrients such as phosphate and sulfate.

The microaerobic state in the symbiosome is maintained by a plant protein called leghemoglobin, which sometimes gives nodules a slight pink coloration. Leghemoglobin is a small protein (MW ~ 16 kDa) similar in structure to myoglobin. It binds oxygen and releases it near the site of the terminal respiratory electron acceptor in the bacterial membrane. This selective release of oxygen enables the symbionts to maintain a high rate of aerobic metabolism and supply the ATP and reduced cofactors needed for nitrogen fixation. The symbiotic bacteria fix dinitrogen and secrete NH_4^+ through specific channels in the symbiosome membrane. Plant enzymes, glutamate synthase and glutamate-oxoglutarate amino transferase, catalyze the formation of glutamine as described previously. Asparagine formation, catalyzed by asparagine synthetase, also occurs. These are the major forms of fixed nitrogen that are transported to the rest of the plant tissues in legumes such as alfalfa and peas (**FIGURE 5.12**). Other legumes, such as soybeans and French bean, transport ureides such as allantoin and citrulline, which are derivatives of urea derived from aspartate.

This elaborate and tightly regulated legume-Rhizobia symbiosis is believed to have evolved from the basic defense mechanisms that plants use to protect themselves from bacterial invasion. It allows legumes to flourish in

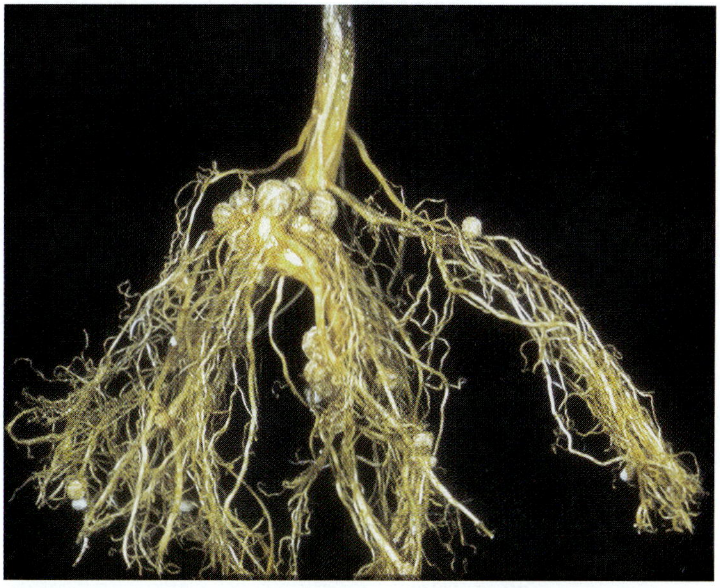

FIGURE 5.11 Nodules on an infected legume root. Nodules are formed after induction of cell division in root cortical cells. The infected cells within these structures house Rhizobia in a microaerophilic environment to promote bacterial nitrogen fixation.

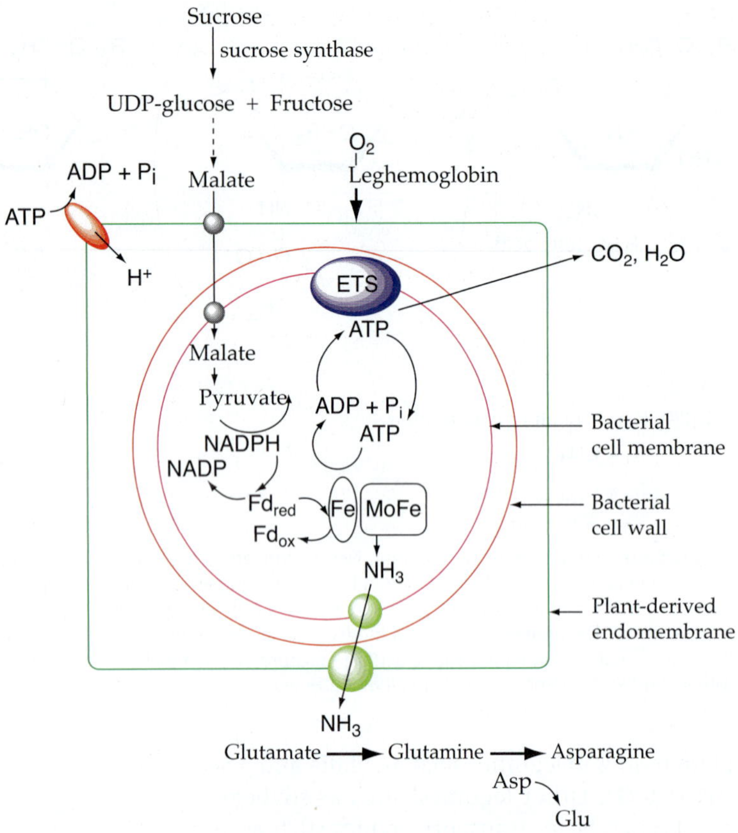

FIGURE 5.12 Generalized diagram of a legume symbiosome. A Rhizobium (red-tinted circles) is shown sequestered in a plant-derived endomembrane of an organelle called a symbiosome (green). The plant supplies fixed carbon to its symbiont in the form of dicarboxylic acids, usually malate. Oxygen is also delivered to the bacterial respiratory system (shown as a blue ellipsoid) in the bacterial cell membrane. The plant leghemoglobin binds oxygen and selectively releases it in the vicinity of this electron transport system. A proton-ATPase in the symbiosome membrane pumps protons into the symbiosome to maintain a slightly acidic pH in the interior that promotes transfer of metabolites across the membrane. The bacterium uses the dicarboxylic acids from the plant as substrates for its citric acid cycle. This cycle provides reduced cofactor that, in turn, reduces ferredoxin needed by the Fe-protein of the nitrogenase complex. The bacterial electron transport system provides ATP to the Fe-protein. The NH_3 generated by the symbiont is transported to the host cytoplasm and used to synthesize glutamine, asparagine, or other organic nitrogen compounds.

nitrogen poor soils and can provide a source of fixed nitrogen to the soil if senescent plant tissues are left to decompose. The bacterial population also increases when released from senescent plants. For many centuries farmers have noticed the ability of leguminous plants to restore the soil and have used crop rotation with beans, peas, clover, and others to restore the productivity of agricultural lands.

Many plants and algae can use organic forms of nitrogen when these are available. Cyanobacteria have an active transport mechanism for urea and the enzyme, urease, to release the ammonia, which is used to produce glutamate:

$$NH_3\text{–}\overset{\overset{\displaystyle O}{\|}}{C}\text{–}NH_3 \rightarrow 2NH_3 + CO_2$$

Urease is also present in some vascular plants, especially in seeds. Most plants can also take up amino acids from the soil and water when these are made available from organic wastes and composting.

Nonbiological Nitrogen Fixation Other natural sources of fixed nitrogen in the biosphere include lightning strikes and volcanic activity. These geological processes supply energy to break the N–N triple bond and promote oxidation to nitrate:

$$N_2 + 3O_2 \rightarrow 2NO_3^-$$

Although the amount of nitrate produced in any single event is small, lightning strikes and active volcanoes are occurring somewhere on the planet at all times and can contribute from 10% to 15% of the total amount of fixed nitrogen found in the biosphere.

Despite the various natural inputs of fixed nitrogen, the amounts required by modern agriculture to sustain the current human population must be supplied by a chemical reaction known as the Haber-Bosch process. The chemical reaction of nitrogen with hydrogen was optimized by Fritz Haber early in the twentieth century:

$$N_2 + 3H_2 \rightarrow 2NH_3$$

The reaction has an overall ΔG of -92 kJ mole^{-1}, but because of the high bond energy of the N_2 molecule (940 kJ mol^{-1}), the above reaction requires a temperature of 300 to 500°C, a pressure of approximately 20 MPa, and an iron catalyst. The Haber process was modified and commercialized by Carl Bosch at BASF in Germany. Despite the high cost of fixing nitrogen into ammonia by this process, it is estimated that 40% to 50% of the fixed nitrogen on the planet is currently produced by this means to satisfy the needs of agriculture and feed most of the human population. The ammonia produced in the Haber-Bosch process can also be oxidized to produce nitrates and used in other industrial processes such as the manufacture of explosives.

5.3 Sulfur Metabolism

The major sulfur-containing metabolites in plants are the amino acids, cysteine and methionine. Cysteine and other compounds such as 3′-phosphoadenosine-5′-phosphosulfate PAPS act as sulfur donors used in the synthesis of the unique sulfolipids found in chloroplast thylakoid membranes and sulfur-containing natural products such as glucosinolates. Plants require relatively large amounts of sulfur, and this element is regarded as a macronutrient. The chief source of assimilated sulfur is sulfate anion (SO_4^{2-}) found in soil; plants can also absorb gaseous sulfur dioxide (SO_2) and hydrogen sulfide (H_2S) from the atmosphere. Unlike fixed nitrogen, sulfur is seldom a limiting nutrient for plant growth.

Sulfate Assimilation

Like nitrate, sulfate is taken up from the soil by an active transport process. Two types of sulfate transporters are present in the membranes of root cells; high affinity transporters bind SO_4^{2-} at 1 to 7 μM, whereas low affinity transporters operate in a range of 100 μM to 1.2 mM sulfate. The *Arabidopsis* genome has 14 genes that code for high affinity transporters. These are proteins of 500 to 700 amino acids with 10 to 12 predicted transmembrane helices. Similar transporters are found in the vacuole membrane for uptake and storage of sulfate ions. Incorporation of sulfate into cysteine occurs mainly in chloroplasts, which implies that the ion is transported in the xylem to photosynthetic tissues for assimilation.

Cysteine Synthesis

The first step in cysteine synthesis is the activation of SO_4^{2-} by reaction with ATP. This reaction, catalyzed by ATP sulfurylase, produces adenosine-5′-phosphosulfate (APS) as shown in FIGURE 5.13. Adenosine phosphosulfate is also the precursor for PAPS. Although PAPS is mainly a sulfur donor in plants, in some organisms, such as cyanobacteria, PAPS can be reduced to sulfite and also provide the precursor for cysteine synthesis. In angiosperms and gymnosperms, the plastid-located APS is the major sulfur donor

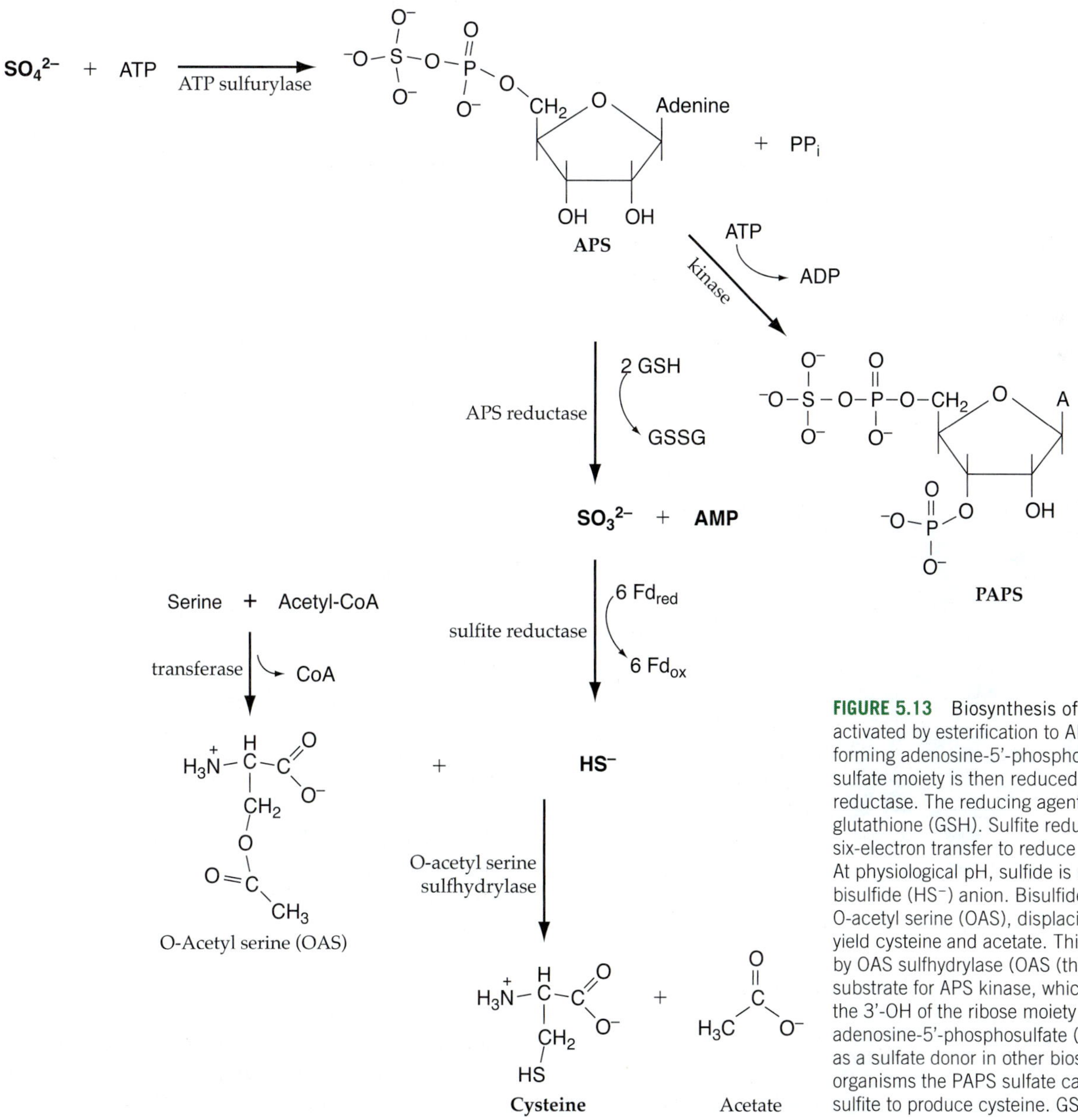

FIGURE 5.13 Biosynthesis of cysteine. Sulfate ion is activated by esterification to AMP by ATP sulfurylase, forming adenosine-5′-phosphosulfate (APS). The sulfate moiety is then reduced to sulfite by APS reductase. The reducing agent for this reaction is glutathione (GSH). Sulfite reductase catalyzes a six-electron transfer to reduce sulfite to sulfide (S^{2-}). At physiological pH, sulfide is mostly present as bisulfide (HS^-) anion. Bisulfide then reacts with O-acetyl serine (OAS), displacing the acetyl group to yield cysteine and acetate. This reaction is catalyzed by OAS sulfhydrylase (OAS (thiol) lyase). APS is also a substrate for APS kinase, which adds a phosphate to the 3′-OH of the ribose moiety to produce 3′-phosphoadenosine-5′-phosphosulfate (PAPS). PAPS serves as a sulfate donor in other biosyntheses; in some organisms the PAPS sulfate can also be reduced to sulfite to produce cysteine. GSSG, oxidized glutathione.

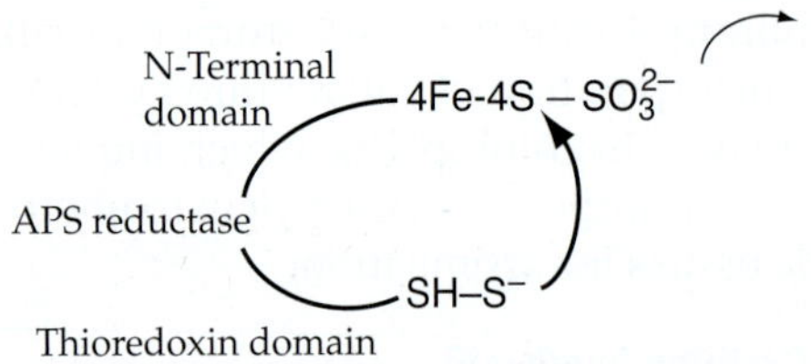

FIGURE 5.14 Diagram of the APS reductase reaction. The enzyme contains a 4Fe-4S cluster in its N-terminal domain. SO_4^{2-} binds to this domain and is reduced to SO_3^{2-}. Sulfite is released from the active site by the action of a thiol group in the thioredoxin-like C-terminal domain of the reductase.

for cysteine synthesis. APS reductase reduces the sulfate moiety to sulfite (SO_3^{2-}) and frees it from the nucleotide carrier. The reductase is a homodimer with a mass of about 90 kDa. It contains one 4Fe-4S cluster bound by a single cysteine residue in the N-terminal domain of the subunits. APS binds to this domain and sulfate is reduced to sulfite, a reaction that requires two electrons. The reducing agent in this reaction is glutathione, which, in turn, is reduced by NADPH and a flavoprotein, glutathione reductase. The C-terminal domain of APS reductase is homologous to the disulfide reductase, thioredoxin. The thioredoxin-like disulfide releases the product, SO_3^{2-}, from the active site (**FIGURE 5.14**).

Sulfite is very reactive, but there is no known sulfite carrier in plants. It is presumably reduced immediately to sulfide (S^{2-}). The enzyme that catalyzes this reaction, sulfite reductase, resembles nitrite reductase. As in nitrite reduction, sulfite reductase must transfer six electrons to the substrate to produce the product. Both enzymes contain a 4Fe-4S cluster and a siroheme (see Figure 5.3). Both plastid-located enzymes use reduced ferredoxin as an electron donor, although some cytoplasmic forms of these enzymes can use NADPH. The tertiary structure and mechanism of action of both enzymes is similar; in fact, sulfite reductase can reduce nitrite with a much lower turnover rate. Although they show only about 20% amino acid identity, both enzymes were probably derived from a common ancestor, perhaps a primitive bifunctional reductase.

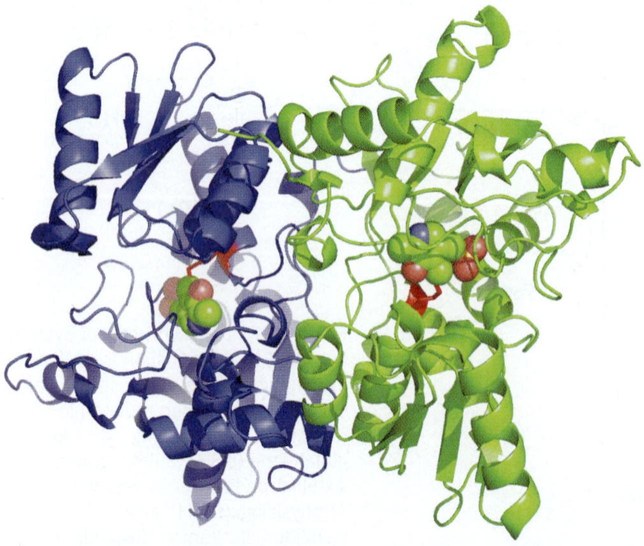

FIGURE 5.15 Structure of the O-acetyl serine sulfhydrylase (OAS (thiol) lyase) from *Arabidopsis*. The enzyme dimer is shown with one subunit in blue and the other in green. Both active sites face the same direction and contain a pyridoximine phosphate cofactor shown as spheres. The cofactor is covalently bound to the ε-amino group of a lysine residue in each monomer, shown as a red line in the green subunit. (Structure from Protein Data Bank 1oas. P. Burkhard, G. S. J. Rao, E. Hohenester, K. D. Schnackerz, P.F. Cook, and J. N. Jansonius, *J. Mol. Biol.* 283 [1998]: 121–133.)

Sulfide is also very reactive and must be used immediately to produce cysteine. At physiological pH, sulfide most likely exists as bisulfide anion (HS^-). The enzyme, O-acetyl serine sulfhydrylase (O-acetyl serine (thiol) lyase), binds HS^- and O-acetyl serine (OAS) as substrates (Figure 5.13). The bisulfide anion displaces the acetyl group to produce cysteine. The enzyme functions as a dimer of subunits with a mass of 35 kDa. Each subunit has a bound pyridoxamine phosphate cofactor. The x-ray crystal structure of the enzyme from *Arabidopsis* has been solved and is shown in **FIGURE 5.15**. As seen in Figure 5.15, the cofactor is bound to a lysine residue in the protein forming a Schiff base (imine). OAS displaces the lysine and binds to the cofactor with release of the acetyl group as acetate. HS^- then binds to the methylene carbon, forming an intermediate aldimine (**FIGURE 5.16**). The enzyme lysine amino group attacks the double bond and restores the original enzymatic Schiff base, releasing cysteine.

Serine is the precursor for the carbon-nitrogen portion of the cysteine molecule. The enzyme that catalyzes acetylation of serine, serine-acetyl-CoA transferase (SAT), is a member of the acyl transferase family of enzymes that utilize acetyl-CoA as an acetylating agent. All these enzymes have an unusual left-handed parallel β-helix domain and are most likely homologous. SAT is composed of 29-kDa monomers that are organized into trimers that then associate as dimers (i.e., it is a dimer of trimers). There are six active sites in the hexamer, each containing a histidine residue that binds to the hydroxyl group of serine and activates it for a nucleophilic attack on the carbonyl carbon of the acetyl group of acetyl-CoA. The SAT hexamer forms a complex with two OAS sulfhydrylase dimers in a supercomplex referred to as cysteine synthase.

SAT is more active in the complex, whereas the active sites of the OAS sulfhydrylases are

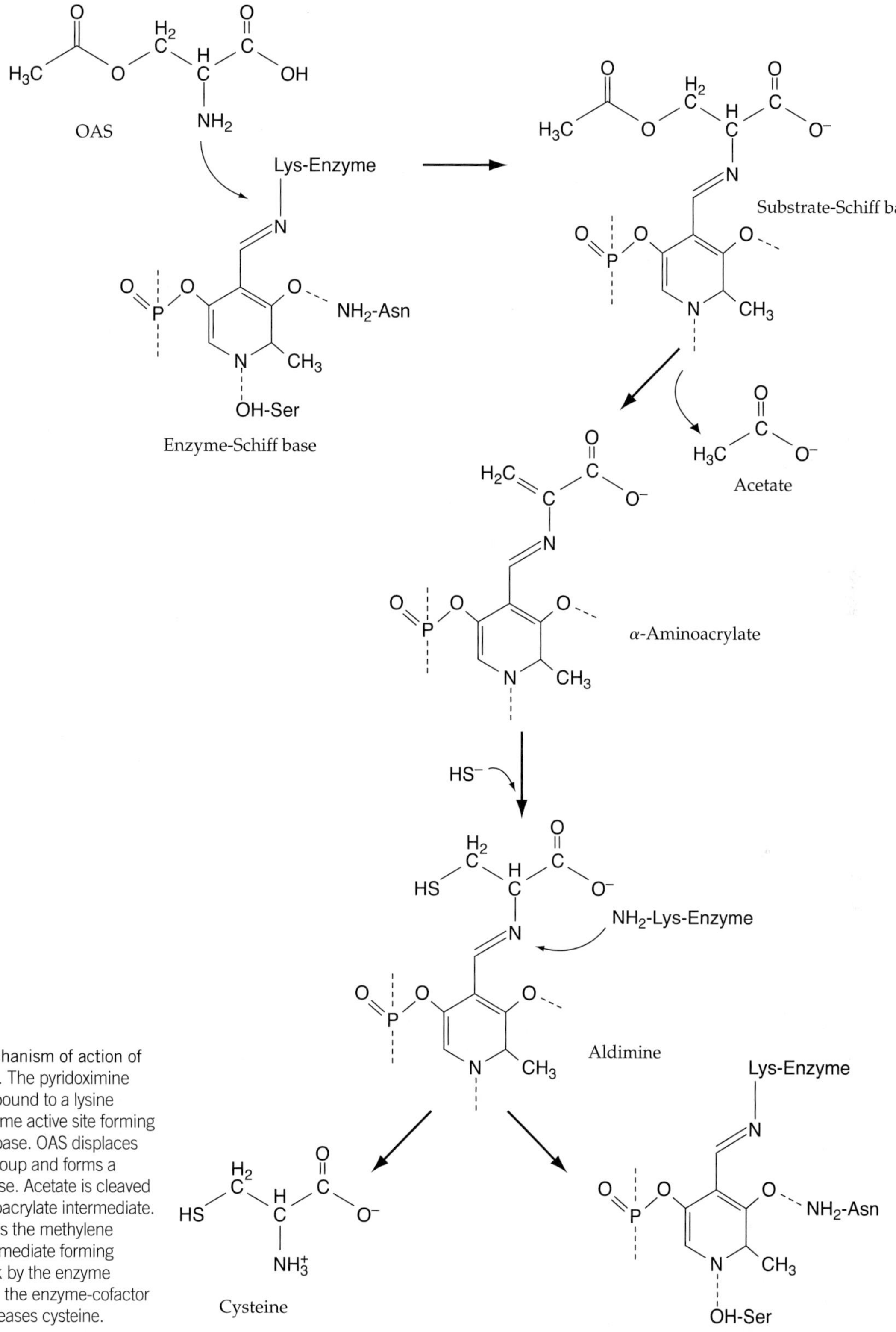

FIGURE 5.16 Mechanism of action of OAS sulfhydrylase. The pyridoximine cofactor is shown bound to a lysine residue in the enzyme active site forming an enzyme-Schiff base. OAS displaces the lysine amino group and forms a substrate-Schiff base. Acetate is cleaved to yield an α-aminoacrylate intermediate. Bisulfide ion attacks the methylene carbon of this intermediate forming an aldimine. Attack by the enzyme lysine-NH_2 reforms the enzyme-cofactor Schiff base and releases cysteine.

blocked and are less active. Under these conditions OAS accumulates, but little cysteine is produced. The increase in concentration of OAS serves as a signal to trigger induction of transcription of the sulfate transporter genes. Increased uptake of sulfate and subsequent processing increases the concentration of HS^-, which acts to dissociate the complex and results in increased synthesis of cysteine. Although the enzymes of cysteine synthesis are produced constitutively, this cysteine-synthase complex formation serves to increase enzyme levels and activity when plants are exposed to oxidative stress and heavy metals.

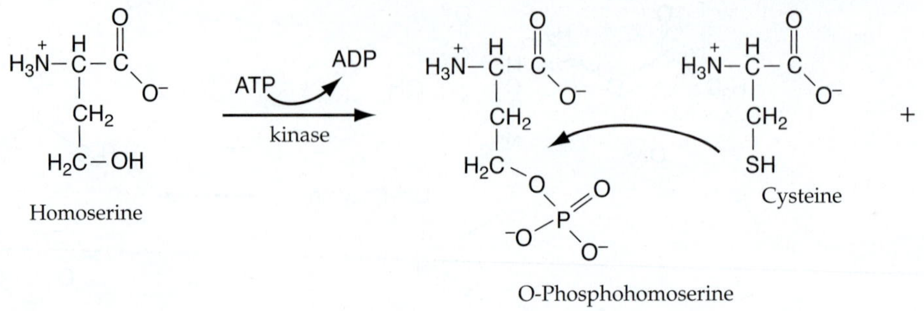
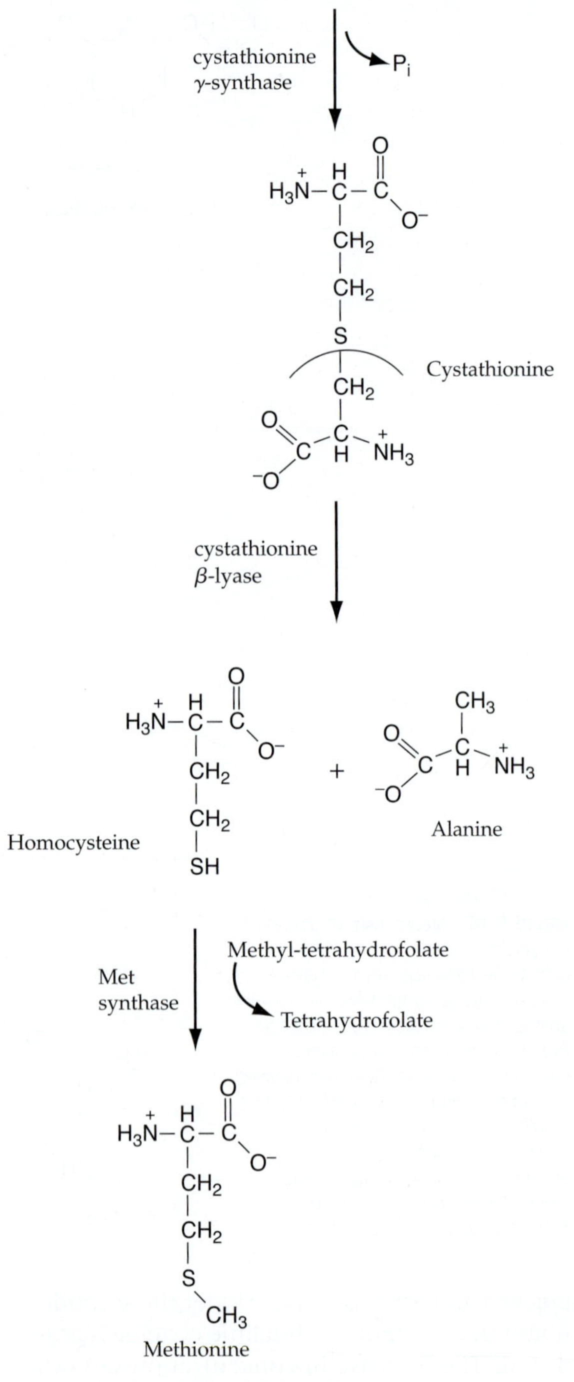

FIGURE 5.17 Synthesis of methionine. Cysteine and homoserine, a product of the reduction of the β-carboxyl group of aspartic acid, are the precursors for methionine synthesis. Homoserine is first activated by phosphorylation. Cysteine then displaces the phosphate to yield cystathionine. A cystathione-β-lyase catalyzes cleavage of cystathionine to produce alanine and homocysteine. Methionine synthase catalyzes the final transfer of a methyl group from methyl-tetrahydrofolic acid to homocysteine to yield methionine.

Cysteine Metabolism

Methionine Synthesis Methionine is the other common amino acid containing sulfur and is synthesized from cysteine. The precursor is homoserine, a product of the reduction of aspartic acid (**FIGURE 5.17**). Homoserine is phosphorylated, and the phosphate is then displaced by the sulfhydryl group of cysteine to form cystathionine. Cleavage of the alanine group from cystathionine yields homocysteine, which is methylated by methyl-tetrahydrofolic acid to yield methionine.

Methionine is used in the synthesis of proteins but has other functions in plants, principally as the major methyl donor in the synthesis of lignin precursors (see Chapter 8) and a variety of additional natural products. Methionine is bonded to the 5′ carbon of adenosine to from S-adenosyl methionine (SAM), which donates its methyl group to other compounds. The resulting S-adenosyl homocysteine is subsequently remethylated in a reaction involving methyl-tetrahydrofolic acid (**FIGURE 5.18**).

Glutathione Synthesis Reduced glutathione is a tripeptide consisting of γ-glutamyl-cysteinyl-glycine. It is readily oxidized and forms a disulfide bond with another glutathione molecule. This couple is the major redox buffer in most cells (**FIGURE 5.19**). Glutathione (GSH) or a similar tripeptide is found in almost all organisms. It provides protection from reactive oxygen species (ROS) (see Appendix 5), reduces/oxidizes the disulfide bonds in proteins, and forms complexes with heavy metals and xenobiotics, such as herbicides, thus providing protection against toxicity.

Two enzymes are required to synthesize glutathione from amino acid precursors. A ligase found in plastids catalyzes the formation of the dipeptide, γ-glutamyl-cysteine (**FIGURE 5.20**). Isoforms of the second enzyme, GSH synthetase, are found in both plastids and cytoplasm. The enzymes of glutathione synthesis are constitutive but can be induced to provide additional GSH upon exposure to heavy metals.

FIGURE 5.18 Methionine as a methyl donor. Methionine is bound to the 5′ C of adenosine to produce S-adenosyl methionine (SAM). SAM is the methyl donor in several biosynthetic pathways in plants, especially lignin synthesis (see Chapter 8). The product after methyl donation, S-adenosyl homocysteine (SAH), is remethylated by methyl-tetrahydrofolate.

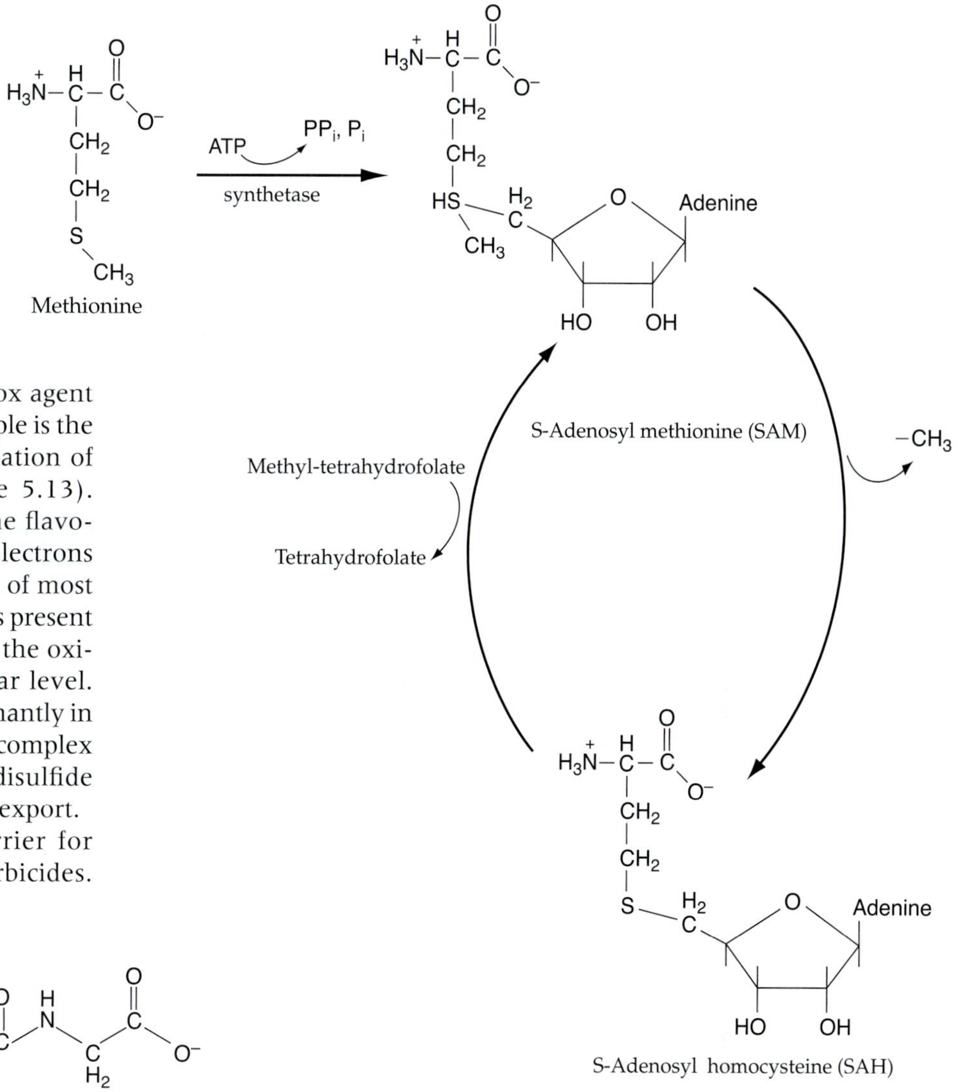

Glutathione can function as a redox agent in some enzymatic reactions. An example is the APS reductase reaction in the assimilation of sulfate described previously (Figure 5.13). Oxidized glutathione is reduced by the flavoprotein, glutathione reductase, with electrons provided by NADPH. In the cytoplasm of most plant cells, reduced glutathione (GSH) is present in millimolar concentrations, whereas the oxidized form is found at the micromolar level. Oxidized glutathione is found predominantly in the endoplasmic reticulum and Golgi complex where it promotes the formation of disulfide bonds in proteins that are destined for export.

Glutathione also provides a carrier for toxic organic compounds such as herbicides.

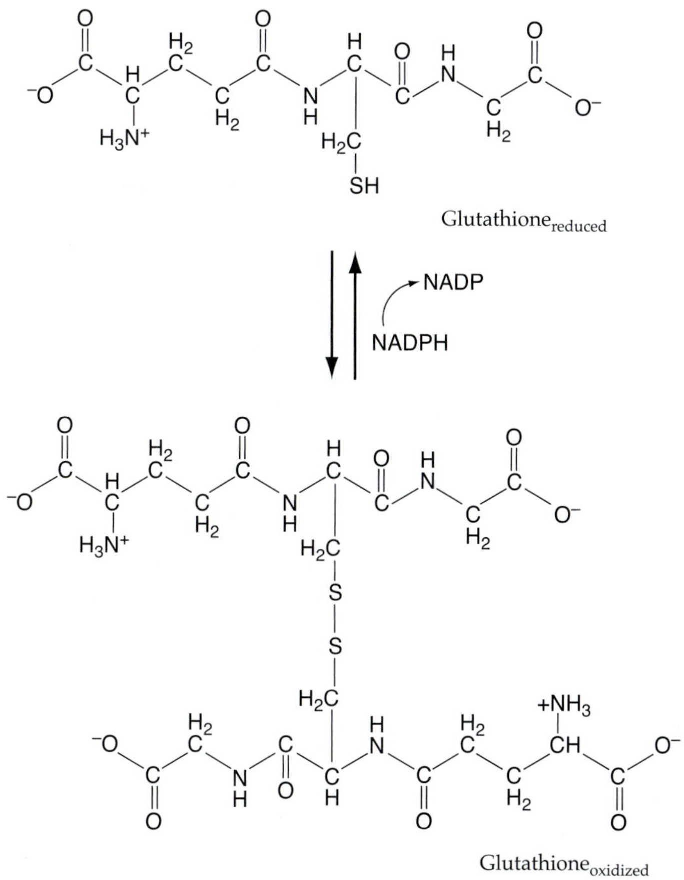

FIGURE 5.19 Structures of reduced/oxidized glutathione. Glutathione is a tripeptide, γ-glutamyl-cysteinyl-glycine. The reduced form (GSH) has a free thiol group. In the oxidized form (GSSG), two glutathione tripeptides are linked by a disulfide bond. The oxidized form is reduced by a flavoprotein, glutathione reductase, that uses NADPH as a reducing agent.

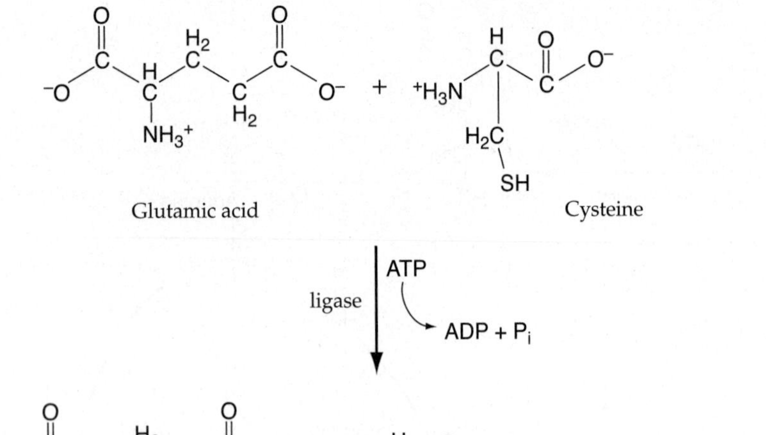

FIGURE 5.20 Synthesis of glutathione. Two enzymes are required for glutathione biosynthesis. The first enzyme is found in plastids and forms a peptide bond between cysteine and the γ-carboxylate of glutamic acid. This ligation requires ATP. The second enzyme, GSH synthetase, catalyzes formation of a peptide bond between the cysteine residue of the γ-glutamyl-cysteine dipeptide and glycine. Isoforms of the γ-glutamyl-cysteine synthetase are found both in chloroplasts and cytoplasm.

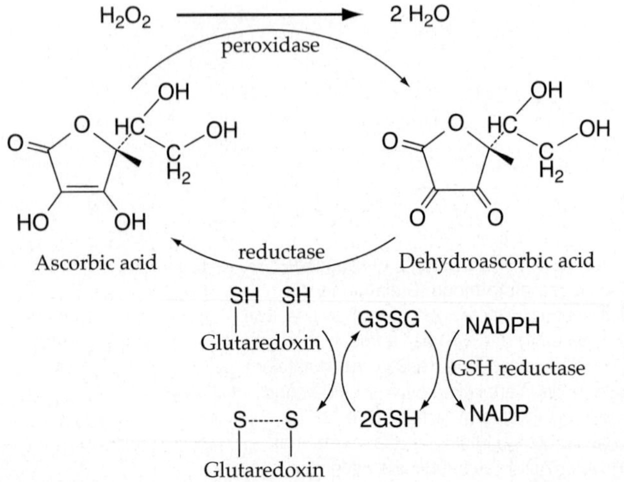

FIGURE 5.21 Glutathione and ascorbate detoxify hydrogen peroxide. Hydrogen peroxide (H_2O_2) is the most abundant reactive oxygen species in plants. It is reduced by ascorbic acid in a reaction catalyzed by ascorbate peroxidase. The dehydroascorbic acid product can be reduced by the disulfide redox protein, glutaredoxin. Glutaredoxin uses glutathione (GSH) as a reducing agent. Oxidized glutathione is, in turn, reduced by glutathione reductase and NADPH. The dithiol form of glutaredoxin can also reduce protein disulfides and repair the damage caused by reactive oxygen species.

Glutathione transferases catalyze the formation of a sulfhydryl link to these xenobiotics. The complex is then transported into vacuoles by multidrug resistance proteins found in the vacuolar membrane. The GSH-xenobiotic complexes can be sequestered in the vacuole where they are harmless and, in some cases, are degraded to cysteine conjugates.

Heavy metals, such as lead and mercury, have a high affinity for sulfhydryl groups and readily attack proteins containing cysteine and methionine. Most plants produce metal-binding thiol peptides called phytochelatins to protect proteins from heavy metals. Phytochelatins are oligomers of γ-glutamyl-cysteine resulting from the degradation of glutathione. Metals form complexes with the thiol groups of the phytochelatins that are stable and can be transported into vacuoles or other vesicles for storage. All enzymes of cysteine, glutathione, and phytochelatin synthesis are up-regulated by the presence of heavy metals in soil and water.

One of the major functions of glutathione is to protect cells from the harmful effects of reactive oxygen species, especially hydrogen peroxide (H_2O_2). Reactive oxygen species are a byproduct of the water-oxidation process in photosystem II, incomplete reduction of O_2 by the mitochondrial electron transport system, and photorespiration. Glutathione acts with ascorbic acid to remove hydrogen peroxide (**FIGURE 5.21**). The enzymes, GSH-ascorbate reductase and ascorbate peroxidase, are found in all organelles and in the cytoplasm. A small protein (~10 kDa), glutaredoxin, can act as a GSH-ascorbate reductase. Glutaredoxin also reduces the disulfide bonds in proteins that often result from oxidative damage. The *Arabidopsis* genome has 31 genes that encode glutaredoxins, which suggests that this protein plays a significant role in protection from oxidative damage. As in the case of phytochelatins, enzyme synthesis increases under conditions of oxidative stress.

Glucosinolate Synthesis A wide variety of natural products contain sulfur and sulfate groups that are derived from cysteine itself or other products of the cysteine synthesis pathway. The best known are the glucosinolates, which are

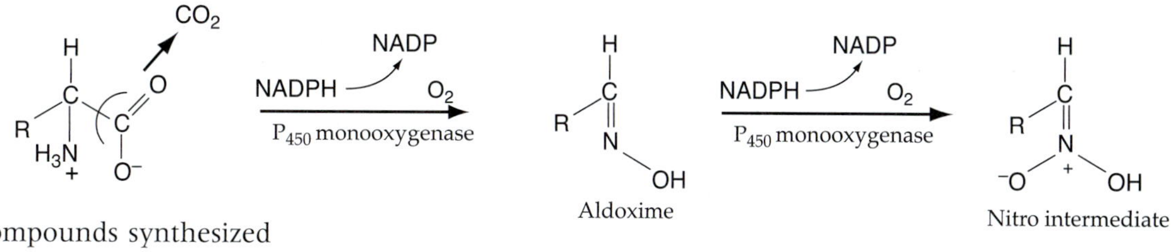

S-sugar conjugates of compounds synthesized from amino acids. The hydrolysis of these compounds releases toxic sulfur-containing compounds that deter herbivores and microorganisms. Glucosinolates have been found only in angiosperms and are mainly produced in the order Capparales, in particular the Brassicaceae family.

Glucosinolates are synthesized from aliphatic amino acids, alanine, valine, leucine, isoleucine, and methionine and from all three aromatic amino acids. These precursors may undergo various modifications, particularly chain elongation, before being incorporated into glucosinolates. Chain elongation involves adding carbons to the amino acid side chain using a complex series of reactions and acetyl-CoA as a methylene donor. Most enzymes and their corresponding genes have been characterized in the model plant, *Arabidopsis thaliana*. A general scheme for the synthesis of glucosinolates is shown in **FIGURE 5.22**. The first step is catalyzed by a cytochrome P450 monooxygenase and yields an intermediate oxime. The second step involves another hydroxylation catalyzed by a P450 monooxygenase to generate a nitro-intermediate. This is followed by conversion of the unstable nitro-intermediate to a thiohydroximate by conjugation to cysteine. The enzyme that catalyzes this step has not been characterized but is believed to be a protein related to the glutathione S-transferase family. A lyase catalyzes removal of alanine from the cysteine moiety producing a thiol intermediate. A specific S-glucosyl transferase adds glucose to the SH group and yields a desulfoglucosinolate. A sulfate anion is then added to the N–OH to complete formation of a glucosinolate. The sulfate donor in this reaction is PAPS.

The basic glucosinolate structure can be further modified after synthesis. This involves changes to the amino acid–derived side chain such as hydroxylation, followed by O-methylation or O-glycosylation, desaturation, and acylation. If the side chain is derived from methionine, the sulfur can be oxidized to a sulfinyl or sulfonyl group. Any one plant species produces several types of glucosinolates. Those derived from methionine or its derivatives are the most common. Glucosinolates are usually part of a plant's constitutive defenses, but their synthesis can be up-regulated or induced by plant hormones, in particular jasmonic acid.

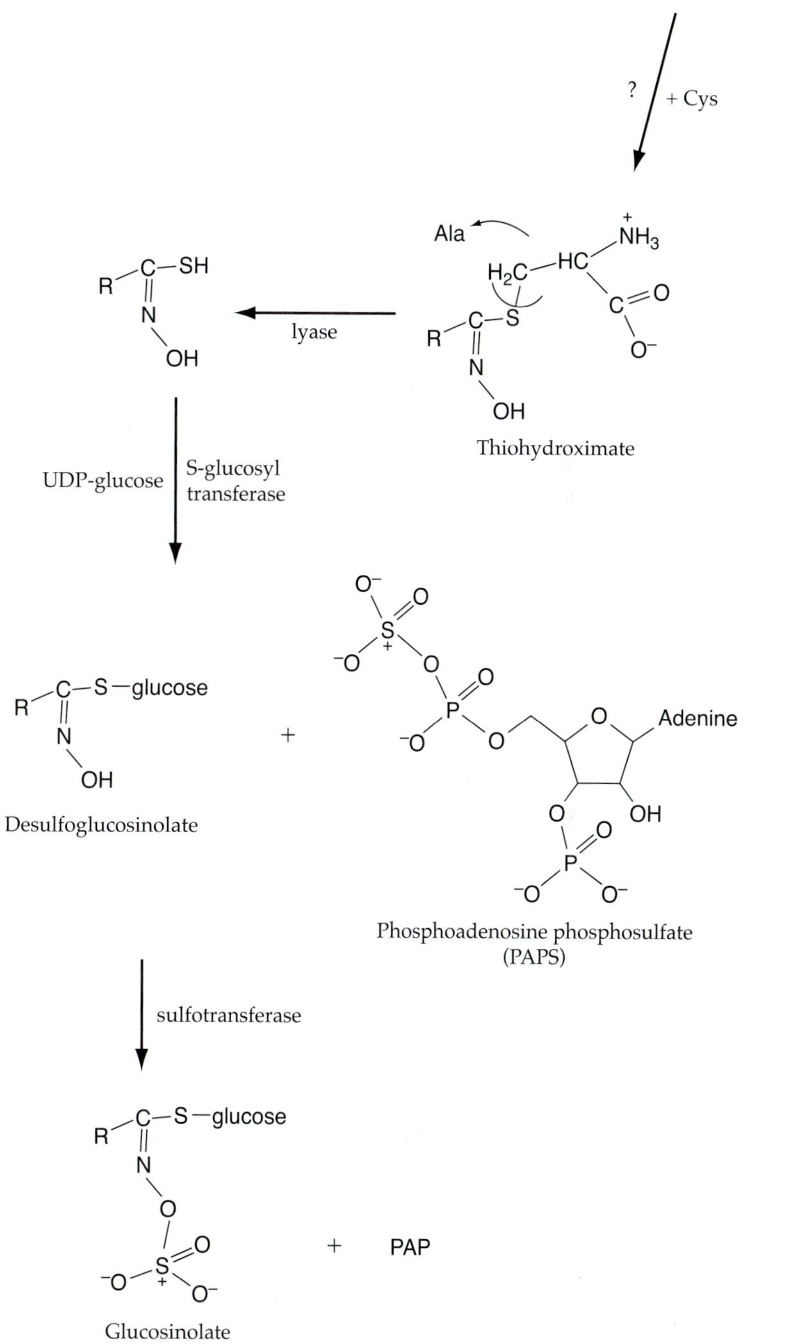

FIGURE 5.22 Synthesis of glucosinolates. Synthesis commences with the decarboxylation of an alpha amino acid that may have been previously modified. A cytochrome P450 monooxygenase adds an OH group to the α-amino nitrogen to generate an aldoxime intermediate. A second monooxygenase adds another OH group to the same N and yields an unstable nitro-intermediate. This is then conjugated to cysteine by an uncharacterized enzyme. The reaction proceeds nonenzymatically *in vitro.* A lyase then catalyzes removal of alanine from the thiohydroximate and produces a thiol intermediate. A specific UDP-glucose-S-transferase catalyzes formation of the unique S-glycosidic bond to the hydroximate intermediate to produce a desulfo-form of the glucosinolate. Sulfate is then transferred from PAPS to generate the final O-sulfated glucosinolate.

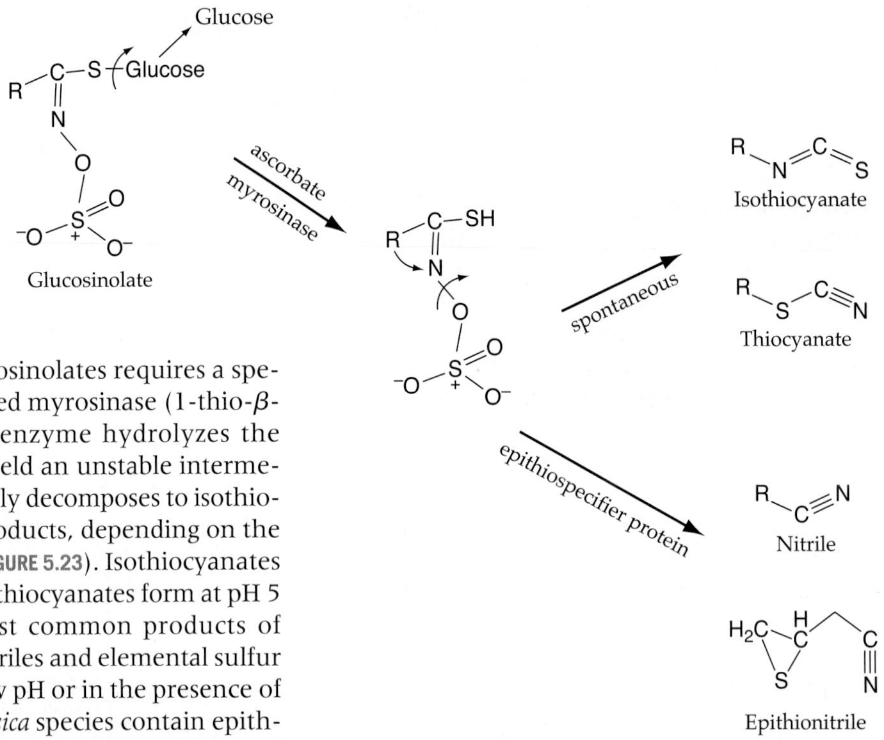

FIGURE 5.23 Decomposition of glucosinolates. A specific S-glucosidase, myrosinase, is required to hydrolyze the sugar from the rest of the glucosinolate molecule. Ascorbate is the cofactor for this enzyme. The resulting aglycone is unstable and will spontaneously rearrange to form isothiocyanates and a small amount of thiocyanates. If the plant contains epithiospecifier proteins, these will guide the breakdown of the aglycone into nitriles.

Hydrolysis of glucosinolates requires a specific S-glycosidase called myrosinase (1-thio-β-D-glucosidase). This enzyme hydrolyzes the thioglucose bond to yield an unstable intermediate that spontaneously decomposes to isothiocyanates and other products, depending on the reaction conditions (**FIGURE 5.23**). Isothiocyanates and lesser amounts of thiocyanates form at pH 5 to 8 and are the most common products of myrosinase action. Nitriles and elemental sulfur are the products at low pH or in the presence of ferric iron. Some *Brassica* species contain epithiospecifier proteins that guide the breakdown of glucosinolates into nitriles, which are toxic to specific herbivores. Thiones (cyclic thioesters) form if the amino acid side chain of the glucosinolate contains a hydroxyl group.

Isothiocyanates are toxic to insect herbivores; they most likely act by binding to proteins and inhibiting the action of digestive enzymes. Some of the glucosinolate decomposition products are volatile and account for the characteristic odor of *Brassica* species. Most insects detect this odor and avoid these plants. However, some insects are attracted to the Brassicaceae by the odor and will feed and lay eggs on the plants. For example, the cabbage butterfly, *Pieris rapae,* can tolerate glucosinolates because it has an insect-produced epithiospecifier protein that guides the production of nitriles during glucosinolate breakdown. Nitriles are less reactive than isothiocyanates and can be detoxified more readily by glutathione S-transferases. This insect–plant interaction is regarded as a good example of coevolution. Attack by herbivores selects for glucosinolate production in the host plants. In turn, selection among herbivores leads to the constitutive synthesis of detoxifying proteins. These specialist herbivores are then able to occupy a specific niche that is not available to other insects.

To be an effective deterrent, the glucosinolates and myrosinase must be stored separately until the plant is wounded. In most cases, myrosinase is stored in specialized cells called idioblasts or myrocin cells (M-cells). The glucosinolates are stored in nearby S-cells. In some species, both components are stored in separate vesicles in the same cell. The synthesis and sequestering of myrosinase has been extensively investigated in *Arabidopsis.* The production of myrosinase and its glucosinolate substrates occurs throughout the plant's life, but the localization is age dependent. In mature plants, myrosinase is produced in the guard cells of the epidermis and in cells of the phloem tissue, but not in roots. The enzyme is also localized to floral stems but is not found in the petals. Unlike many other *Brassica* species, myrosinase in *Arabidopsis* is not found in seeds but is highly concentrated in cells of the siliques (seed pods). In seedlings, the enzyme is produced only in the cotyledons. Glucosinolates are concentrated in separate cells in these same tissues.

Myrosinase is a glycoprotein that belongs to a large superfamily of glycosidases that catalyze hydrolysis of an O-sugar bond. The enzyme from various *Brassica* species has a mass of approximately 50 kDa and functions as a dimer in which the subunits are held together by a zinc ion. The active enzyme presumably has two independent active sites. Although myrosinase has 40% to 60% amino acid similarity with O-glycosidases and shows almost equivalent tertiary structure, it specifically catalyzes the hydrolysis of S-glycosidic bonds. Unlike the O-glycosidases, myrosinase has only one glutamic acid residue in the active site that functions as an acid/base residue during

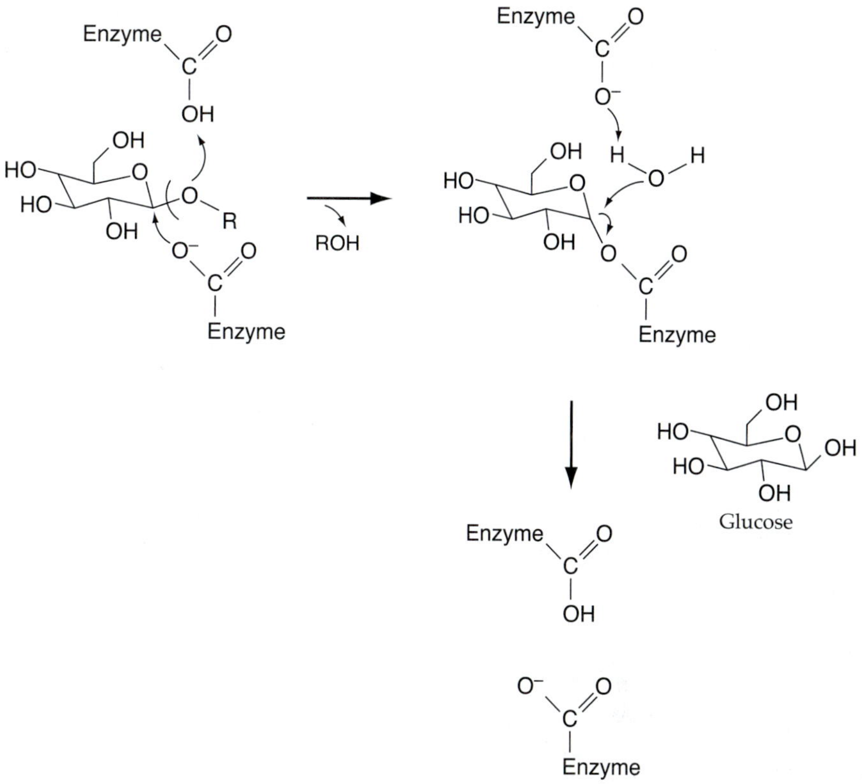

FIGURE 5.24 Mechanism of action of O-glycosidases. A glycosylated substrate (R-O-glucose) binds to the active site of the enzyme that contains two glutamic acid residues on either side of the substrate-binding cavity. One of these glutamate carboxyl groups acts as a base and forms a covalent bond to the glucose residue. The second glutamate carboxyl group acts as an acid and transfers a proton to the R-O group, which results in release of the ROH product from the active site. In the next step, a water molecule enters the active site, and the second glutamic acid now acts as a base and removes a proton from H_2O. This leaves the hydroxyl group of water to attack the anomeric carbon of the sugar and cleave its bond to the enzyme. The second product of the reaction, glucose, then diffuses out of the active site of the enzyme, which is now ready to catalyze another hydrolysis. This mechanism of action is common to "retaining" glycosidases (i.e., the glucose product has the same configuration at the anomeric carbon as in the substrate).

catalysis. A typical glycosidase has two glutamic acid residues in the active site to catalyze sugar cleavage, as shown in **FIGURE 5.24**. Kinetic studies of myrosinase showed that it has a slower turnover than a typical O-glycosidase but could be activated by ascorbic acid at concentrations up to 1 mM. Addition of higher concentrations of ascorbate to *in vitro* assays resulted in competitive inhibition with glucosinolates, suggesting that ascorbate binds to the enzyme's active site. Further kinetic studies combined with structural analysis demonstrated that ascorbate is a cofactor for myrosinase and substitutes for the second glutamate residue that is missing in the active site of this glycosidase (**FIGURE 5.25**).

The structures of two common glucosinolates are shown in **FIGURE 5.26**. Although one type of compound may be more abundant, most *Brassica* species produce several different glucosinolates. Glucoraphanin is the major compound found in broccoli. The hydrolysis product, sulforaphane, may protect humans against certain forms of cancer. The isothiocyanates produced upon hydrolysis are detoxified by

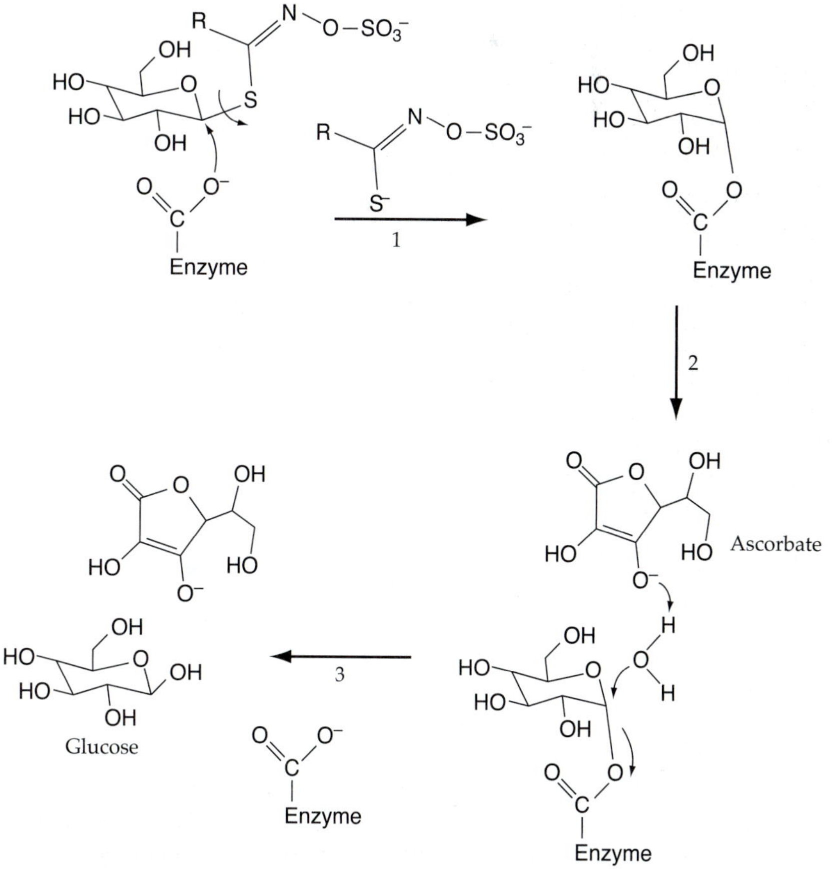

FIGURE 5.25 Mechanism of action of the S-glycosidase, myrosinase. As in the O-glycosidase reaction shown in Figure 5.24, the glucosinolate substrate enters the active site and an essential glutamate carboxyl group attacks the anomeric carbon of the sugar, forming an enzyme-sugar bond. The glucosinolate side chain is not protonated because myrosinase lacks the second acidic residue in the active site. The side chain is released from the active site as a thiolate anion as shown in step 1. The glucose residue remains bound to the active site, which accounts for the low turnover rates for myrosinase observed *in vitro*. In step 2, ascorbic acid and a water molecule enter the active site. Ascorbate acts as a base and removes a proton from water. The resulting hydroxyl group attacks the anomeric carbon of the sugar and displaces it from the active site glutamate carboxyl group. The free sugar and the ascorbate now diffuse out of the active site and the enzyme is able to bind another glucosinolate substrate. Note that the oxidation state of ascorbic acid does not change; it acts as a base, not as a redox cofactor in this reaction.

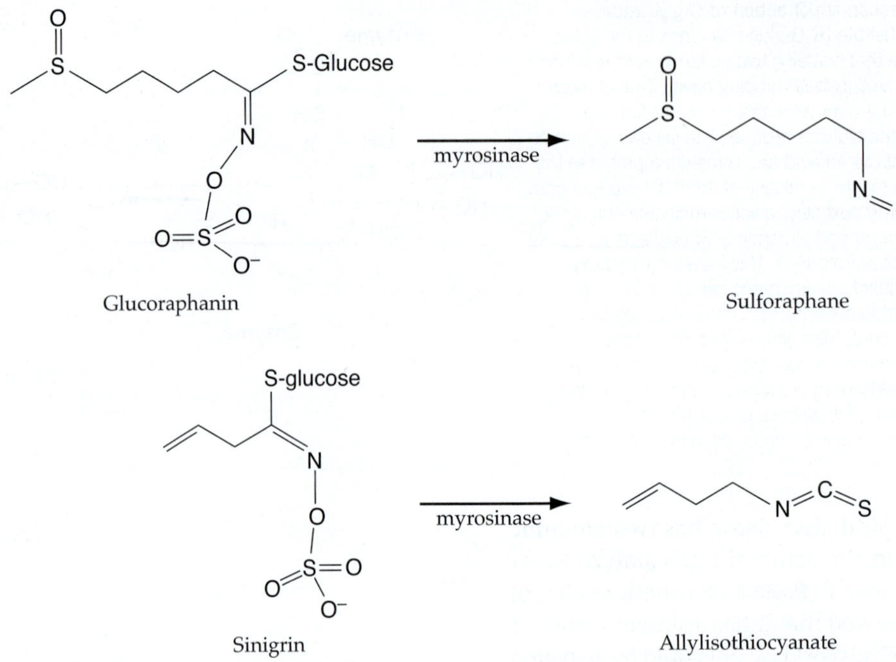

FIGURE 5.26 Examples of two common glucosinolates. Glucoraphanin is a glucosinolate commonly produced in varieties of *Brassica oleracea*. The side chain is synthesized from a methionine starter unit that has been elongated by addition of two methylene groups. The methionine sulfide has been oxidized to a sulfinyl group. Sinigrin is the major glucosinolate found in mustard seeds (*Brassica nigra, B. juncea*). The allyl side chain is synthesized from an aliphatic amino acid, probably leucine.

conjugation to glutathione in the human liver. Diets rich in glucoraphanin induce Phase II detoxification enzymes, such as glutathione S-transferases, probably by interacting with transcription factors. The breakdown products also inhibit the cell cycle and induce apoptosis of human cells grown in culture. The available evidence suggests that eating broccoli and other *Brassica* may be beneficial to human health. There is little evidence, however, to show that sufficient amounts of the active components are absorbed from the gastrointestinal tract to be effective in the whole organism.

Sinigrin (Figure 5.26) is a glucosinolate that is predominant in mustard seeds (*Brassica nigra*). Its hydrolysis product, allyisothiocyanate, in this condiment gives mustard its characteristic "bite." Isothiocyanates can down-regulate the inflammatory response in mammals by interacting with signaling molecules such as prostaglandins. Isothiocyanates in general can inhibit the growth of *Helicobacter pylori*, a common inhabitant of the human stomach that may cause ulcers.

Progoitrin (**FIGURE 5.27**) is produced by wild-type varieties of *Brassica napus*. The amino acid side chain has a hydroxyl group in a position that promotes the formation of a cyclic thione after hydrolysis of glucose. Goitrin inhibits the uptake of iodine from the intestinal tract in humans and other mammals and suppresses the formation of the hormone, thyroxine. This leads to a condition called goiter, which is characterized by swelling and enlargement of the thyroid gland and other symptoms of hypothyroidism, such as abnormal glucose metabolism.

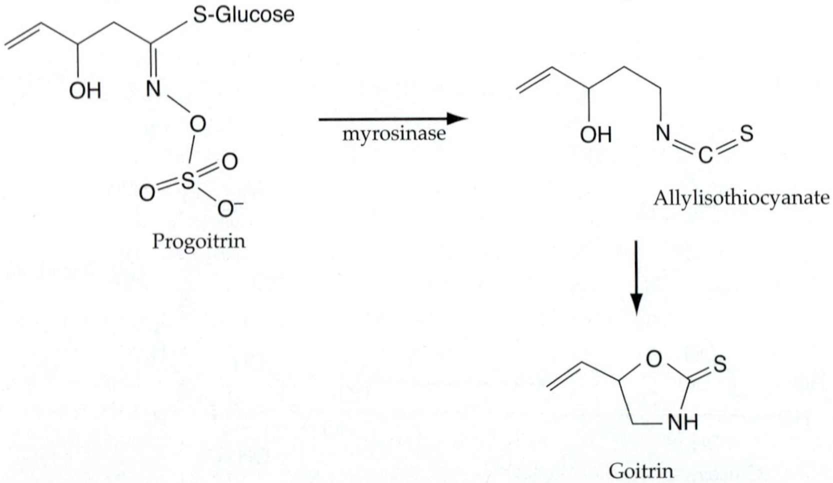

FIGURE 5.27 Structure of progoitrin and its decomposition to goitrin. Progoitrin is the glucosinolate produced by wild type varieties of *Brassica napus* in which the allyl side chain has been hydroxylated. The resulting allyl isothiocyanate spontaneously cyclizes to form a thione ring of goitrin. Goitrin inhibits uptake of iodine in the mammalian gastrointestinal tract.

Alliin Another sulfur-containing natural product that is common in the human diet is alliin (**FIGURE 5.28**) found in *Allium* species such as onions and garlic. The alliins are synthesized from amino acids, principally methionine, and

FIGURE 5.28 Structure and decomposition of alliin. The enzyme alliinase catalyzes removal of the alanine moiety from alliin to produce a three-carbon sulfenic acid, pyruvate, and ammonia. Further enzymatic or spontaneous reactions produce a variety of sulfur-containing products. Allicin is the precursor for the formation of propanethial S-oxide. In aqueous solution, this compound further decomposes to propionaldehyde, hydrogen sulfide, and sulfuric acid. The latter is responsible for the lacrimator properties (tear-causing) of peeled onions.

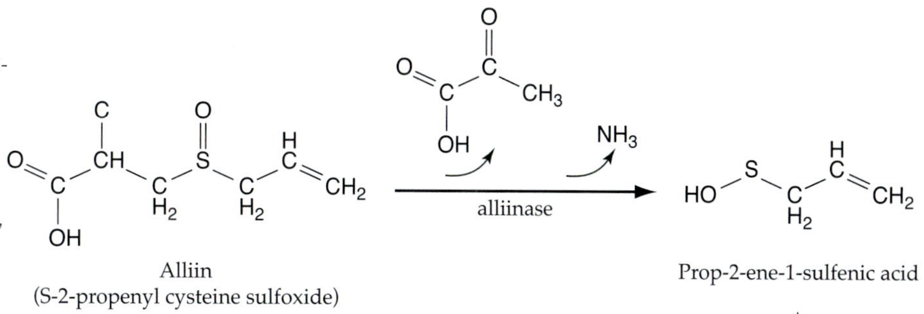

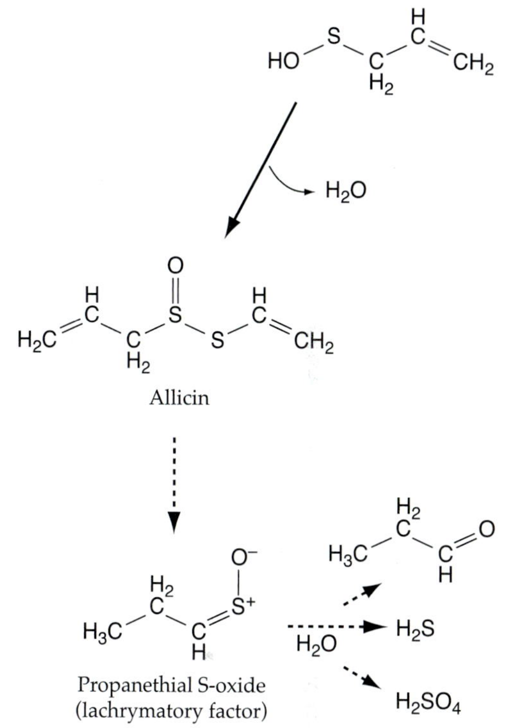

are inactivated by glycosylation. Wounding of plant cells mixes the alliins with a specific alliinase that leads to hydrolysis of the sugar and the breakdown of these compounds into a variety of odoriferous products with antibacterial activity. The alliin in onions decomposes to propane thiol S-oxide, a potent lachrymator or tearing agent (Figure 5.28). The alliins are produced by *Allium* species to deter herbivory and have activity against microorganisms. These plants, however, have been components of the human diet for millennia and may have some health benefits. Allicin and its derivatives inhibit the cell cycle and induce apoptosis in human cells in culture and may protect against cancer. Ajoene, which is responsible for the odor of garlic, prevents platelet aggregation and formation of blood clots and has been suggested as a preventative for stroke. Like the glucosinolates, the alliins are produced to deter herbivores and are effective against some humans but may actually be good for you!

CHAPTER SUMMARY

Nitrogen and sulfur are two of the major elements required by plants and other photosynthetic organisms such as algae and cyanobacteria. Photosynthetic autotrophs obtain these elements mainly in the form of nitrate (NO_3^-) and sulfate (SO_4^{2-}) ions taken up from the soil or water. Uptake is followed by the process of assimilation into organic compounds, principally amino acids.

Nitrate assimilation consists of reduction to nitrite, further reduction to ammonia, and incorporation into the amino acid, glutamate. Because of the high energy demand, assimilation occurs mainly in photosynthetic tissues. Glutamate is used in protein synthesis but is also the major amino donor for the synthesis of other amino acids and essential nitrogen-containing metabolites such as nucleotides and cofactors. Only prokaryotes, including many cyanobacteria, can use atmospheric nitrogen for biosyntheses. These microorganisms produce a nitrogenase enzyme complex that binds and reduces N_2 to NH_3. In addition to being a high energy requiring process, the nitrogenase complex is inactivated by oxygen, which restricts the conditions under which nitrogen can be fixed. Many vascular plants have taken advantage of the N_2-fixing organisms by forming symbiotic relations. These associations may be rather loose, such as those between the water fern, *Azolla*, and species of cyanobacteria in the genus *Anabaena*. Some associations are very intimate and highly regulated, such as those between legumes and bacteria in the genus *Rhizobia*. In this latter symbiosis, the plants attract and encourage infection by specific species of *Rhizobia*. The bacteria are sequestered in special root cells where the plant maintains an environment favorable for bacterial nitrogen fixation.

Fixed forms of nitrogen are often limiting for plant growth. Agricultural needs for fixed nitrogen in developed

CHAPTER SUMMARY, CONT.

countries are supplied by a chemical reaction, the Haber-Bosch process.

Sulfate ions are actively taken up by plants and reduced to sulfide, which is incorporated mainly into the amino acid cysteine and to a lesser extent into the sulfate donor, 3′-phosphoadenosine-5′-sulfate (PAPS). Cysteine is the precursor for the synthesis of methionine, the other sulfur-containing amino acid. In addition to their incorporation into proteins, cysteine and methionine serve many other essential functions. Cysteine is a component of the tripeptide, glutathione, which is the major redox buffer in cells and also serves to detoxify xenobiotic compounds. Likewise, methionine is required for the synthesis of the methyl donor, S-adenosylmethionine, which is essential for lignin precursor and natural product syntheses. Both cysteine and PAPS are required for glucosinolate synthesis, which are important defensive natural products in the Brassicaceae. Methionine is the precursor for alliin and other sulfur-containing natural products in *Allium* species. These are only a few examples of the plant-specific metabolites that contain sulfur and illustrate the importance of this element in plant survival.

REFERENCES FOR FURTHER INFORMATION

Campbell, W.H. (2001). Structure and function of eukaryotic NAD(P)H:nitrate reductase. *Cell. Mol. Life Sci.* **58:**194–204.

Forde, B.G. (2000). Nitrate transporters in plants: Structure, function and regulation. *Biochim. Biophys. Acta* **1465:**219–235.

Halkier, B.A. & Gershenzon, J. (2006). Biology and biochemistry of glucosinolates. *Annu. Rev. Plant Biol.* **57:**303–333.

Jones, K.M., Kobayashi, H., Davies, B.W., Taga, M.E., & Walker, G.C. (2007). How rhizobial symbionts invade plants: The *Sinorhizobium-Medicago* model. *Nat. Rev. Microbiol.* **5:**619–633.

Leustek, T. & Saito, K. (1999). Sulfate transport and assimilation in plants. *Plant Physiol.* **120:**637–643.

MacKintosh, C. & Meek, S.E.M. (2001). Regulation of plant NR activity by reversible phosphorylation, 14-3-3 proteins and proteolysis. *Cell. Mol. Life Sci.* **58:**205–214.

Oldroyd, G.E.D. & Downie, J.A. (2008). Coordinating nodule morphogenesis with rhizobial infection in legumes. *Annu. Rev. Plant Biol.* **59:**519–546.

Rausch, T. & Wachter, A. (2005). Sulfur metabolism: a versatile platform for launching defence operations. *Trends Plant Sci.* **10:**503–509.

Seefeldt, L.C., Hoffman, B.M., & Dean, D.R. (2009). Mechanism of Mo-dependent nitrogenase. *Annu. Rev. Biochem.* **78:**701–722.

PLANT BIOCHEMISTRY *on the web*

Visit the *Plant Biochemistry* student companion Web site at http://biology.jbpub.com/gleason/plantbiochemistry/ for additional online study resources on these topics.

CHAPTER 6

Lipids

CHAPTER OUTLINE

6.1 Overview

Plant lipids were experimentally defined by chemists in the beginning of the twentieth century as a fraction of the oily cellular substances that were not soluble in water. They could be extracted with organic solvents such as ether that obviously yielded a fraction containing all the other nonpolar constituents of the cell such as chlorophylls. Lipids are now chemically defined as esters of fatty acids and glycerol and their derivatives. Most organisms store energy as triacylglycerols (TAGs). Lipids are a more efficient storage form than carbohydrates. Fatty acids are more reduced; that is, they have more carbon-hydrogen bonds than starch or glycogen. Therefore, TAGs provide more energy upon oxidation than these carbohydrates, approximately 38 versus 17 kJ/g. In addition, because lipids are not hydrated like carbohydrates, they can be densely packed into smaller volumes, as in the oil bodies found in seed tissues. Phospholipids, which consist of two fatty acids esterified to two of the hydroxyl groups of glycerol (diacylglycerols [DAGs]) and a phosphate-containing compound in the third position, are the primary components of most biological membranes. In addition, lipids and fatty acids are often used in the cell to modify other molecules, such as proteins (lipoproteins) and carbohydrates (glycolipids). Plants synthesize a large number of unusual lipids that are important in maintaining membrane fluidity in the thylakoids and can function in defense against herbivores and pathogens in seeds. Plants also produce a number of lipid-based signaling molecules such as green-leaf volatiles and jasmonates, which serve as stress hormones.

Plant lipids are an essential component of the human diet. Most mammals are unable to produce polyunsaturated fatty acids (PUFAs; i.e., those having more than one double bond) and must obtain these from their diet. This includes eating plants and plant products. Polyunsaturated fats are required for synthesis of mammalian hormones such as prostaglandins and leukotrienes, which mediate inflammation and immune responses. In addition, humans have used plant oils for a variety of nonfood applications, such as lighting, lubricants, and soaps. Plant oils are now being utilized and further explored as a source of hydrocarbons to replace petroleum in fuels and manufacturing of plastics and other materials.

6.2 Fatty Acid Structure and Nomenclature

Plants synthesize over 300 different kinds of fatty acids, more than any other group of organisms. However, this diversity is not uniform; the most common fatty acids found in plants are 16-carbon, palmitate, and 18-carbon, stearate, and their unsaturated derivatives.

The structures of some of these common fatty acids are shown in **FIGURE 6.1**.

As shown in Figure 6.1, it is common to use a numerical designation to indicate the structures of fatty acids. For example, stearic acid is 18:0; 18 is the total number of carbons in the acid and the number after the colon indicates the number of double bonds. Stearic acid has no double bonds and is referred to as a saturated fatty acid. Oleic acid, 18:1 Δ9c, is an 18-carbon acid that contains one double bond (i.e., a monounsaturated fatty acid). The position of the double bond is indicated after the delta (Δ) as determined from the carboxyl group as position 1. In some cases the Greek letter omega (ω) is used to designate the position of the double bond, and the numbering proceeds from the methyl end of the fatty acid. Using this latter system, linoleic acid in Figure 6.1 is 18:2 ω6c, 9c. Rotation is restricted around a double bond, and two positional isomers are possible. The letter "c" indicates that the configuration around the double bond is *cis* (*Z*, in organic chemical nomenclature). Most fatty acids found in living organisms have the *cis* configuration; *trans* (t, or *E*) fatty acids occur in some seeds' oils and pollen grains. *Trans* fatty acids that are found in commercially processed oils and foods are produced by rearrangements that occur during chemical hydrogenation of polyunsaturated fatty acids. Diets high in *trans* fatty acids have been shown to cause heart disease in experimental animals.

The number of carbons in a fatty acid and the number and configuration of the double bonds affect its physical properties. The rigid structure of double bonds produces a "kink" in the molecule and prevents close packing of the molecules, making the corresponding lipids more fluid than in the case of saturated fatty acids. This kinking is more pronounced with *cis* double bonds (Figure 6.1). The *trans* configuration produces a more linear structure similar to saturated acids. These molecular variations are most readily apparent in the differences in melting temperatures shown in **TABLE 6.1**.

CHAPTER 6
Lipids

CHAPTER OUTLINE

6.1 Overview

Plant lipids were experimentally defined by chemists in the beginning of the twentieth century as a fraction of the oily cellular substances that were not soluble in water. They could be extracted with organic solvents such as ether that obviously yielded a fraction containing all the other nonpolar constituents of the cell such as chlorophylls. Lipids are now chemically defined as esters of fatty acids and glycerol and their derivatives. Most organisms store energy as triacylglycerols (TAGs). Lipids are a more efficient storage form than carbohydrates. Fatty acids are more reduced; that is, they have more carbon-hydrogen bonds than starch or glycogen. Therefore, TAGs provide more energy upon oxidation than these carbohydrates, approximately 38 versus 17 kJ/g. In addition, because lipids are not hydrated like carbohydrates, they can be densely packed into smaller volumes, as in the oil bodies found in seed tissues. Phospholipids, which consist of two fatty acids esterified to two of the hydroxyl groups of glycerol (diacylglycerols [DAGs]) and a phosphate-containing compound in the third position, are the primary components of most biological membranes. In addition, lipids and fatty acids are often used in the cell to modify other molecules, such as proteins (lipoproteins) and carbohydrates (glycolipids). Plants synthesize a large number of unusual lipids that are important in maintaining membrane fluidity in the thylakoids and can function in defense against herbivores and pathogens in seeds. Plants also produce a number of lipid-based signaling molecules such as green-leaf volatiles and jasmonates, which serve as stress hormones.

Plant lipids are an essential component of the human diet. Most mammals are unable to produce polyunsaturated fatty acids (PUFAs; i.e., those having more than one double bond) and must obtain these from their diet. This includes eating plants and plant products. Polyunsaturated fats are required for synthesis of mammalian hormones such as prostaglandins and leukotrienes, which mediate inflammation and immune responses. In addition, humans have used plant oils for a variety of nonfood applications, such as lighting, lubricants, and soaps. Plant oils are now being utilized and further explored as a source of hydrocarbons to replace petroleum in fuels and manufacturing of plastics and other materials.

6.2 Fatty Acid Structure and Nomenclature

Plants synthesize over 300 different kinds of fatty acids, more than any other group of organisms. However, this diversity is not uniform; the most common fatty acids found in plants are 16-carbon, palmitate, and 18-carbon, stearate, and their unsaturated derivatives.

The structures of some of these common fatty acids are shown in **FIGURE 6.1**.

As shown in Figure 6.1, it is common to use a numerical designation to indicate the structures of fatty acids. For example, stearic acid is 18:0; 18 is the total number of carbons in the acid and the number after the colon indicates the number of double bonds. Stearic acid has no double bonds and is referred to as a saturated fatty acid. Oleic acid, 18:1 Δ9c, is an 18-carbon acid that contains one double bond (i.e., a monounsaturated fatty acid). The position of the double bond is indicated after the delta (Δ) as determined from the carboxyl group as position 1. In some cases the Greek letter omega (ω) is used to designate the position of the double bond, and the numbering proceeds from the methyl end of the fatty acid. Using this latter system, linoleic acid in Figure 6.1 is 18:2 ω6c, 9c. Rotation is restricted around a double bond, and two positional isomers are possible. The letter "c" indicates that the configuration around the double bond is *cis* (*Z*, in organic chemical nomenclature). Most fatty acids found in living organisms have the *cis* configuration; *trans* (t, or *E*) fatty acids occur in some seeds' oils and pollen grains. *Trans* fatty acids that are found in commercially processed oils and foods are produced by rearrangements that occur during chemical hydrogenation of polyunsaturated fatty acids. Diets high in *trans* fatty acids have been shown to cause heart disease in experimental animals.

The number of carbons in a fatty acid and the number and configuration of the double bonds affect its physical properties. The rigid structure of double bonds produces a "kink" in the molecule and prevents close packing of the molecules, making the corresponding lipids more fluid than in the case of saturated fatty acids. This kinking is more pronounced with *cis* double bonds (Figure 6.1). The *trans* configuration produces a more linear structure similar to saturated acids. These molecular variations are most readily apparent in the differences in melting temperatures shown in **TABLE 6.1**.

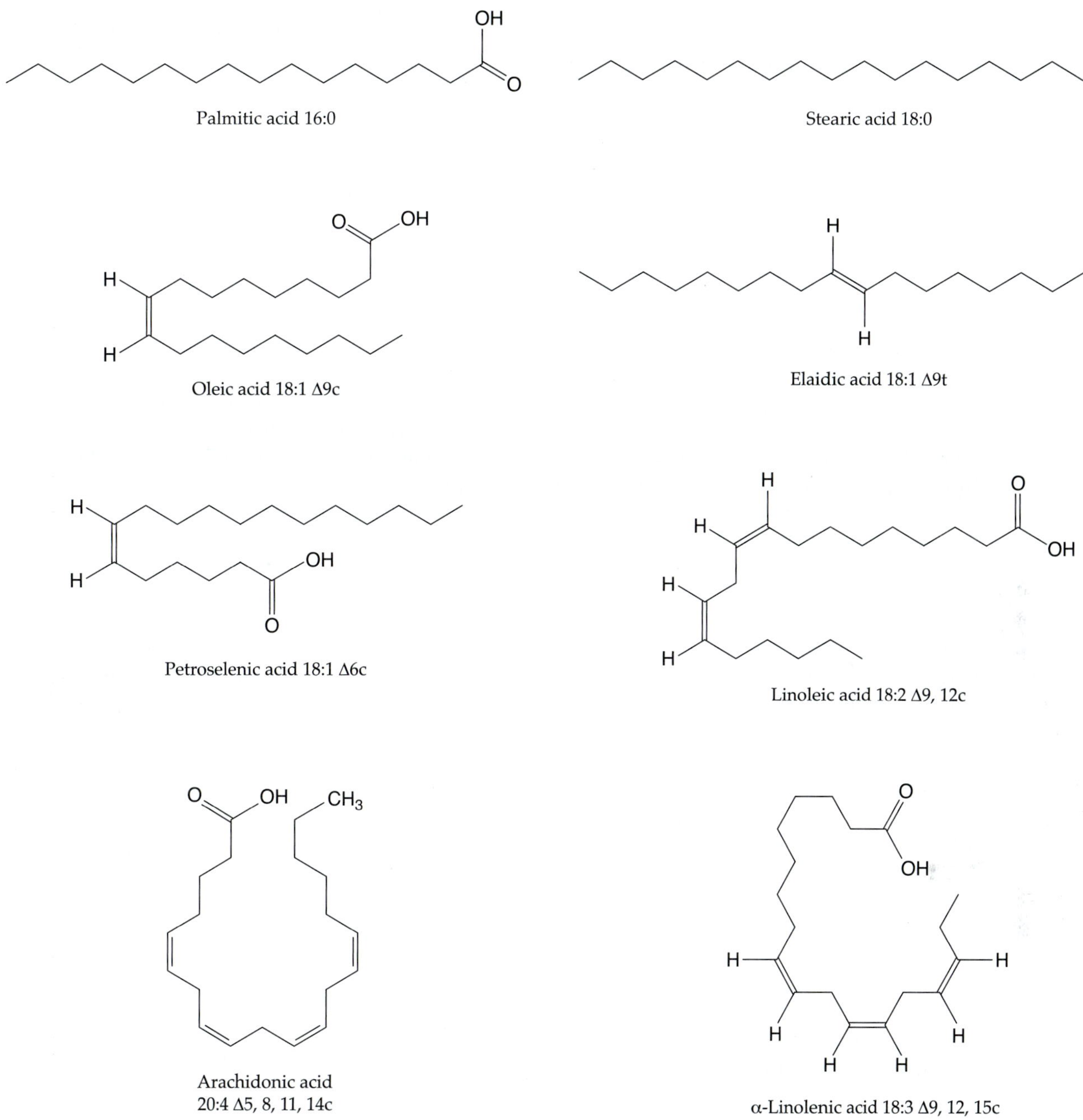

FIGURE 6.1 Structures of some fatty acids produced by plants. Palmitic (16:0) and stearic (18:0) acids are saturated fatty acids; they have no double bonds. Oleic, elaidic, and petroselenic acids are 18-carbon monounsaturated fatty acids; they each have one double bond. Oleic acid (18:1 Δ9c) is the most common of the three acids; it is the major component of olive oil. The double bond is at position 9, starting from the carboxyl end of the molecule and the configuration around the double bond is *cis* (*Z*). Elaidic acid is a rare fatty acid with a *trans* (*E*) configuration around the double bond. It is produced in some pollen grains. Petroselenic acid, found in parsley, has an uncommon *cis* double bond between C6 and C7. Linoleic and linolenic acids are polyunsaturated fatty acids, having two or three *cis* double bonds.

Table 6.1 Melting Temperatures of Some Common Fatty Acids

Fatty Acid	Melting Point (°C)	Fatty Acid	Melting Point (°C)
Lauric acid (12:0)	44	Elaidic acid (18:1, Δ9t)	45
Myristic acid (14:0)	52	Petroselinic acid (18:1, Δ6c)	32
Palmitic acid (16:0)	63	Linoleic acid (18:2, Δ9,12c)	−9
Stearic acid (18:0)	70	α-Linolenic acid (18:3, Δ9, 12, 15c)	−17
Oleic acid (18:1, Δ9c)	13	Arachidonic acid (20:4, Δ5,8,11,14c)	−50

6.3 Fatty Acid Biosynthesis

In the 1950s E. Conn and P. Stumpf and coworkers initiated labeling studies to trace the *de novo* synthesis of fatty acids. Using techniques similar to the Calvin group working on the C-3 pathway (see Chapter 2), these researchers fed $^{14}CO_2$ and were able to demonstrate that labeled carbon was directly incorporated by plants into acetate and then into 16- and 18-carbon fatty acids. Separation and identification of fatty acids were done by thin-layer chromatography (TLC) on glass plates coated with silica gel. This is still a useful technique for initial characterization of small, hydrophobic molecules. The development of gas chromatography to separate fatty acids, coupled to mass spectrometry for identification, greatly facilitates such investigations today (see Appendix for description of gas chromatography-mass spectrometry [GC-MS]).

6.4 Acetate-Malonate Pathway

In plants, *de novo* fatty acid synthesis originates in the stroma of plastids. One of the first ^{14}C-labeled compounds is acetyl-CoA. Labeling with radioactive and stable isotopes further indicates that acetate can be made from phosphoglyceric acid, an intermediate in the C-3 cycle. This synthesis involves reducing phosphoglyceraldehyde to glycerol, dehydration to pyruvate, and oxidative decarboxylation to acetyl-CoA by a chloroplast pyruvate dehydrogenase complex similar to the multienzyme complex found in the mitochondria (see Chapter 3). An alternate source of acetyl-CoA is by transport of acetate from the mitochondrion and its conversion to acetyl-CoA by a synthetase in the stroma.

The initial step in fatty acid synthesis is addition of bicarbonate ion to acetyl-CoA to produce a three-carbon intermediate, malonyl-CoA:

$$\text{acetyl-CoA} + HCO_3^- + \text{ATP} \rightarrow \text{malonyl-CoA} + \text{ADP} + \text{Pi}$$

This reaction is catalyzed by the acetyl-CoA carboxylase (ACC) complex (**FIGURE 6.2**). This protein complex consists of three components: (1) the biotin carboxyl carrier protein (BCCP), (2) biotin carboxylase (BC), and (3) a heterodimeric protein, carboxyltransferase (CT). The cofactor, biotin, is covalently bound to the epsilon amino group of a lysine residue in BCCP. The biotin carboxylase component catalyzes the ATP-dependent transfer of bicarbonate ion to the cofactor as shown in Figure 6.2. ATP is required to dehydrate the bicarbonate ion to CO_2, which is a more efficient carboxylating agent. The carboxyl transferase binds acetyl-CoA and catalyzes transfer of the carboxyl group from BCCP-biotin to acetyl-CoA to produce the three carbon acid, malonyl-CoA. The biotin cofactor remains bound to BCCP. Biotin acts like a "swinging arm," moving the carboxyl group between the other two components of the complex.

There are two different forms of the ACC complex. In plastids of plants and many bacteria, the complex consists of three separate protein components and is referred to as the prokaryotic variant. The β subunit of carboxyl transferase is encoded by a gene in the chloroplast genome, whereas genes for the α subunit of carboxyltransferase, the BCCP, and biotin carboxylase are found in the nuclear genome and the proteins

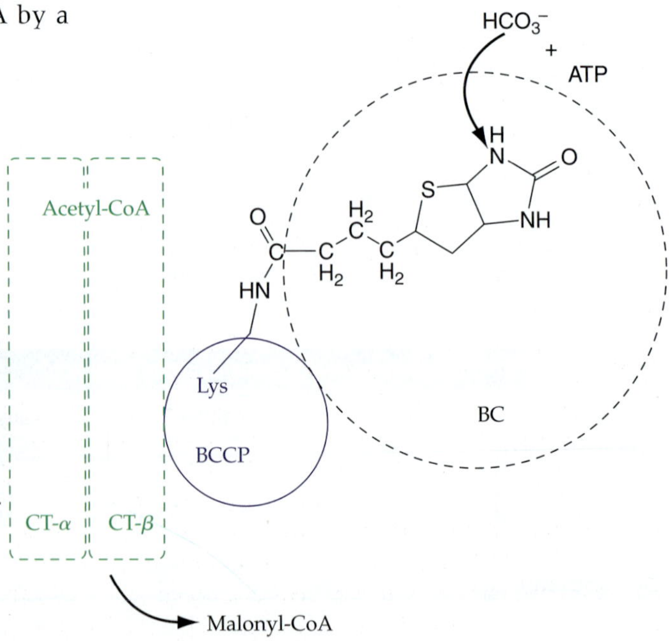

FIGURE 6.2 Acetylcoenzyme A carboxylase (ACC) complex. The ACC complex in the chloroplast stroma consists of three types of proteins held together by noncovalent interactions. The biotin carboxyl carrier protein (BCCP) occupies the center of the complex; a biotin cofactor is bound to the epsilon amino group of a lysine residue. Biotin carboxylase (BC) catalyzes the addition of CO_2 to an amino group of the cofactor. ATP is required to "activate" the substrate, HCO_3^-, in this reaction. Carboxyl transferase (CT), a heterodimeric enzyme, catalyzes the transfer of the bound CO_2 from biotin to acetyl-CoA, producing the three-carbon product, malonyl-CoA.

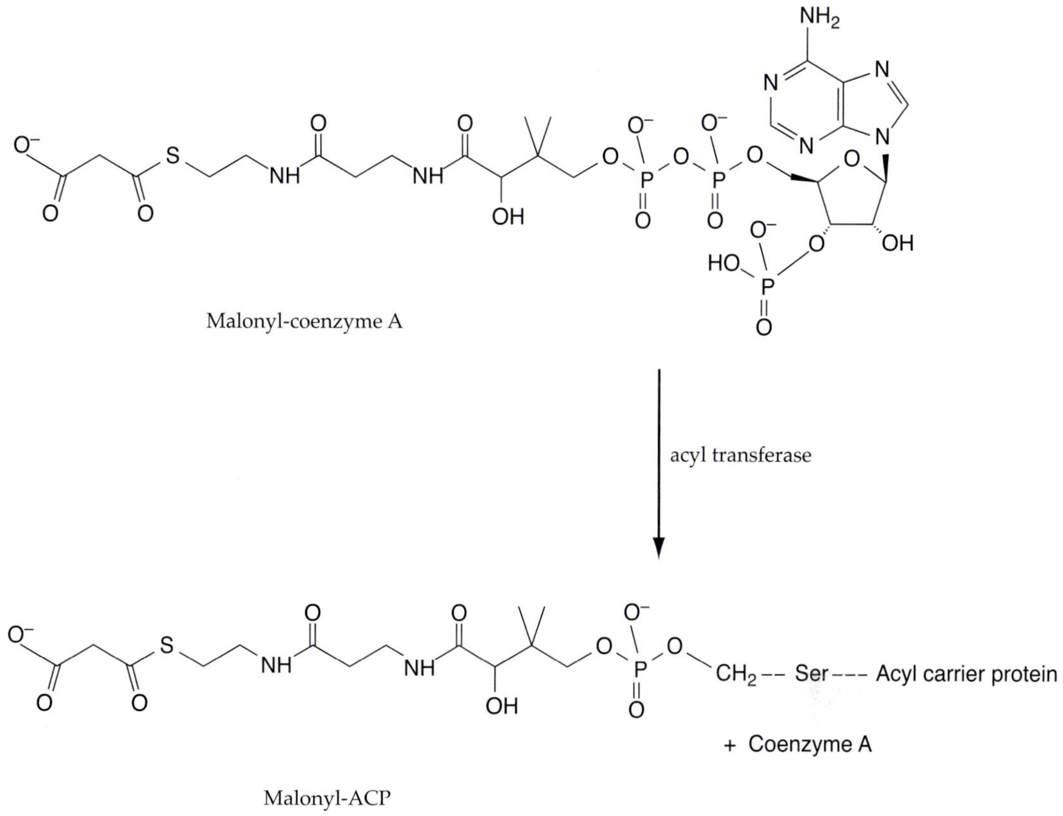

FIGURE 6.3 Reaction catalyzed by acyl carrier protein (ACP) transferase (malonyl CoA:ACP transacylase). The enzyme, in this example, specifically binds malonyl-CoA. It catalyzes cleavage of the thioester-malonyl bond of the phosphopantetheine group in CoA and transfers the malonyl moiety to a similar group that is covalently bound to a serine residue in ACP.

are produced in the cytoplasm. The eukaryotic variant is a large fusion protein of all four subunits. A eukaryotic type of ACC occurs in both mitochondria and cytoplasm of mammalian and avian cells. Plants also have a eukaryotic variant of ACC associated with the endoplasmic reticulum (ER).

The three-carbon malonyl moiety synthesized by the ACC complex is transferred from CoA to an acyl carrier protein (ACP). ACP is a small acidic protein with a phosphopantetheine cofactor covalently bonded to a serine residue. The malonyl transfer is catalyzed by an acyl transferase (malonyl CoA:ACP transacylase) (**FIGURE 6.3**).

Malonyl-ACP then docks with the fatty acid synthase (FAS) complex, a large, multienzyme complex found in the chloroplast stroma consisting of several enzymes that produce saturated fatty acids. As in the case of ACC, FAS complexes are found in more than one form. Fatty acid synthase complex I is a large fusion protein found in the cytoplasm and possibly the mitochondria in mammalian and avian cells. Complex II consists of separate enzyme components held together by noncovalent interactions. It is found in bacteria and plastids of plants. A modified form of complex II is found in fungi such as yeast. Complex II is the only type of FAS described to date in plants and the initial reactions of *de novo* fatty acid synthesis presumably occur only in plastids. The steps are illustrated in **FIGURE 6.4**.

The first step in fatty acid synthesis is condensation of malonyl-ACP with acetyl-CoA. This is catalyzed by β-ketoacyl-ACP synthase (KAS). Three isoforms of this enzyme occur in the FAS complex. The initial condensation is catalyzed by the form designated β-ketoacyl-ACP synthase III (KASIII). This initial "priming" reaction is a variation of a Claisen condensation in organic chemistry. Loss of CO_2 from malonate drives the reaction in the direction of acetoacetyl synthesis (**FIGURE 6.5**). Note that although the ACC complex fixes CO_2 into malonate, this CO_2 is lost in the FAS and there is no net carbon dioxide fixation during fatty acid synthesis.

The product of the β-ketoacyl-ACP synthase condensation remains attached to acyl carrier protein ACP throughout the following series of steps. The next reaction is a reduction catalyzed by β-ketoacyl-ACP reductase, which requires NADPH. Water is then removed, resulting in a double bond between C2 and C3 (refer to Figure 6.4). This reaction is stereospecific and results in a double bond with the *trans* (*E*) configuration. In the final step, the double bond is reduced and a product with two more carbons than the starting material

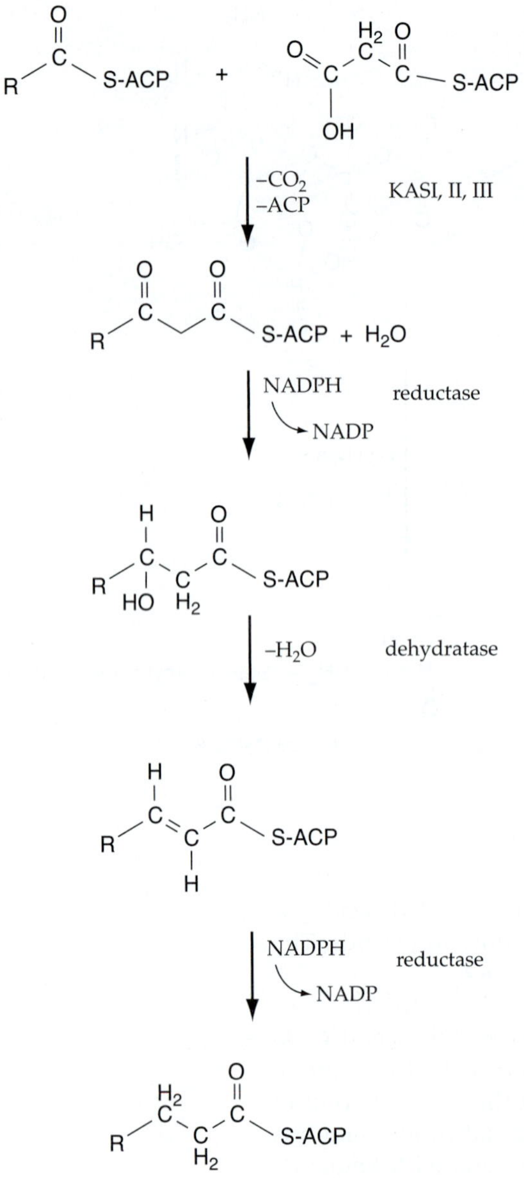

FIGURE 6.4 Fatty acid synthesis by the fatty acid synthase (FAS) complex. In plants, the type II FAS consists of an association of several enzymes found in the plastid stroma. The first reaction is a condensation of malonyl-ACP with a fatty acyl-ACP (acetyl-CoA in the first step of synthesis), catalyzed by one of three forms of β-ketoacyl-ACP synthase (KAS). In the second step, the keto group on C3 is reduced to an alcohol. The reductase uses NADPH as reducing agent. The third step is a stereo-specific dehydration. The resulting double bond has the *trans* configuration. A final reductase, using NADPH, catalyzes a reduction of the double bond and the product is a fatty acid attached to ACP that is two carbons longer than the starting acyl-ACP. The product then binds to another KAS isoform and is further elongated by this series of four reactions.

is produced. The resulting fatty acyl-ACP returns to the first step where it condenses with another molecule of malonyl-ACP, catalyzed by β-ketoacyl-ACP synthase I or II, and the cycle repeats. The fatty acid chain continues to grow by two carbon units until it is 16 to18 carbons in length. It is then released from the FAS complex by the action of a fatty acyl-ACP thioesterase. Several isoforms of this enzyme are found in the chloroplast with specificity for different chain lengths, usually 16 or 18 carbons. The resulting palmitic and stearic acids are then condensed with CoA by acyl-CoA synthetase or ACP and processed further in the chloroplast or transported to the endoplasmic reticulum.

6.5 Desaturation of 18:0 (Stearic Acid) in the Chloroplast

A large fraction of the fatty acids found in most plant lipids contain one or more double bonds. Two forms of desaturase enzymes introduce double bonds into saturated fatty acids. These enzymes require molecular oxygen and NADPH and belong to a large superfamily of enzymes classified as monooxygenases. A C–H bond is very stable (approximately 410 kJ/mol), and breaking this type of bond usually requires a metal ion at the active site and an unusual reaction mechanism. The chloroplast stroma contains a unique soluble desaturase that uses stearoyl-ACP as its substrate. The more common desaturases are membrane-bound complexes embedded in the ER that usually bind fatty acids after they are esterified to glycerol.

Because it is a soluble protein, the chloroplast desaturase has been studied extensively, and crystal structures of the protein from castor bean are available. The protein is active as a homodimer; each monomer consists mainly of 11 α-helices with a few β-strands near the C-terminal. The active site is located in a channel between four of the helices and contains a di-iron cluster bonded to glutamic acid and histidine residues (**FIGURE 6.6**). The channel between these four helices is approximately the correct size to accommodate the stearoyl moiety of the ACP-bound substrate with its C9−C10 bond near the Fe atoms at the active site. The protein is usually associated with a ferredoxin-NADP

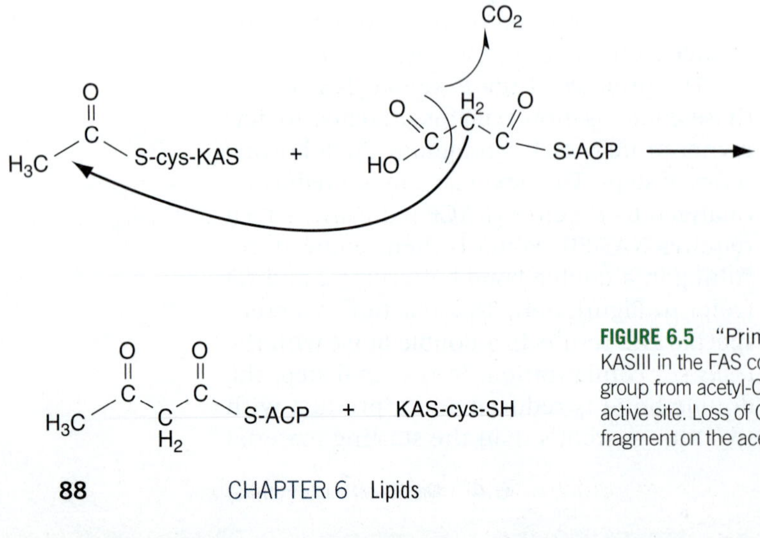

FIGURE 6.5 "Priming" reaction catalyzed by β-ketoacyl-ACP synthase III (KASIII). KASIII in the FAS complex catalyzes the initial condensation of malonyl-ACP to an acetyl group from acetyl-CoA. The acetyl group is transferred to a cysteine residue in the enzyme active site. Loss of CO_2 from malonate is followed by attack of the remaining two-carbon fragment on the acetyl carbonyl group and release of acetoacetyl-ACP from KASIII.

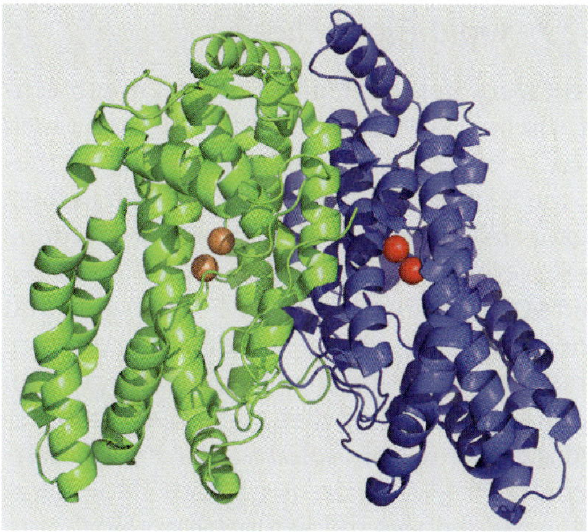

(a)

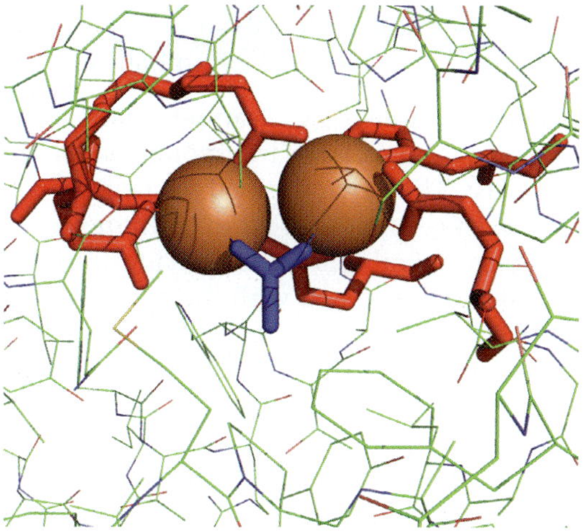

(b)

FIGURE 6.6 Structure of the soluble Δ9 stearoyl-ACP desaturase found in the chloroplast stroma. (a) The enzyme is active as a dimer. Each monomer consists mainly of 11 α-helices that enclose the active site. The two catalytic Fe atoms are shown as red balls. (b) The active site of the enzyme is shown in detail. The two Fe atoms are coordinated to histidine and glutamic acid residues shown as red tubes. The stearoyl moiety of the substrate binds in a groove between the helices that is occupied by a bound acetate molecule (blue) in this structure. (Structure in part a from Protein Data Bank 1oq4. M. Moche, A. K. Ghoshal, J. Shanklin, and Y. Lindqvist, *J. Mol. Biol.* 278 [2003]: 25072–25080. Structure in part b from Protein Data Bank 1oq9. M. Moche, A. K. Ghoshal, J. Shanklin, and Y. Lindqvist, *J. Mol. Biol.* 278 [2003]: 25072–25080.)

reductase and ferredoxin. Ferredoxin is reduced by ferredoxin-NADP reductase and transfers electrons to the di-iron cluster in the desaturase. In a typical monooxygenase, the reduced Fe cluster binds O_2 and transfers one oxygen atom to the substrate and reduces the other O to H_2O. In the chloroplast desaturase, this mechanism is modified to produce a *cis* double bond in stearic acid (**FIGURE 6.7**). The product, 18:1 Δ9c, remains attached to ACP and can be used for lipid synthesis in the chloroplast or transported to the cytoplasm for further processing in the ER.

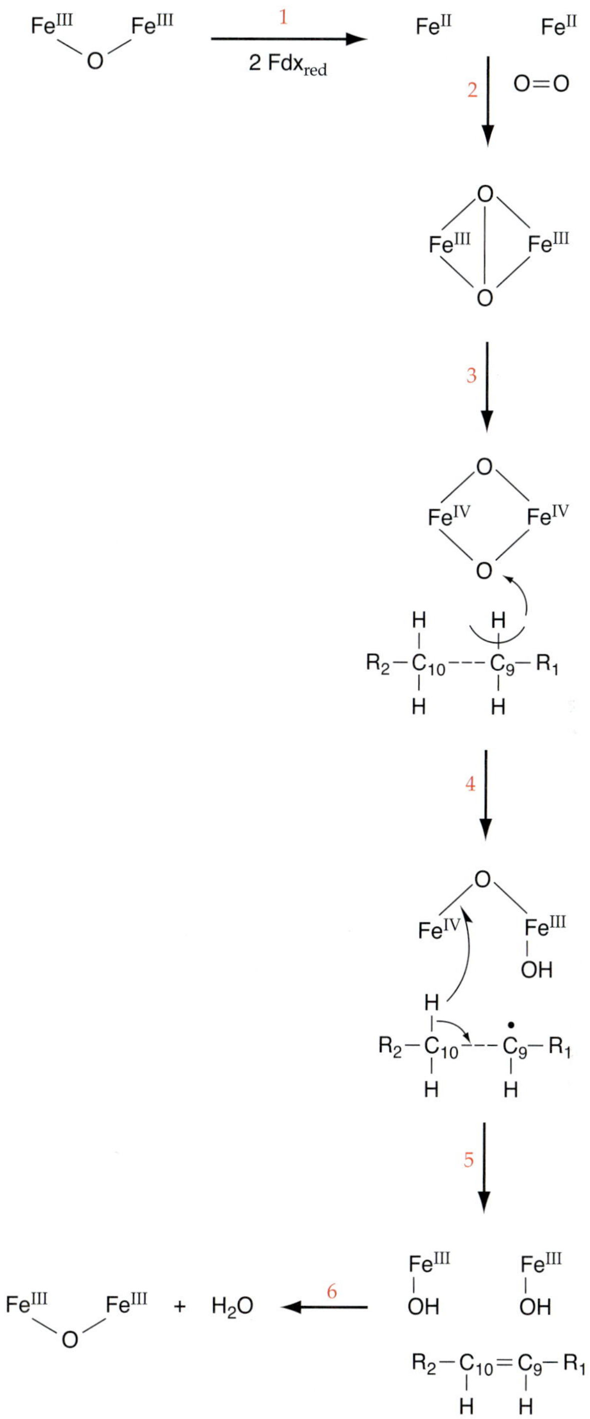

FIGURE 6.7 Proposed reaction mechanism of the soluble Δ9 stearoyl-ACP desaturase. In step 1 the di-iron center at the active site is reduced to its Fe^{II} state by two electrons from two ferredoxins. The reduced desaturase then binds molecular O_2 in step 2. In step 3 the irons are oxidized to a transient Fe^{IV} state by the bound oxygen molecule. The $Fe^{IV}-O_2$ complex abstracts an H atom from the stearoyl substrate, producing a transient radical on C9 in step 4. Abstraction of a second H atom from the substrate results in a C9–C10 *cis* double bond in the product and a hydroxylated di-iron center in the enzyme (step 5). The iron then loses water and returns to the initial oxidized Fe^{III} oxidized state in step 6.

6.6 Desaturation of Fatty Acids in the Endoplasmic Reticulum

The desaturases in the ER are integral membrane proteins found in a variety of organisms, including plants. Analysis of membrane proteins is more difficult than soluble proteins, and no crystal structure is available as yet for these desaturases. Gene analyses have provided information on these enzymes and their probable mechanism of action. Similar to the soluble desaturase, the ER protein contains two Fe atoms that are coordinated to histidine residues. The ER desaturases contain eight histidine residues, seven of which are strictly conserved and occur in three clusters referred to as His boxes. These boxes are separated by hydrophobic regions that presumably span the ER membrane and bring the boxes and Fe ions into proximity on the cytoplasmic face of the membrane (**FIGURE 6.8**). Membrane desaturases are usually found in a complex with a cytochrome b_5 and an NADPH-cytochrome b_5 reductase. In some organisms this complex consists of a fusion protein of the cytochrome at the N-terminal region of the desaturase. As in the soluble desaturase the iron atoms are reduced, in this case by the cytochrome, and then bind molecular oxygen. ER-desaturases act on a variety of fatty acids that are esterified to glycerol. The mono- and polyunsaturated products form triacylglycerols (TAGs), which are stored as oil bodies. Alternatively, they are further processed into phospholipids (diacylglycerols [DAGs]) for membrane synthesis. In addition, some DAGs are transported back to the plastids for further modification in these organelles.

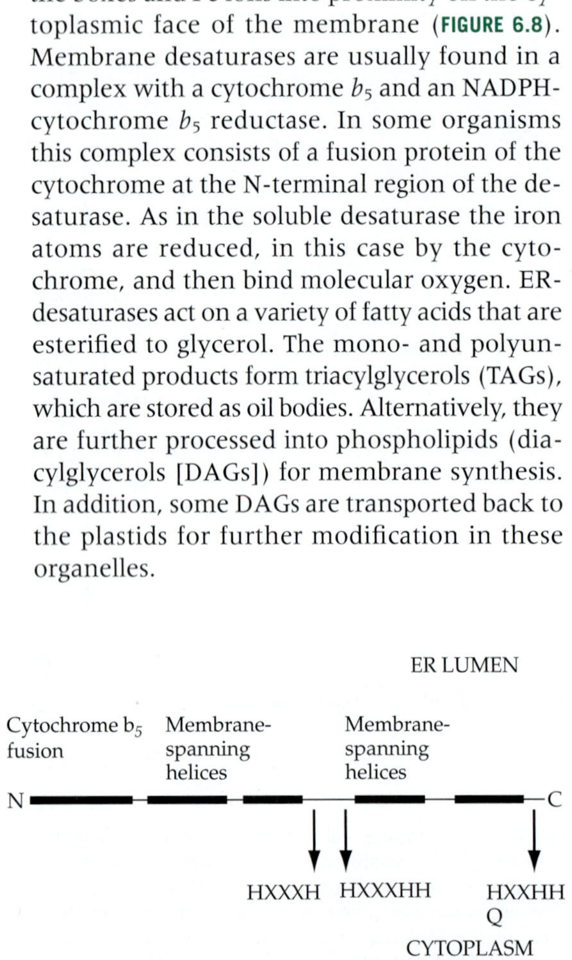

FIGURE 6.8 Diagram of the proposed structure of an integral membrane fatty acyl desaturase. The enzyme's active site consists of three conserved histidine-rich sequences (His boxes) separated by putative membrane-spanning helices. These helices traverse the ER bilayer and bring the His boxes and associated iron atoms together on the cytoplasmic surface of the membrane. ER desaturases are often synthesized as fusion proteins with a *b*-type cytochrome at the N-terminus. The fused cytochrome transfers electrons from NADPH to reduce the iron atoms at the active site.

6.7 Lipid Biosynthesis

Early work in the middle of the twentieth century on lipid synthesis was done in bacteria such as *Escherichia coli*. The basic steps of TAG synthesis, the Kennedy pathway, are common to most organisms. The initial work in plants was done by the group of P. Stumpf in the 1960s. Most lipid synthesis occurs in the ER in plants and other organisms, although plants also produce some membrane lipids in plastids. The glycerol required for lipid synthesis is derived from dihydroxyacetone phosphate (DHAP), an intermediate in glycolysis. An NADH-dependent reductase reduces the ketone to an alcohol to yield glycerol-3-P.

$$\text{glycolysis} \rightarrow \text{DHAP} + \text{NADH} \rightarrow \text{glycerol-3-phosphate} + \text{NAD}$$

Saturated and monounsaturated fatty acids are cleaved from ACP and transported out of the chloroplast. They are esterified to coenzyme A by an acyl-CoA synthetase located in the outer chloroplast membrane. The enzymes of lipid synthesis are associated with the outer face of the ER. An acyl-CoA:glycerol-3-phosphate acyltransferase usually adds an 18:0 or 18:1 to the sn-1 position of glycerol-3-P. An acyl-CoA:lysophosphatidic acid acyl transferase esterifies an unsaturated fatty acid to the sn-2 position, resulting in the synthesis of phosphatidic acid (**FIGURE 6.9**).

Two possible modifications to phosphatidic acid yield lipids with different functions. The phosphate can be removed and the third position esterified to a third fatty acid to yield TAGs, or the phosphate can be removed and replaced with a "head" group such as choline. In this latter pathway, the phosphate moiety in phosphatidic acid is hydrolyzed by a phosphatase. This hydrolysis is coupled to addition of an activated head group such as CDP-choline to yield phosphatidylcholine (Figure 6.9).

$$\text{Phosphatidic acid} \xrightarrow{\text{phosphatase}}$$

$$\text{DAG} + \text{CDP-choline} \xrightarrow{\text{synthase}} \text{phosphatidylcholine} + \text{CDP}$$

Alternatively, diacylglycerol can be activated by addition of CDP and then react with a "head" group.

$$\text{DAG} + \text{CTP} \rightarrow \text{CDP-DAG} + \text{choline} \rightarrow \text{phosphatidylcholine} + \text{CDP}$$

These modified DAGs—phosphatidylcholine, phosphatidylethanolamine, and phosphatidylinositol (Figure 6.9)—are amphipathic lipids,

FIGURE 6.9 Structures of some common phospholipids found in plants. Initial DAG synthesis in most plants results in phosphatidic acid (PA), which consists of ester linkages to two fatty acids at the sn-1 and sn-2 positions in glycerol and a phosphoester at sn-3. PA, which is synthesized in the ER, usually has an 18:0 or 18:1 fatty acid at the sn-1 position and an 18:1 moiety at sn-2. When synthesized in the plastid, PA generally has an 18:1 acid at sn-1 and a 16-carbon fatty acid at sn-2. The third position is then modified by other groups such as ethanolamine, choline, or inositol. Phosphatidylethanolamine (PE) is found in cellular membranes other than the plastid, whereas phosphatidylcholine (PC) and phosphatidylinositol (PI) are common in all membranes. Sulfolipids and galactolipids are unique to thylakoids and the inner membranes of chloroplasts. The nonglycerol sphingolipid is found in plasma membranes.

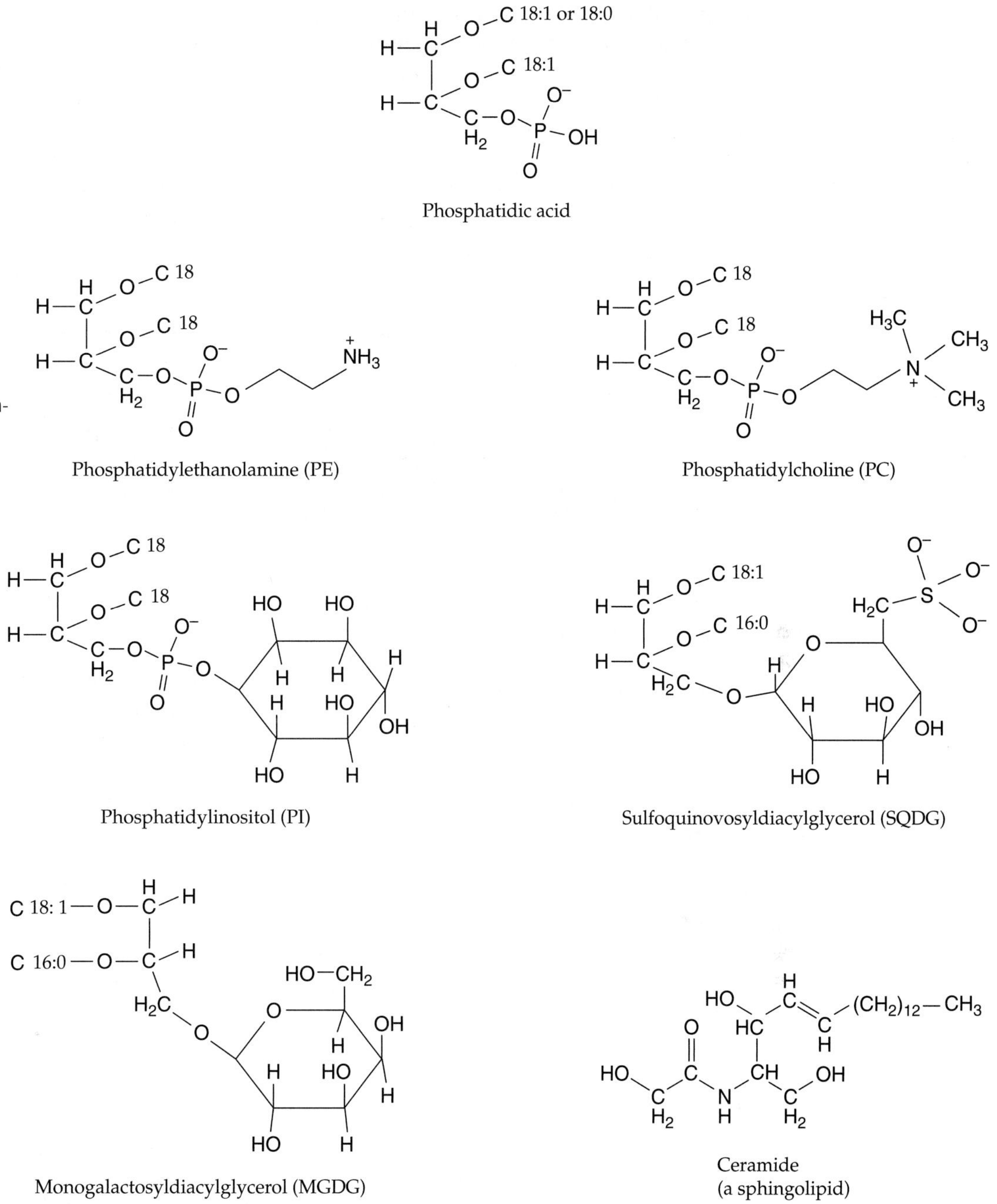

having both a polar, charged moiety at the sn-3 position and two nonpolar hydrophobic fatty acid moieties. These modified DAGs make up most of the cell's membrane bilayers. Further modification of the fatty acids in these lipids can subsequently be catalyzed by ER desaturases, elongases, and other enzymes.

Lipid synthesis also occurs in plastids in a slightly different fashion. The plastid acyl transferases utilize the ACP conjugates of fatty acids and esterify these to glycerol-3-P.

In plastids, an 18:1 fatty acid is usually esterified to the sn-1 position of glycerol-phosphate, and a 16-carbon fatty acid is esterified to sn-2. Galactolipids are unique to thylakoid membranes and are synthesized from CDP-activated DAG and UDP-galactose to yield monogalactosyldiacylglycerol (MGDG, Figure 6.9).

$$\text{UDP-glucose} \xrightarrow{\text{epimerase}}$$

$$\text{UDP-galactose} + \text{CDP-DAG} \xrightarrow{\text{MGDG synthase}}$$

$$\text{1,2-diacyl-3-O-}\beta\text{-galactosylglycerol} + \text{CDP}$$

A second galactose can be added to this galactose moiety by an α-1,6 bond to produce a digalactosyl lipid. Galactolipids comprise up to 85% of the bilayer of thylakoid membranes. They are also abundant in the inner chloroplast membrane.

Sulfolipids are another unusual membrane lipid produced in plastids. Sulfite is added to UDP-glucose to yield the sugar-nucleotide, UDP-sulfoquinovose, which is then added to the sn-3 position of activated diacyl glycerol.

$$SO_3^{2-} + \text{UDP-glucose} \xrightarrow{\text{synthase}}$$

$$\text{UDP-sulfoquinovose} + \text{CDP-DAG} \xrightarrow{\text{SQDG synthase}} \text{1,2-diacylglycerol-3,6-sulfo-}\alpha\text{-quinovosyl-glycerol (SQDQ)}$$

Both sulfolipids and galactolipids help to maintain the high degree of fluidity found in photosynthetic membranes. They also serve to minimize the amount of phosphate ions that must be bound into chloroplast membranes.

Small amounts of other lipids, such as cardiolipins and sphingolipids, are found in plant membranes. Cardiolipin (diphosphatidylglycerol) is characteristic of the inner membrane of the mitochondrion and may be synthesized in this organelle. It is the result of esterification of two molecules of CDP-DAG to glycerol. Sphingolipids are found mainly in the outer cell (plasma) membranes. They are adducts of fatty acids and sphingosine rather than glycerol (Figure 6.9).

6.8 Oil Composition of Seeds

TAGS are the major form of stored carbon synthesized in seeds and fruits. In TAGS the sn-3 position in glycerol is esterified to a third fatty acid, usually an 18:0 or 18:1. The fatty acid moieties of TAGS can then be further modified by the ER desaturases or other enzymes. TAGS are packaged and stored in lipid bodies. The membranes of these organelles consist of a single layer of a DAG such as phosphatidyl choline. Small proteins called oleosins are targeted to the lipid surface of the oil bodies. These proteins prevent oil bodies from coalescing, thus maintaining smaller packages that are more readily processed during seed germination.

Most seed lipids contain high contents of unsaturated fatty acids, but the composition can be highly variable (**TABLE 6.2**). For example, the "other" in coconut oil is lauric acid (12:0); developing seed tissues in coconut have a specific acyl-ACP thiolase that releases this fatty acid from the fatty acid synthase complex after six rounds of condensation. This high content of 12:0 makes coconut oil particularly fluid and a commercially important material for the manufacture of soaps.

Castor bean (*Ricinis communis*) is unique in that most fatty acids in its seeds are hydroxylated to produce ricinoleic acid (12OH 18:1 Δ9c), the major ingredient in castor oil (**FIGURE 6.10**).

$$\text{TAG-18:1 } \Delta 9c + O_2 + \text{NADPH} \rightarrow \text{TAG-12-OH-18:1 } \Delta 9c + H_2O + \text{NADP}$$

The enzyme that catalyzes this hydroxylation is a modified desaturase that acts on oleic acid moieties bound to glycerol in TAGs. Because this modified lipid occurs mainly in seeds, it must be specifically transported from the site of modification in ER membranes of maternal tissues to oil bodies in seeds. Castor oil is used industrially in the manufacture of paints and lubricants. Unfortunately, large-scale cultivation of *R. communis* to obtain this oil is problematic because the plants also produce toxic compounds such as the protein, ricin. Attempts to transfer the gene for the modified desaturase to other plants have not resulted in higher yields of ricinoleic acid in transgenic seeds. These results indicate that more than one protein is required for effective synthesis, transport, and storage of the modified lipids.

Minor changes in the amino acid residues in the active site of ER desaturases also lead to the production of a variety of modified fatty acids found in TAGS that are useful to humans. For example, an acetylenase found in some plants modifies linoleic acid to crepenynic acid.

$$\text{Linoleic acid (18:2 } \Delta 9,12c) \rightarrow \text{Crepenynic acid (18:2 } \Delta 9c\text{, 12 acetylene)}$$

Table 6.2 Fatty Acid Composition of Common Storage Oils (% of Total Lipids)

Plant	16:0	18:0	18:1	18:2	18:3	Other
Avocado	20	1	60	18	0	1
Canola	4	1	54	23	10	8
Castor bean	1	1	3	4	Trace	91
Cocoa butter	27	34	35	3	Trace	1
Coconut	9	2	7	2	0	80
Maize	14	2	30	50	2	2
Olive	12	2	72	11	1	2
Soybean	11	4	22	53	8	2
Sunflower	6	5	19	68	Trace	2

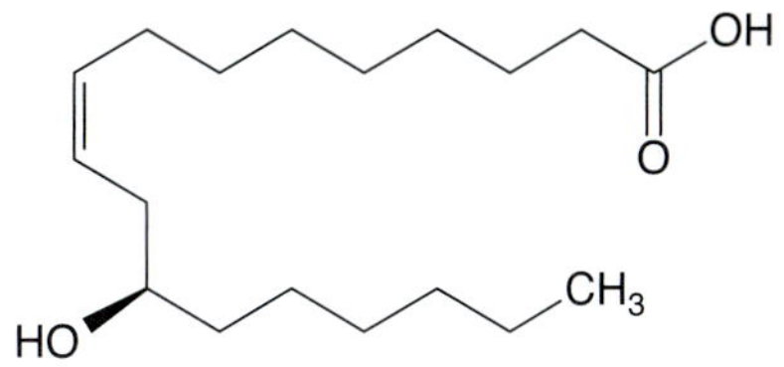

Ricinoleic acid
12OH-18:1 D9c

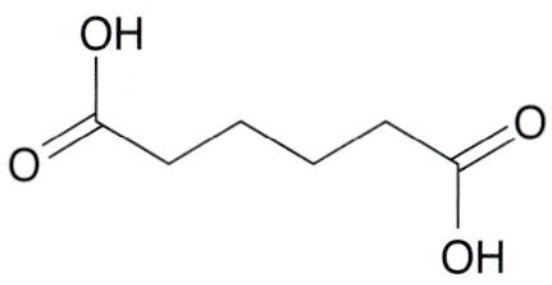

Adipic acid

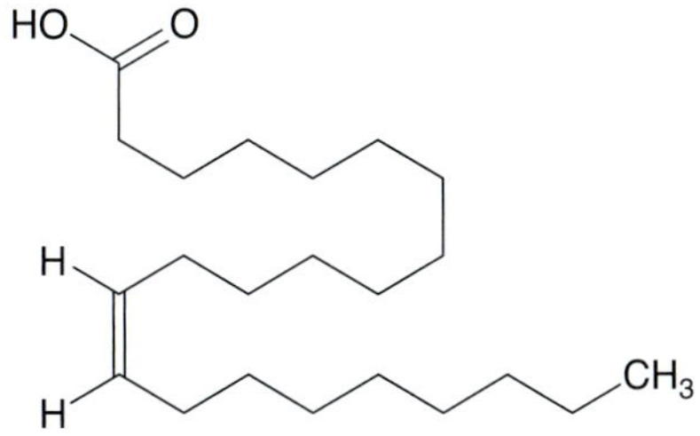

Erucic acid
22:1 D13c

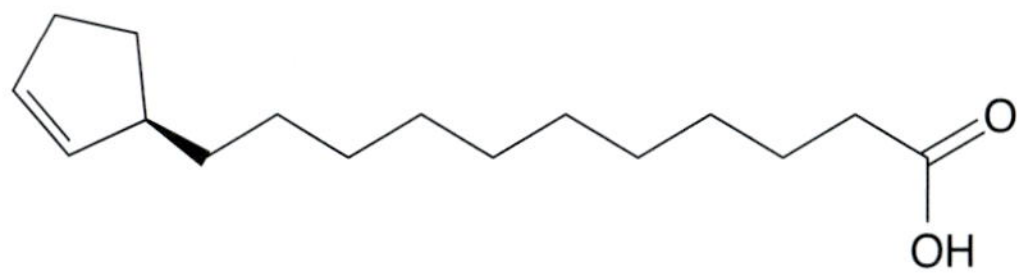

Hydnocarpic acid
11-(2-cyclopenten-1-yl) 11:0

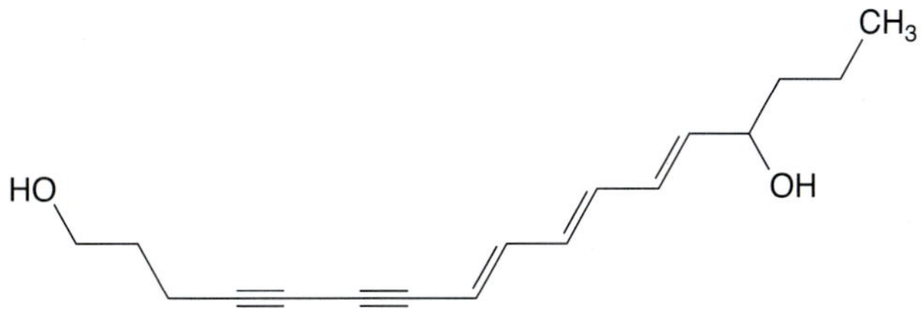

Cicutoxin

FIGURE 6.10 Some unusual fatty acids and derivatives found in plants. Desaturases have undergone various mutations in their active site residues to produce the C12-hydroxylated ricinoleic acid from oleic acid. Triple bonds as in cicutoxin (water hemlock, *Cicuta virosa*) are the product of a desaturase modified to add a triple bond to an unsaturated fatty acid (acetylenase). Cyclases produce the cyclopentyl ring seen in hydnocarpic acid found in chaulmoogra oil (*Hydnocarpus wightiana*). Elongases catalyze the formation of the 20-carbon erucic acid from oleic acid and malonyl CoA. Erucic acid is found in the seeds of some cultivars of *Brassica napus*. It is used as a lubricant and precursor of adipic acid, which is used commercially in the manufacture of plastics.

Crepenynic acid is a common constituent in the seeds of goat's beard (*Crepis rubra*). It can be converted into a highly effective cold-tolerant lubricant for machinery that must operate at low external temperatures. Epoxydases introduce an epoxide group in linoleic acid moieties in triglycerides.

Linoleic acid (18:2 Δ9,12c) →
Vernolic acid (12,13 epoxy 18:1 Δ9c)

Vernolic acid is found in seeds of *Veronica* sp. and in some varieties of soybeans (*Glycine max*). It can be used in the manufacturing of plasticizers. Additional unusual fatty acid derivatives include conjugated and cyclopentyl modifications and are shown in Figure 6.10. Increased cost and depletion of petroleum supplies have stimulated major efforts to engineer over-production of these types of fatty acids in plants. A more notorious acetylenic derivative is cicutoxin (1,14OH 17:5 Δ4a, 6a, 8c, 10c, 12c), the toxic component found in the storage organs of water hemlock (*Cicuta virosa*). In this case the original linoleic acid has undergone several different modifications. The synthesis of these unusual modified fatty acids may be fortuitous in some plants. However, the toxicity of derivatives such as cicutoxin suggests that lipid modifications may make the storage organs and seeds resistant to herbivory and attack by pathogens.

6.9 Very-Long-Chain Polyunsaturated Fatty Acids

Very-long-chain polyunsaturated fatty acids (VLC-PUFAs) are fatty acids with 20 or more carbons and three or more double bonds. They are normal constituents of membranes in some algae and fungi. Because algae form the base of the food chain in aquatic systems, VLC-PUFAs are commonly found in fish oils. These omega fatty acids are important in human health and include omega-3 eicosapentaenoic acid (20:5 Δ5,8,11,14,17c) and omega-6 arachidonic acid (20:4 Δ5,8,11,14c). PUFAs are the precursors for the hormones prostaglandins, leukotrienes, and thromboxanes, which mediate fever, inflammation, blood pressure, and heart function. Humans can synthesize these from linoleic and linolenic fatty acids that, in turn, must be obtained from eating plants. Dietary sources of PUFAs such as fatty fish are also recommended.

The VLC-PUFAs are synthesized from 18-carbon fatty acids by sequential action of elongases and modified desaturases. Elongases are ER-bound enzymes with a mechanism of action similar to enzymes in *de novo* fatty acid synthesis. The elongases add malonyl-CoA to existing fatty acids to produce 20 or more carbon chains that are subsequently modified. For example, synthesis of eicosapentaenoic acid (20:5 Δ5,8,11,14,17) from α-linolenic acid (18:3) occurs in three steps:

$$18{:}3 \xrightarrow{\Delta 6 \text{ desaturase}} 18{:}4 \xrightarrow{\text{elongase + malonyl-CoA}} 20{:}4 \xrightarrow{\Delta 5 \text{ desaturase}} 20{:}5$$

PUFA supplements sold in health food stores and added to certain processed foods are made by fungal (*Mortieriella alpina*) cultures. Attempts have been made to engineer crop plants with genes for elongases and desaturates from microorganisms such as cyanobacteria. Genetically modified plants can produce PUFAs, but these oils do not accumulate in seeds. These results again emphasize a lack of basic information on lipid transport into developing seeds. Synthesis and transport engineering are needed to make VLC-PUFA synthesis in plants into a viable alternative to fish oils.

One crop plant that does produce a long-chain fatty acid in its seed oils is *Brassica napus*, rapeseed. The seeds accumulate high levels of monounsaturated erucic acid (22:1 Δ13c) produced by elongases in the ER. Erucic acid is used commercially in the production of lubricants and adipic acid but is toxic in oils consumed by animals. In this case the elimination of the elongase gene in *B. napus* led to the production of a valuable seed-oil crop called canola.

6.10 Hydroperoxy Fatty Acids and Oxylipin Synthesis

Attack by herbivores or pathogens that penetrate the plant cell membrane initiates the synthesis of hydroperoxy fatty acids from unsaturated fatty acids. The hydroperoxide formation is catalyzed by lipoxygenases, which are a type of dioxygenase. Like monooxygenases, dioxygenases bind molecular oxygen but incorporate both oxygen atoms into the substrate to yield the product. Plant lipoxygenases (LOXs) have a mass of approximately 100 kDa and, like the desaturases, have nonheme iron in the active site that is reduced by an NADPH-requiring reductase. The iron atoms are liganded to four histidine residues. Lipoxygenases have two domains, an N-terminal beta barrel domain that binds to the periphery of the lipid bilayer and an alpha helical catalytic domain that binds, linoleic (18:2), or α-linolenic acid (18:3). As in the desaturases, a free radical intermediate is generated. Both atoms of O_2 are added to C9 in linoleic or C13 in linolenic acid to generate a hydroperoxy fatty acid. The hydroperoxy group is unstable, and this fatty acid derivative undergoes enzymatic cleavage by a hydroperoxide lyase. Spontaneous chemical rearrangements then generate compounds called oxylipins (**FIGURE 6.11**). Several products are possible, such as the volatile compounds hexenal and hexanol, which account for the "green odor" of cut leaves and stems (**FIGURE 6.11a**). Green leaf volatiles are signaling molecules that attract both herbivorous insects and carnivorous insects that prey on the herbivores. It has also been reported that these volatile products can stimulate defense responses in other tissues of the wounded plant and may even alert nearby plants to the presence of herbivores and initiate defense responses in these neighbors.

A series of enzymatic reactions also generates the plant hormone, jasmonic acid from the hydroperoxy derivative (**FIGURE 6.11b**). Jasmonates are known to function in plant immunity. The initial enzyme involved in the synthesis of this oxylipin, allene oxide synthase, is another type of monooxygenase. The enzymes that act on the hydroperoxy derivatives belong to a large family of cytochrome P450 monooxygenases that contain iron in a heme group that is covalently bound to a cysteine residue in the protein. The heme group has an absorption maximum at 450 nm when carbon monoxide is bound to the iron. The P450 enzymes in oxylipin synthesis are unusual in that they utilize the oxygen atoms that have already been incorporated into the fatty acid by the lipoxygenase reaction and therefore do not require addition of molecular oxygen or a reducing agent. In the jasmonic acid pathway, the epoxide intermediate produced by the allene oxide monooxygenase is then cyclized at C12–C13 by a specific cyclase and reduced in a reaction that requires NADPH. Three cycles of β-oxidation, which remove two carbons at a time from the carboxylic acid end of the intermediates, result in a 12-carbon isojasmonic acid. Isojasmonate is subsequently rearranged to form the biologically active *R,S* isomer. The side chain

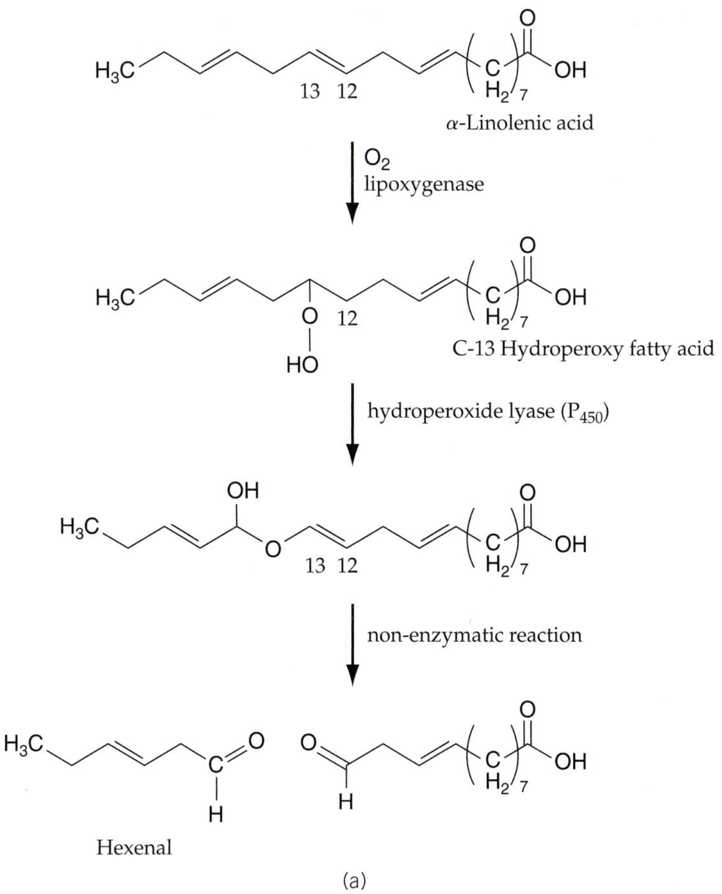

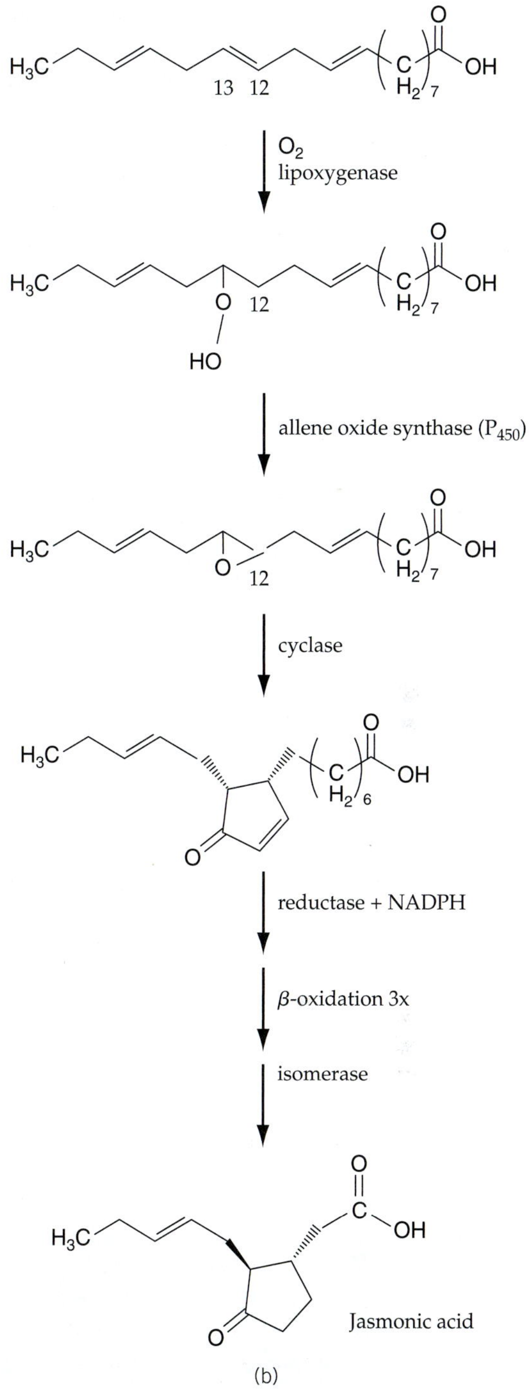

FIGURE 6.11 Synthesis of oxylipins from a-linolenic acid in membrane lipids. (a) Green leaf volatiles. After cleavage from a membrane phospholipid, α-linolenic acid is modified by a ΔC13 lipoxygenase (a dioxygenase), which adds a molecule of O_2 to C13 of the C12–C13 double bond, producing an unstable peroxy intermediate. This intermediate is a substrate for hydroperoxide lyase, a cytochrome P450 monooxygenase. Both oxygen atoms are incorporated into an unstable product that spontaneously decomposes to a number of short chain compounds such as hexenal. These fatty acid derivatives are known as green leaf volatiles. (b) Jasmonic acid. As in A, a lipoxygenase catalyzes the addition of O_2 to α-linolenic acid that has been cleaved from a membrane phospholipid by a lipase. Allene oxide synthase, a cytochrome P450 monooxygenase, catalyzes a C12–C13 epoxidation of the fatty acyl peroxy intermediate. A specific cyclase then catalyzes formation of a cyclopentyl ring, followed by reduction of a double bond in the ring. Three cycles of β-oxidation (see Figure 6.14) remove six carbons from the side chain to yield a 12-C intermediate. Isomerization of this product results in one of the isomers of the plant stress hormone, jasmonic acid.

of jasmonates is usually modified by methylation, sulfation, or addition of a sugar or amino acid. The conjugate with isoleucine has been shown to interact with a jasmonate receptor and is most likely the biologically active form.

6.11 Cutin and Suberin

Plant cuticles are hydrophobic layers of polymeric material that is insoluble in organic solvents. The cuticle is synthesized by nonphotosynthetic epidermal cells that cover leaves, stems, and other organs. It constitutes a barrier to pathogens and controls water movement. The cuticle consists mainly of two types of fatty acid derivatives, cutin and cuticular wax.

Fatty acid synthesis in epidermal cells is initiated in the nongreen plastids. Subsequent processing of the fatty acids into cuticle polymers takes place on the ER. Elongases act on 18-carbon fatty acids to yield products with chain lengths from 24 to 32 carbons. Some of these elongated acids are then reduced to their corresponding alcohols. These alcohols are esterified to fatty acids and form long-chain polymers that constitute cuticular waxes. The long-chain fatty acids are also the precursors for alkane biosynthesis. A decarbonylation reaction (loss of CO) results in a chain with an odd number of carbon atoms. In some plants these alkanes are exported for cutin biosynthesis. In other plants they are modified by oxygenases to yield secondary alcohols that may be further oxidized to ketones.

It is assumed that these various monomers are transported to the cell exterior in Golgi vesicles that fuse with the cell membrane. The hydrophobic monomers, however, may remain bound to the inner face of the cell membrane, and enzymes called "flippases" are required to release them to the outside. Alternatively, the fatty acid derivatives may remain attached to specific sites in the ER that subsequently fuse with the plasma membrane and the acids are then exported via transport proteins in the cell membrane. Polyester synthesis then would take place on the outer face of the primary cell wall; however, the enzymes required for extracellular transport and processing have not been characterized. The final cutin layer consists of random ester linkages among these various fatty acid derivatives. This provides an additional barrier to microbial and insect attack and water loss (**FIGURE 6.12**).

Suberin is a polymer similar to cutin, but in addition to cutin polymers and waxes, it contains phenolic compounds such as ferulic acid and other monomers characteristic of lignin, which is a component of secondary plant walls. Wounding induces suberin polymerization outside the cell walls on the plant surface, but suberin is also found normally on some internal plant tissues, such as the root endoderm. Suberin biosynthesis also requires elongation of the 18-carbon fatty acids before modification. In addition to esterases that catalyze cross-links among the fatty acids, feruloyl transferases are active in making bonds to the phenolic monomers. In some plants suberin seems to be covalently bonded to the cell wall, although a specific linkage has been difficult to characterize.

6.12 Catabolism of TAGs

The biochemical steps in the breakdown of lipids for energy have been studied extensively in germinating seeds such as castor bean and soybean. Recent genetic analysis in *Arabidopsis* has facilitated the characterization of the enzymes and transport proteins involved in this process. Lipases, which are associated with membranes of oil bodies, ER, and glyoxysomes, cleave the ester bond between fatty acids and glycerol. The free glycerol is phosphorylated by a cytoplasmic kinase to glycerol-3-phosphate and oxidized to dihydroxy acetone-phosphate, an intermediate in glycolysis. In developing seeds the glycolysis pathway is reversed to generate sugars such as sucrose that are transported to the embryo to support heterotrophic growth.

After cleavage from glycerol, free fatty acids are esterified to CoA by a long-chain acyl-CoA synthetase.

$$\text{Fatty acid} + \text{CoA} + \text{ATP} \rightarrow \text{Fatty acyl-CoA} + \text{AMP} + \text{PPi}$$

The reaction requires the activation of the fatty acid by esterification to AMP with release of pyrophosphate that is then cleaved to two inorganic phosphate ions. Hydrolysis of the fatty acyl-AMP phosphoester bond facilitates transfer to the thiol of CoA.

The steps of fatty acyl-CoA catabolism occur in glyoxysomes. These organelles are peroxisomes specifically modified to degrade fatty acids and transport the resulting products to the cytoplasm for sugar synthesis or to the mitochondria for energy generation via aerobic respiration. Glyoxysomes are often seen adjacent to oil bodies in the seed endosperm (**FIGURE 6.13**).

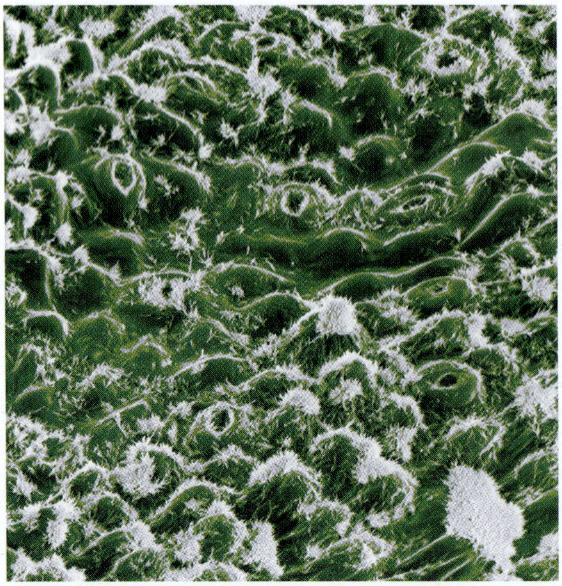

FIGURE 6.12 Scanning electron micrograph of the surface of a *Eucalyptus* leaf. Accumulations of waxes are observed on the surface of the leaf above the cutin layer. The waxes and fatty alcohol precursors are secreted by leaf epidermal cells and form random three-dimensional polymers. These wax formations provide protection from insect herbivores and fungi. Magnification ~9,200x.

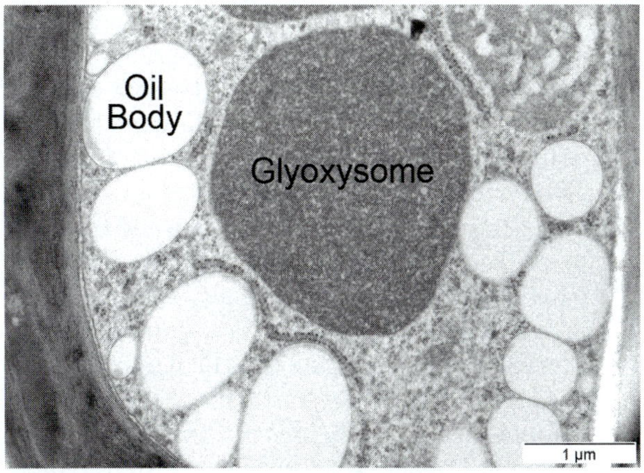

FIGURE 6.13 Transmission electron micrograph of an *Arabidopsis* seed section showing a glyoxysome in proximity to several oil bodies.

In the glyoxysome, the fatty acyl-CoA esters are oxidized by a specific acyl-CoA dehydrogenase that uses an FAD cofactor.

Acyl-CoA + FAD → 2-*trans*-enoyl-CoA + $FADH_2$

The reduced FAD cofactor generated in this reaction is reoxidized by molecular oxygen, producing hydrogen peroxide (H_2O_2). Two enzyme systems are present in glyoxysomes that detoxify peroxide. Catalase is a soluble enzyme that converts H_2O_2 to H_2O and O_2. Ascorbate peroxidase is a membrane-bound enzyme that oxidizes peroxide and reduces ascorbic acid to dehydroascorbate. A membrane-associated NADH-dehydroascorbate reductase then reoxidizes ascorbate.

$$\text{Ascorbic acid} + H_2O_2 \xrightarrow{\text{ascorbate peroxidase}} \text{Dehydroascorbate} + 2\ H_2O$$

$$\text{Dehydroascorbate} + \text{NADH} \xrightarrow{\text{ascorbate dehydrogenase}} \text{Ascorbate} + \text{NAD}$$

The resulting double bond in the 2-*trans*-enoyl-CoA ester of the oxidized fatty acid is hydrated to an alcohol and then oxidized to a ketone by a multifunctional protein with hydratase and dehydrogenase activities. A thiolase breaks the C–C bond, liberating acetyl-CoA, and esterifies the resulting acyl fragment to another CoA. The sequence of reactions resembles the *de novo* synthesis of fatty acids in reverse (**FIGURE 6.14**). Each round of β-oxidation results in one molecule of acetyl-CoA, one NADH, and a fatty acid that is two carbons shorter than the original. The sequence of reactions is repeated until the entire fatty acid is converted to acetyl-CoA.

The product of β-oxidation, acetyl-CoA, can be transported to the mitochondria and catabolized in the citric acid cycle. In germinating

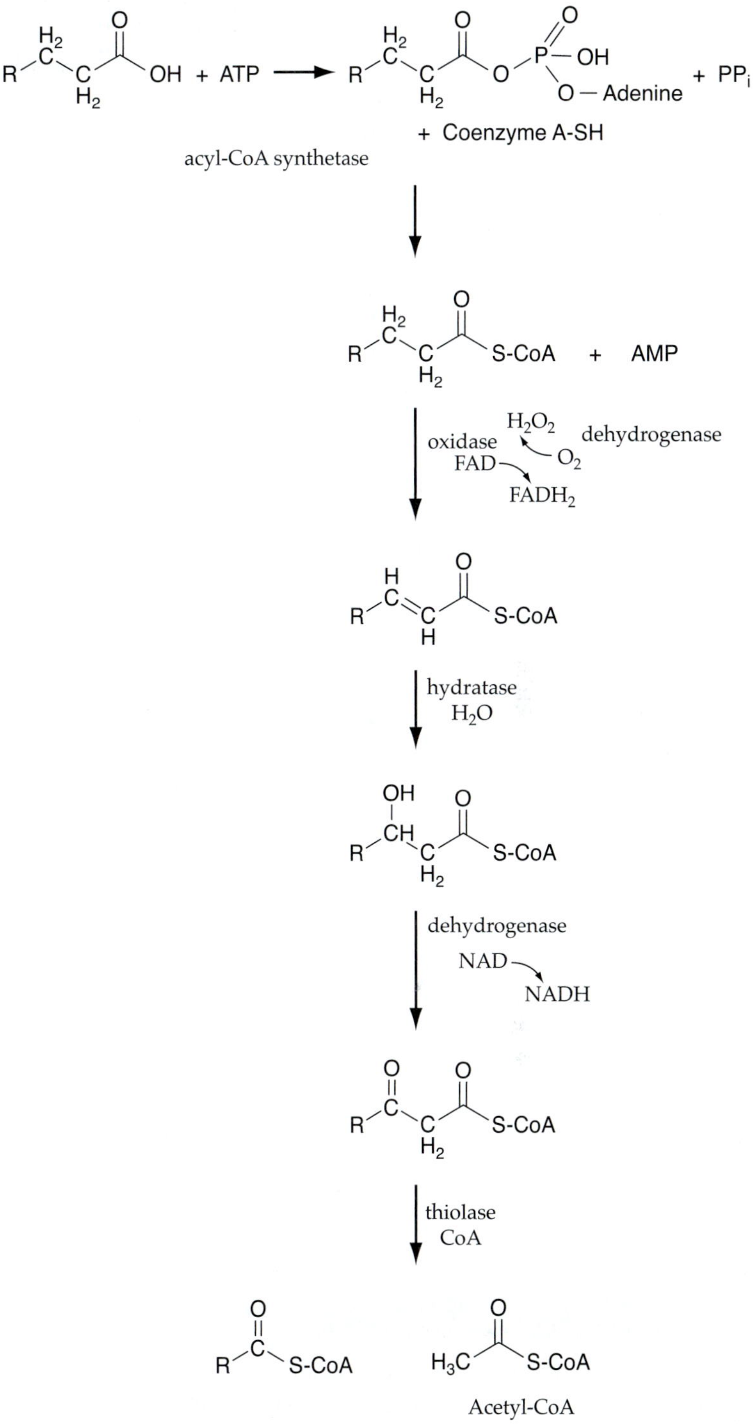

FIGURE 6.14 β-Oxidation of saturated fatty acids. In the glyoxysome, an acyl-CoA synthetase catalyzes the activation of free fatty acids to their CoA derivatives. This reaction is endothermic and proceeds *via* an adenylated intermediate. The pyrophosphate product is cleaved by a pyrophosphatase into two molecules of inorganic phosphate. The fatty acyl-CoA is then oxidized to a 2-*trans*-enoyl intermediate. The oxidase that catalyzes this reaction uses an FAD cofactor. Reoxidation of the cofactor generates hydrogen peroxide, which is removed by enzymes in the glyoxysome. A multifunctional enzyme then acts on the double bond of the acyl intermediate. The hydratase activity of the enzyme adds H_2O to the double bond and the dehydrogenase activity oxidizes the alcohol to a ketone. A thiolase cleaves a two-carbon fragment from the carboxyl end of the fatty acid and esterifies the remainder to another CoA molecule. This results in one molecule of acetyl-CoA and a fatty acyl-CoA that is two carbons shorter than the starting acid. The cycle repeats until the fatty acid is completely converted to acetyl-CoAs.

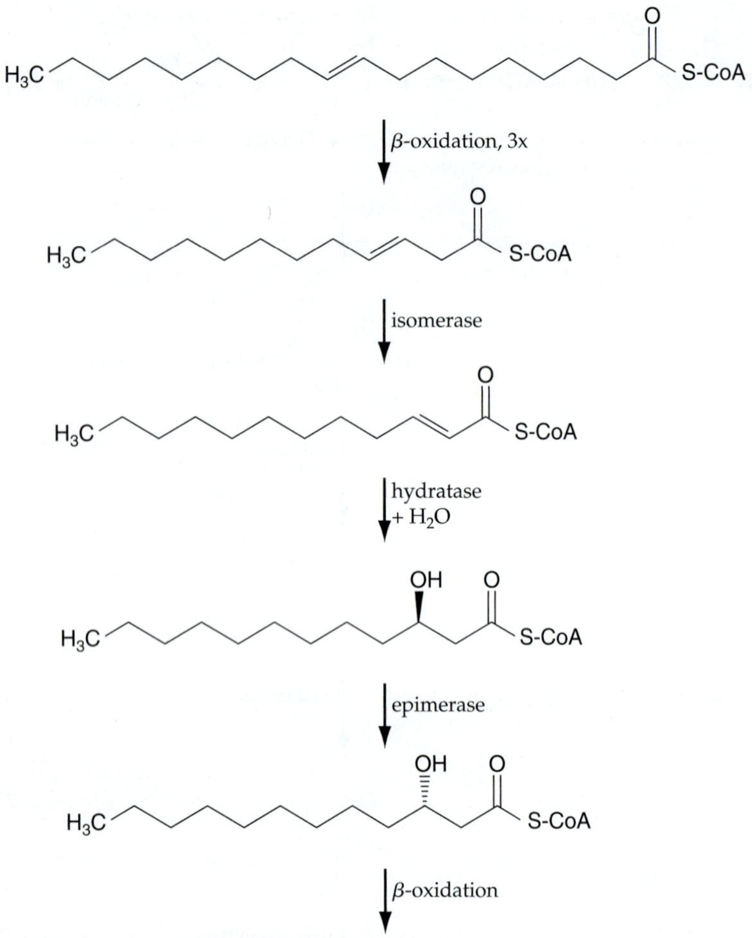

FIGURE 6.15 Catabolism of unsaturated oleic acid (18:1 D9c). Catabolism proceeds initially by β-oxidation as outlined in Figure 6.14 to cleave the first six carbons to produce three acetyl-CoAs. The double bond at C3 in the remaining 12 carbon intermediate is moved to C2 by an isomerase. A hydratase catalyzes the formation of an OH group on C3 that must be epimerized to the configuration recognized by the dehydrogenase in the β-oxidation sequence. After epimerization, β-oxidation proceeds to completion as shown in Figure 6.14.

seeds, however, acetyl-CoA can be converted to sugars (gluconeogenesis), a process that does not occur in animal cells. A malate synthase enzyme is unique to glyoxysomes. It catalyzes the condensation of acetyl-CoA from β-oxidation with glyoxylate to produce malate. This four-carbon acid is transported to the cytoplasm and oxidized to oxaloacetic acid, generating reduced NAD. The oxaloacetate can reenter the glyoxysome and is now reduced to malate, a reaction that is required to regenerate oxidized cofactor for the dehydrogenase step in β-oxidation. Malate in the cytoplasm can also be decarboxylated to phosphoenolpyruvate. A reversal of the steps of glycolysis from phosphoenolpyruvate leads to the production of glucose. The soluble sugars are then used for sucrose synthesis and transported to the germinating embryo.

Catabolism of unsaturated fatty acids requires a modification of the basic scheme for β-oxidation to remove the double bonds. The simplest example is the catabolism of oleic acid, 18:1 Δ9c (**FIGURE 6.15**). Initially, the fatty acid undergoes three cycles of β-oxidation as described above to yield a 12-carbon intermediate, 12:1 Δ3c. An isomerase then converts this to 12:1 Δ2c. Action of a hydratase produces the corresponding hydroxy intermediate that must be acted on by an epimerase to generate the correct *trans* configuration. The resulting 12-C fragment can then undergo complete catabolism by β-oxidation. Multiple activities of these enzymes are required to remove double bonds in PUFAs. The overall energy yield is somewhat lower than for breakdown of saturated fatty acids. The products are the same as in the catabolism of saturated fatty acids, however, and are generally used in gluconeogenesis.

CHAPTER SUMMARY

Lipids are esters of fatty acids and glycerol. Modified DAGs are essential components of membranes, and TAGs are the main energy storage compounds in living organisms. In plants, the *de novo* synthesis of fatty acids and lipids involves a series of enzymes found in the stroma of the plastids and integral membrane proteins in the ER and requires transfers of components and coordination of activities between organelles. Plants synthesize both saturated and unsaturated fatty acids. The latter requires catalysis by modified monooxygenases called desaturases. In addition to the formation of double bonds in fatty acids, modified plant desaturases also catalyze the production of a number of unusual fatty acids containing hydroxyl, epoxy, acetylenic, and other chemical groups not found in the fatty acids of animal cells. The enzymes that catalyze these reactions are all related to the basic ER desaturases and provide a particularly compelling example of protein evolution. Gene duplication, accompanied by minor changes in a few active site residues, provides an enzyme that acts on a primary metabolite, usually oleic acid, and produces a divergent product.

In addition to modification of fatty acids in DAGs and TAGs, the plant oxygenases catalyze the formation of plant-signaling molecules, the oxylipins. These include the green leaf volatiles and isomers of jasmonic acid, which initiate further cellular changes in plant defense pathways. Structural defensive compounds, cutin, suberin, and waxes are also synthesized from fatty acids.

Lipids are catabolized by a process called β-oxidation that resembles the synthesis of fatty acids in reverse. The product is mainly acetyl-CoA, which can be metabolized via the citric acid cycle to provide ATP and various intermediates for biosynthesis. In plants, β-oxidation occurs mainly in specialized organelles called glyoxysomes. Malate synthesis and export from these organelles allow the formation of sugars that are essential to nourish the embryos in germinating seeds.

Humans and many other animals require the PUFAs made in plants for proper nutrition. Genetic engineering of plant enzymes has provided some success in generating new sources of fatty acids, such as canola, but in general has not led to major changes in the composition of seed TAGs. Economic commercialization of plant lipids both for food and to replace petroleum products will require additional work, not just in genetic engineering but in providing more detailed information about the biochemistry and cell physiology of lipid synthesis.

REFERENCES FOR FURTHER INFORMATION

Dyer, J.M., Stymne, S., Green, A.G., & Carlsson, A.S. (2008). High-value oils from plants. *Plant J.* **54:**640–655.

Graham, I.A. (2008). Seed storage oil mobilization. *Annu. Rev. Plant Biol.* **59:**115–142.

Mosblech, A., Feussner, I., & Heilmann, I. (2009). Oxylipins: Structurally diverse metabolites from fatty acid oxidation. *Plant Physiol. Biochem.* **47:**511–517.

Samuels, L., Kunst, L., & Jetter, R. (2008). Sealing plant surfaces: Cuticular wax formation by epidermal cells. *Annu. Rev. Plant Biol.* **59:**683–707.

Wallis, J.G., Watts, J.L., & Browse, J. (2002). Polyunsaturated fatty acid synthesis: What will they think of next? *Trends Biochem. Sci.* **27:**467–473.

PLANT BIOCHEMISTRY ***on the web***

Visit the *Plant Biochemistry* student companion Web site at http://biology.jbpub.com/gleason/plantbiochemistry/ for additional online study resources on these topics.

CHAPTER 7

Isoprenoid Compounds (Terpenes)

CHAPTER OUTLINE

7.1 Overview

Plants produce thousands of compounds from a five-carbon precursor, isopentenyl diphosphate (IPP). IPP resembles isoprene, and compounds derived from it are referred to as isoprenoids. Plant isoprenoids are also known as terpenes from turpentine, which is produced from conifer resins and contains several small, volatile hydrocarbons most of which are derivatives of IPP.

The best-known compounds in the isoprenoid class are the sterols. These 30-carbon, cyclic isoprenoids are vital components of cell membranes in most living organisms and are precursors for the synthesis of hormones and other signaling molecules. The IPP used in sterol synthesis is derived from acetyl-CoA via the mevalonic acid pathway located in the cytoplasm of plants, animals, and some microorganisms. In photosynthetic organisms a second pathway generates IPP from pyruvate and glyceraldehyde-3-phosphate. The enzymes for this methyl-erythritol-phosphate pathway are found in the stroma of plastids. Plant-specific products produced in the plastids include carotenoids, the phytol "tail" of chlorophylls, and precursors for plant hormones such as gibberellins and abscisic acid. Many terpenes, however, are unique to a particular plant species and are not essential for plant survival. This has led to their traditional classification as secondary metabolites. Given their role in helping plants thrive in their natural environment by attracting pollinators and combating herbivores and pathogens, however, these compounds are certainly essential and the term "natural products" is more appropriate.

7.2 Biosynthesis of IPP Isopentenyl Diphosphate

Acetate-Mevalonate Pathway

Isopentenyl diphosphate and its isomer, dimethylallyldiphosphate (DMAPP), are the basic building blocks of isoprenoids and are produced from acetyl-CoA by enzymes located in the cytoplasm. Some of these enzymes are soluble proteins, whereas others are integral membrane proteins embedded in the endoplasmic reticulum. Acetyl-CoA is produced in the mitochondria via the pyruvate dehydrogenase complex, but this precursor can also be synthesized in peroxisomes. In many organisms, citrate is exported from the mitochondria by a specific transporter. In the cytoplasm, an ATP-dependent citrate lyase catalyzes the cleavage of citrate to acetyl-CoA and oxaloacetate.

$$\text{Citrate} + \text{ATP} + \text{CoA} + H_2O \rightarrow \text{Oxaloacetate} + \text{acetyl-CoA} + \text{ADP} + \text{Pi}$$

The acetyl-CoA is used in the cytoplasm in several pathways, such as fatty acid elongation (see Chapter 6) and the synthesis of isoprenoid compounds. There is also evidence for IPP synthesis in peroxisomes in some plants.

The acetate-mevalonate pathway (MEV or MVA pathway) starts with the condensation of two molecules of acetyl-CoA to produce acetoacetyl-CoA, catalyzed by acetoacetyl-CoA thiolase (**FIGURE 7.1**). In the next step, the enzyme, 3-hydroxy-3-methylglutaryl-CoA synthase, adds a molecule of acetyl-CoA to a cysteine residue in its active site. The enzyme then binds acetoacetyl-CoA and catalyzes an aldol condensation to produce a six-carbon intermediate, 3-hydroxy-3-methylglutaryl-CoA (HMG-CoA). HMG-CoA is then reduced to mevalonic acid. The HMG-CoA reductase is a tightly regulated enzyme in many organisms, including most plants. It is subject to feedback inhibition by several of the products of the MEV pathway such as sterols. In humans with high cholesterol levels, drugs called statins are prescribed to inhibit sterol synthesis at this step.

Two additional steps catalyzed by mevalonate kinases use ATP and add two phosphate groups to mevalonic acid to produce mevalonate diphosphate. A subsequent decarboxylation and dehydration, both catalyzed by 5-diphosphomevalonate decarboxylase, yield IPP. ATP hydrolysis provides the energy to drive this reaction.

Isopentenyl diphosphate can be isomerized to DMAPP by an isomerase, and these two compounds are the precursors for further terpene biosyntheses that occur in the cytoplasm, including sterol synthesis. The MEV pathway is common to plants, animals, fungi, and many microorganisms. Labeling experiments done in the 1980s with bacteria, however, suggested that some terpenes are not derived from acetate. For example, labeled acetate is incorporated very slowly into carotenoids in plants. Likewise, ^{13}C-labeled glucose does not yield the labeling pattern in carotenoids that would be expected from IPP synthesized by the MEV pathway. Further research was done on green algae using ^{13}C- and ^{2}H-labeled precursors. Characterization of carotenoids by gas chromatography–mass spectroscopy (GC-MS) and nuclear magnetic resonance spectroscopy (NMR) made it possible to precisely determine

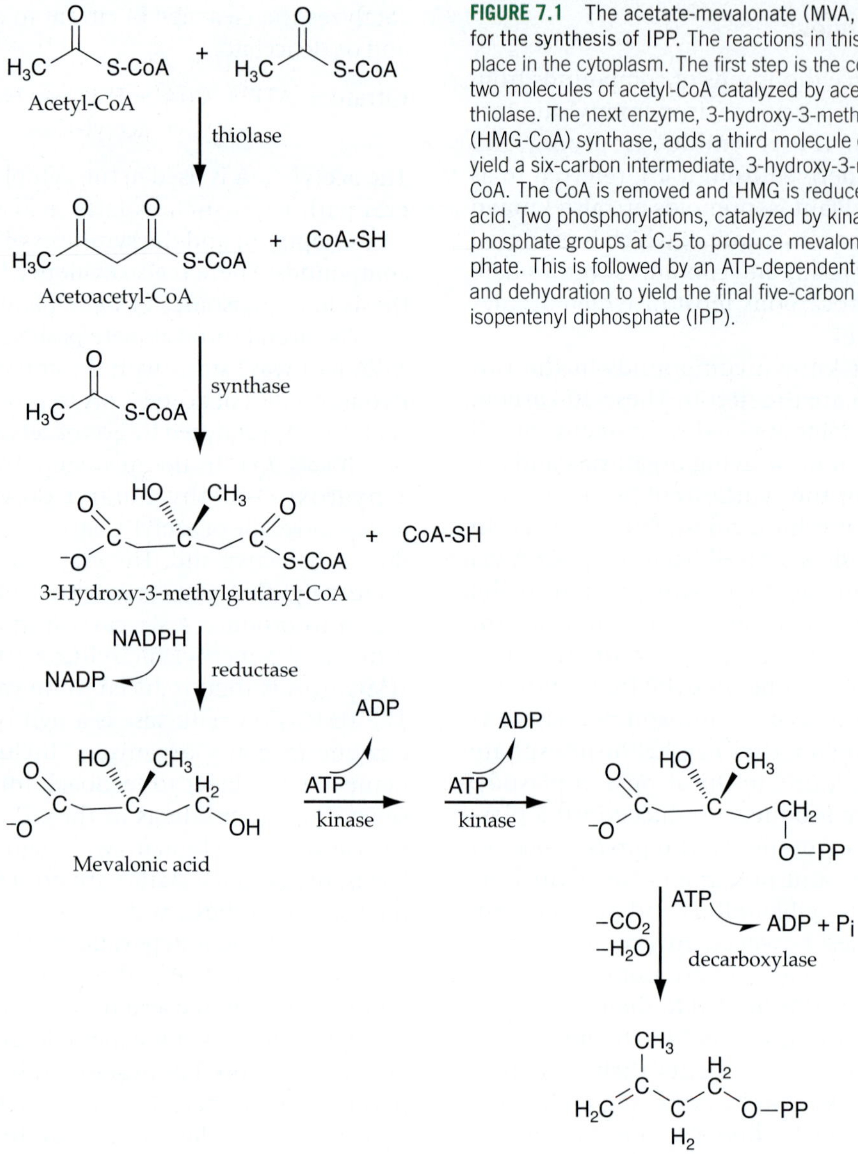

FIGURE 7.1 The acetate-mevalonate (MVA, MEV) pathway for the synthesis of IPP. The reactions in this pathway take place in the cytoplasm. The first step is the condensation of two molecules of acetyl-CoA catalyzed by acetoacetyl-CoA thiolase. The next enzyme, 3-hydroxy-3-methylglutaryl-CoA (HMG-CoA) synthase, adds a third molecule of acetyl-CoA to yield a six-carbon intermediate, 3-hydroxy-3-methylglutaryl-CoA. The CoA is removed and HMG is reduced to mevalonic acid. Two phosphorylations, catalyzed by kinases, add phosphate groups at C-5 to produce mevalonate diphosphate. This is followed by an ATP-dependent decarboxylation and dehydration to yield the final five-carbon product, isopentenyl diphosphate (IPP).

the location of the isotopic labels. This led to the proposal that some organisms can produce isoprenoids from a five-carbon sugar, deoxy-xylulose-5-P.

Methyl-Erythritol Phosphate Pathway (MEP or DOXP Pathway)

A second pathway, the methyl-erythritol phosphate pathway (MEP or DOXP pathway), to produce IPP is found in the chloroplasts of gymnosperms and angiosperms. In some green algae, cyanobacteria, and apicomplexan protozoans, it is the only pathway for the synthesis of IPP and terpenes. A combination of biochemical analyses and gene knock-out mutations done in the late 1990s led to a complete elucidation of the steps and enzymes involved in plastid-based synthesis of IPP. Although all of the reactions in this pathway take place in plastids, the genes for the enzymes are encoded in the nuclear genome.

The pathway starts with pyruvate and glyceraldehyde-3-phosphate. Glyceraldehyde-3-P is an intermediate in the C-3 cycle, and pyruvate can be imported from the cytoplasm or synthesized by a pyruvate dehydrogenase type complex in plastids. The first enzyme in the pathway, 1-deoxy-xylulose-5-phosphate (DOXP) synthase, catalyzes the decarboxylation of pyruvic acid and binds the remaining two-carbon fragment to its thiamine-PP (thiamine diphosphate) cofactor. This is a transketolase-like reaction similar to that in the pyruvate dehydrogenase complex (see Chapter 3). Transfer of this two carbon fragment from thiamine-PP to glyceraldehyde-3-P yields the five-carbon sugar, 1-deoxy-xylulose-5-P (DOXP)

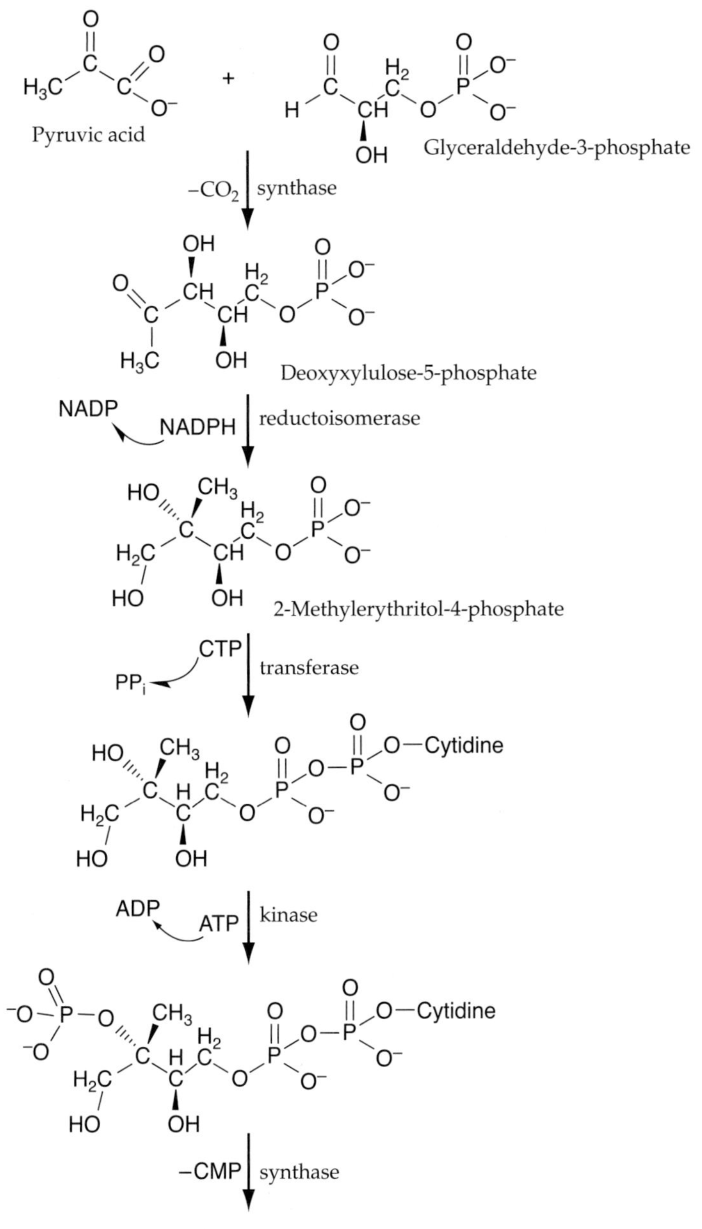

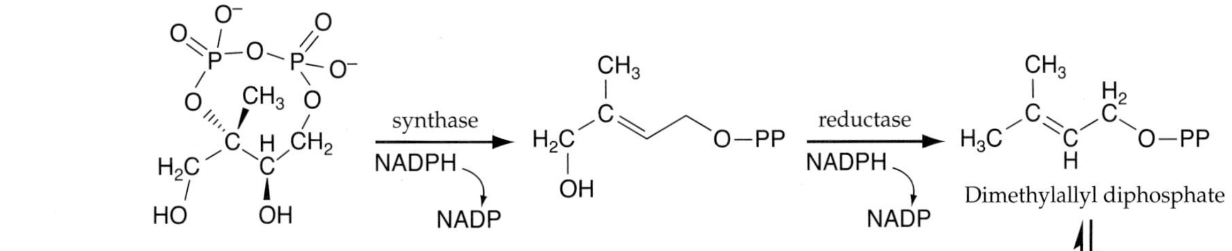

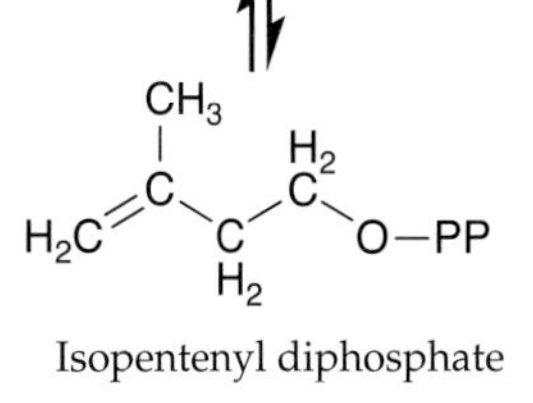

FIGURE 7.2 Methyl erythritol phosphate (MEP, DOXP) pathway for the synthesis of IPP. The reactions of this pathway take place in the plastids of plants. Pyruvic acid is decarboxylated, and the resulting two-carbon fragment is condensed with glyceraldehyde-3-phosphate. This reaction is catalyzed by 1-deoxy-xylulose-5-phosphate (DOXP) synthase. The product, DOXP, is then isomerized and reduced by DOXP reductoisomerase. The resulting sugar alcohol, 2-methyl-erythritol-4-phosphate (MEP), is transferred to cytidine triphosphate (CTP) with the loss of pyrophosphate. A kinase catalyzes the addition of a phosphate to the C-2 hydroxyl group to produce 4-diphosphocytidyl-2-methylerythritol. In the next step, a synthase removes cytidine monophosphate (CMP) and catalyzes the formation of an unusual 2,4-cyclodiphosphate intermediate. A 1-hydroxyl-2-methyl-2-butenyl-4-phosphate synthase opens the diphosphate ring and also introduces a double bond between C-2 and C-3. In the final step, a 4-hydroxy-3-methyl but-2-enyl diphosphate reductase reduces the hydroxyl group and catalyzes the rearrangement to IPP and, its isomer, dimethyl allyl diphosphate (DMAPP).

(**FIGURE 7.2**). A reductoisomerase rearranges DOXP to 2-methyl-erythrose-4-phosphate and then reduces the ketone to yield 2-methyl-erythritol-4-phosphate (MEP). This step of the pathway is inhibited by the antibiotic, fosmidomycin. Because the MEP pathway is the only route to sterol synthesis in apicomplexan parasites, this inhibitor may be useful in treating protozoan-caused diseases such as malaria.

The MEP is subsequently attached to cytidine monophosphate (CMP) by a transferase and a kinase adds a phosphate to C-3. The enzyme, 2-methyl-erythritol-2,4-cyclodiphosphate synthase, then catalyzes removal of CMP and produces an unusual cyclic phosphodiester. A 1-hydroxyl-2-methyl-2-butenyl-4-diphosphate synthase catalyzes ring opening and double bond formation between C-2 and C-3. A final reduction of the alcohol by a reductase yields IPP and its isomer, DMAPP.

7.3 Prenyl Transfer

Isopentenyl diphosphate and DMAPP are the precursors for the synthesis of a large number of compounds. The enzymes that initiate synthesis belong to a general class of prenyl transferases. These enzymes require Mg^{2+}, bind DMAPP, and then catalyze the removal of the diphosphate, which is a good leaving group. The resulting dimethyl allyl cation is resonance stabilized (**FIGURE 7.3**) and can be transferred to a variety of different compounds by electrophilic addition.

The simplest reaction of the allyl cation is to lose a proton and generate isoprene (Figure 7.3). Some plants, such as poplars, have an isoprene synthase in leaf chloroplasts that catalyzes this reaction. Isoprene is volatile (boiling point ~33 degrees) and can be released from the leaves on hot days. Oak, willow, and poplar trees emit isoprene, giving rise to the haze often seen over forests on hot, windless summer days. The physiological benefit of this isoprene emission is unknown, but it may serve to cool the leaf tissues and dissipate reactive oxygen species produced under high light intensity. Some mosses, ferns, and gymnosperms can also emit isoprene, but this ability is not found in all plants despite their common need to deal with heat stress.

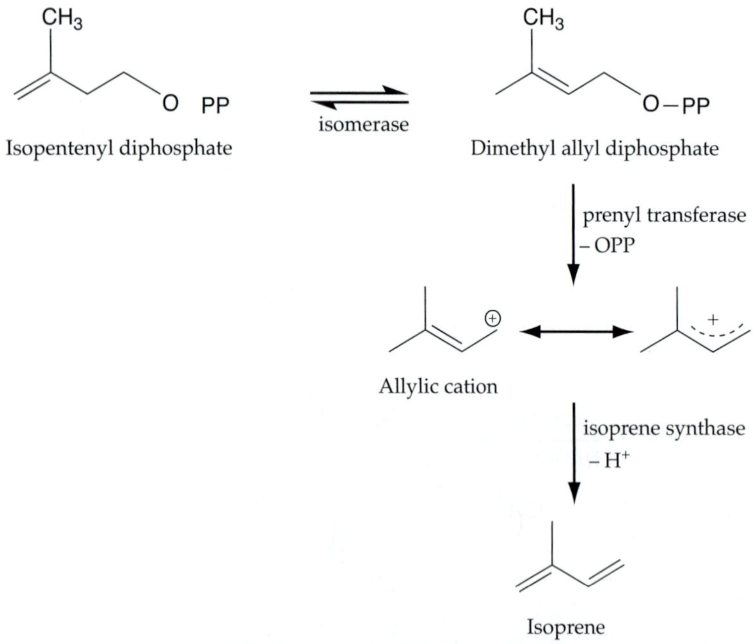

FIGURE 7.3 Isomerization of isopentenyl diphosphate (IPP) to dimethyl allyl diphosphate (DMAPP) and synthesis of isoprene. An isomerase catalyzes the inter-conversion of IPP and DMAPP. Enzymes in the prenyl transferase family facilitate the removal of the diphosphate moiety of DMAPP, resulting in an enzyme-bound dimethylallyl cation. This carbocation is resonance stabilized and can be transferred to electron-rich sites in other molecules. In some plants, an isoprene synthase located in plastids catalyzes the removal of an H^+ from the cation and releases isoprene.

Prenyl groups can also be transferred to other molecules such as nucleotides. Transferases catalyze the prenylation of some nucleotides found in transfer RNAs, such as the isopentenyl adenosine that is often found in the anticodon loop. The growth-promoting cytokinin, isopentenyl adenosine, is a product of transferases found in both plastids and cytoplasm. Prenyl groups are often added to various natural products such as flavonoids. This makes the compounds more hydrophobic and may facilitate their transport through membranes. Proteins can also be post-translationally modified by prenyl groups. Transferases can add a five-carbon or longer prenyl group to a cysteine residue in membrane-associated proteins. Prenylated proteins can then be anchored in the lipid bilayer by means of this hydrophobic hydrocarbon. Prenylated proteins are often involved in signal transduction pathways that are initiated at membrane receptors.

Monoterpenes (10-Carbon Isoprenoids)

Prenyl transferases can catalyze the addition of an IPP to the DMAPP allylic cation to generate longer chain terpenes. The reaction of DMAPP and IPP yields compounds that are 10 carbons in length. Most of the transferases catalyze a "tail to head" addition as seen in **FIGURE 7.4**. The product is geranyl diphosphate (GPP).

Terpene synthases, sometimes called cyclases, can now catalyze ring closure, leading to a variety of different products. The synthases use the same basic mechanism as the prenyl transferases; removal of the diphosphate generates a resonance stabilized carbocation that attacks a double bond in the same molecule. However, as seen in **FIGURE 7.5**, several different resonance forms are possible, including different configurations around the double bonds. Terpene synthases from different plants are highly homologous, but small changes in amino acid residues surrounding the active site will lead to predominantly one product. For example, in *Mentha* sp., GPP synthase produces 94% limonene and small amounts of other

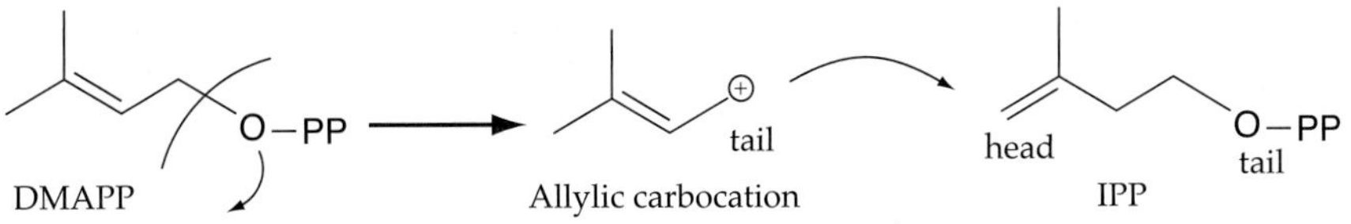

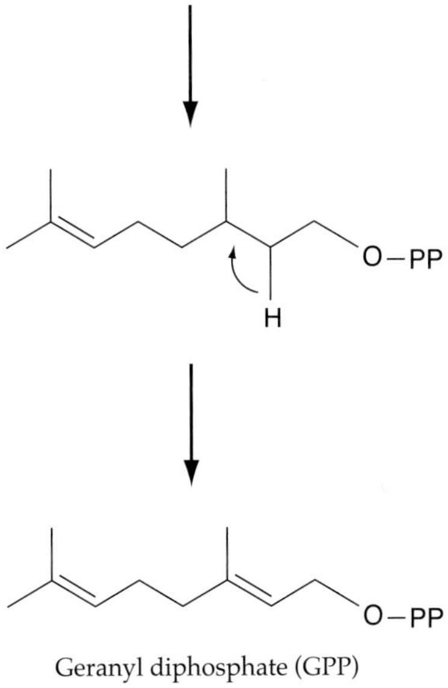

FIGURE 7.4 Synthesis of geranyl diphosphate (GPP). Specific prenyl transferases, the monoterpene transferases, facilitate removal of the diphosphate from DMAPP and the production of an allylic cation. This is designated the "tail" and can be added to the double-bonded "head" of an IPP molecules to yield the linear, 10-carbon product, GPP. (Note: There is a lack of consistency among authors as to which ends of the IPP-DMAPP isomers are designated "head" and "tail." Some authors prefer to designate the DMAPP cation the "head," which is then added to the "tail" of an IPP molecule. The head–tail designations used in this chapter follow the usage by P. M. Dewick in *Medicinal Natural Products*, Wiley, 2002. Dewick designates the diphosphate end of the molecule as the tail.)

terpenes such as pinenes and myrcene. Conversely, in *Pinus* sp., the conifer synthase produces predominantly pinenes.

The structures of some common monoterpenes are shown in **FIGURE 7.6**. These are relatively small molecules and most of them are volatile and odoriferous. They give many plants, such as spices, their characteristic odor and have been referred to as "essential oils," meaning "producing an essence" rather than referring to any vital physiological function. The stereochemistry is not shown for all these structures, but the configuration can have a major impact on the odor of a terpene. For example, (+)carvone is produced in caraway (*Carum carvi*), whereas (−)carvone is the spearmint odor of *Mentha spicata*.

Limonene is found in a number of different plants and can be modified by the addition of "decorating" groups. Addition of a hydroxyl group to the ring is the most common modification as shown in **FIGURE 7.7**, which illustrates the synthesis of carvone from limonene. Hydroxylation of monoterpenes is catalyzed by monooxygenases, mostly of the cytochrome P450 type. Terpenes with OH groups can be further modified by glycosylation, usually with one or more glucose residues. Glycosylation facilitates transport and storage of terpenes. In many plants, terpenes are produced in epidermal cells and stored in special structures called trichomes, which can be seen protruding from stems and

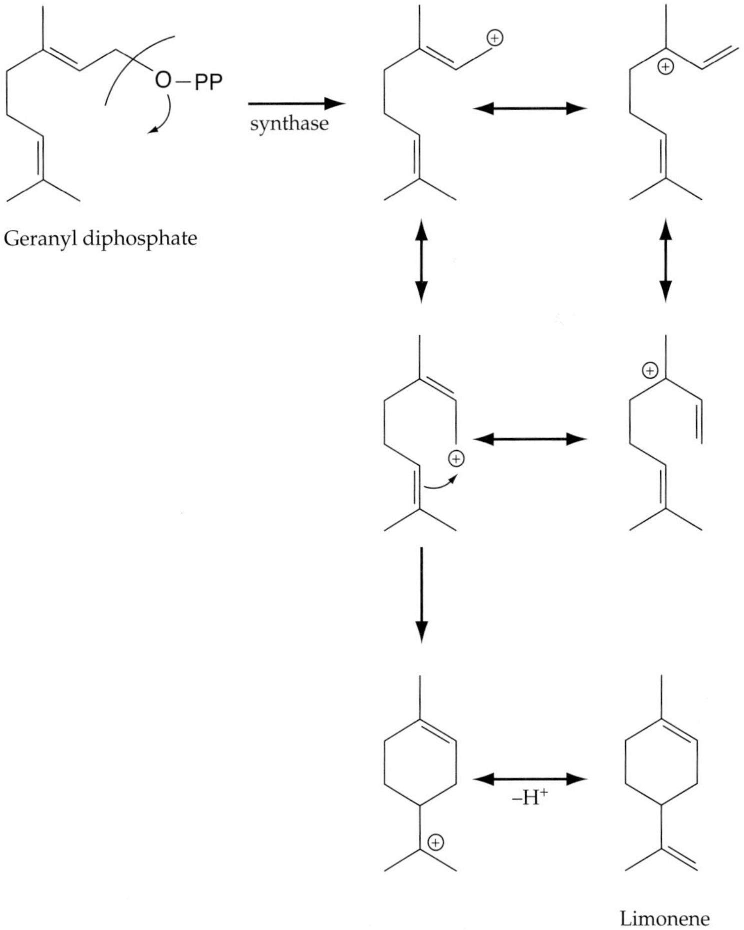

FIGURE 7.5 Synthesis of limonene. The limonene synthase (also called a cyclase) facilitates removal of the diphosphate group from GPP. The resulting carbocation is resonance stabilized and can relocate to several positions in the enzyme-bound intermediate. Limonene synthase produces the six-membered ring of limonene, a common volatile terpene, but small amounts of other 10-C products are also made.

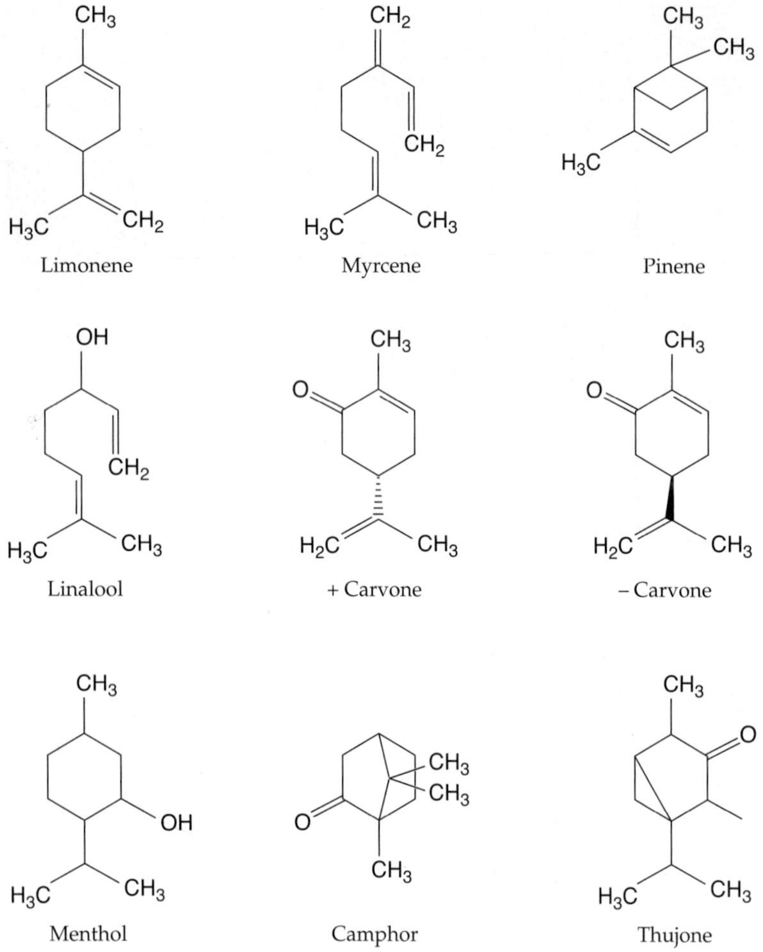

FIGURE 7.6 Structures of some common monoterpenes. All the compounds shown in this figure are products of specific monoterpene synthases. Mutations in residues near the active sites of the synthases direct the synthesis of predominantly one product in a particular plant species. Limonene is found in many angiosperms, whereas menthol is specific to *Mentha* sp. The stereochemistry may also differ as shown for the two isomers of carvone.

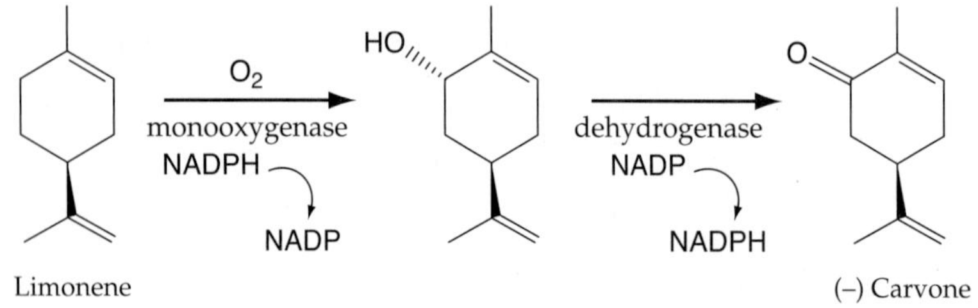

FIGURE 7.7 Synthesis of (−)carvone in *Mentha spicata.* A monooxygenase catalyzes the addition of a hydroxyl group to limonene. This OH is then oxidized by a specific dehydrogenase to yield (−)carvone, the essence of spearmint. The (+)carvone isomer is made similarly in *Carum carvi* and has the odor characteristic of caraway.

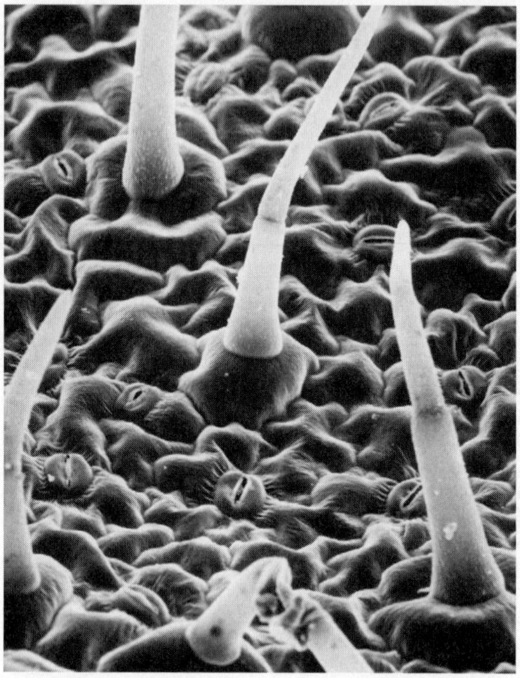

FIGURE 7.8 Epidermal trichomes on the surface of a tomato leaf. The small hairs are extensions of epidermal cells that protrude through the cuticle. The hairs are easily broken on contact, and this causes the release of volatile compounds that are stored in the basal cell and produce the characteristic "tomato odor." (Magnification ~900×.)

leaves (**FIGURE 7.8**). Mechanical damage, such as herbivory, releases the terpenes. Many monoterpenes are a feeding deterrent to insect herbivores, such as the pinenes, which are toxic to bark beetles. Others, such as linalool, are often produced in epidermal cells in flower petals and serve to attract and guide pollinators.

Humans generally find the monoterpene odors attractive, and plants that produce them are used to flavor foods, such as myrcene in bay leaves or menthol in peppermint (Figure 7.6). Myrcene is also a common component of perfumes. Some monoterpenes, such as camphor from *Cinnamomum camphora,* are made as an insect deterrent by the plant but are used as an analgesic by humans. Likewise, thujone from *Artemesium* sp. has a bitter taste, but terpenes from this plant produce the unique flavors in a variety of the liqueurs such as absinthe. Absinthe was a popular drink in Europe and North America in the beginning of the twentieth century. The terpenes, particularly thujone, make absinthe extremely bitter so the liqueur was generally diluted with water that was poured through a slotted spoon holding a sugar cube (**FIGURE 7.9**). Because terpenes are generally rather hydrophobic, the thujone starts to precipitate when water is added, giving the drink an attractive opalescence. Thujone was subsequently shown to be toxic to laboratory animals, and the sale of absinthe was banned in many countries. The amount of thujone required to produce any toxic effects, however, is exceedingly large;

most cases of reported human "absinthe poisoning" were more likely due to excess consumption of ethanol!

Although tail to head condensation is the most common addition catalyzed by monoterpene transferases, other reactions are possible. A tail to middle transferase is found in chrysanthemums that catalyzes the reaction shown in **FIGURE 7.10**. The product, chrysanthemic acid, is subsequently modified in a number of different ways to produce a class of compounds called pyrethrins. These compounds are toxic to a variety of insects and are mass produced for use by organic farmers as natural insecticides.

FIGURE 7.9 How to drink absinthe. The terpenes, mostly thujone, in the liqueur give it a very bitter flavor. To make the liqueur palatable, sugar cubes, placed on a slotted spoon, are dissolved in absinthe by pouring water through the spoon. The addition of water causes the hydrophobic terpenes to precipitate, giving the drink its distinctive milky opalescence as seen in the tumbler next to the bottle in the photo.

Sesquiterpenes (15-Carbon Isoprenoids)

Tail-to-head addition of IPP to a geranyl cation gives rise to a 15-carbon terpene, farnesyl diphosphate (FPP). Farnesyl diphosphate and its derivatives are referred to as sesquiterpenes. Sesquiterpene synthase (cyclase) action is mechanistically similar to that of the monoterpene synthases, but the sesquiterpene enzyme active site has an increased capacity to accommodate the larger FPP substrate. Because of the expanded number of cation resonance forms, there are obviously a larger number of possible products (**FIGURE 7.11**).

The crystal structure of the sesquiterpene synthase, aristolochene synthase from tobacco, has been solved to 2.2-Å resolution. Although the primary structure of this enzyme has very low (~15%) sequence identity to sesquiterpene synthases from animal and bacterial sources, the three-dimensional structures of the cyclases are very similar in most organisms. These enzymes are composed of 10 α-helices with five of the helices forming a large open cavity to bind the linear terpene substrate. All synthases have a conserved active site motif of aspartic acid residues that bind Mg^{2+} ions. In angiosperms, this is usually a sequence of DDXXD residues (**FIGURE 7.12**). Hydrophobic residues surrounding the active site aspartic acids guide the formation of a particular intermediate and thereby result in one predominant product. In the case of the tobacco synthase, one of the final products of aristolochene synthesis is the phytoalexin, capsidiol (**FIGURE 7.13**). Phytoalexins are plant defense compounds synthesized in response to wounding or signal transduction from other tissues in the plant that

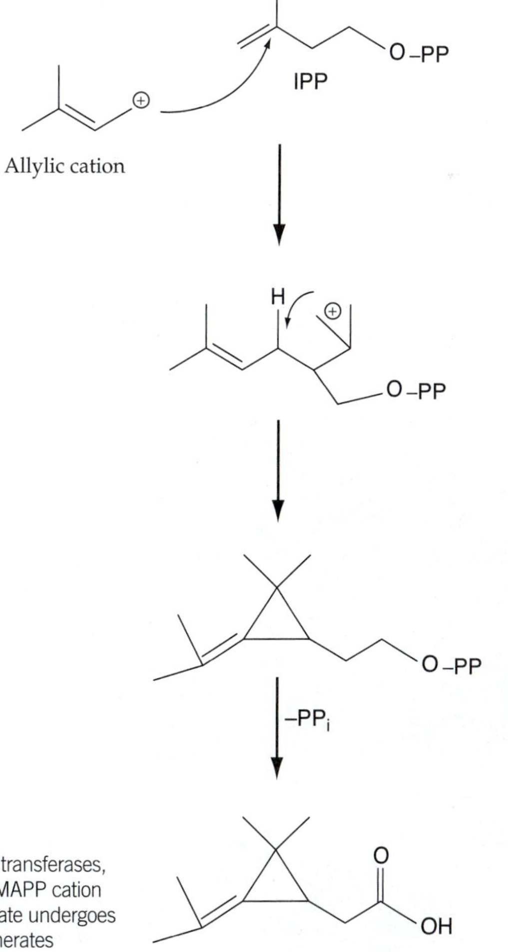

Chrysanthemic acid

FIGURE 7.10 Synthesis of chrysanthemic acid. Unlike the usual prenyl transferases, the enzyme in some chrysanthemum species catalyzes addition of the DMAPP cation to C-3 in the middle of the IPP molecule. The resulting cationic intermediate undergoes ring closure to form a cyclopropyl ring. Loss of the diphosphate group generates chrysanthemic acid, the precursor for a class of natural insecticides called pyrethrins.

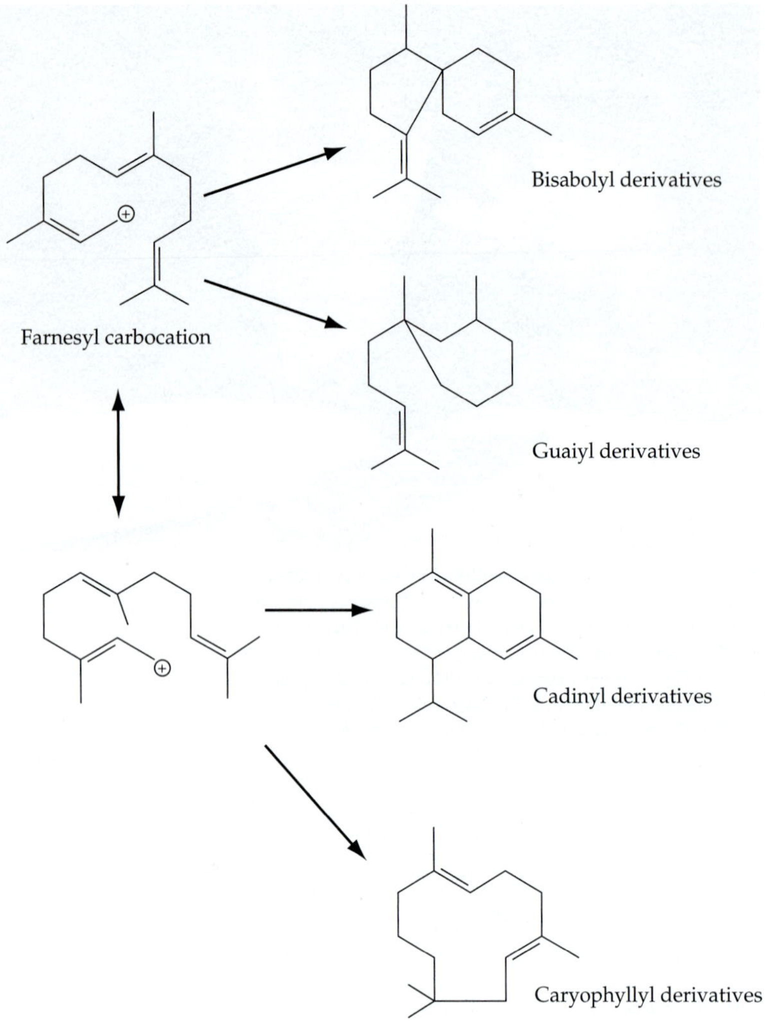

FIGURE 7.11 Activity of sesquiterpene synthases. This synthase generates a farnesyl carbocation from FPP that can assume a number of resonance forms, two of which are shown on the left of the diagram. These intermediates, in turn, can generate ring closure in a variety of different ways, giving rise to a large number of sesquiterpenes.

are under attack from pathogens. Phytoalexins can be toxic to the producing plants, and their synthesis is seldom constitutive. As seen in Figure 7.13, other plants also use a similar synthase to produce slightly different final products. In potato, vetispiradiene is the sesquiterpene synthase product that is further modified to produce a unique potato phytoalexin, lubimin.

Structures of some representative sesquiterpenes are shown in **FIGURE 7.14**. Cadinenes are common in many angiosperms and gymnosperms, and some are produced constitutively. Alpha-cadinene is the bitter compound present in juniper berries. It is also the starting material for the synthesis of artemisinin in *Artemisia annua*. This sesquiterpene was first identified in certain traditional Chinese herbal preparations. It is currently the most effective treatment for malaria. The protozoan, *Plasmodium falciparum*, has become immune to previously used drugs such as chloroquine, which is an alkaloid-based compound. Artemisinin kills the parasites as they reproduce in red blood cells. Delta-cadinene in cotton seeds is the precursor to an unusual aromatic sesquiterpene dimer, gossypol. Several additional enzymes are involved in its synthesis. Monooxygenases add hydroxyl groups and reductases lead to double bond formation and, finally, to aromatic rings, which are uncommon in terpenes. Gossypol reacts readily with amino groups in proteins and is toxic to fungi and animals.

Additional sesquiterpenes with a caryophyll ring are shown in Figure 7.14. Humulene is the major bitter-tasting component in hops. It is added to beer as a flavoring agent and a preservative because of its antimicrobial properties. Parthenolide is a sesquiterpene lactone made by feverfew (*Tanacetum parthenium*). Humans have found it useful in treatment of migraines.

Geosmin is not readily recognized as a sesquiterpene because it has only 12 carbons (**FIGURE 7.15**). Its synthesis from FPP is catalyzed by a unique bifunctional synthase that appears to be the result of gene duplication and subsequent

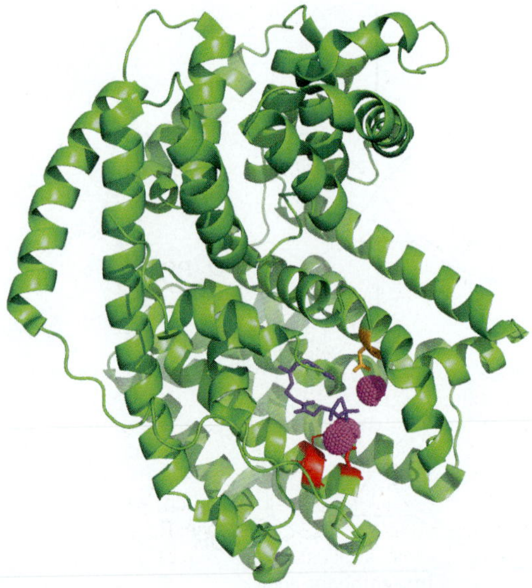

FIGURE 7.12 The three-dimensional structure of 5-epi-aristolochene synthase from tobacco. The enzyme is composed of α-helices connected by short loops. The active site is located in a hydrophobic pocket formed by five of the helices. In this structure a substrate analog, 2-fluoroFPP, shown as a purple stick figure, is bound. The common active site motif of terpene synthases, DDXXD, is composed of Asp-301, Asp-302, and Asp 304 in the tobacco enzyme. These residues are shown in red on the right side of the active site pocket coordinating two Mg^{2+} ions shown as magenta spheres that protrude into the pocket. A third Mg^{2+} ion is coordinated to Asp-444 in orange; it lies on the opposite side of the active site pocket. The Mg^{2+} ions bind the phosphates of FPP and facilitate their removal and generation of the intermediate cation. (Structure from Protein Data Bank 3LZ9. J. P. Noel, N. Dellas, J. A. Faraldos, M. Zhao, B. A. Hess Jr., L. Smentek, R. M. Coates, and P. E. O'Maille, *Acs. Chem. Biol.* 5 [2010]: 377–392.)

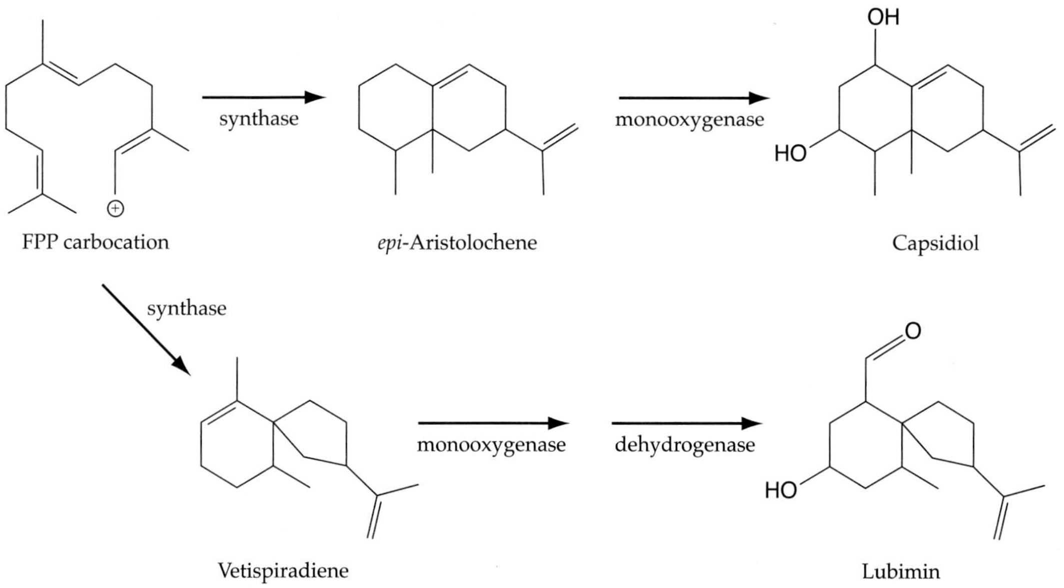

FIGURE 7.13 Synthesis of phytoalexins from farnesyl diphosphate. Aristolochene synthase generates a cadinyl intermediate called *epi*-aristolochene. Cytochrome P450 monooxygenases introduce two hydroxyl groups to one of the rings yielding capsidiol, a phytoalexin produced in tobacco plants. Vetispiradiene synthase catalyzes the formation of the five- and six-membered rings from FPP to yield an intermediate called vetispiradiene. Two hydroxyl groups are added to the intermediate by monooxygenase action. One of the hydroxyl groups is then oxidized to an aldehyde to yield lubimin, a phytoalexin produced in potatoes.

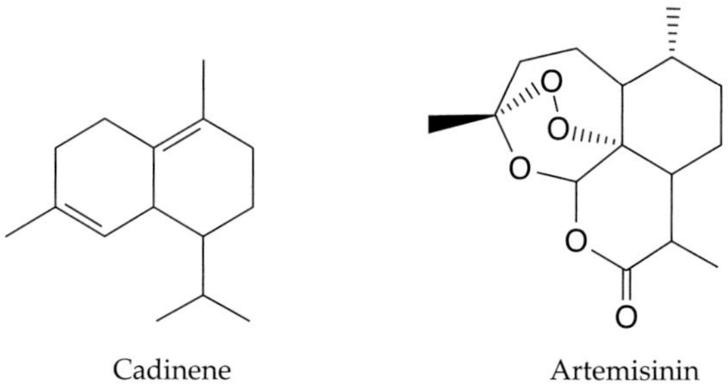

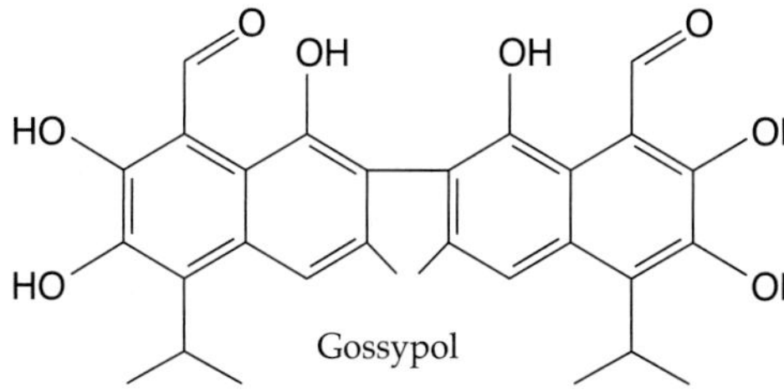

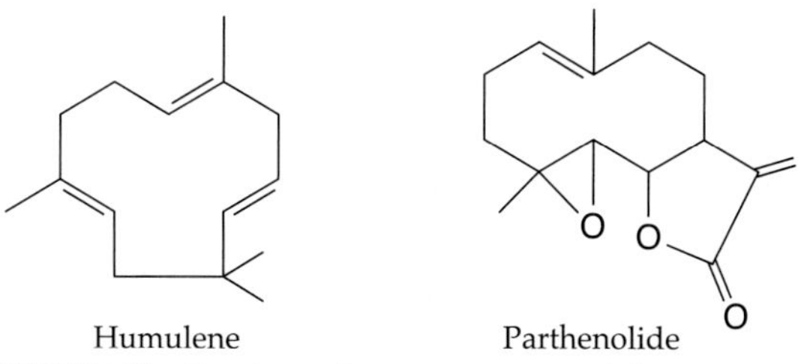

FIGURE 7.14 Structures of some common sesquiterpenes. Cadinene is an intermediate in the synthesis of many inducible phytoalexins but is produced constitutively in some plants such as juniper. Gossypol is an unusual example of an aromatic sesquiterpene dimer found in cotton seeds. Humulene, from hops, is an example of a caryophyllyl derivative (see Figure 7.11). Artemisinin and parthenolide are sesquiterpene lactones.

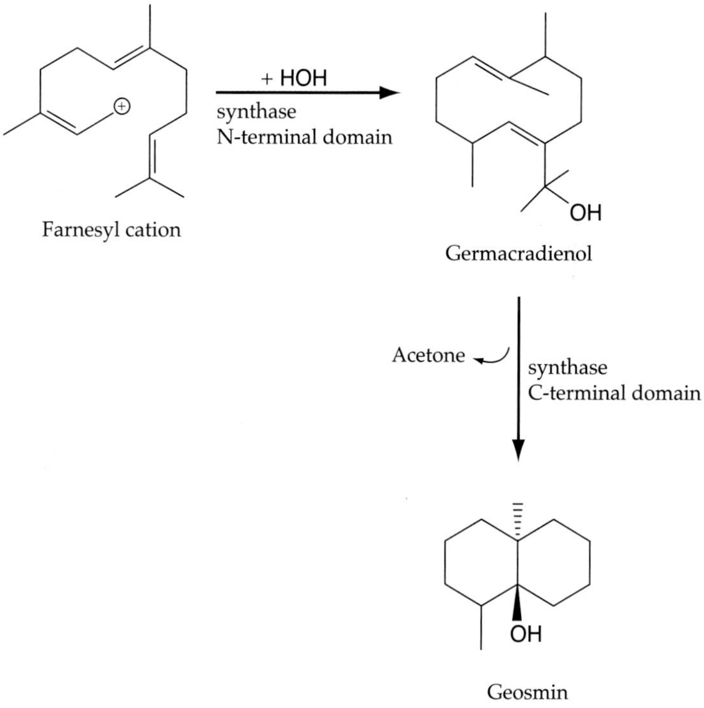

FIGURE 7.15 Synthesis of geosmin, an unusual 12-carbon sesquiterpene derivative. The geosmin synthase is the apparent product of an ancient gene duplication and subsequent fusion. This bifunctional enzyme has two synthase active site domains. The domain located near the N-terminal catalyzes ring closure to produce the intermediate, germacradienol. An OH-group, derived from water, is added at the same time. Germacradienol can be released from the enzyme but is more likely to be transferred to the active site in the C-terminal domain. Three carbons are cleaved from germacradienol and released as acetone. Ring closure then generates geosmin, a 12-carbon sesquiterpene derivative.

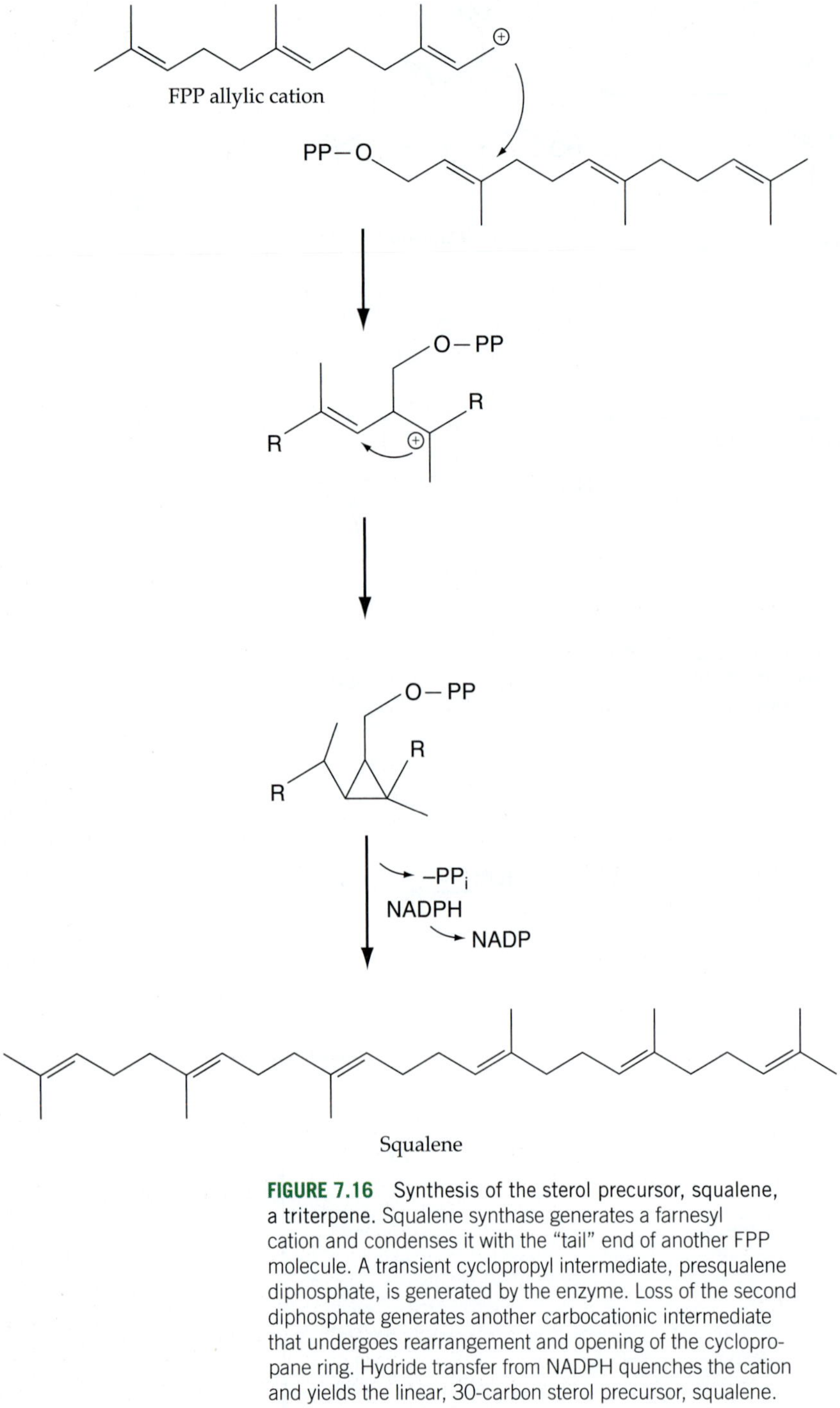

Squalene

FIGURE 7.16 Synthesis of the sterol precursor, squalene, a triterpene. Squalene synthase generates a farnesyl cation and condenses it with the "tail" end of another FPP molecule. A transient cyclopropyl intermediate, presqualene diphosphate, is generated by the enzyme. Loss of the second diphosphate generates another carbocationic intermediate that undergoes rearrangement and opening of the cyclopropane ring. Hydride transfer from NADPH quenches the cation and yields the linear, 30-carbon sterol precursor, squalene.

fusion. The FPP is bound to the N-terminal domain of the enzyme and the intermediate, germacradienol, is produced. Addition of an OH group from water to the intermediate is the source of the hydroxyl group in this reaction. This intermediate is then transferred to the C-terminal synthase domain that catalyzes the removal of an isopropyl group and releases the 12-C product, geosmin. Geosmin is produced by many bacteria, fungi, cyanobacteria, and some cacti. It is moderately volatile and imparts an earthy smell to soil or water.

Triterpenes (30-Carbon Isoprenoids)

Prenyl transferases in the cytoplasm condense two FPP units to produce 30-carbon squalene, a reaction that is common in most organisms. In this case the transferase catalyzes a tail-to-tail condensation (**FIGURE 7.16**), and a phosphocyclopropane intermediate is generated. As in other isoprenoid reactions, subsequent loss of the diphosphate produces a cation intermediate that can then rearrange. The cation is quenched by hydride transfer from NADPH to yield squalene. Note that the double bond configuration changes in the two halves of the product; this type of arrangement is characteristic of a tail-to-tail prenyl transferase. A cytochrome P450 monooxygenase adds an OH group to squalene, followed by a synthase (cyclase) catalyzed reaction to produce the common sterol ring system.

Various enzymes can then modify the basic sterol structure to produce several kinds of products. Some sterol structures are shown in **FIGURE 7.17**. Relatively few plants produce cholesterol, which is a characteristic product in animals. Sitosterol is the major sterol component found in most plant membranes. Analysis of *Brassica napus* showed that campesterol can be further modified to produce brassinolide, which contains a seven-membered lactone ring. This compound is a growth-promoting hormone that has subsequently been found in many angiosperms, gymnosperms, and even some ferns and algae.

Sterols can be further modified by glycosylation with one or more glucose residues. These are referred to as steroidal saponins because of their foaming, soap-like characteristics. Such compounds are often toxic to insect herbivores but may be useful to humans. The saponin, diosgenin (**FIGURE 7.18**), from Mexican yams (*Dioscorea* sp.), is the starting material for the synthesis of female contraceptives. Saponins produced by foxglove (*Digitalis purpurea*) are known as cardenolides because they can inhibit the Na^+-K^+ ATP-dependent transporter in animal cells. Cardenolides and synthetic analogs are used to treat irregular heart beating in humans. The cardenolides are produced by a number of additional plants and are found in a milky secretion called the latex. These cardenolides are generally toxic to herbivores, including humans. However, some insects, such as the Monarch butterfly, have evolved a mechanism to sequester the toxic compounds in their cuticle. This then makes the insects inedible to carnivores that prey on them, such as birds (**FIGURE 7.19**).

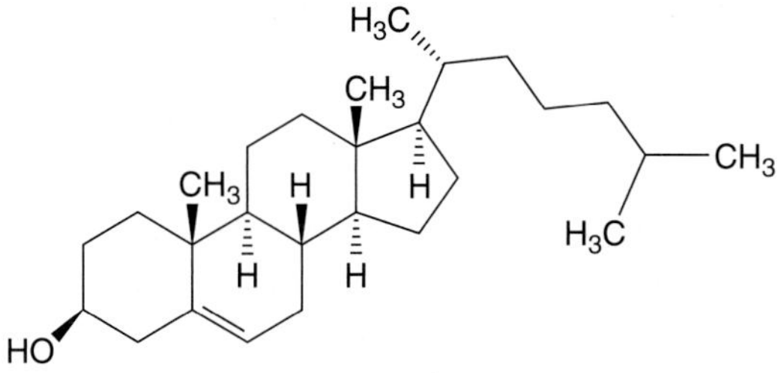

Cholesterol

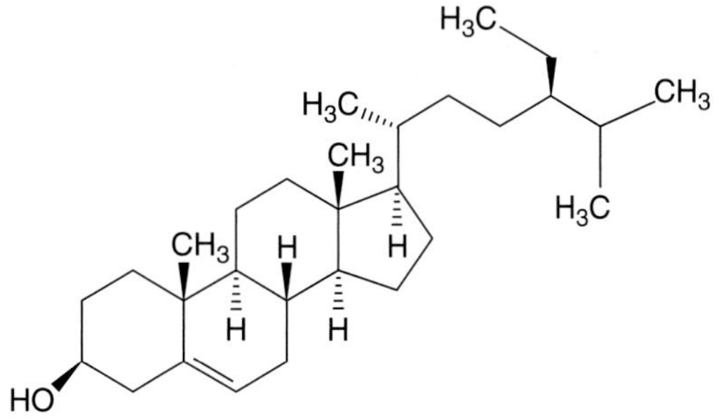

Sitosterol

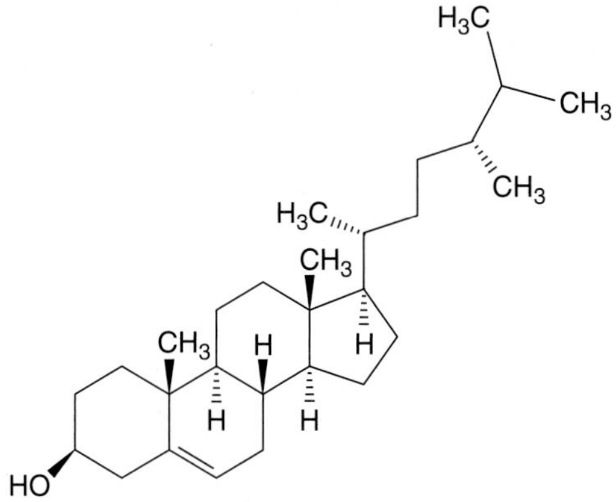

Campesterol

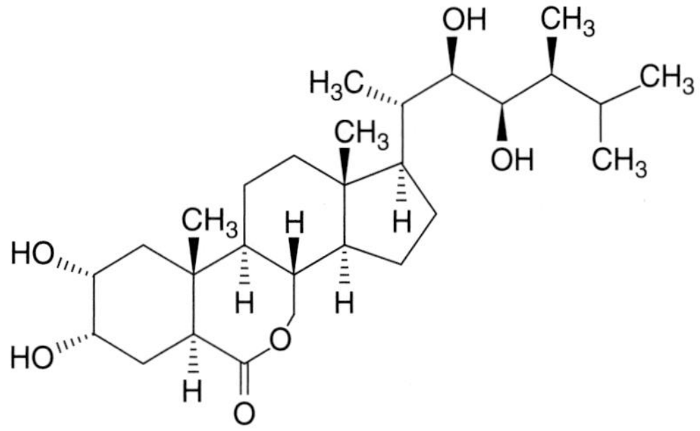

Brassinolide

FIGURE 7.17 Structures of some common sterols. Cholesterol is rare in plants but is found in some tissues, such as avocado fruits. Sitosterol is the most common sterol in angiosperm cell membranes. Campesterol is the biosynthetic precursor of the plant hormone, brassinolide. Although discovered initially in the Brassicaceae, brassinosteroids are growth hormones that are widespread in the plant kingdom.

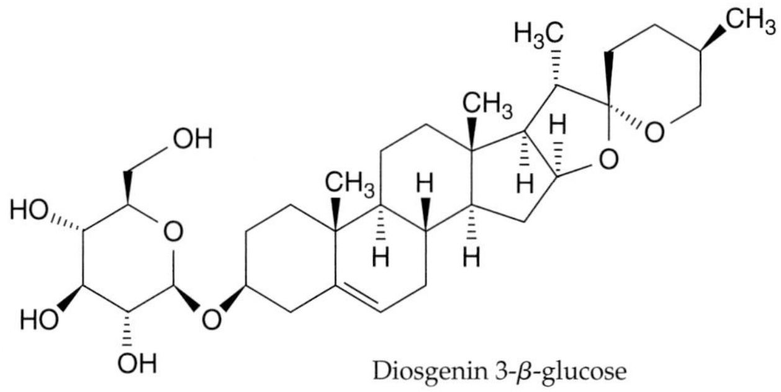

Diosgenin 3-β-glucose

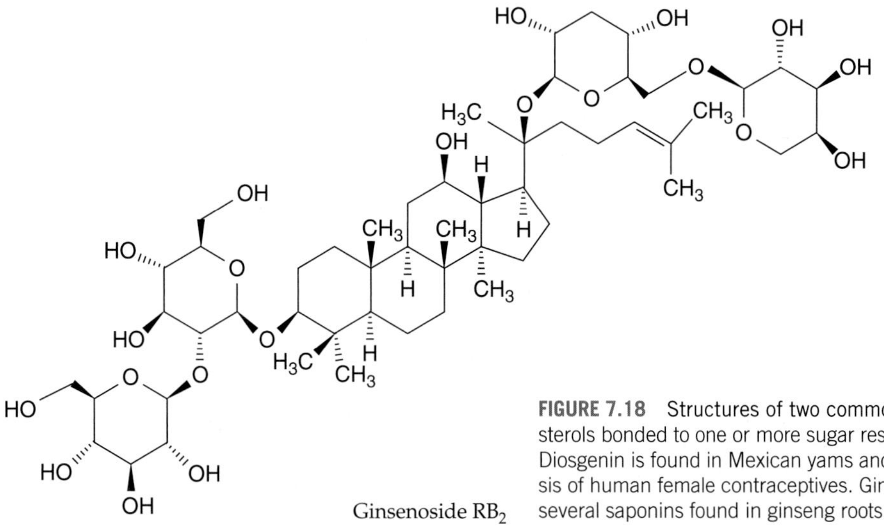

Ginsenoside RB_2

FIGURE 7.18 Structures of two common saponins. Saponins are sterols bonded to one or more sugar residues, usually glucose. Diosgenin is found in Mexican yams and is used in the semisynthesis of human female contraceptives. Ginsenoside RB_2 is one of several saponins found in ginseng roots.

FIGURE 7.19 Adaptation of the monarch butterfly to cardenolides. Although the cardenolides (cardiac glycosides) are toxic to most insects, the monarch caterpillar (a) can feed on otherwise toxic milkweed (*Asclepias* sp.) plants without harm. The saponins in the plants are ingested and stored in the outer layer of cells of the caterpillar and remain in place in the adult (b), thus protecting both stages of the insect's life cycle from carnivores.

Other examples of saponins used by humans are the ginsenosides from *Panax ginseng*. These are found in the plant roots and are marketed as "adaptogens" (i.e., they enable the body to cope with stress). Ginsenosides have a variety of physiological effects on humans, but these are often contradictory. For example, ginseng tea is presumably a relaxant but also causes sleeplessness. To date, there is no clinical evidence for their physiological effectiveness in advertised benefits such as lowering blood pressure or stimulating immune responses.

Diterpenes (20-Carbon Isoprenoids)

A typical tail to head transferase adds IPP to a farnesyl cation to produce the 20-carbon compound, geranyl geranyl diphosphate (GGPP). A prenyl transferase in chloroplasts can then catalyze transfer of GGPP to chlorophyllide, the tetrapyrrole precursor of chlorophyll. Most of the double bonds in the geranyl geranyl side chain are then reduced to yield the characteristic phytol side chain of chlorophylls (see Chapter 1). A specific transferase also adds GGPP units to hydroxybenzoic acid to produce phylloquinones. A similar enzyme in the cytoplasm, or possibly the mitochondria, catalyzes the production of ubiquinone, which is a component of the mitochondrial electron transport system (see Chapter 3).

Geranyl geranyl diphosphate in the chloroplast can undergo cyclization to a sterol-like ring compound called *ent*-kaurene, the starting material for the synthesis of gibberellins (**FIGURE 7.20**). *Ent*-kaurene is then transported out of the plastid. In the cytoplasm, *ent*-kaurene is modified by enzymes in the endoplasmic reticulum to produce over 120 modified gibberellins. Only a few of these compounds have been shown to have growth-stimulating activity in plants and also promote and sustain germination in seeds. The other gibberellins may be hormone precursors or degradation products.

Tetraterpenes (40-Carbon Isoprenoids)

The major tetraterpenes in plants are carotenoids, and these are produced exclusively in plastids. A bifunctional enzyme, phytoene

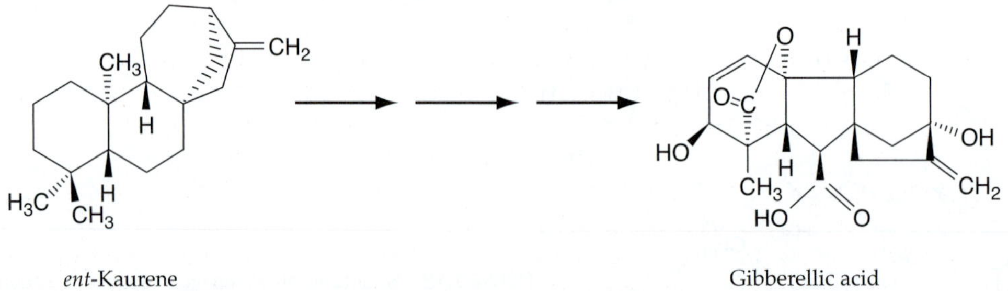

FIGURE 7.20 Abbreviated synthesis of gibberellic acid. Gibberellin synthesis is initiated in plastids from GGPP. A specific synthase then catalyzes ring closure to generate *ent*-kaurene, which is transported out of the plastid. Gibberellin synthesis then continues in the cytoplasm with several modifications to the basic *ent*-kaurene structure. Over 120 different structures of gibberellins have been reported; all have the core ring structure resembling gibberellic acid.

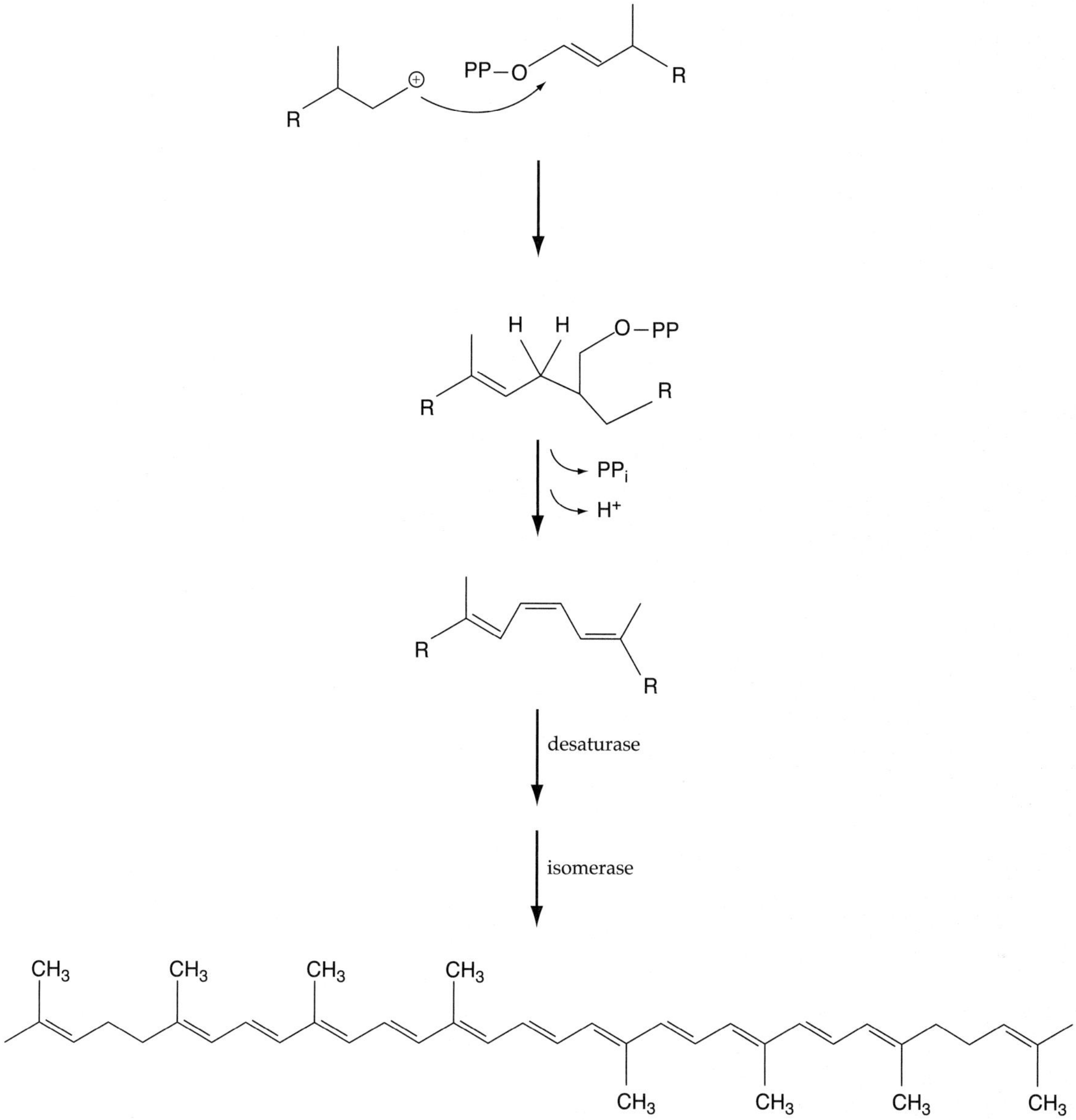

all-*trans* Lycopene

FIGURE 7.21 Synthesis of lycopene. Geranylgeranyl transferase catalyzes a "tail-to-tail" addition of GGPP to a GGPP cation. Loss of the second diphosphate group and a proton produces a double bond in the resulting product, phytoene. Desaturation of this 40-carbon intermediate and isomerization yields all-*trans* lycopene, a precursor of carotenoids.

synthase, catalyzes a tail-to-tail condensation of two GGPP units (**FIGURE 7.21**). As in squalene synthesis a diphosphate intermediate is generated, and loss of the diphosphate generates a cation. In phytoene synthesis, however, this leads to formation of a double bond and release of *cis*-phytoene from the enzyme. After a desaturation and isomerization, the final product is all-*trans* lycopene. Note that this molecule is not fully conjugated. It is the principal pigment responsible for the red color of ripe tomatoes. Lycopene is then acted on by alpha- and/or beta-cyclases, which give rise to rings at the ends of the lycopene chain, producing carotenes (**FIGURE 7.22**). The rings are further modified by addition of oxygen. Both nonheme type oxygenases, resembling the soluble fatty acid desaturase in chloroplasts, and membrane-bound cytochrome P450 monooxygenases catalyze these additions. Epoxidases then catalyze the formation of xanthophylls. These epoxidases generally have a FAD cofactor and are reduced by ferredoxin.

Carotenoids are highly conjugated molecules (Figure 7.22). They absorb visible light in the 420- to 500-nm range and appear yellow, orange, or red. As noted above, they are the pigments found in many fruits and flowers.

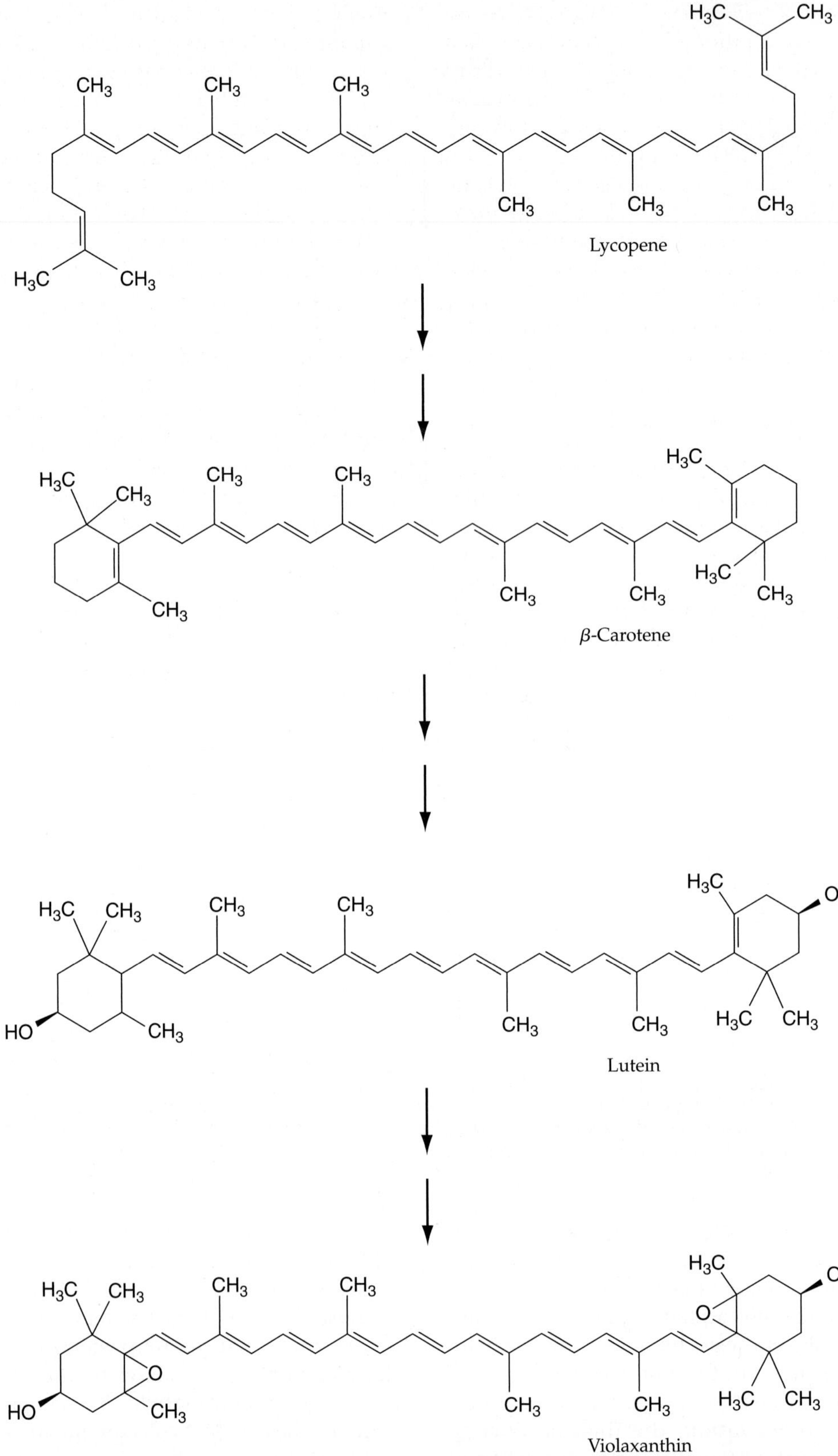

FIGURE 7.22 Synthesis of carotenoids. This figure shows one branch of the carotenoid synthesis pathway. The formation of β-carotene from lycopene is catalyzed by a β-carotene cyclase. A β-carotene hydroxylase (monooxygenase) then adds hydroxyl groups to the rings. Lutein epoxidase catalyzes formation of one epoxide to yield lutein and a second epoxidation produces violaxanthin.

Carotenoids are also present in high concentrations in leaves, although their color is masked by chlorophyll. They function as accessory antenna pigments in photosynthesis and protect the photosynthetic reaction centers by scavenging free radicals. The xanthophyll cycle also functions under high light intensity to dissipate excess light energy as heat. A de-epoxidase catalyzes the conversion of violaxanthin to zeaxanthin with consequent loss of energy, a process called nonphotochemical quenching. In the dark, epoxidases restore zeaxanthin to violaxanthin (see Chapter 1).

7.4 Carotenoid Cleavage

Dioxygenases in the chloroplast can catalyze the cleavage of carotenoids into smaller molecules with incorporation of the oxygen atoms into the products (**FIGURE 7.23**). These enzymes are nonheme iron proteins with the iron coordinated to histidine residues at the active site. A similar retinal synthase-dioxygenase occurs in mammals to produce the visual pigment, retinal, from dietary carotenoids. Specific dioxygenases are found in plants that can cleave carotenoids at every possible C–C bond, leading to a variety of products called apocarotenoids. Some of the products are small, volatile molecules, such as β-cyclocitral and β-ionone; these produce the characteristic odors when cutting plants and vegetables. Although compounds such as β-ionone are widespread, some cleavage products are unique and occur only in certain parts of a plant. An example is crocetin, a dialdehyde produced by cleavage of zeaxanthin. Crocetin is produced only in the stigmas of the autumn crocus to attract pollinators and is responsible for the unique odor (flavor) of saffron. In some algae, these carotenoid cleavage products produce unpleasant odors in water supplies. They may function as allelopathic agents for the producing plants.

Carotenoid cleavage enzymes in the chloroplast also initiate the synthesis of the plant stress hormone, abscisic acid (ABA). As shown in **FIGURE 7.24**, neoxanthin synthase catalyzes

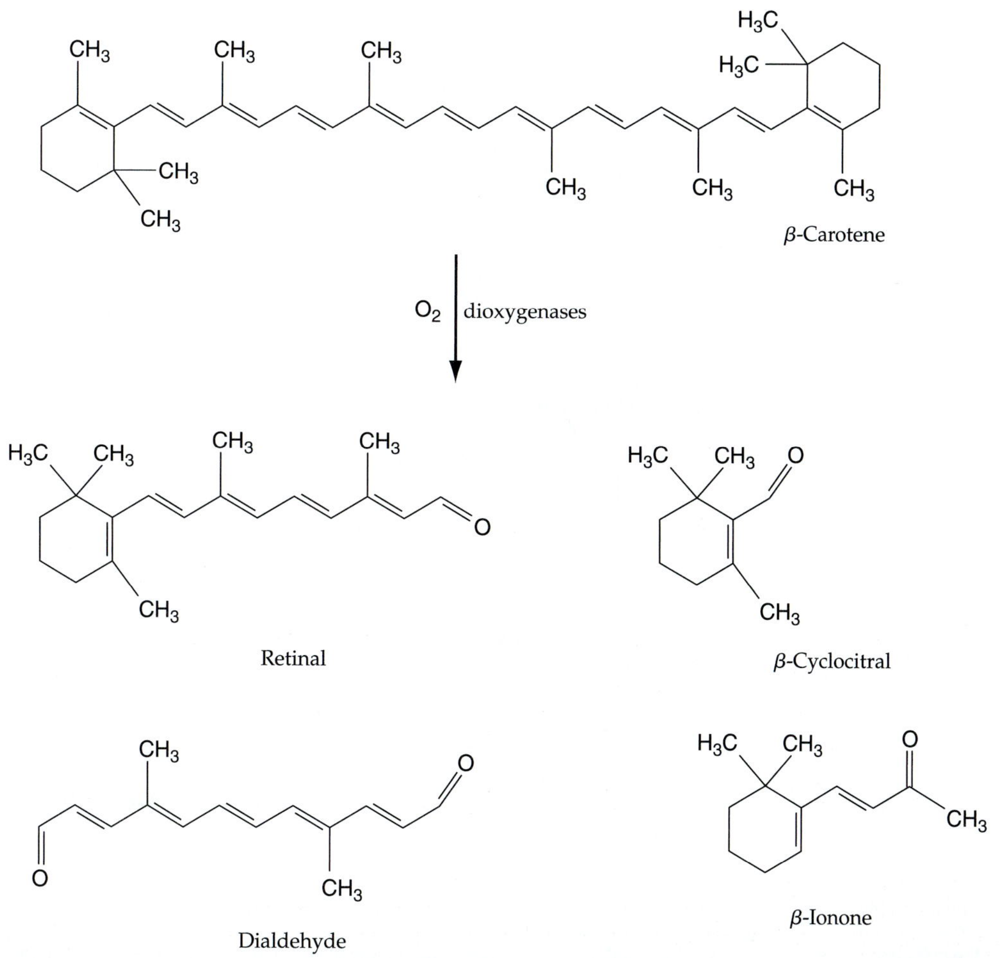

FIGURE 7.23 Cleavage of carotenoids to apocarotenoids. Specific dioxygenases catalyze C–C bond cleavage at several positions in the β-carotene molecule. This yields a variety of shorter carotenoid derivatives containing oxygen. Many of the cleavage products are volatile and contribute to the odor of injured plants.

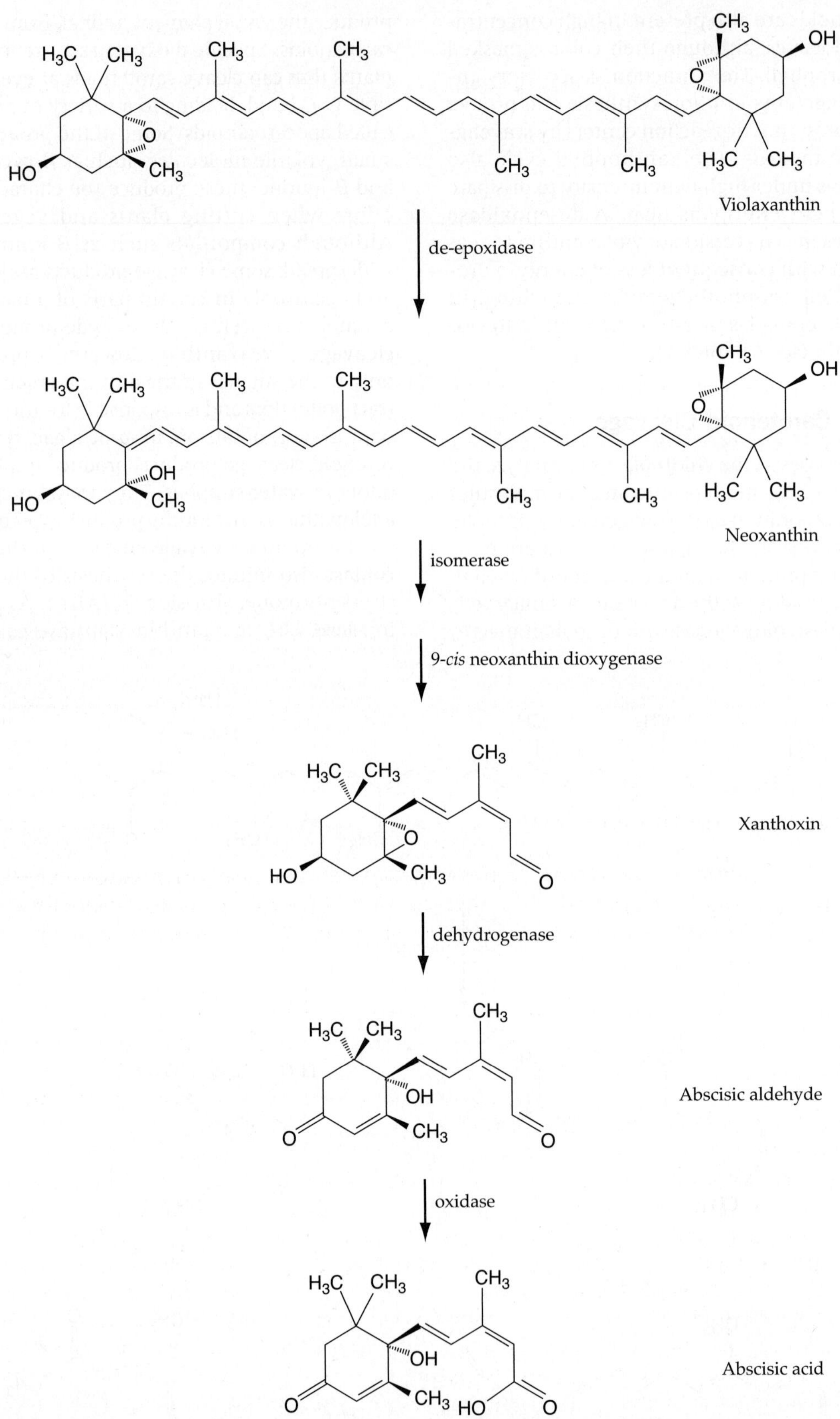

FIGURE 7.24 Synthesis of abscisic acid. Abscisic acid biosynthesis starts in plastids from xanthophyll precursors, violaxanthin and neoxanthin. A specific carotenoid cleavage dioxygenase, 9-*cis* neoxanthin dioxygenase, generates the 15-carbon abscisic acid precursor, xanthoxin. This precursor is transported out of the plastids, and abscisic acid synthesis is completed by enzymes found in the cytoplasm. The last step, oxidation of abscisic aldehyde, requires the action of a specific oxidase containing a molybdenum cofactor.

de-epoxidation and rearrangement of the carotenoid, violaxanthin, to neoxanthin. After isomerization to 9-*cis* neoxanthin, a specific dioxygenase (9-*cis* neoxanthin dioxygenase) cleaves neoxanthin and adds oxygen to produce a 15-carbon product, xanthoxin. Xanthocin is subsequently transported across the chloroplast membranes to the cytoplasm. A dehydrogenase then oxidizes the hydroxyl group of xanthoxin to a ketone. Abscisic oxidase is a unique molybdenum-containing enzyme that catalyzes the final oxidation to abscisic acid.

Abscisic acid is produced in leaves under heat and drought stress. It promotes closing of the stomates and conservation of water. It is also found in high concentrations in seeds and tubers where it serves to maintain dormancy. Subsequent breakdown of abscisic acid and synthesis of gibberellins stimulate germination.

It has recently been reported that another plant hormone is derived from cleavage of carotenoids. The strigolactones were initially described as seed germination promoters in the parasitic plants, *Striga* and *Orobanche*. They are exuded from the roots of many angiosperms, and promote a symbiotic association with mycorrhizal fungi. The strigolactones also inhibit shoot branching and maintain bud dormancy in angiosperms.

Polymers

Many plants contain prenyl transferases that catalyze the polymerization of IPP into chains containing several thousand units. These polymers remain as linear molecules and are exported to the apoplast and transported in the xylem. They are the major organic compound in the sap of many plants, principally the Euphorbiaceae. Especially long polymers of *cis*-isoprene are attached to small proteins and transported in the xylem of *Hevea brasiliensis*, the Brazilian rubber tree, which is the major source of natural rubber. The all-*cis* configuration gives the molecules a slight helical twist and accounts for the elasticity of this particular latex.

CHAPTER SUMMARY

Isoprenoid compounds are produced from the five-carbon compound, isopentenyl diphosphosphate (IPP), and its isomer, dimethylallyl diphosphate (DMAPP). These precursors are synthesized from acetyl-CoA in the cytoplasm by the mevalonic acid (MEV) pathway. The principal end products of this pathway are sterols, a major component of cellular membranes and the precursors for brassinosteroid hormones. Inducible enzymes also utilize FPP as a substrate for phytoalexin synthesis.

A second pathway to IPP is found in the chloroplast. The methyl-erythritol phosphate MEP pathway starts with pyruvate and glyceraldehyde-3-phosphate. Carotenoids are major products of isoprenoid synthesis in plastids and also hormone precursors for gibberellins, abscisic acid, and strigolactones (**FIGURE 7.25**).

Terpene synthesis relies mainly on enzymes that use the same basic mechanism (i.e., the removal of a diphosphate from DMAPP). The resulting carbocation can be added to a number of different molecules, including IPP itself to produce longer chain terpenes. These can, in turn, be extended by addition to IPP. The linear terpenes are cyclized by synthases that also catalyze loss of diphosphate and formation of a cation. A large family of plant enzymes can catalyze these reactions. The enzymes vary in the size of the active site and often have changes in amino acid residues near the active center that lead to the production of products specific to a particular plant or plant class. Cyclic terpenes are further modified by additional enzymatic reactions, most often by monooxygenases that add OH groups and glycosyl transferases that add glucose.

Isoprenoid compounds comprise a large number of primary metabolites as noted above but also produce unique plant natural products. These include monoterpenes, which are volatile and serve to attract pollinators or deter herbivores, and sesquiterpene-derived phytoalexins, which are synthesized in response to pathogen invasions.

The presence of two seemingly redundant pathways to IPP synthesis in plants may be a result of their mixed evolutionary origin from a nonphotosynthetic cell with the MEV pathway and a cyanobacterium with the MEP pathway. Although each pathway produces precursors for specific end products, there is evidence for some transport of IPP between the cytoplasm and the chloroplast. However, one pathway cannot completely substitute for the functions of the other. The most dramatic illustrations are the knock-out mutants in the MEP pathway in the chloroplast. These mutants cannot produce carotenoids, and cytoplasmic IPP does not substitute for the loss of the MEP product. MEP mutants are albinos, and the mutation is lethal.

CHAPTER SUMMARY, CONT.

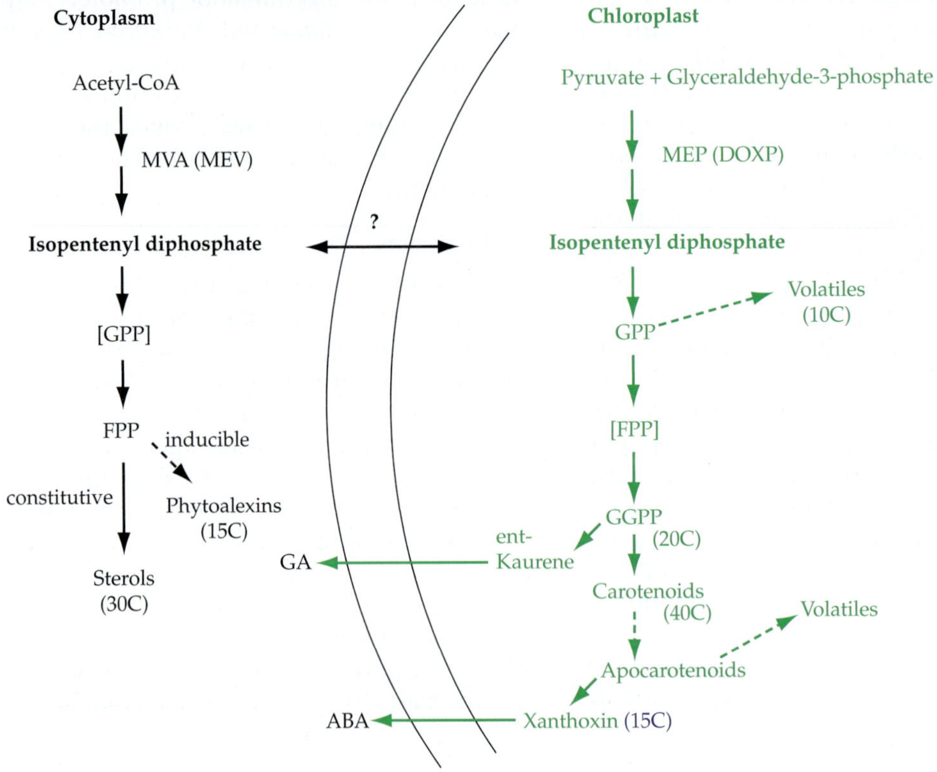

FIGURE 7.25 Summary of isoprenoid metabolism in plants. The MVA (MEV) pathway in the cytoplasm produces IPP that is used to maintain a pool of FPP for sterol synthesis. FPP is also used for induced phytoalexin synthesis. The MEP (DOXP) pathway yields IPP in the chloroplast stroma, which maintains a pool of GPP for monoterpene synthesis. GPP can also be used for GGPP synthesis, normally bypassing FPP production in the plastid. GGPP is used mainly to synthesize carotenoids. Isoprenoid hormone precursors, *ent*-kaurine and xanthoxin, are also produced in plastids and transported to the cytoplasm for the final steps of hormone synthesis. Although there is evidence for transport of IPP between compartments, plants require both pathways to be fully functional.

REFERENCES FOR FURTHER INFORMATION

Bouvier, F., Rahier, A., & Camara, B. (2005). Biogenesis, molecular regulation and function of plant isoprenoids. *Prog. Lipid Res.* **44:**357–429.

Eisenreich, W., Bacher, A., Arigoni, D., & Rohdich, F. (2004). Biosynthesis of isoprenoids via the non-mevalonate pathway. *Cell. Mol. Life Sci.* **61:**1401–1426.

Gershenzon, J. & Dudareva, N. (2007). The function of terpene natural products in the natural world. *Nat. Chem. Biol.* **3:**408–414.

Kirby, J. & Keasling, J.D. (2009). Biosynthesis of plant isoprenoids: Perspectives for microbial engineering. *Annu. Rev. Plant Biol.* **60:**335–355.

Moise, A.R., vonLintig, J., & Palczewski, K. (2005). Related enzymes solve evolutionarily recurrent problems in the metabolism of carotenoids. *Trends Plant Sci.* **10:**178–186.

PLANT BIOCHEMISTRY *on the web*

Visit the *Plant Biochemistry* student companion Web site at http://biology.jbpub.com/gleason/plantbiochemistry/ for additional online study resources on these topics.

CHAPTER 8

Aromatic and Phenolic Compounds

CHAPTER OUTLINE

8.1 Overview

Plants are autotrophic organisms and must synthesize all their metabolites from the intermediates of the C-3 cycle and other pathways of primary carbon and nitrogen metabolism. This includes the 20 common amino acids used in protein synthesis. Many of these amino acids are synthesized directly from precursors derived from primary metabolism. For example, the amino acid glutamic acid is produced in a single reaction: transfer of an amino group to 2-oxoglutarate (α-ketoglutaric acid), an intermediate in the citric acid cycle. Synthesis of the aromatic amino acids, phenylalanine, tyrosine, and tryptophan, is more complex. In addition to their function in protein synthesis, most amino acids serve as precursors for the synthesis of other metabolites. This is especially true for phenylalanine, which is the precursor for the phenylpropanoid pathway. Removal of the amino group from phenylalanine yields *trans*-cinnamic acid followed by addition of an OH and coenzyme A, which produces *p*-coumaroyl-CoA. This key intermediate is essential in synthesis of lignin precursors and is considered a primary metabolic pathway in terrestrial plants. Coumaroyl-CoA is also the initial compound for a variety of natural products, including flavonoid synthesis.

Flavonoids are found in most plants and give rise to colored compounds such as anthocyanins. They function to protect plants from intense light and ultraviolet (UV) radiation and can act as antioxidants. A vast number of additional natural products are synthesized from the aromatic amino acids and their precursors. Only a small number of these are presented in this chapter.

8.2 Shikimic Acid Pathway

All three aromatic amino acids are derived from intermediates of the same series of reactions, the shikimic acid pathway. Synthesis starts with erythrose-4-phosphate and phosphoenolpyruvic acid (PEP). The pathway is named after the first intermediate that is unique to aromatic amino acid synthesis, shikimic acid. The reactions in this pathway are similar in plants, fungi, and bacteria but are absent in animals. The component enzymes are homologous in all these kingdoms. In plants, the enzymes are found in plastids and are presumably all soluble in the stroma. The genes for these enzymes are encoded in the nucleus. An overview of the pathway is shown in **FIGURE 8.1**. In addition to the three aromatic amino acids, the intermediates of the shikimic acid pathway are also used in the synthesis of a variety of other metabolites, including the plant hormone salicylic acid, vitamins such as folic acid, and a number of species-specific natural products.

The erythrose-4-phosphate precursor is an intermediate in the C-3 cycle and the pentose phosphate pathway in the chloroplast stroma (see Chapters 2 and 3). PEP is an intermediate in glycolysis and may be imported into chloroplasts. Alternatively, it can be a product of chloroplast pyruvate metabolism via pyruvate kinase and pyruvate, Pi dikinase. The first reaction in the pathway is an aldol condensation that results in the synthesis of a 7-carbon ketose, 3-deoxy-arabino-heptulosonic acid-7-phosphate (DAHP), shown in **FIGURE 8.2**.

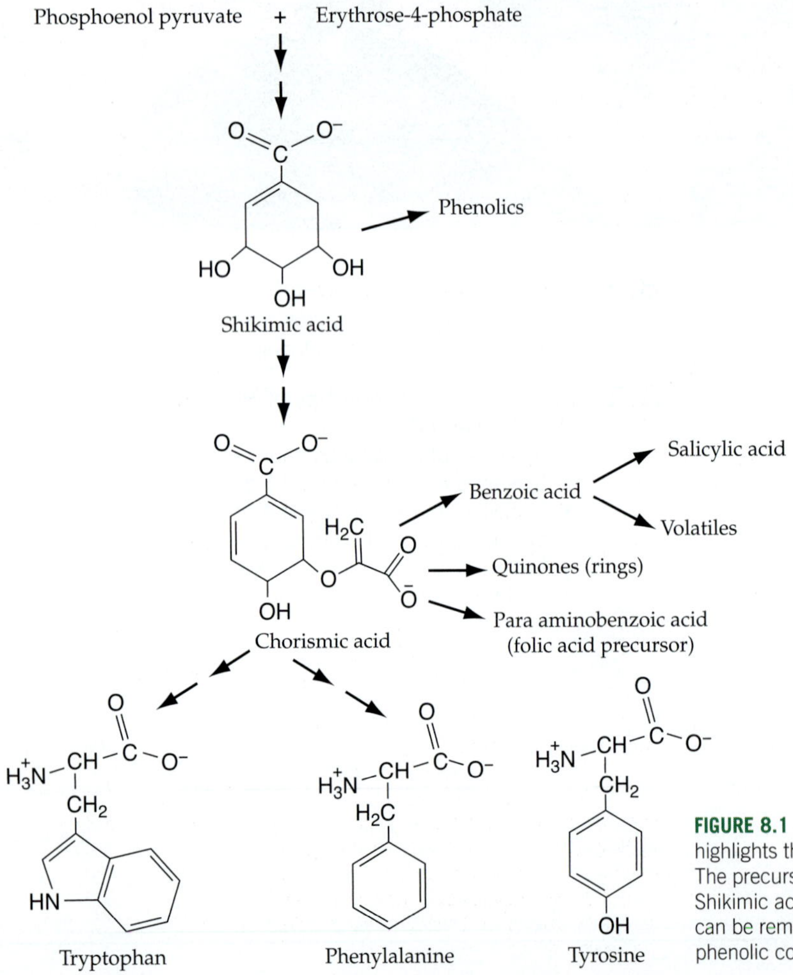

FIGURE 8.1 Simplified overview of the shikimic acid pathway. This diagram highlights the main intermediates and products of the shikimic acid pathway. The precursors are phosphoenolpyruvate (PEP) and erythrose-4-phosphate. Shikimic acid and chorismic acid are key intermediates. These intermediates can be removed from the pathway and used to synthesize a variety of additional phenolic compounds.

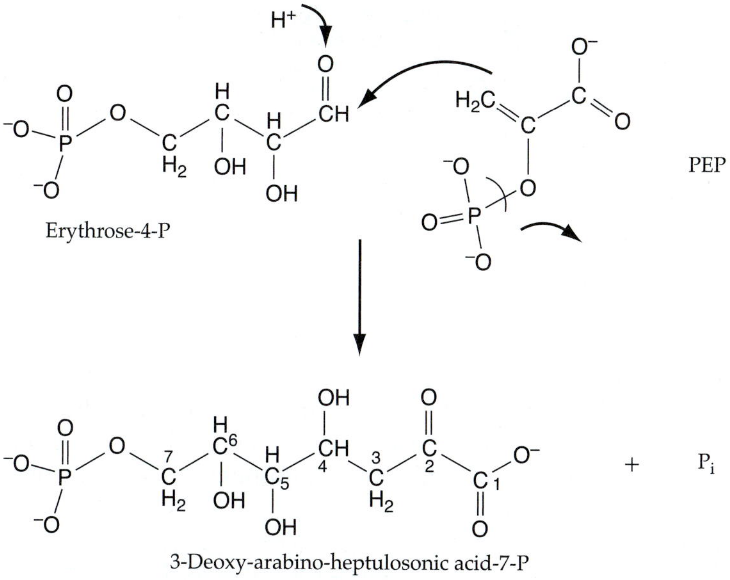

FIGURE 8.2 Synthesis of 3-deoxy-arabino-heptulosonic acid-7-phosphate (DAHP). The DAHP synthase catalyzes an aldol-type condensation. The phosphate is removed from PEP, and its methylene carbon then forms a bond with the carbonyl carbon of erythrose-4-phosphate.

DAHP synthases in plants and bacteria show only about 20% amino acid identity. The plant enzyme, however, complements bacterial mutants lacking this synthase. The bacterial enzyme exhibits feedback inhibition by the aromatic amino acid end products, phenylalanine and tyrosine. This does not appear to be true of the plant synthases. Expression of different isozymes of DAHP synthase in plants, however, is influenced by environmental factors such as high light intensity or wounding and the presence of hormones, gibberellic acid and jasmonic acid. The intermediates and end products of the shikimic acid pathway are often intermediates in the production of phytoalexins and other components of plant defense reactions.

The second step in the pathway is the formation of a cyclic intermediate, 3-dehydroquinic acid, from DAHP (**FIGURE 8.3**). The enzyme, dehydroquinate synthase, is an oxidoreductase that requires NAD as a cofactor. The enzyme in this case catalyzes both an oxidation followed by a reduction, thereby regenerating the oxidized cofactor.

Dehydroquinic acid is the substrate for a dehydratase that catalyzes removal of water and introduces a double bond into the ring. This is followed by reduction of the ketone to an alcohol forming shikimic acid, the first unique intermediate (**FIGURE 8.4**). In some plants, shikimic acid is the starting material for the synthesis of a variety of phenolic natural products. The most common are the water-soluble gallotannins, which are complexes of phenolics with sugars, usually glucose. Plants make these compounds to protect their tissues from UV damage and to deter herbivores. Tannins can bind irreversibly to proteins and inhibit enzyme activity. Humans have traditionally used tannic acid–rich plant extracts to tan and preserve animal hides. Some humans also like the bitter flavor of gallotannins in beverages such as tea.

Shikimic acid is further metabolized to chorismic acid, another key intermediate in this pathway. A kinase adds a phosphate group to one of the meta hydroxyls to produce shikimic acid-3-phosphate. This intermediate is then condensed with another molecule

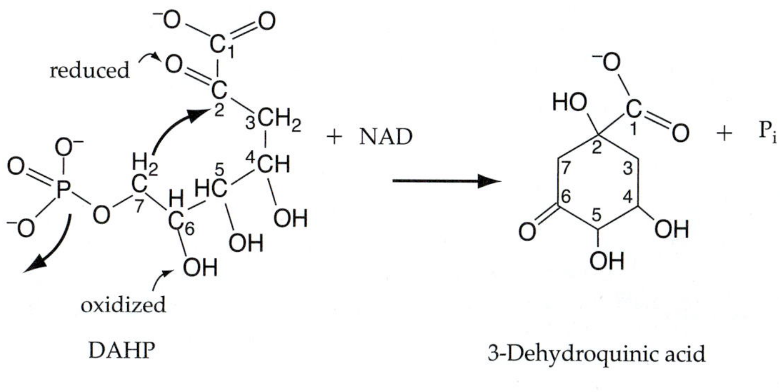

FIGURE 8.3 Reactions catalyzed by dehydroquinate synthase. The phosphate is removed from C-7 of DAHP and the resulting carbocation forms a bond with C-2. The enzyme then catalyzes the oxidation of the hydroxyl (OH) group on C-6 to a ketone and a reduction of the ketone on C-2 to an OH to yield 3-dehydroquinic acid. The redox cofactor for this enzyme is NAD, which is regenerated during the reaction.

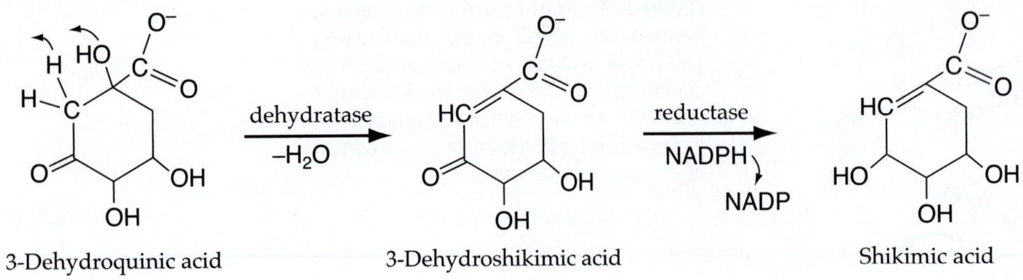

FIGURE 8.4 Synthesis of shikimic acid. A dehydratase-reductase binds dehydroquinic acid. It catalyzes removal of the hydroxyl next to the carboxyl group on C-1 and a proton from the adjacent C-6 carbon, generating a double bond in the ring. The NADPH-dependent reductase activity then reduces the keto group to a hydroxyl group to generate shikimic acid.

of PEP to produce 5-enolpyruvyl-shikimate-3-phosphate (EPSP) (**FIGURE 8.5**). The enzyme, EPSP synthase, is inhibited by an amino acid analog, *N*-phosphonomethyl glycine, also known as glyphosate. Although the structures do not appear to be similar, glyphosate competes with PEP for the same binding site on the synthase. As shown in **FIGURE 8.6**, the synthase consists of two globular domains. Binding of shikimate-3-phosphate triggers a global

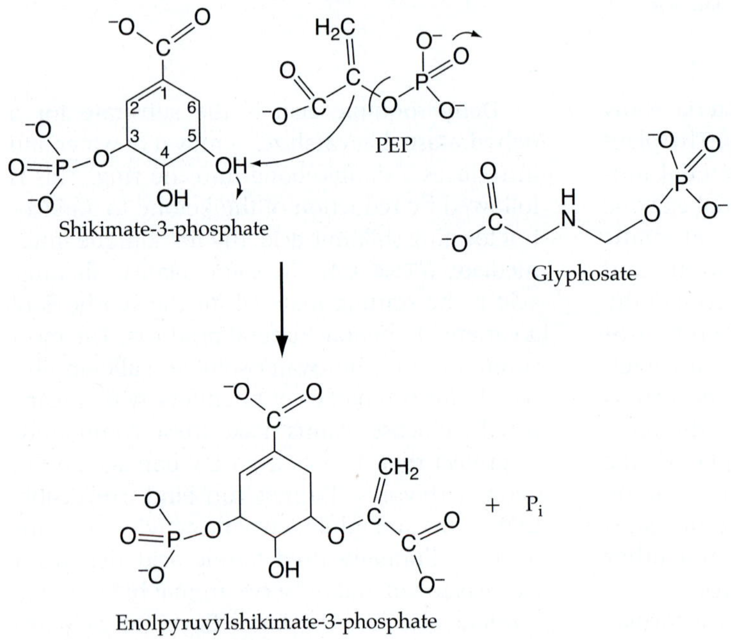

FIGURE 8.5 Synthesis of 5-enol-pyruvyl-shikimate-3-phosphate (EPSP). The EPSP synthase catalyzes removal of the phosphate from PEP. The resulting three-carbon fragment is added to the hydroxyl group on C-3 of shikimate-3-phosphate. The amino acid analog, glyphosate (inset), is a competitive inhibitor of PEP.

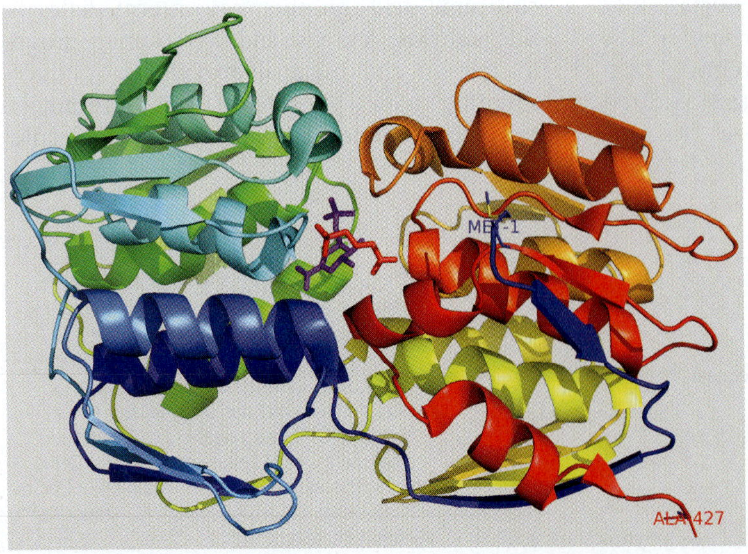

FIGURE 8.6 Structure of EPSP synthase. The enzyme consists of two domains shown in blue-green and red-yellow. In the absence of substrate, a wide gap separates the two domains. After binding shikimate-3-phosphate (shown as purple sticks), the enzyme undergoes a large conformational shift to narrow the gap between the domains as shown. The constricted active site can now bind pyruvate or glyphosate, shown as red sticks. The EPSP synthase is a good example of "induced fit" where binding of one substrate induces a conformational change in the protein to accommodate the second substrate and subsequent catalysis. (Structure from Protein Data Bank 1G6S. E. Schonbrunn, S. Eschenburg, W. Shuttleworth, J. V. Schloss, N. Amrhein, J. N. S. Evans, and W. Kabsch, *Proc. Natl. Acad. Sci.* 98 [2001]: 1376–1380.)

conformational change to the more closed structure seen in Figure 8.6. PEP then binds to the active site formed at the interface of the two domains. Both kinetic and crystal structure analyses show that glyphosate competes with PEP for this same site. The unique nature of this structural change is further illustrated by the fact that glyphosate does not compete with PEP in any of the many other reactions that use PEP as a substrate. Glyphosate is a highly effective, broad-spectrum herbicide and is the active component in Monsanto's Roundup®.

EPSP is the substrate for a synthase that removes the phosphate from C-3 and introduces a second double bond into the ring. This results in the production of chorismic acid (**FIGURE 8.7**). Chorismic acid can be a precursor for benzoic acid and its derivatives and also salicylic acid in some plants. Methylated benzoic and salicylic acids are volatile and are responsible for many floral scents that attract pollinators. Salicylic acid is also a signaling molecule that initiates responses to abiotic stress and pathogen invasion, a process referred to as systemic acquired resistance (SAR). Additionally, chorismic acid in some plants provides the ring portion of quinones and *para*-amino benzoic acid, the precursor for folic acid synthesis. In many plants some of these compounds can also be derived from phenylalanine.

Phenylalanine and Tyrosine Synthesis

Chorismic acid is the branch point between synthesis of phenylalanine and tyrosine or tryptophan (Figure 8.1). Chorismate mutase catalyzes the transfer of the pyruvyl side chain to C-1 to produce prephenic acid, the precursor of phenylalanine and tyrosine (Figure 8.7). This mutase is allosterically regulated by the three aromatic amino acids. Allosteric regulators are generally small molecules that can either activate or inhibit enzyme activity, but these regulators do not bind to the enzyme's active site. Most allosterically regulated enzymes are composed of two or more subunits, and the regulators generally act by changing the subunit association. In the case of chorismate mutase, the enzyme's activity is inhibited by phenylalanine and tyrosine and activated by tryptophan, thereby ensuring balanced pools of all three amino acids.

In the next step, an amino group is added to prephenic acid to produce arogenic acid (Figure 8.7). The transaminase that catalyzes

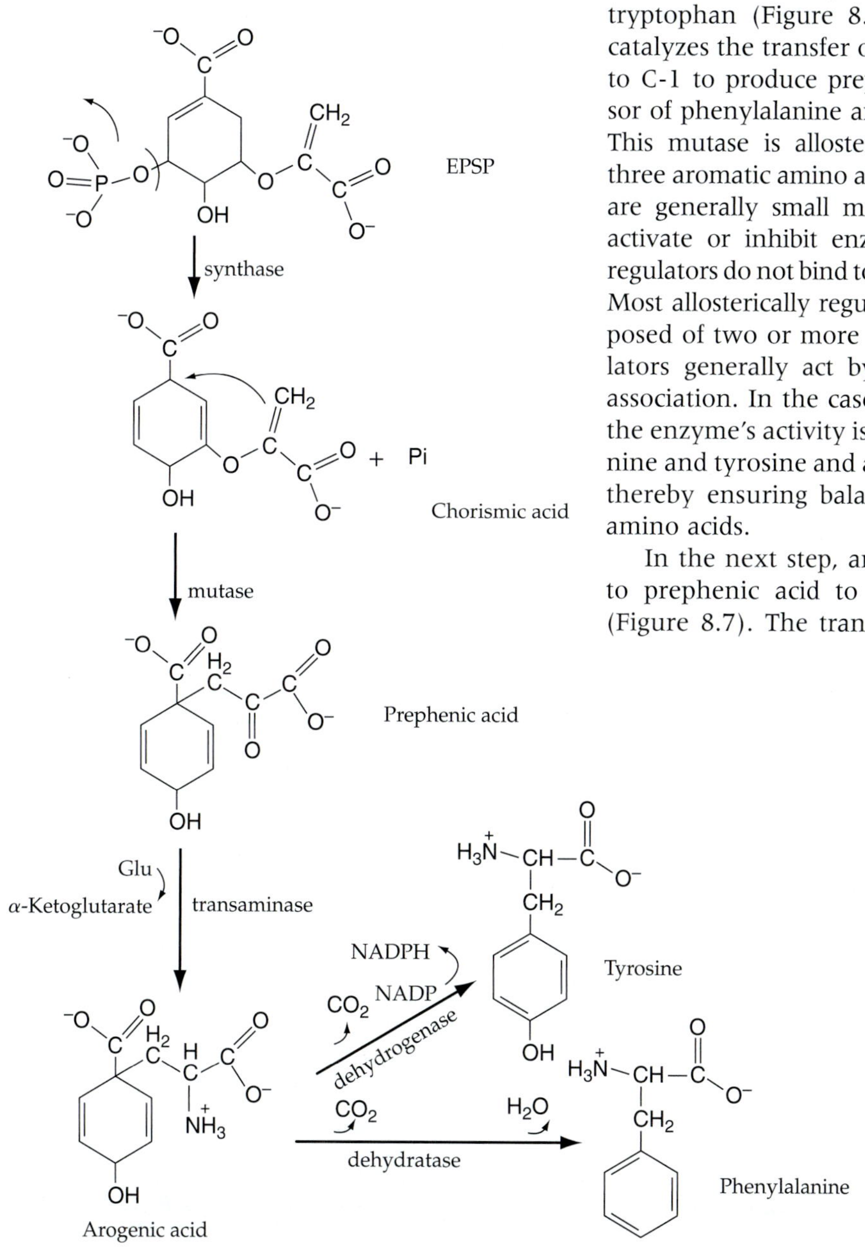

FIGURE 8.7 Synthesis of tyrosine and phenylalanine from chorismic acid. Chorismate synthase catalyzes the removal of the phosphate group from EPSP and introduces a second double bond into the ring to generate chorismic acid. The prephenate mutase reaction results in the relocation of the pyruvyl moiety to C-1 of the ring producing prephenic acid. A transaminase then catalyzes transfer of an amino group from glutamic acid to C-2 of the side chain resulting in the amino acid, arogenic acid. In most plants, arogenic acid is the precursor for both tyrosine and phenylalanine. To produce tyrosine, an NADP-dependent dehydrogenase oxidatively decarboxylates C-1 of the ring and catalyzes formation of a third double bond to generate an aromatic ring. In phenylalanine synthesis, a dehydratase removes both the C-1 carboxyl group and a water molecule from the ring, thereby generating the aromatic amino acid.

this reaction uses glutamic acid as the amino donor. Oxidative-decarboxylation of arogenic acid by an NADP-dependent dehydrogenase yields tyrosine. To synthesize phenylalanine, decarboxylation of arogenic acid is followed by a dehydration to produce the aromatic ring.

Tryptophan Synthesis

Chorismic acid is also the precursor for the synthesis of tryptophan (Figure 8.1). A synthase catalyzes removal of the pyruvyl side chain and a transamination to produce anthranilic acid (**FIGURE 8.8**). The amino donor in this reaction is glutamine. The next step involves a glycosylation, that is, addition of 5-phospho-ribosyl-1-diphosphate to the amino group of anthranilic acid. The sugar-diphosphate is a good leaving group, and the sugar is bonded to anthranilate by an *N*-glycosidic bond as seen in nucleosides. An isomerase then catalyzes the formation of the ketose derivative. Decarboxylation of the aromatic ring and dehydration and cyclization produce the five-membered ring of indole glycerol-phosphate.

The final step in this branch of the pathway is catalyzed by tryptophan synthase. This enzyme cleaves the glycerol-P moiety from the indole ring and replaces it with serine (Figure 8.8). Tryptophan synthase is an $\alpha_2\beta_2$ heterotetramer in plants. It is an unusual example of a bifunctional enzyme. Indole glycerol-phosphate binds to an active site located on the α subunit. Binding causes a conformational change that closes the active site. After removal of glyceraldehyde-3-phosphate, the indole ring diffuses through a long protein tunnel (~25Å) between subunits to the active site in the β subunit. Serine then binds and closes the active site in the β subunit. The hydroxyl group is removed from serine, and the beta-carbon is bonded to the indole ring. Formation of tryptophan opens the active sites on both subunits, and the products are released. Although crystal structures of several tryptophan synthases with bound intermediates are available, it is still not known how the two active sites communicate to coordinate their activity, but this enzyme does not produce free indole. Some plants, such as maize, produce indole-based phytoalexins. In this case an enzyme resembling only the α subunit of tryptophan synthase is found in the cytoplasm. A similar enzyme is the most likely source of the volatile indole released from many plants.

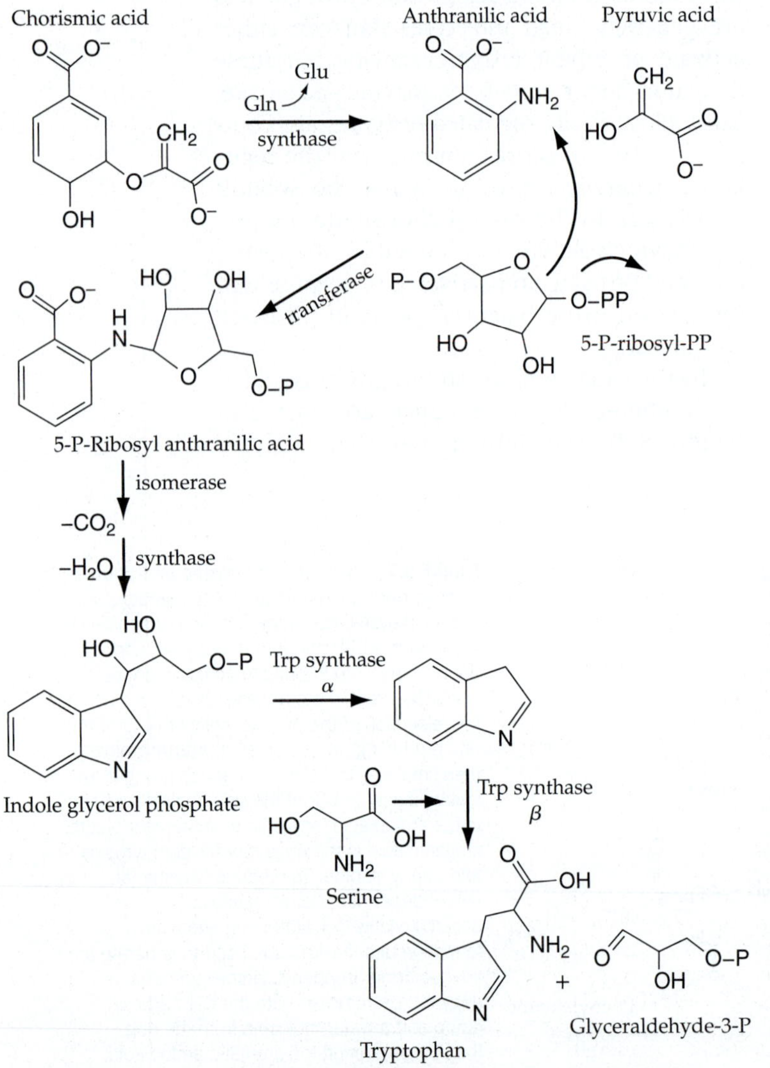

FIGURE 8.8 Synthesis of tryptophan. Anthranilate synthase binds chorismic acid and catalyzes removal of the pyruvyl moiety followed by transfer of an amino group from glutamine to the ring to form anthranilic acid. The sugar, 5-phosphate-ribosyl-diphosphate, is coupled to the amino group in anthranilate by an *N*-glycosidic bond and pyrophosphate is released. The 5-phospho-ribosyl moiety then undergoes isomerization to a ketose. The ring of the intermediate is decarboxylated, and a third enzyme catalyzes a ring closure of the ketose to generate the five-membered ring in indole glycerol-3-phosphate. This intermediate is the substrate for the α subunit of $\alpha_2\beta_2$ tryptophan synthase. Binding of indole glycerol-phosphate to the α subunit closes this subunit active site, and the enzyme cleaves the glyceryl-phosphate moiety from the indole ring. The products are not released from the enzyme. The indole intermediate is transported through a channel in the enzyme to the β subunit. Serine also binds to the β subunit, and binding closes this second active site. The hydroxyl group is removed from serine, and the remaining three-carbon fragment is coupled to the indole ring to generate tryptophan. Opening of the active sites on both subunits releases tryptophan from the β subunit and glyceraldehyde-3-phosphate from the α-subunit site.

8.3 Phenylpropanoid Pathway

Many complex plant products are synthesized from the aromatic amino acids, including both primary and secondary metabolites (natural products). Phenylalanine occupies a particularly prominent position in plant metabolism as it is the starting material for the synthesis of a large number of aromatic compounds. These metabolites are referred to as phenylpropanoids and include primary products such as the monolignol precursors of lignin and a large number of species-specific natural products. Lignin is a major component of secondary cell walls and the second most abundant polymer in the biosphere after cellulose. Its synthesis is vital to the survival of all terrestrial plants. Lignin is a polymer of phenylpropanoid precursors that are exported to the cell wall matrix of mature plant tissues. In the wall the precursors are polymerized and form cross-links to components of the primary cell wall. Phenylpropanoids are also the precursors for phytoalexins in some plants and volatiles such as eugenol found in cloves. Another major role of phenylpropanoids is in the production of polyketides, including flavonoids, which are found in most plants and enable them to interact and cope with environmental factors. All this complex variety of end products starts with a phenylalanine precursor.

Synthesis of *Trans*-Cinnamic Acid

The synthesis of phenylpropanoids starts with the removal of the amino group of phenylalanine to produce *trans*-cinnamic acid. The enzyme that catalyzes this initial step is phenylalanine ammonia lyase (PAL). PAL is found in angiosperms and gymnosperms and also in some mosses, algae, fungi, and bacteria. Several PAL isozymes occur in angiosperms. The plant enzymes are found in the cytoplasm, most likely associated with the outer face of the endoplasmic reticulum. PAL from various plant sources is a tetramer with a mass of 225 to 330 kDa. In dicotyledonous plants, PAL is specific for phenylalanine. Some dicots also have another enzyme that deaminates tyrosine (tyrosine ammonia lyase). In monocots, PAL accepts both phenylalanine and tyrosine as substrates, although the catalytic efficiency is lower for tyrosine.

Plant PAL belongs to the amino acid–ammonia–lyase family of enzymes. The plant enzyme is homologous to the bacterial histidine ammonia lyase (HAL) and has the same mechanism of action. The crystal structure of PAL from parsley (*Petroselinum crispum*) has been determined. The protein is mainly composed of α-helices, and its overall tertiary structure is very similar to the PAL from *Rhodosporidium toruloides* and histidine ammonia lyase from the bacterium, *Pseudomonas syringae*. All these enzymes are unique in that they generate their own cofactor at the active site by cyclization and dehydration of an alanine-serine-glycine sequence to generate 4-methylidene-imidazole-5-one (MIO) (**FIGURE 8.9**). The methylidene group of the MIO cofactor binds the amino group of the substrate, phenylalanine, and weakens the C–H bond on the β-carbon of the substrate

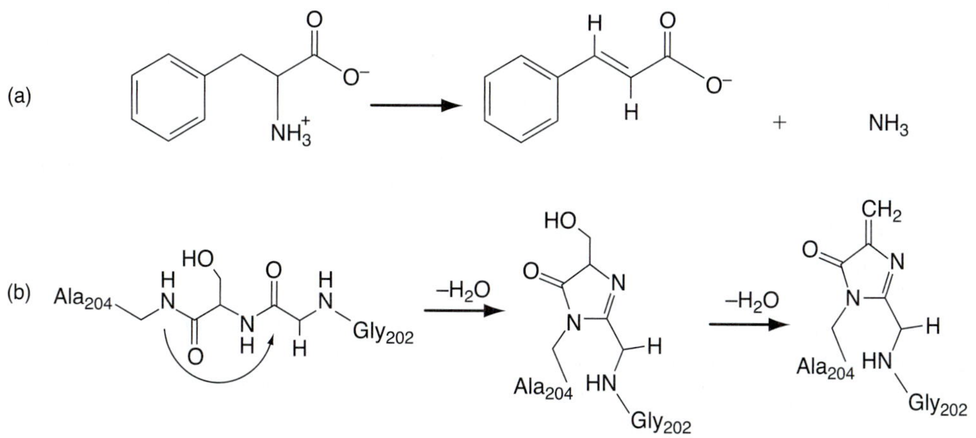

FIGURE 8.9 Activity of phenylalanine ammonia lyase (PAL). (a) PAL catalyzes the removal of the amino group from the α-carbon and a hydrogen from the β-carbon of phenylalanine to produce *trans*-cinnamic acid and ammonia. (b) The enzyme generates a unique cofactor at its active site by cyclization of three residues, alanine-204, serine-203, and glycine-202 (residue numbering is from the bacterial enzyme), followed by a dehydration. The amino group in the peptide bond between ala-204 and ser-203 attacks the carbonyl carbon in the peptide bond between ser-203 and gly-202. Subsequent removal of two H_2O molecules generates the 4-methylidene-imidazole-5-one (MIO) cofactor that binds the substrate, phenylalanine, and facilitates removal of the amino group to generate NH_3 (shown in Figure 8.10).

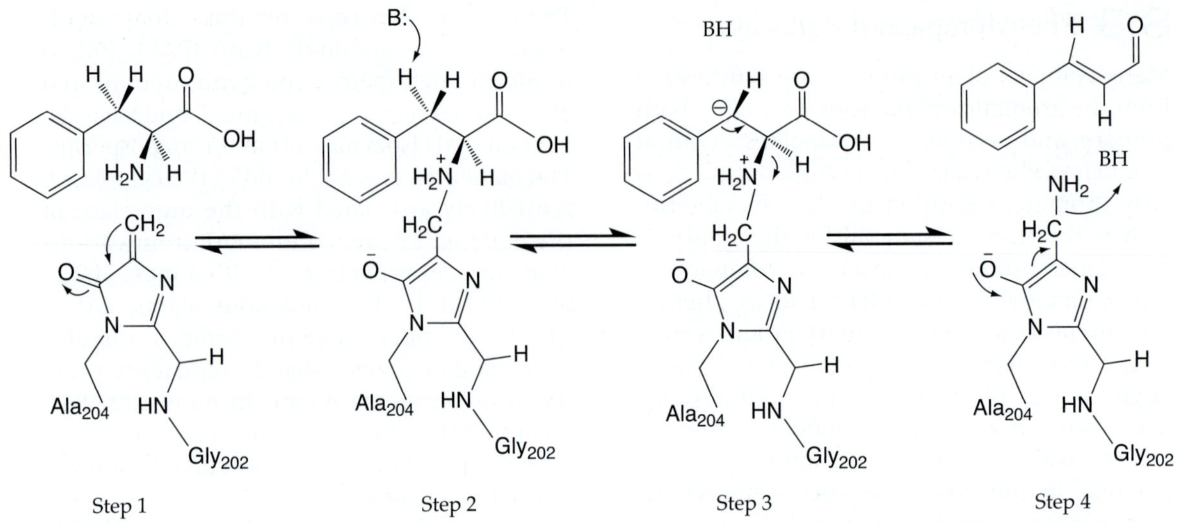

FIGURE 8.10 Mechanism of action of phenylalanine ammonia lyase (PAL). This enzyme is the gateway to the phenylpropanoid pathway, and its mechanism has been thoroughly investigated. Step 1: The methylidene carbon of the MIO cofactor (Figure 8.9B) in PAL binds to the amino group of phenylalanine. Step 2: A basic group in the enzyme, B, stereospecifically removes one of the protons from the β-carbon of phenylalanine. Step 3: Rearrangement of the resulting anion generates a double bond in the substrate and facilitates cleavage of the bond between the α-carbon of phenylalanine and its amino group. Step 4: *Trans*-cinnamic acid is released from the enzyme. The protonated BH residue in the active site donates its H^+ to the aminated cofactor and facilitates removal of the NH_3 and regeneration of MIO. (*Mechanism adapted from*: Calabrese, J. et al., "Crystal Structure of Phenylalanine Ammonia Lyase: Multiple Helix Dipoles Implicated in Catalysis," *Biochemistry* 43 [2004]: 11403–416.)

(**FIGURE 8.10**). A proton is removed from this β-carbon with the help of a basic group in the active site, most likely a histidine residue. Breaking of the very stable C–H bond in this reaction is facilitated by the electrophilic environment around the active site created by the aligned helix dipoles of the enzyme. Removal of the proton generates an anion that rearranges and results in the cleavage of the C–N bond between phenylalanine and the MIO cofactor. The product, *trans*-cinnamic acid, is released from the active site. The cofactor is regenerated by H^+ donation from an acidic residue in the enzyme and ammonia is produced. *Trans*-cinnamic acid is the precursor for most of the phenylpropanoids, including the monolignols, used to synthesize lignin.

Lignin Synthesis

Lignin is a highly irregular polymer formed from aromatic alcohols referred to as monolignols. The major monolignols are coumaroyl, coniferyl, and sinapyl alcohols. These precursors are synthesized from *trans*-cinnamic acid by enzymes associated with the endoplasmic reticulum. The monolignols are exported to the cell wall, and polymerization of the monolignols then occurs outside the cell membrane.

Synthesis of *p*-Coumaroyl Alcohol The aromatic ring of *trans*-cinnamic acid can be hydroxylated by a specific cytochrome P450 monooxygenase. The monooxygenases in this pathway are often referred to as hydroxylases. The hydroxyl (OH) group is added at the *para*-position with respect to the side chain to produce 4-coumaric acid (*p*-coumaric acid). Coumaric acid can also be produced in some plants by direct deamination of tyrosine in a TAL-catalyzed reaction (**FIGURE 8.11**).

Coumaric acid can be ligated to coenzyme A by a ligase; this reaction requires two equivalents of ATP to activate coumaric acid and catalyze the formation of a thioester bond. Coumaroyl-CoA is a major branch point in the synthesis of monolignols and other phenylpropanoids. There is experimental evidence that all the monolignols are present as CoA esters until a final reduction to the alcohols, but some of the modifying enzymes can use the free acids or other derivatives as substrates. In addition, recent analyses have also shown that monolignols are present as complexes with shikimic or quinic acids, which are intermediates in the shikimic acid pathway (**FIGURE 8.12**). Because these acids are synthesized in plastids, it is assumed that they must be specifically transported to the cytoplasm to serve as monolignol carriers.

Para-coumaroyl-CoA, or other conjugate or the free acid, undergoes reduction to coumaroyl alcohol by two reductases that use NADPH as a cofactor. The mechanism of action of these two enzymes is slightly different. The first enzyme, coumaroyl-CoA reductase (CCR),

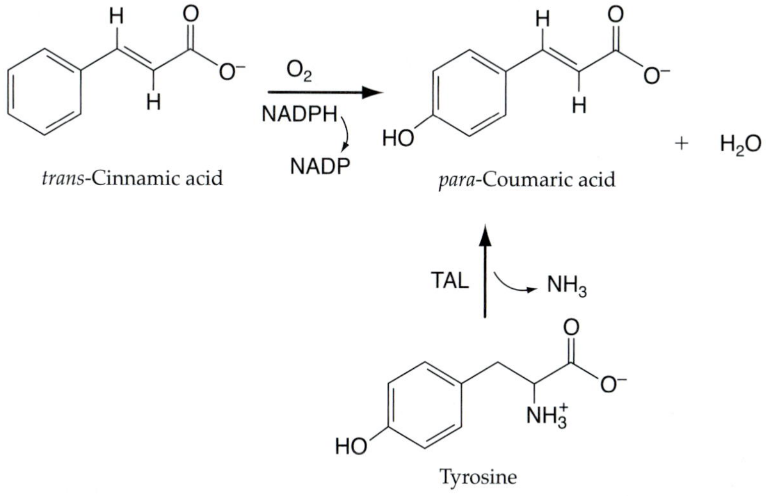

FIGURE 8.11 Synthesis of *p*-coumaric acid. A hydroxyl (OH) group is added to *trans*-cinnamic acid by a cytochrome P450 monooxygenase (hydroxylase). The enzyme adds the OH-group at the *para* position with respect to the side chain. Coumaric acid can also be synthesized by direct deamination of tyrosine catalyzed by tyrosine ammonia lyase (TAL) in some plants.

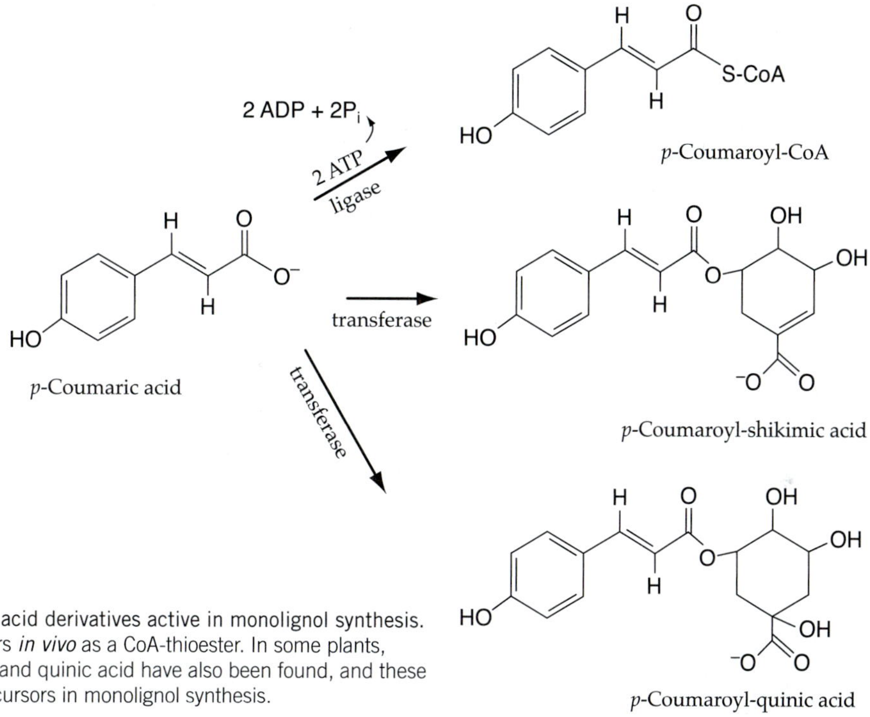

FIGURE 8.12 *Para*-coumaric acid derivatives active in monolignol synthesis. Coumaric acid generally occurs *in vivo* as a CoA-thioester. In some plants, conjugates with shikimic acid and quinic acid have also been found, and these may function as alternate precursors in monolignol synthesis.

reduces the acid to the aldehyde. It is referred to as a type B reductase because it transfers the HB hydride from NADPH (**FIGURE 8.13**). The second reductase, coumaroyl aldehyde dehydrogenase (CAD), is a type A reductase. It transfers the HA hydride from the cofactor and reduces the aldehyde to an alcohol. These reductases are highly specific and cannot substitute for each other. This has been demonstrated in genetically altered plants designed to generate different pools of monolignols.

Coumaroyl alcohol is one of the monolignol precursors. It can be polymerized into a complex referred to as *p*-hydroxyphenyl lignin, or H-lignin. H-lignin is common in some monocots, but in most plants it is often a minor component found in complexes with other types of lignin.

Synthesis of Coniferyl and Sinapyl Alcohols Coumaric acid, either as the free acid or a derivative, is hydroxylated by a second monooxygenase to produce the dihydroxy derivative, caffeic acid. This second OH group in the aromatic ring is then methylated by an O-methyl transferase that requires S-adenosyl methionine (SAM) as the methyl donor. Monohydroxy substituted rings are usually not good substrates for the methyl transferases, but they can bind a number of different hydroxylated substrates. The methylated caffeic acid is ferulic acid (**FIGURE 8.14**). Ferulic acid is then

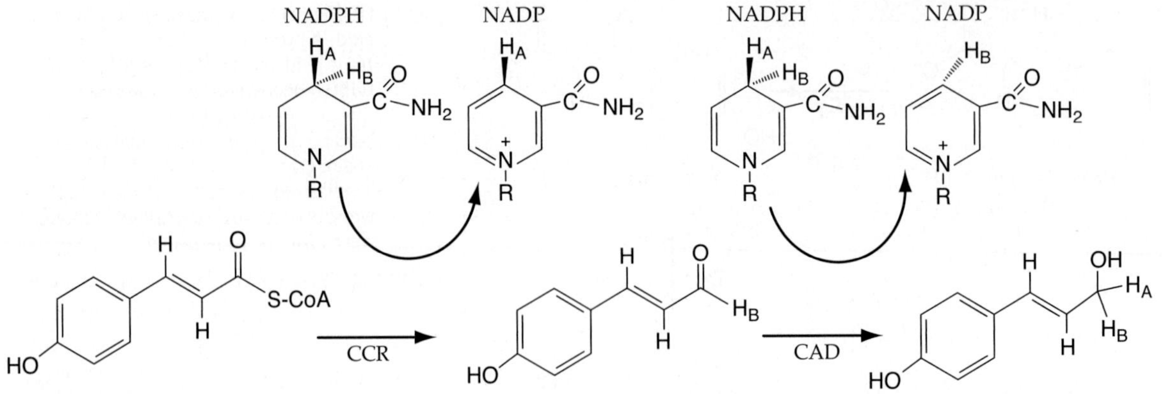

FIGURE 8.13 Reductases that reduce *p*-coumaroyl-CoA and other monolignol precursors. The first enzyme, *p*-coumaroyl-CoA reductase (CCR, also called cinnamate-CoA reductase), transfers the HB hydride from NADPH and reduces the thioester to the aldehyde. It is referred to as a type B reductase. The second enzyme, coumaroyl aldehyde dehydrogenase (CAD), transfers the HA hydride from the cofactor to produce coumaroyl alcohol; it is a type A reductase. Both CCR and CAD can also catalyze reduction of other monolignol precursors; NADPH is the reducing agent in both reactions.

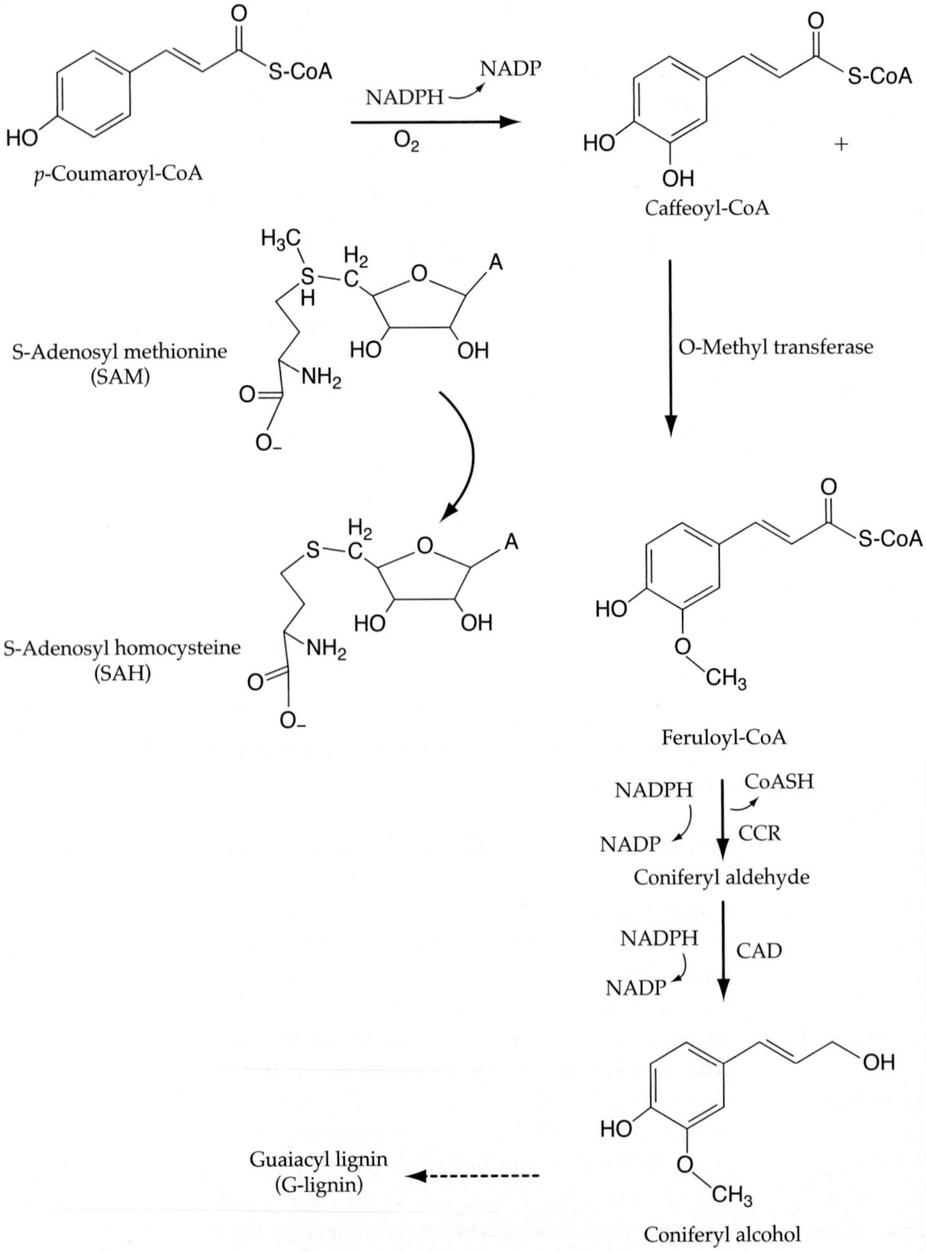

FIGURE 8.14 Synthesis of coniferyl alcohol. Coumaroyl-CoA is hydroxylated by a specific cytochrome P450 monooxygenase at the *meta* position (with respect to the side chain) to generate caffeic acid or a caffeoyl conjugate. A caffeoyl O-methyl transferase then adds a methyl group to this *meta*-hydroxyl group to produce ferulic acid. The methyl donor is S-adenosyl methionine (SAM). Removal of the methyl group from SAM yields S-adenosyl homocysteine (SAH), which can be remethylated. Ferulic acid and its conjugates are reduced by the *p*-coumaroyl-CoA reductase (CCR) and coumaroyl aldehyde dehydrogenase (CAD) reductases described in Figure 8.13 to yield coniferyl alcohol. This product is a lignin precursor that can be polymerized to guaiacyl or G-lignin, which is the major lignin type found in gymnosperms, especially conifers.

reduced to the alcohol by the same redox enzymes, CCR and CAD, described in Figure 8.13 to yield coniferyl alcohol. This alcohol is the major component of guaiacyl lignin (G-lignin).

Another cytochrome P450 monooxygenase introduces a third hydroxyl group into the aromatic ring to produce 5-hydroxy ferulic acid. This third hydroxyl group is methylated by a SAM-dependent transferase, as shown in **FIGURE 8.15**, to produce sinapic acid. Sinapic acid is then reduced to sinapyl alcohol, a major monolignol made in dicotyledonous angiosperms and the precursor of syringyl lignin (S-lignin).

Polymerization of Monolignols

The monolignol precursors, coumaroyl, coniferyl, and sinapyl alcohols, are exported to the exterior of the cell. They may be packaged into Golgi vesicles or transported directly. In some mature plant tissues monolignols are found as glycosylated derivatives, presumably for storage until secondary wall synthesis is induced. Polymerization of the monolignols occurs outside the cell membrane, between it and the primary cell wall. The polymerization process is nonenzymatic but is initiated by an enzymatic free radical formation. Two types of oxidases, peroxidases and laccases, are found

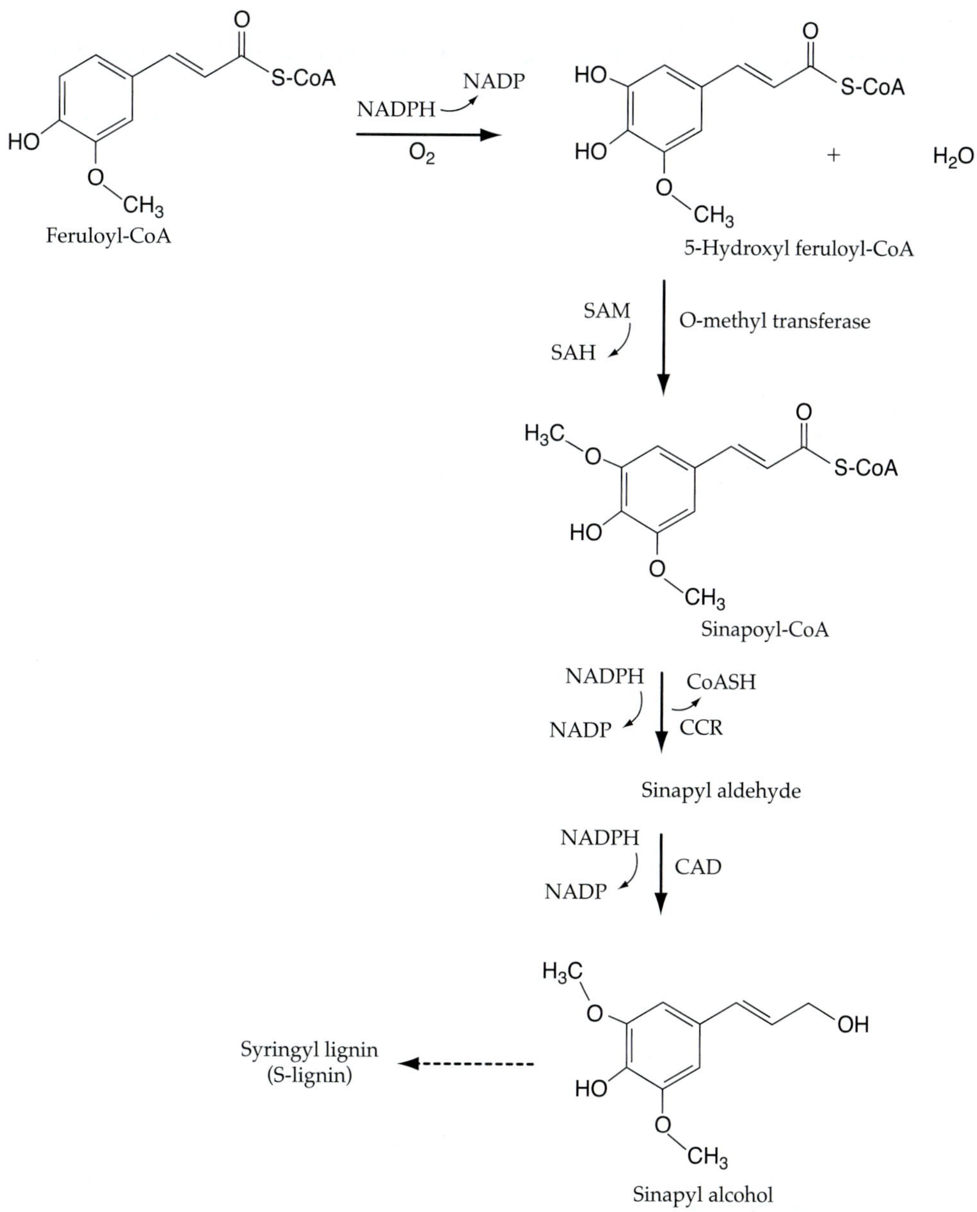

FIGURE 8.15 Synthesis of sinapyl alcohol. Ferulic acid, or a derivative, is hydroxylated by a cytochrome P450 monooxygenase to 5-hydroxy ferulic acid. An O-methyl transferase adds a methyl group to this third hydroxyl group to produce sinapic acid, which is reduced to the alcohol by *p*-coumaroyl-CoA reductase (CCR) and coumaroyl aldehyde dehydrogenase (CAD). Sinapyl alcohol is a primary monolignol that is polymerized into syringyl or S-lignin, a major component of secondary cell walls in dicotyledonous plants.

in the cell wall that can act as free radical initiators. Peroxidases are heme-Fe containing enzymes that use a variety of substrates and hydrogen peroxide to generate a radical:

$$2AH + H_2O_2 \rightarrow 2A\bullet + 2H_2O$$

Plants contain a number of isoforms of these enzymes; the peroxidases in the cell wall are designated class III peroxidases. They can generate reactive oxygen species (ROS, see Appendix 5) associated with the wound response. Peroxidases are also capable of generating monolignol radicals. The source of the hydrogen peroxide in the cell wall is unknown but could presumably be generated by the peroxidase itself from molecular oxygen.

The second type of radical initiator, the laccases, are Cu^{+}-containing proteins belonging to the phenol oxidase enzyme family. They generate radicals from molecular oxygen:

$$4AH + O_2 \rightarrow 4A\bullet + 2H_2O$$

An unpaired electron that is introduced into a monolignol can be delocalized over the aromatic ring and side chain and subsequently couple with another radical to form a bond in a number of different linkages. Some of the possible linkages between monolignols are shown in **FIGURE 8.16**. A similar random radical mechanism links lignin to carbohydrates in the matrix of the primary cell wall (see Chapter 4).

Unlike the carbohydrates of the primary cell wall, lignin is a relatively inflexible polymer. It makes cell walls rigid and provides support for plant stems and vascular tissues. It is especially important in xylem vessels where it maintains the vessel structure under negative water pressure. Lignin also provides a barrier to pathogen invasion. The lignin structure with its random chemical bonds is highly resistant to both enzymatic and chemical degradation. Only a few organisms, mostly fungi, can degrade lignin; they use a free radical mechanism to generate random breaks in the polymer, essentially reversing the polymerization process.

Genetic Engineering of Lignin

The properties of lignin in plants depend on the concentration of monolignols exported to the wall, which determines the different types, H, G, or S. Gymnosperm wood is mainly G-lignin; the monolignol pool contains mostly coniferyl alcohol with smaller amounts of sinapyl alcohol. This pool gives rise to a type of lignin with mainly C–C linkages that are more stable than the ether linkages that result from a higher concentration of sinapyl alcohol precursors (**FIGURE 8.17**). Plants containing a high content of G-lignin are more difficult for herbivores, such as domestic cattle and sheep, to digest. G-lignin also requires harsher chemical treatment in the pulping process to make paper and presents a barrier to the production of ethanol-based biofuels.

Many research groups have attempted to genetically alter the composition of lignin with the goal of producing a product that is easier to degrade, either enzymatically or chemically. Examples of these manipulations include a gene knock-out or down-regulation of the enzyme that methylates caffeoyl-CoA (CCoAOMT). This should result in a pool of monolignols with a higher concentration of coumaroyl alcohol (use **FIGURE 8.18** as a guide). The engineered lignin that results from this manipulation should have less of both the G- and S-lignin components

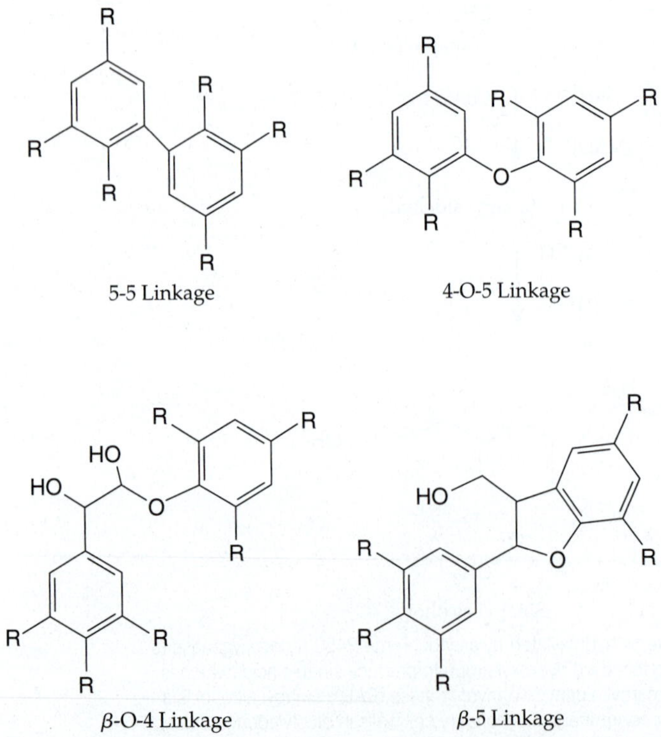

FIGURE 8.16 Some common linkages found in lignin. The types of linkages found in lignin depend on the pool of monolignols that are synthesized and exported from the cell. In conifers, coniferyl alcohol is the most abundant precursor, and the resulting G-lignin contains mainly C–C bonds such as the 5-5 linkage shown. Dicotyledonous angiosperms produce mainly sinapyl alcohol precursors and the resulting S-lignin contains more ether linkages such as the 4-O-5 and β-O-4 linkages. Because most plants produce some of each type of monolignol, all the above linkages and many mixed linkages such as the β-5 configuration with both C–C and C–O bonds are possible.

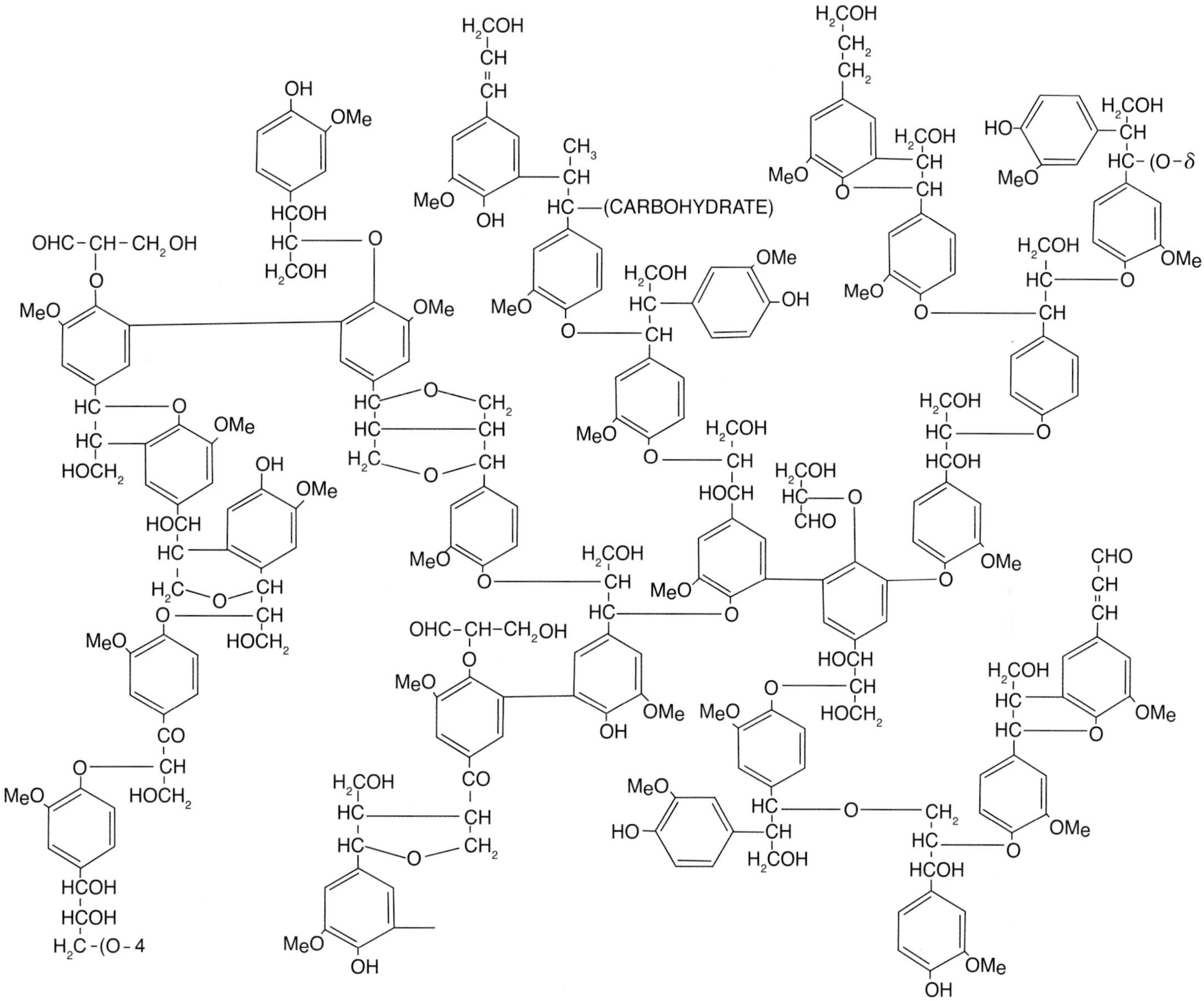

FIGURE 8.17 Model of linkages found in conifer lignin. These are some of the major linkages that can be found in G-lignin formed predominantly from coniferyl alcohol. Both C–C and C–O bonds are formed, but the C–C linkages are more common in this type of secondary cell wall polymer. (*Reproduced from*: M. Baucher, B. Monties, M. Van Montagu, et al. *Crit. Rev. in Plant Sci.* 17 [1998]: 125–197, Figure 7. Courtesy of Marc Van Montagu, Ghent University.)

and therefore fewer recalcitrant C–C bonds. The lignin in transgenic tobacco and poplar plants, however, actually had a higher content of 5-hydroxyguaiacyl units (i.e., more coumaroyl alcohol was incorporated into the polymer), but additional 5-hydroxyferulic alcohol was also found. Apparently, the other O-methyl transferases can adequately substitute for CCoAOMT in these plants, and the resulting modified lignin still contains a high proportion of linkages that are resistant to degradation.

Naturally occurring mutations in the monolignol enzyme genes were found in maize (*Zea mays*) and were designated *bm1,2,3,4* (*brown mid-rib*). The *bm3* mutant was found to have reduced O-methyl transferase (COMT) activity (see Figure 8.18). The mutant plants contain less S-lignin in their walls as expected but again compensated with a higher content of hydroxyferulic alcohol. The mutant maize leaves and stalks are more digestible by ruminants. Grain yield is lower in mutant plants, however, and they are more susceptible to lodging (i.e., stems break easily in the wind).

In the model plant, *Arabidopsis*, mutants with knock-out mutations in the gene for ferulic acid monooxygenase (F5H) produce no S-lignin. Conversely, overexpression of the gene for this monooxygenase leads to production of cell walls that are predominantly S-lignin, suggesting that manipulation of the expression of the ferulate-5-hydroxylase gene

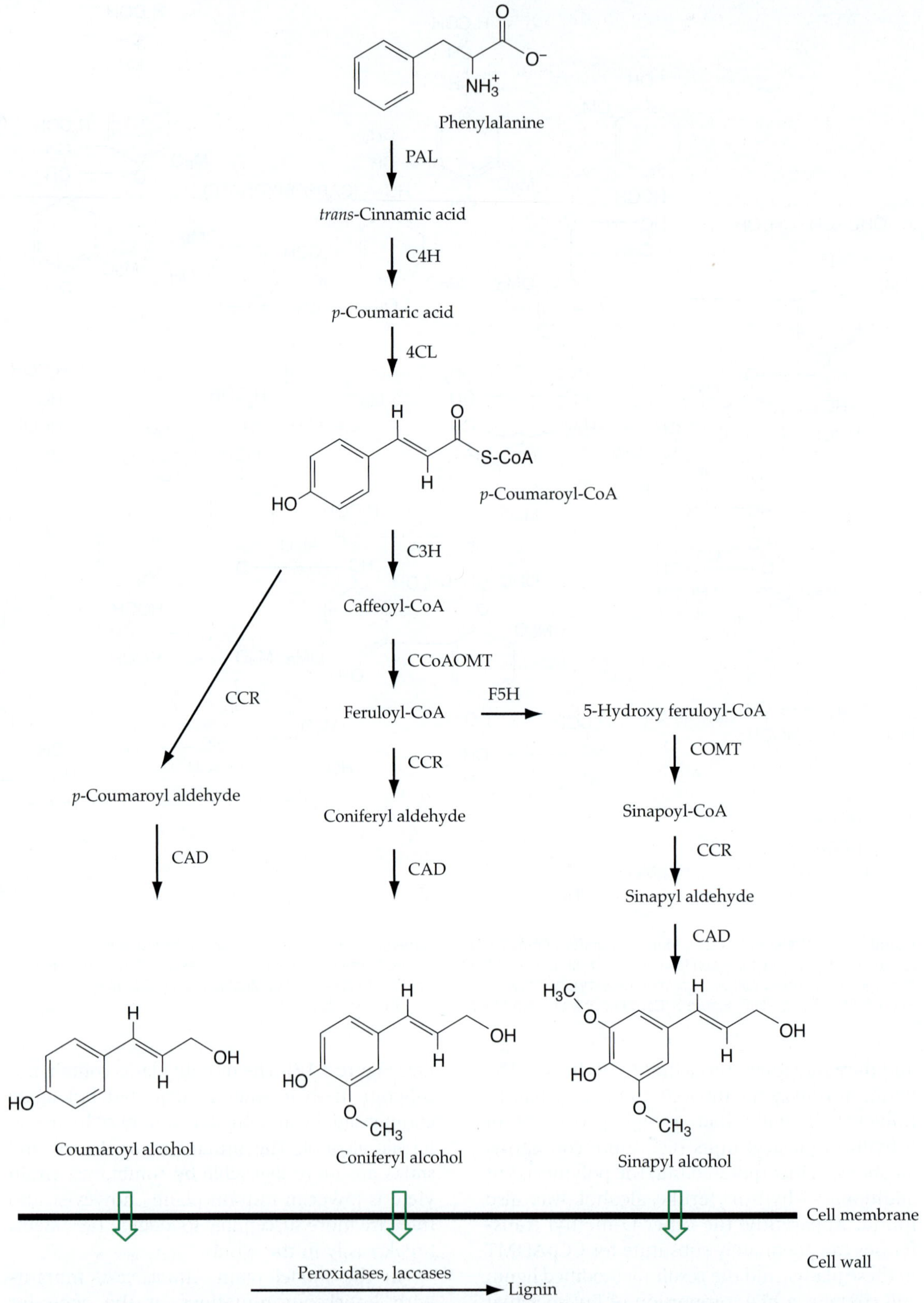

FIGURE 8.18 Simplified map of monolignol and lignin synthesis in vascular plants. The most direct route to the three major monolignols synthesized from *p*-coumaric acid and some of the enzymes required are shown. In most plants, alternate reactions/enzymes are present, providing additional routes to the major monolignols and other compounds that may be found in lignin. The *in vivo* pathway for lignin synthesis is more grid-like than this simple linear diagram. The mechanism by which monolignols are transported across the cell membrane is unknown. CAD, coumaroyl alcohol dehydrogenase; CCoAOMT, caffeoyl-CoA-O-methyl transferase; C3H, *p*-coumarate-3-hydroxylase (monooxygenase); C4H, coumaroyl-4-hydroxylase; CCR, coumaroyl-CoA reductase; 4CL, 4-coumarate CoA ligase; COMT, caffeic acid/5-hydroxy ferulic acid O-methyltransferase; F5H, ferulate-5-hydroxylase; PAL, phenylalanine ammonia lyase.

may be a key step in increasing the digestibility of forage crops.

The results in any one plant system, however, may not be applicable to other species. Down-regulation of the gene for 4-coumarate CoA ligase (4CL) in tobacco produced plants with less overall lignin in the cell walls and a polymer with a higher content of coumaroyl alcohol. Conversely, in *Arabidopsis* the total lignin content in 4-coumarate CoA ligase mutant plants was reduced but the walls still contained high levels of S-lignin. This suggests that alternate pathways for monolignol synthesis are present in some species and can compensate for alterations in the main sequence of monolignol production, but these alternatives may be species specific. The striking results in poplar (*Populus tremuloides*) illustrate the flexibility of plants in producing lignins. Down-regulation of the aldehyde reductase (CAD) synthesis results in plants with a higher content of aldehydes in their cell walls. Aldehyde incorporation produces a pronounced red coloration in the vascular tissues of the woody stems. Although genomics and genetic engineering techniques have greatly increased our knowledge of lignin synthesis, application of these techniques to manipulate the quantity and properties of lignin is still problematic because of our overall lack of understanding of how cell walls are made. Methods for analyzing lignin and linkages, such as nuclear magnetic resonance (NMR) spectroscopy, are a valuable asset to lignin engineers and are currently used to provide a more directed path to lignin manipulation.

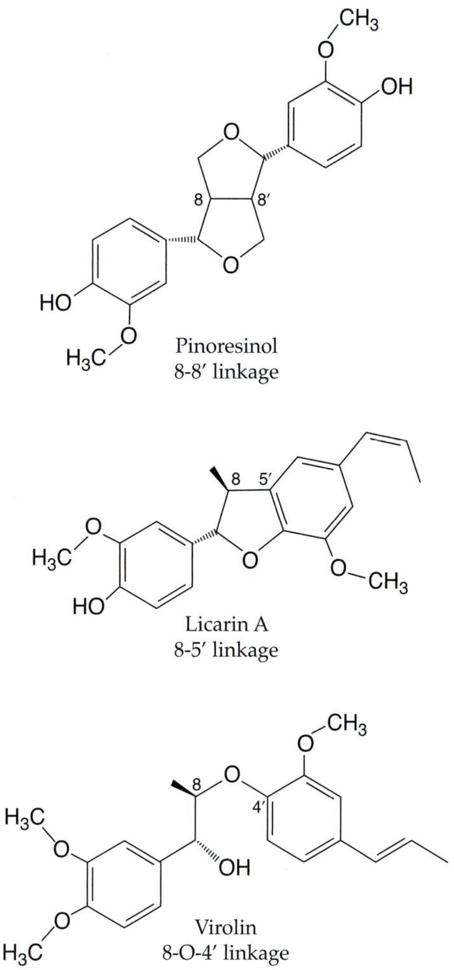

FIGURE 8.19 Structures of some common lignans. Lignans are dimers of monolignols with precise linkages formed between the precursors. They are synthesized by free radical–initiated bond formation guided by dirigent proteins. The dilignol bonds in these compounds correspond to linkages found in lignin (see Figure 8.16).

8.4 Natural Products Derived from the Phenylpropanoid Pathway

Natural Products from Monolignols

Lignans Lignans are dimeric phenylpropanoids that are synthesized from monolignols, mainly coniferyl alcohol. They are found in the seed coats, flowers, and epidermal layers of stems and leaves where they serve to defend plant tissues against pathogens. They are also precursors for a variety of complex natural products. Lignans are synthesized by a free radical–coupling mechanism, but the linkages are specific rather than random as in lignin. The structures of some common lignans are shown in **FIGURE 8.19**. The monolignol substrates are held in position by proteins called dirigents (guide proteins). These proteins have no enzymatic activity; they serve to bind the lignols and guide the radical-mediated bond formation (**FIGURE 8.20**). As in the case of lignin synthesis, the free radical initiator is a peroxidase.

Lignan dirigent proteins are found in the cytoplasm, but similar dirigents have been localized in the cell walls of some plants. It has been proposed that these extracellular dirigent proteins may be active in lignin formation. The binding of monolignols to dirigents with different sites could account for the regular bonding patterns found in some lignin polymers. Although a large number of different dirigent proteins and their genes have been predicted from genomic analyses, an even larger number would presumably be required to guide total lignin synthesis. The dirigent proteins in cell walls may determine the bonding in the initial products of polymerization to which additional monolignols are then added in a random, radical-coupled process.

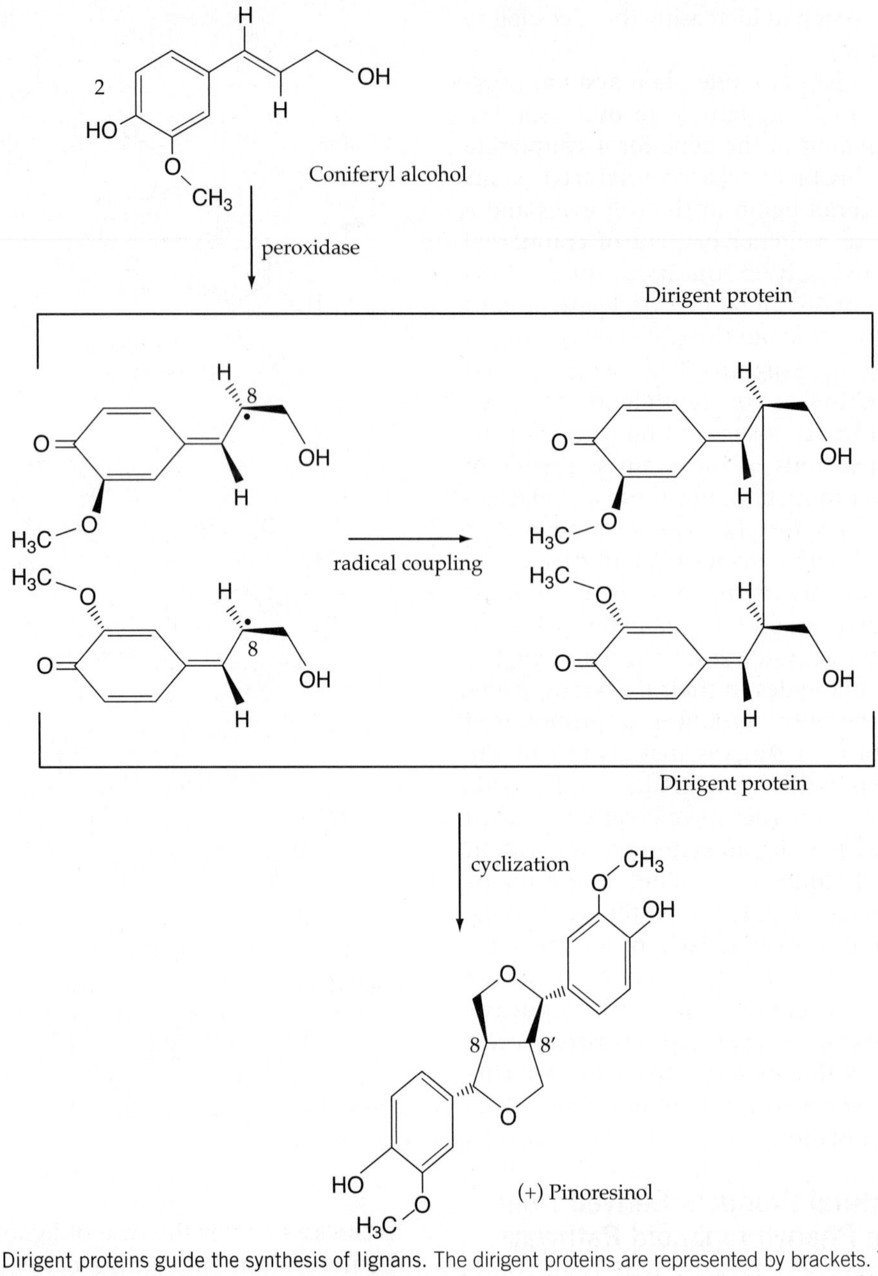

FIGURE 8.20 Dirigent proteins guide the synthesis of lignans. The dirigent proteins are represented by brackets. The proteins bind two molecules of coniferyl alcohol in this example and hold them in a defined position. A radical is generated by a cytoplasmic peroxidase. Because of binding to the dirigents, radical coupling occurs only between the 8-8′-positions. The product is the lignan, pinoresinol, a compound that is common in the Oleaceae (olive family).

In some plants, lignans are modified to produce species-specific natural products. Some of these modified lignans are useful to humans as anticancer agents and nutritional supplements. The structures of three examples are shown in **FIGURE 8.21**. Sesamin is an antioxidant found in the seed oil of the sesame plant (*Sesamum indicum*). Gomisin A, from *Schisandra chinensis*, a tropical vine from Asia, is used in traditional herbal medicines as a treatment for liver disorders. Podophyllotoxin from May apple (*Podophyllum peltatum*) inhibits cell division by binding to tubulin and preventing the formation of the spindle apparatus. It is, therefore, a potential anticancer agent but is too toxic to be taken internally. It is used topically to treat some skin cancers. A synthetic derivative, etoposide, is less toxic and can be taken internally to treat certain types of lung and testicular cancers. Etoposide acts by inhibiting topoisomerase activity and DNA synthesis rather than binding to tubulin.

Volatiles Many plants produce volatile compounds from monolignols. Volatiles in flowers and fruits are attractive to pollinators and her-

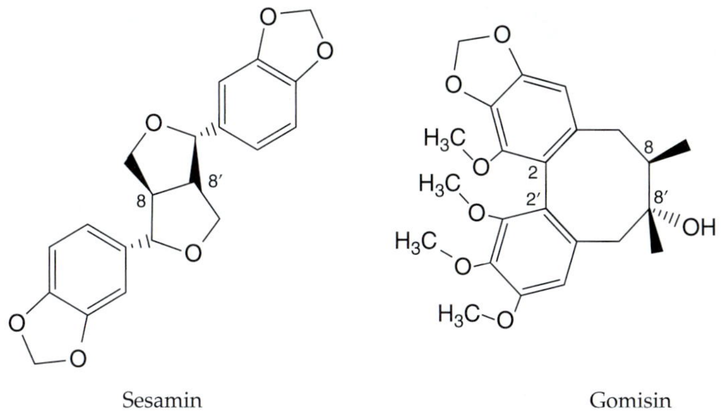

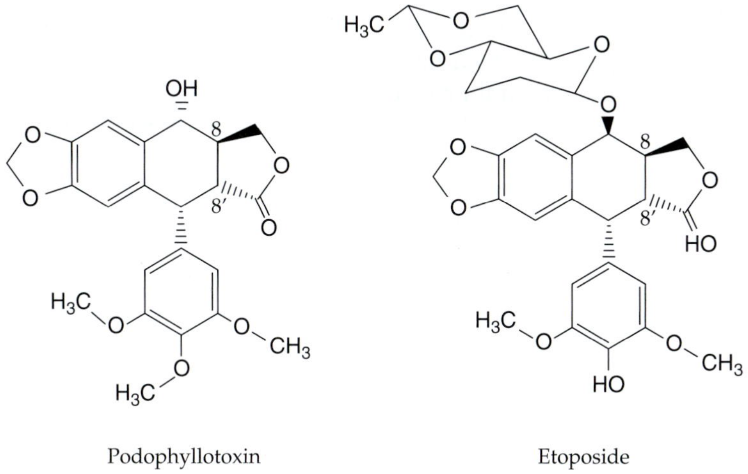

FIGURE 8.21 Structures of modified lignans that are useful to humans. Sesamin is an antioxidant found in sesame seed oil that is marketed as a neutraceutical. Gomisin-A has an unusual eight-membered ring. It is the active ingredient in some traditional herbal medicines used to treat liver disorders. Podophyllotoxin inhibits tubulin polymerization and prevents cell division. Although podophyllotoxin is too toxic for therapeutic use, it is the natural model for etoposide, a synthetic anticancer drug.

bivores, while volatile compounds in leaves and stems repel insects or attract insect predators. A common pathway for the synthesis of some volatiles is shown in **FIGURE 8.22**. The alcohol group of monolignols such as coniferyl alcohol can be removed by a reductase to generate a resonance-stabilized carbocation intermediate. Reduction of the allylic cation yields eugenol, the aromatic compound found in cloves (*Syzgium aromaticum*), which are used by humans as both a flavoring agent and an analgesic (pain reliever). Similarly, reduction of the corresponding propenyl cation and dehydration produces anethole, a volatile found in many plants such

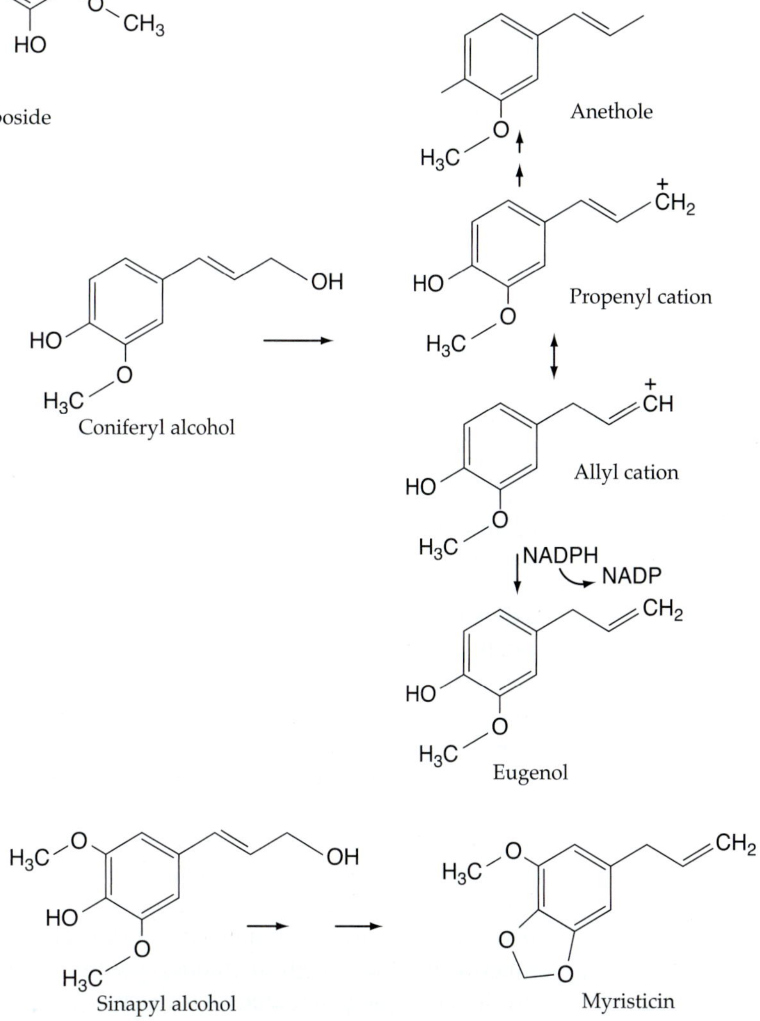

FIGURE 8.22 Examples of volatile natural products produced from monolignols. Removal of the hydroxyl group from coniferyl alcohol generates two resonance- stabilized carbocations. Reduction of the allylic cation generates the methylene group in eugenol, the volatile compound in cloves. Reduction of the propenyl cation and subsequent dehydration generates anethole, the volatile found in many plants such as fennel. Removal of the hydroxyl group of sinapyl alcohol and reduction of the resulting allylic cation to a methylene group initiates the formation of myristicin, which also requires ring closure to form the methylene dioxy ring. Myristicin is the volatile that adds to the fragrance of nutmeg seeds.

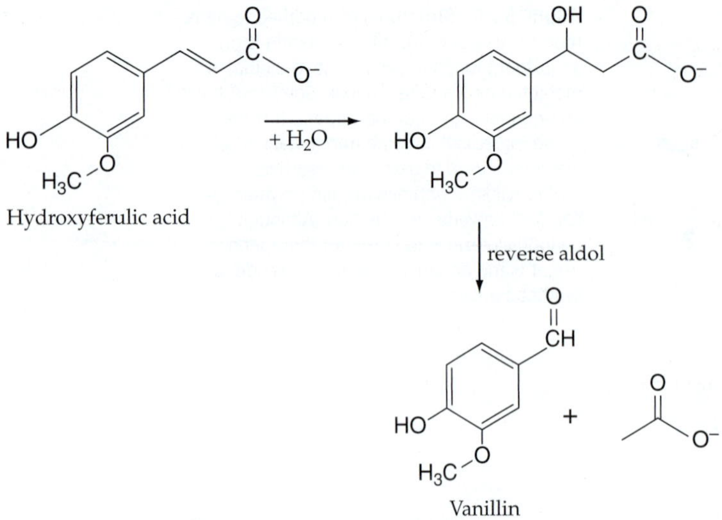

FIGURE 8.23 Synthesis of vanillin. Vanillin is produced by a fermentation process that alters ferulic acid derivatives found in secondary cell walls. Hydration of the side chain of hydroxyferulic acid is followed by a reverse aldol reaction to generate the aldehyde, vanillin.

as fennel (*Foeniculum vulgare*). If the starting material is sinapyl alcohol, a similar reduction and subsequent formation of a methylene dioxy ring yields myristicin (Figure 8.22), the volatile compound found in nutmeg (*Myristica fragrans*).

Vanillin is a natural product derived from hydroxyferulic acid. In the pathway illustrated in **FIGURE 8.23**, the double bond in hydroxyferulic acid is hydrated followed by a reverse aldol reaction that cleaves the side chain to yield vanillin. In some plants an alternate route uses the CoA ester of ferulic acid as a precursor. Commercial vanilla extract from the pods of the vanilla orchid (*Vanilla planiflora*) is produced after a long fermentation period. The pods are then dried and extracted with ethanol. Real vanilla extract contains vanillin and a number of additional compounds. Imitation vanilla extracts can be produced by fermentation of the ferulic acid found in some woody stems.

Salicylic Acid The plant hormone, salicylic acid, can be synthesized by a variety of pathways. It is derived in some plants from coumaroyl-CoA. In bacteria and other plants, such as *Arabidopsis*, salicylic acid can be produced from chorismic acid, an intermediate in the shikimic acid pathway (Figures 8.1 and 8.7). Salicylic acid is a hormone produced after pathogen invasion. It is transported to undamaged plant tissues where it mediates systemic acquired resistance (SAR). It is also the active ingredient in aspirin (acetyl salicylic acid), although the modern commercial product is made by chemical synthesis.

8.5 Synthesis and Properties of Polyketides

As described in the previous section, the phenylpropanoid pathway furnishes monolignols that are essential for lignin synthesis and, therefore, can be considered part of the primary metabolism of plants. This same pathway also provides precursors for many different aromatic compounds that are unique to a particular plant, such as eugenol and myristicin (Figure 8.22), and are generally regarded as secondary metabolites or natural products. In addition, coumaroyl-CoA from the phenylpropanoid pathway is the starting material for the synthesis of polyketides such as the flavonoids, which have traditionally been considered natural products.

Many flavonoids are unique to a particular family of plants. Both their synthesis in a specific plant and molecular details have been used to construct biochemically based phylogenies. Most land plants and many aquatic species, however, contain some class of flavonoids that function to protect the organisms against UV radiation, free-radical damage, pathogens, and herbivores or to attract pollinators, mycorrhizal fungi, and symbiotic microorganisms. Given their essential roles in enabling plants to interact with both biotic and abiotic factors in their environment, flavonoids and other polyketides are definitely of primary importance in plant metabolism.

Synthesis of Chalcones

Coumaroyl-CoA is a key intermediate in the phenylpropanoid pathway, and its synthesis from phenylalanine is described in Figures 8.11 and 8.12. It is a substrate for the enzyme, chalcone synthase (CHS). CHS is found in most plants and provides the precursor for a large number of flavonoids. A cysteine thiolate anion in the enzyme's active site attacks the carbonyl carbon of coumaroyl-CoA, displacing the CoA and binding the coumaroyl moiety (**FIGURES 8.24** and 8.26). The second substrate is malonyl-CoA, which is produced by an acyl-CoA carboxylase (ACC) complex found in the cytoplasm. Unlike the plastid enzyme complex, the ACC complex in the cytoplasm is one large fusion protein. Its mechanism of action is the same as the multisubunit ACC found in chloroplasts; it uses bicarbonate, ATP, and acetyl-CoA to produce the three-carbon acid, malonyl-CoA (see Chapter 6).

Malonyl-CoA coordinates with an asparagine (Asn) residue in the CHS active site,

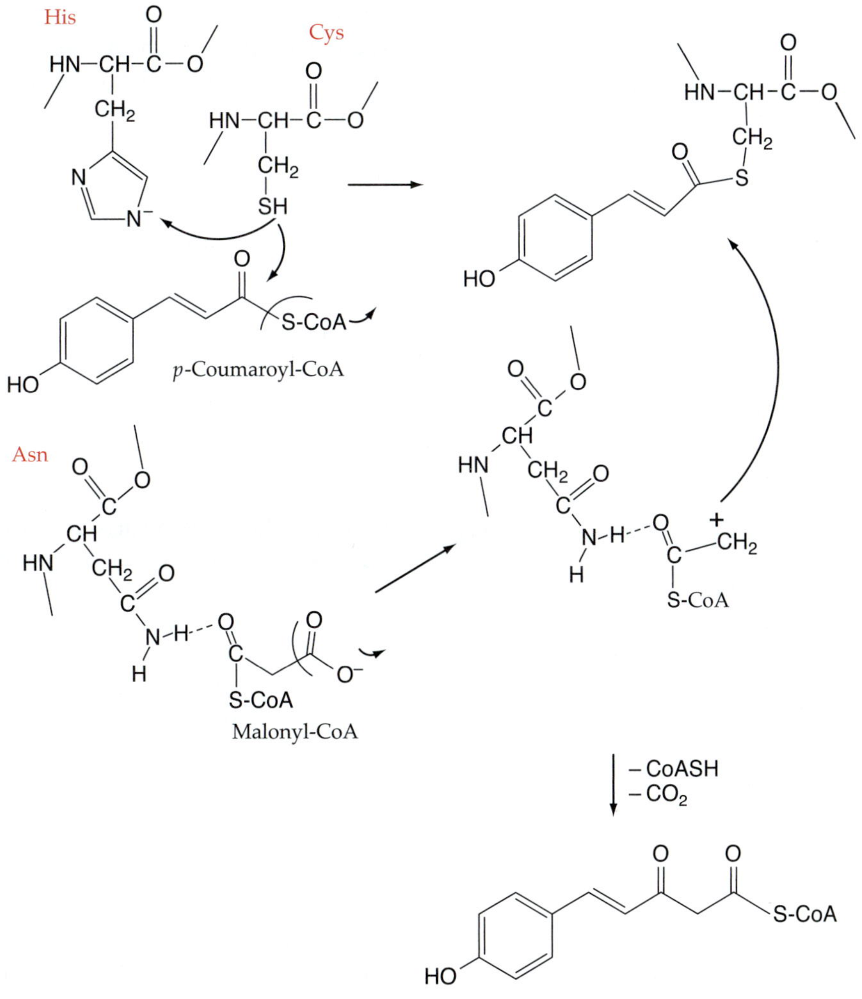

FIGURE 8.24 Mechanism of action of chalcone synthase (CHS). All polyketide synthases have a conserved catalytic triad consisting of histidine (His), cysteine (Cys), and asparagine (Asn) residues in their active sites. The histidine imidazole-N acts as a base and removes a proton from the cysteine sulfhydryl group generating a reactive thiolate anion. This anion attacks the carbonyl carbon of coumaroyl-CoA and displaces the CoA. The second substrate, malonyl-CoA, binds near the asparagine residue. The Asn amide-NH_2 forms an H bond with the carbonyl oxygen of malonyl-CoA and facilitates its decarboxylation. The resulting acetyl cation attacks the carbonyl carbon of the coumaroyl–cysteine complex and produces a diketide product.

which facilitates its decarboxylation. The remaining two-carbon fragment, an acetyl-CoA cation, attacks the carbonyl carbon of the coumaroyl moiety and displaces it from the cysteine active site. The product is coumaroyl diketide (Figure 8.24). This diketide intermediate is reattached to the cysteine thiolate in the active site of CHS and a second equivalent of malonyl-CoA is coordinated to the asparagine residue, decarboxylated, and added to the diketide. The resulting triketide intermediate undergoes another round of binding to the active site cysteine and acetate addition to yield a tetraketide. This final ketide intermediate undergoes cyclization and aromatization while still bound to CHS to produce the product called a chalcone (**FIGURE 8.25**). The overall reaction catalyzed by CHS is as follows:

$$\text{Coumaroyl-CoA} + 3\ \text{Malonyl-CoA} \rightarrow \text{Chalcone} + 4\ \text{CoA-SH} + 3CO_2$$

Chalcone synthases are found in all vascular plants and some bryophytes. They also occur in fungi and certain classes of algae. Most CHSs have a monomer molecular weight of approximately 40 kDa and function as dimers. The CHSs found in plants are part of a large family of enzymes that can catalyze the synthesis of an aromatic ring from fatty acid precursors, usually malonyl-CoA. The plant synthases are part of a larger superfamily, the polyketide synthases. Polyketides are produced by many organisms, including bacteria such as *Streptomyces*, and

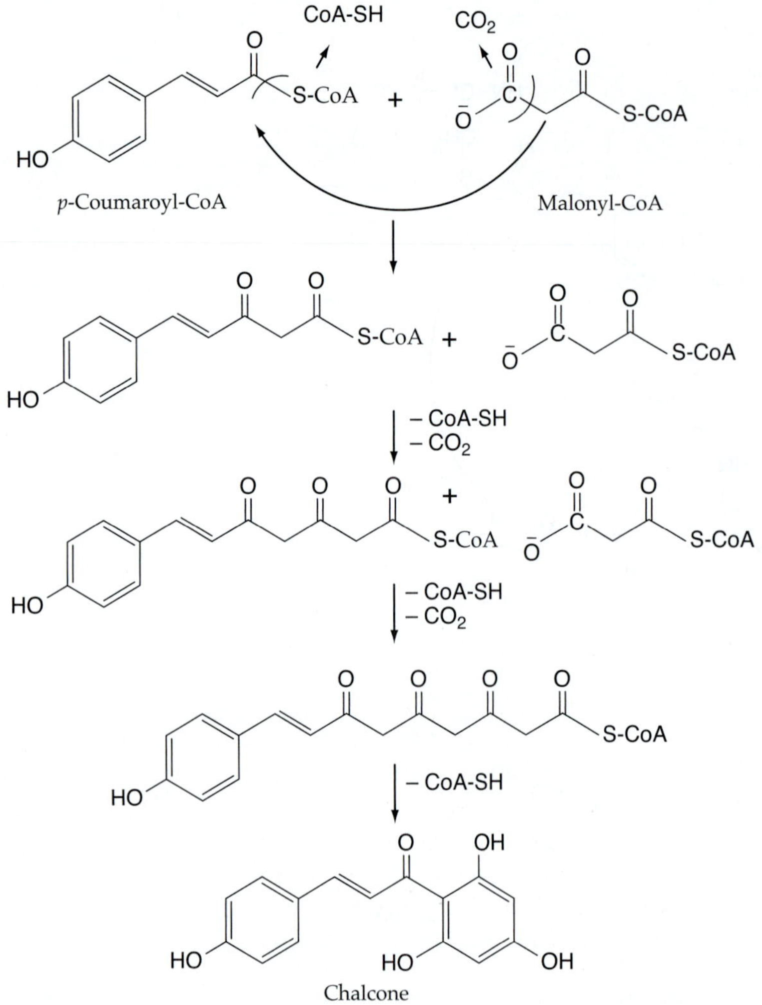

FIGURE 8.25 Chalcone synthesis catalyzed by chalcone synthase (CHS). CHS binds *p*-coumaroyl-CoA first and then adds a two-carbon fragment resulting from the decarboxylation of malonyl-CoA (see Figure 8.24). Loss of CO_2 drives the reaction toward synthesis. The cycle repeats two more times to generate a tetraketide intermediate. The tetraketide undergoes cyclization and generation of double bonds to produce a second aromatic ring in the final product, a tetrahydroxy-chalcone.

form the core structures of certain antibiotics, such as erythromycin. Although they catalyze similar reactions, the polyketide synthases of plants have very little amino acid sequence similarity to their bacterial counterparts. The plant enzymes probably evolved from fatty acid synthases (see Chapter 6).

The x-ray crystal structure of the CHS from alfalfa (*Medicago sativa*) has been solved to 1.5-Å resolution (**FIGURE 8.26a**). The homodimer has two structurally separate active sites. Coumaroyl-CoA binds first in a cleft that excludes other phenolics due to its size and the charged residues in the vicinity of the active site. As shown in **FIGURE 8.26b**, the active site consists of a catalytic triad of histidine, cysteine, and asparagine residues. The histidine acts as a general base and removes a proton from the cysteine sulfhydryl group to produce an active thiolate anion that binds coumarate by displacing the CoA. Malonyl-CoA enters the active site near the asparagine residue, which provides hydrogen bonds to this substrate by way of its amide-NH_2 (γ-amino group). Decarboxylation of malonyl-CoA produces the cation that attacks the coumaroyl moiety, producing a CoA-bound diketide. This cycle is repeated two more times to yield a tetraketide intermediate that then folds and is desaturated to produce an aromatic ring. Naringenin chalcone diffuses out of the active site, which is not large enough to allow further condensation to longer polyketides. The CHS from alfalfa has a high degree of sequence similarity to other polyketide synthases found in plants, including those that produce different products, such as stilbenes and pyrones. Most of the plant synthases seem to have evolved from a common ancestor.

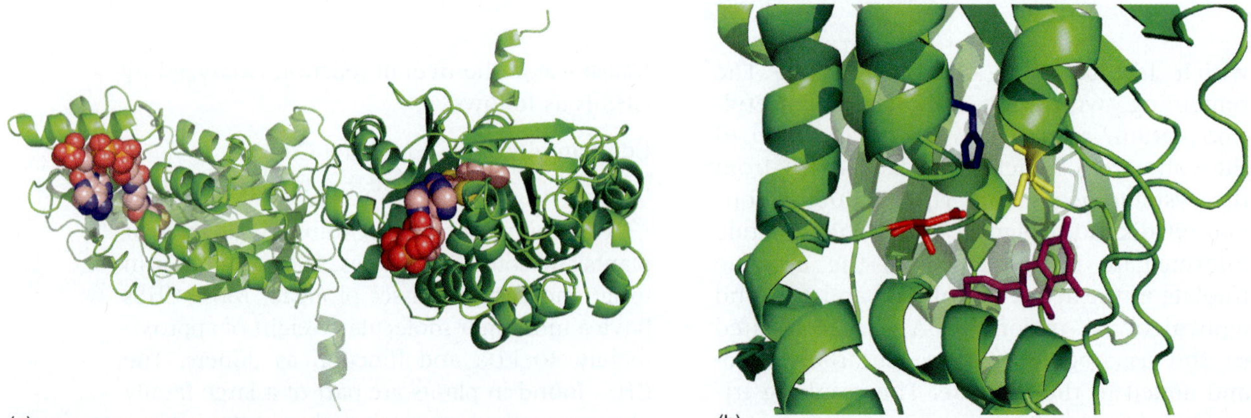

(a) (b)

FIGURE 8.26 X-ray crystal structure of CHS from alfalfa. (a) CHS occurs as a homodimer with two independent active sites. The tunnel into the active site is occupied by a coenzyme A analog shown as spheres. (b) The active site of CHS is shown with a bound narigenin chalcone in purple sticks. The active site histidine-303 (blue) is placed at the top of the binding cavity. It acts as a base to remove a proton from the cysteine-164 shown in yellow. The ketide intermediates are bound to this cysteine residue. Asparagine-336 in red sticks is the binding site for malonyl-CoA. (Structure a from Protein Data Bank 1CHW. J.-L. Ferrer, J. Jez, M. E. Bowman, R. Dixon, and J. P. Noel, *Nat. Struct. Biol.* 6 [1999]: 775–784. Structure b from Protein Data Bank 1CHGK. J.-L. Ferrer, J. Jez, M. E. Bowman, R. Dixon, and J. P. Noel, *Nat. Struct. Biol.* 6 [1999]: 775–784.)

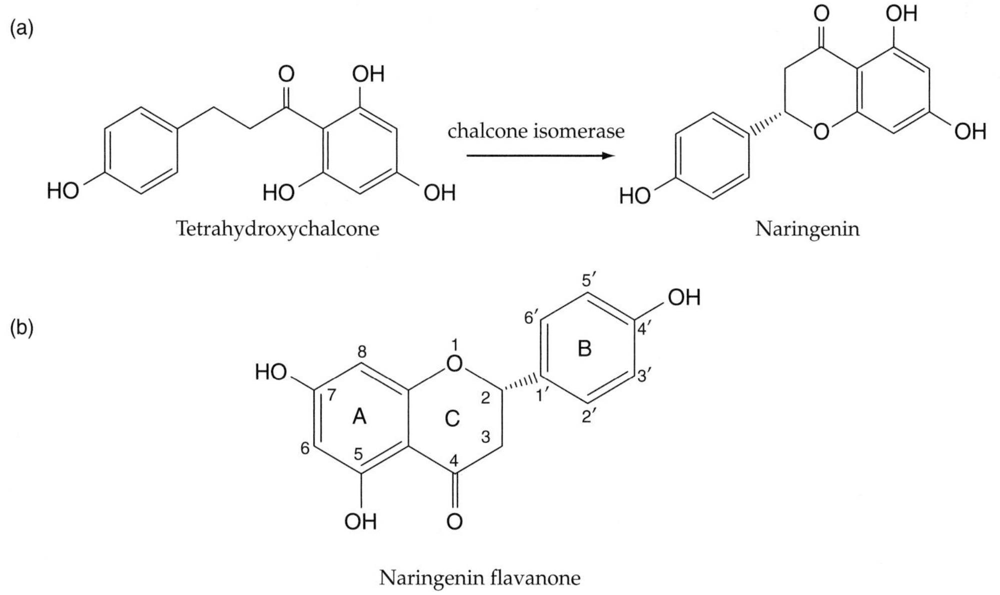

FIGURE 8.27 Synthesis of naringenin flavanone. (a) A specific chalcone isomerase (CHI) catalyzes formation of the pyran ring in the *S*-configuration. (b) The letter designations and numbering system for the general flavonoid molecule are illustrated.

■ Synthesis of Flavanones and Derivatives

The chalcones are the precursors for most flavonoid compounds. Further ring closure to produce a pyran ring in naringenin occurs spontaneously. Plants, however, have an isomerase that catalyzes this reaction. The enzymatic ring closure is at least 100 times faster than the uncatalyzed reaction and also produces the biologically active *S*-isomer (**FIGURE 8.27**). The uncatalyzed reaction results in a racemic mixture. The product of isomerization is classified as a flavanone and is the key intermediate in the synthesis of other classes of flavonoids. The major differences in these compounds are changes in the oxidation state of the pyran (C) ring (**FIGURE 8.28**).

Many of the modifications shown in Figure 8.28 are catalyzed by oxygenases that add hydroxyl groups or a double bond to the C ring of naringenin flavanone. These enzymes are associated with the outer face of the endoplasmic reticulum. Additional enzymatic reactions can also modify the A and B rings to produce a large array of derivatives that are collectively referred to as flavonoids.

Most flavonoids are glycosylated for storage or transport. The sugar attached to the flavonoid is usually glucose but can also be galactose, arabinose, xylose, or rhamnose. These reactions are catalyzed by glycosyl transferases that use a UDP-sugar substrate and make an ether bond between the sugar and an OH group in the flavonoid. Additional sugars are often added to the first one.

Some flavonoids are acylated, usually with methyl or acetyl groups, but other additions are possible. The acyl groups are added to an OH in one of the rings or to the sugar moiety in glycosylated flavonoids. Acylation can change the solubility of the flavonoids and cause a shift in their absorption spectra.

A sulfotransferase can catalyze the sulfation of the B ring of some flavonoids. The sulfur

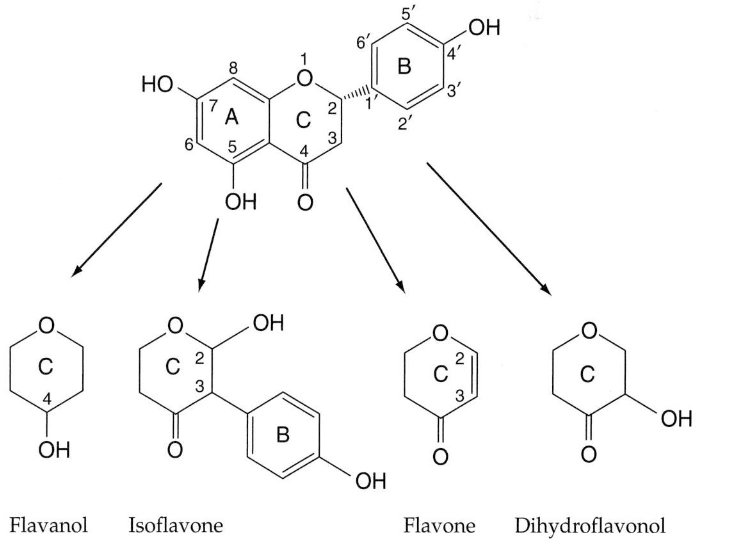

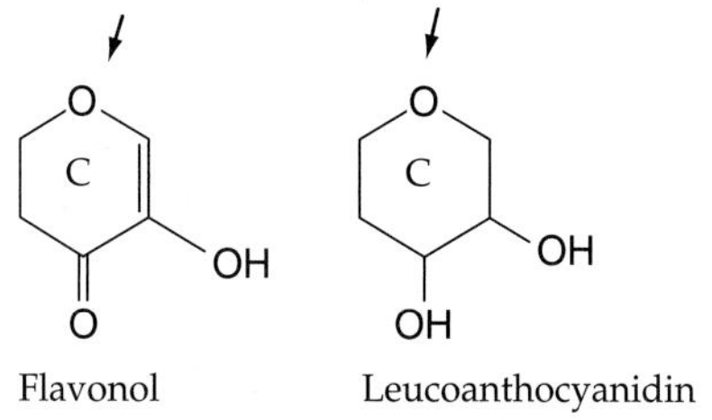

FIGURE 8.28 Structures of the various flavonoid classes. Flavonoids are placed in different classes, flavonols, isoflavones, flavones, and dihydroflavonols, based on the modifications shown in the C ring. Within each class, further additions are often made to the A and B rings.

donor is phosphoadenosine phosphosulfate (PAPS). Sulfated flavonoids are often signaling molecules used by plants such as legumes to attract the appropriate species of symbiotic bacteria (see Chapter 5).

Flavonoids can be prenylated, usually with dimethylallyldiphosphate but occasionally with longer terpene precursors. This increases the hydrophobicity of the products. Prenylated flavonoids generally function as phytoalexins.

■ Synthesis and Properties of Flavones

Flavones are made from naringenin flavanone by addition of a double bond to the C-ring between C-2 and C-3 (Figure 8.28). Two different flavone synthases (FNS) can catalyze this desaturation. In many angiosperms, flavone synthase II (FNS II) is a cytochrome P450 monooxygenase. In others, such as some species in the Apiaceae (parsley, carrot), the synthase (flavone synthase I [FNS I]) is an oxoglutarate-dependent dioxygenase (ODD). The ODDs are nonheme iron–containing enzymes that can bind O_2 and transfer one oxygen atom to oxoglutarate (α-ketoglutarate) and the other oxygen atom to a substrate or catalyze the formation of a double bond as in the fatty acid desaturases.

As illustrated in **FIGURE 8.29**, the iron in flavone synthase I is ligated to histidine and aspartic acid residues in the active site. This Fe^{3+} is reduced by ascorbic acid. The ferrous form of the enzyme then binds oxoglutarate and molecular oxygen. The oxoglutarate undergoes oxidative decarboxylation to succinate, which is released from the enzyme. This reaction generates an active ferryl intermediate, and the enzyme can now bind and catalyze the formation of a double bond in the substrate, naringenin flavanone. Additional reactions in flavonoid modification are catalyzed by ODDs, and genomic analyses suggest that this type of dioxygenase is more common than previously thought. The *Arabidopsis* genome has approximately 64 genes for putative ODDs, and some oxygenase-catalyzed reactions may use this type of enzyme rather than cytochrome P450 monooxygenases.

Flavones are colorless compounds with an absorption maximum in the UV region of the spectrum. They are generally stored in cells as 7-O-glycosides and can be found in all tissues of most plants, except for the Brassicaceae, which do not produce flavones. Flavones function to protect plant cells from UV radiation and may also act as allelochemicals (i.e., compounds that inhibit growth of competing photosynthetic organisms). They may also serve to attract and guide pollinators. As seen in **FIGURE 8.30**, colorless flavones fluoresce under UV light, and the patterns can be detected by bees and some other potential pollinators. The flavones serve as a nectar guide but also ensure that pollen is transferred onto the insect.

■ Synthesis and Properties of Anthocyanidins

Anthocyanidins are the precursors of plant pigments collectively referred to as anthocyanins. Their colorless precursors, leuco-anthocyanidins, are synthesized from dihydroflavonols, which are the products of another ODD that introduces an OH group at C-3 in the C ring of naringenin flavanone. The keto group in the

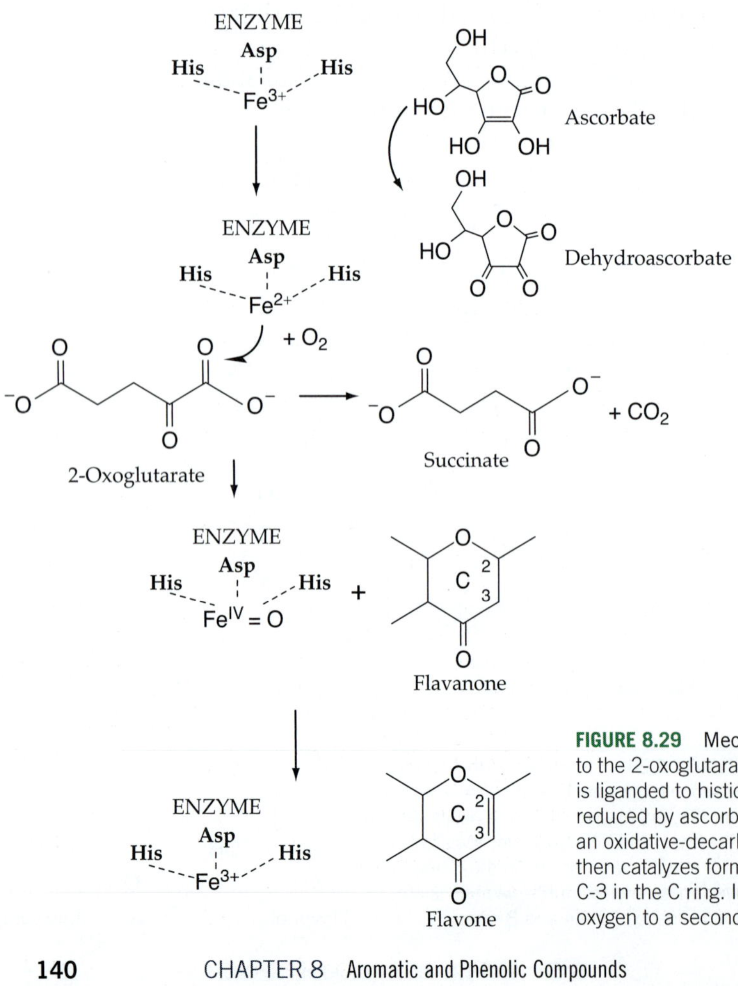

FIGURE 8.29 Mechanism of action of flavone synthase I (FNS I). FNS I belongs to the 2-oxoglutarate–dependent nonheme iron family of dioxygenases (ODD). Iron is liganded to histidine and aspartic acid residues in the protein. The ferric iron is reduced by ascorbate. The Fe^{2+} form of the enzyme binds oxoglutarate and catalyzes an oxidative-decarboxylation to succinate. The resulting $Fe^{IV} = O$ form of the enzyme then catalyzes formation of a double bond in the flavanone substrate between C-2 and C-3 in the C ring. In a more generalized reaction, an ODD transfers the Fe-bound oxygen to a second substrate to generate a hydroxylated product.

(a) (b)

FIGURE 8.30 Colorless flavones are visible under UV light. Flower (*Coreopsis leavenworthii*, top; *Bidens mitis*, bottom) images are shown photographed under visible (a) and UV (b) light. The UV images are transduced into blue. UV-absorbing flavones are concentrated in the center of the flowers. The patterns are visible to some insects, such as honeybees, and may increase attraction to the flowers and help guide the pollinator after it lands.

C ring is then reduced, and a second ODD, anthocyanidin synthase, introduces two double bonds into the ring to produce the anthocyanidin flavylium cation (**FIGURE 8.31**). The final steps in anthocyanin synthesis include glycosylation of the OH group in the C ring and further modifications of the A and B rings.

The unmodified anthocyanidin flavylium cation can be generated *in vitro* and exhibits a reddish hue at neutral pH. The color is influenced both by the type of solvent and pH. In flower petals the color of the anthocyanins can vary depending on the chemical modifications, including further glycosylation, acylation, and prenylation. Anthocyanins are often stored in special vacuoles modified for pigment storage. These vacuoles are acidic and generally give rise to a blue color. Changes in the vacuolar pH, such as raising the pH to 7 or higher, can shift the color toward the red end of the spectrum. Changes in cell shape influence stacking of the aromatic rings of the anthocyanins and lead to variations in the perceived colors. The presence of cations, mainly Mg^{2+} and Al^{3+}, can also influence the final color of flower pigments. A particularly dramatic example are the colors of

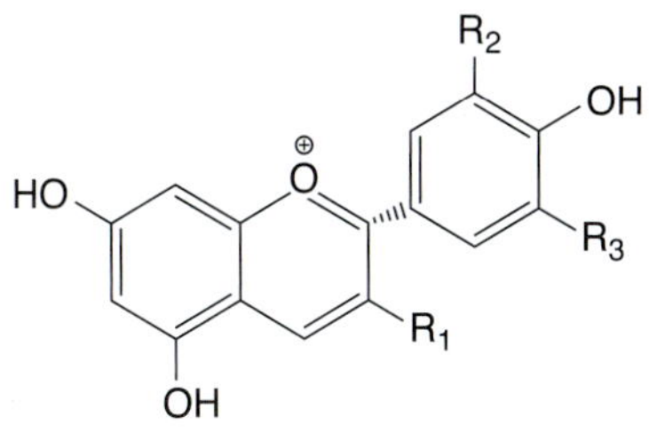

Pelargonidin: R_1 = OH, R_2 = H, R_3 = H
Luteolinidin: R_1 = H, R_2 = OH, R_3 = H
Delphinidin: R_1 = OH, R_2 = OH, R_3 = OH
Cyanidin: R_1 = OH, R_2 = OH, R_3 = H

FIGURE 8.31 Examples of anthocyanidins. The product of anthocyanidin synthase is the flavylium cation shown at the top that is further modified to produce different colored products. The hydroxyl (OH) group at R_1 in most anthocyanidins is glycosylated to stabilize the ring structure. Hydroxyl groups in the A and B rings are also often glycosylated. The colors shown above can be generated *in vitro* from purified compounds. The actual colors of anthocyanidins *in vivo* depend on a number of physiological parameters such as pH of the compartment in which the pigments are stored and presence of metal ions and copigments.

hydrangea flowers, which are generally pink when grown in alkaline soils but can change to blue after application of an acidic mixture of soluble aluminum salts to the soil (see photo on page 119). The presence of other flavonoids, such as the colorless flavones, act as copigments and can also modulate color. Anthocyanins in flowers can obviously have a variety of colors and are a major factor in the attraction of pollinators. Genetic engineering techniques have been used to try to alter flower colors for commercial purposes. Given the number of factors that can influence the final petal colors, this is not readily accomplished by modifications of a few genes.

In contrast to flower petals, anthocyanins in leaves are usually a red color and are responsible for the bright red of autumn leaves, particularly in some maple species. Anthocyanins in photosynthetic tissues are stored in vacuoles and protect cells against the damaging effects of intense sunlight. Their synthesis can be increased during periods of high light intensity and low temperatures, such as early spring in the northern hemisphere. Under these conditions, the light absorption by anthocyanins reduces the rate of the light absorption and transfer in photosystem II. In cool conditions, CO_2 fixation is limiting and cannot recycle the cofactors required in the light reaction. The attenuation of oxygen evolution in photosystem II by anthocyanins helps prevent the formation of oxygen free radicals. The anthocyanins themselves are also very effective antioxidants. They can readily dissipate reactive oxygen species such as the hydroperoxy radical and other radical species produced in peroxisomes (see Appendix 5).

Anthocyanidins can be polymerized into oligomers called proanthocyanidins or condensed tannins. The precursors accumulate in vacuoles and are exported to the cell wall. Subsequent oxidation by laccases produces linear polymers with a deep purple to brown color characteristic of some seeds, fruits, and tree bark. These compounds inhibit proteolytic enzymes in herbivores and also slow the growth of bacterial and fungal pathogens.

Synthesis and Properties of Isoflavonoids

Isoflavonoids are synthesized from naringenin flavanone by a cytochrome P450 monooxygenase. Isoflavone synthase (IFS) catalyzes the cleavage of the C–H bond at C-3 in the C ring and generates a radical intermediate. The B ring then migrates to C-3 and the enzyme adds an OH group to C-2 (**FIGURE 8.32**). A dehydratase can remove H_2O from this initial product to yield a specific isoflavone such as genistein. Isoflavone synthase can also act on deoxyflavonones that lack the hydroxyl group on C-5 of ring A to produce an additional class of isoflavonoids. As in the other flavonoids, modifying groups such as sugars can then be added to the basic isoflavonoid ring system.

Isoflavonoids are not widespread among plants but are commonly found in legumes. Isoflavonoid synthesis is inducible, and the products act as phytoalexins in leaf tissues to protect against pathogens. Isoflavonoids are excreted from legume roots and serve as attractants for nitrogen-fixing symbiotic bacteria (see Chapter 5).

The isoflavone ring system bears a resemblance to the mammalian sex hormones, the estradiols (**FIGURE 8.33**). Isoflavonoids are able to bind weakly to mammalian estrogen

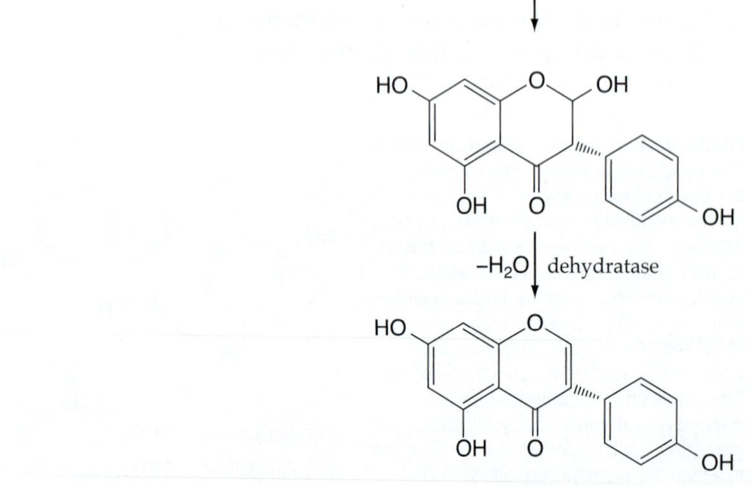

FIGURE 8.32 Synthesis of genistein isoflavone. The enzyme, isoflavone synthase, is a specific cytochrome P450 monooxygenase. It binds the S-isomer of naringenin flavanone and catalyzes removal of an H atom at C-3 in the C ring, producing a radical intermediate. Migration of the aromatic B ring to C-3 and hydroxylation of C-2 yield an isoflavone intermediate. In soybean (*Glycine max*), a dehydratase removes H_2O from the C ring, producing the isoflavone, genistein.

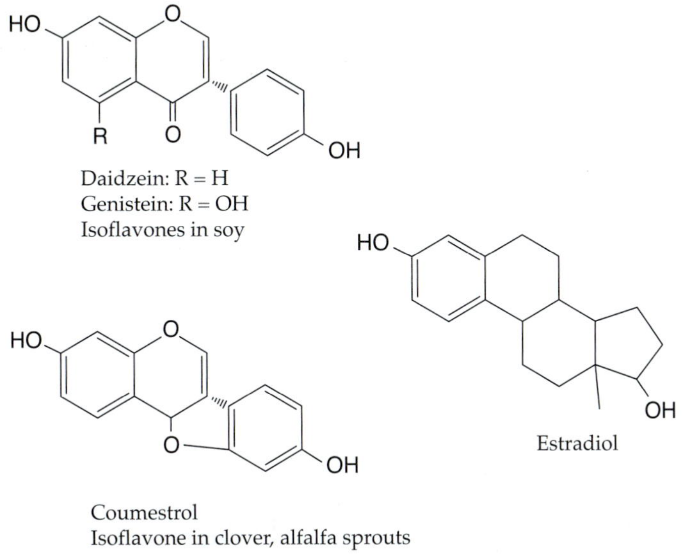

FIGURE 8.33 Some isoflavonoids resemble estradiol. The ring structure of isoflavones in many legumes resembles the mammalian sex hormone, estradiol, and can bind to estrogen receptors. The critical property for receptor binding is the distance between the hydroxyl groups.

receptors and can function as phytoestrogens. The phytoestrogen in clover, coumestrol, is known to have negative effects on the reproductive cycles of ruminants such as sheep. The phytoestrogens in soy-rich foods, such as tofu, may have beneficial effects in humans. Clinical trials have shown that a soy-rich diet can reduce blood cholesterol levels and the risk of heart attacks in postmenopausal women. The U.S. Food and Drug Administration recommends an intake of 25 g per day of soy protein to diminish the effects of menopause and reduce cholesterol levels. The effects are variable and may depend on the activity of gut flora that can modify the phytoestrogens and promote uptake from the intestines.

Examples of Other Plant Polyketide Synthases

CHS is ubiquitous in plants and provides the naringenin precursor for a wide variety of flavonoids. Other types of polyketide synthases have been described that are often species specific, leading to the production of unique natural products. Only a few of these are described here to illustrate how small changes in enzyme structure can lead to a variety of different products. One example is the synthesis of curcumin in *Curcuma longa*, a plant in the Zingiberaceae (ginger) family. In *C. longa*, a unique diketide synthase adds malonyl-CoA to the CoA ester of ferulic acid (**FIGURE 8.34**). A second feruloyl-CoA is added to the diketide product to yield curcumin, the main ingredient in the spice, turmeric. Curcumin is produced mainly in the roots of the plant as a phytoalexin. Humans use the roots to provide a pungent taste and yellow color to foods. Curcumin also inhibits the secretion of mucus in mammalian lungs and may be useful in treatment of the symptoms of asthma and cystic fibrosis.

Pyrone Synthases The pyrone synthases are another type of polyketide synthase with restricted distribution among plants. The basic pyrone synthase uses acetyl-CoA as its initial substrate and adds malonyl-CoA in a condensation reminiscent of fatty acid synthesis (**FIGURE 8.35**). Pyrone synthases catalyze the production of a triketide intermediate and its subsequent cyclization to 6-methyl-4-hydroxy-pyrone.

Like the CHSs, the pyrone synthases are dimers with two active sites. The chalcone and basic pyrone synthases exhibit 30% to 50% amino acid similarity and the same catalytic triad of histidine, cysteine, and asparagine. A comparison of the x-ray crystal structures of CHS from *Medicago sativa* and the pyrone synthase from *Gerbera hybrida* is shown in **FIGURE 8.36**. The structural data clearly illustrate that production of different products is a result of minor residue mutations that affect the size and hydrophobicity of the ketide binding cavity.

Stilbene Synthase An especially perplexing example of polyketide synthesis is the production of stilbenes.

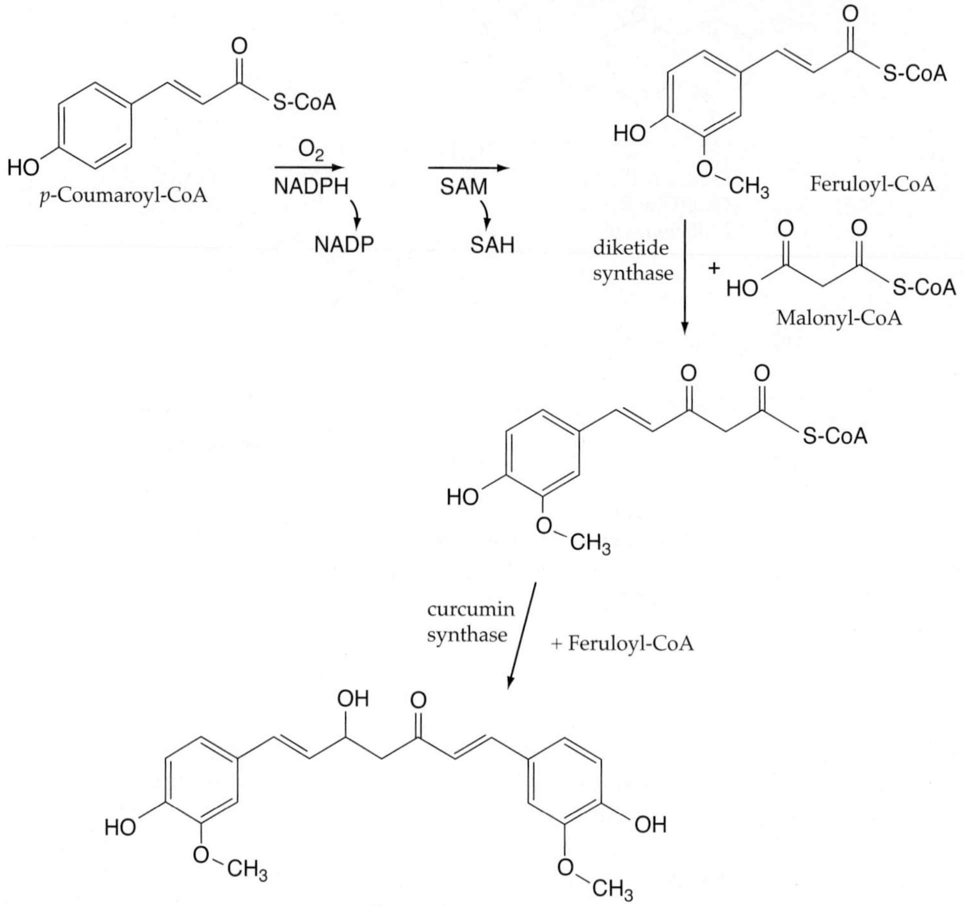

Curcumin

FIGURE 8.34 Synthesis of curcumin. A unique diketide synthase binds the CoA ester of ferulic acid (see Figure 8.14) and catalyzes the addition of malonyl-CoA. The feruloyl-CoA diketide intermediate is further modified by a specific polyketide synthase that adds a second feruloyl moiety to the side chain to yield curcumin. This particular polyketide synthase is unique to some plants of the Zingiberaceae (ginger) family.

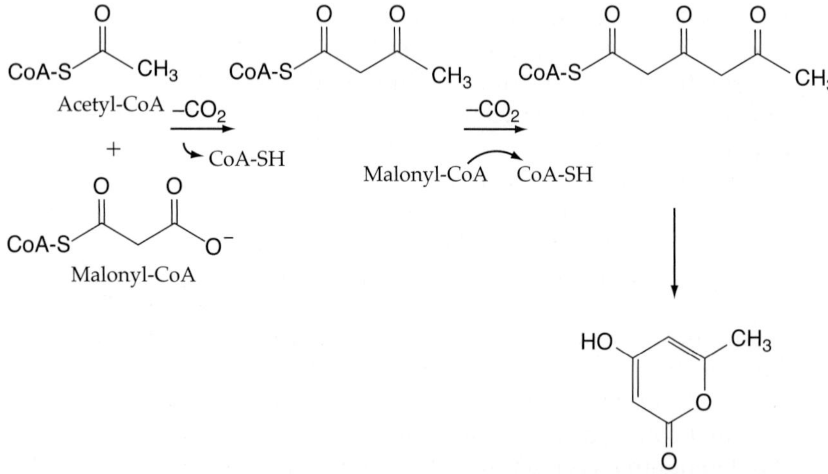

6-Methyl-4-hydroxy-2-pyrone

FIGURE 8.35 Activity of pyrone synthase. The enzyme catalyzes the sequential condensation of two molecules of malonyl-CoA with an acetyl-CoA starter unit instead of the *p*-coumaroyl-CoA starter in CHS. A triketide intermediate is produced in this reaction and then cyclized into an unsaturated δ-lactone called a pyrone.

Stilbene synthases utilize coumaroyl-CoA as their initial substrate and add three units of malonyl-CoA to produce the same tetraketide intermediate that is found in chalcone synthesis. The stilbene synthases, however, catalyze a subsequent decarboxylation of the intermediate and a C-2 to C-7 cyclization instead of the C-1 to C-6 cyclization in chalcone synthases (**FIGURE 8.37**). Both types of synthases exhibit over 50% amino acid similarity and have identical residues in the active site and surrounding substrate entrance cavity. Yet they consistently catalyze the formation of different products from the same tetraketide intermediate. Subtle differences in the active site region apparently favor the formation of a larger molecule by CHSs and decarboxylation of the tetraketide and synthesis of a smaller product by stilbene synthases.

As in the case of the flavonoids, the stilbenes can be further modified, mainly by glycosylation.

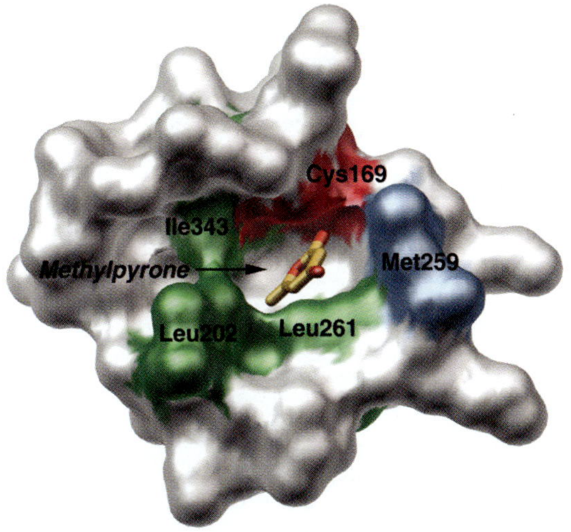

FIGURE 8.36 Comparison of the active site structures of 2-pyrone synthase (2-PS) (left) and chalcone synthase (CHS) (right). The major difference between these homologous, dimeric enzymes is the size of the binding cavity, which is considerably smaller in the case of 2-PS. In addition, 2-PS has an active site entrance that is constricted compared with CHS and lined with several hydrophobic residues. Relatively small molecules have access to the active site cavity and result in the condensation and exit of relatively hydrophobic products. This is in contrast to the more open and hydrophilic cavity of CHS (see also Figure 8.26). (*Reproduced from*: J. M. Jez, M. B. Austin, J.-L. Ferrer, M. E. Bowman, J. Schröder, and J. P. Noel, *Chemistry & Biology* 7 [2000]: 919–930, Figure 6. Courtesy of Joseph P. Noel, Howard Hughes Medical Institute.)

They are widespread but not universally distributed in flowering plants. Thcy have antibacterial and antifungal properties and act as phytoalexins. Resveratrol, the stilbene shown in Figure 8.37, is abundant in grapes. Based on experiments *in vitro* and experiments in animals, it acts as an antioxidant and can inhibit blood platelet aggregation (clotting) in mammals. Resveratrol is the compound in red wine that is presumably responsible for the "French paradox," that is, how to consume a diet rich in saturated fats and cholesterol and still maintain a low risk for atherosclerosis and heart attack. Experiments in mice have recently shown that consumption of resveratrol protects mice from the adverse effects of a high-calorie diet even though the animals were relatively obese compared with controls. Larger quantities of resveratrol than are present in a normal diet, however, must be consumed to obtain these health benefits. Resveratrol is available in pure form and clinical trials are currently being done in humans suffering from type 2 diabetes.

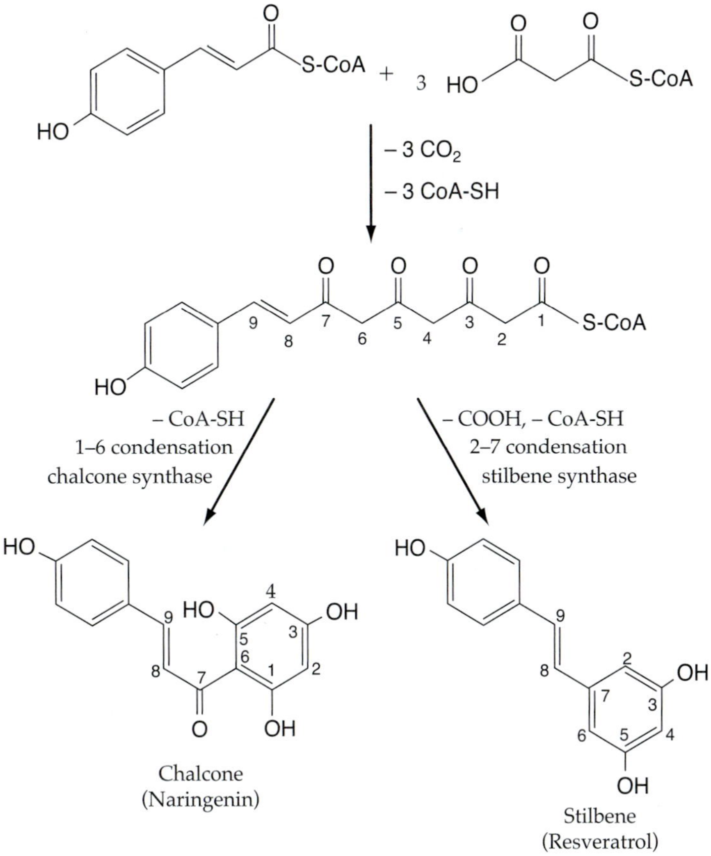

FIGURE 8.37 Comparison of the activity of chalcone and stilbene synthases. Both enzymes catalyze the condensation of three molecules of malonyl-CoA with a *p*-coumaroyl-CoA staring unit to generate the same tetraketide intermediate. In chalcone synthase CHS, the intermediate undergoes a C-1 to C-6 cyclization and aromatization to produce naringenin chalcone. In stilbene synthase, the tetraketide intermediate undergoes a decarboxylation before cyclization of C-2 to C-7 to produce the second aromatic ring to yield resveratrol.

■ Synthesis and Activity of Coumarins

The coumarins, like other products of the phenylpropanoid pathway, start with the synthesis of the key intermediate, coumaroyl-CoA. Coumarin synthesis requires a specific hydroxylase (cytochrome P450 monooxygenase) that adds an OH to the aromatic ring at the *ortho* position. The next step is a nonenzymatic isomerization of

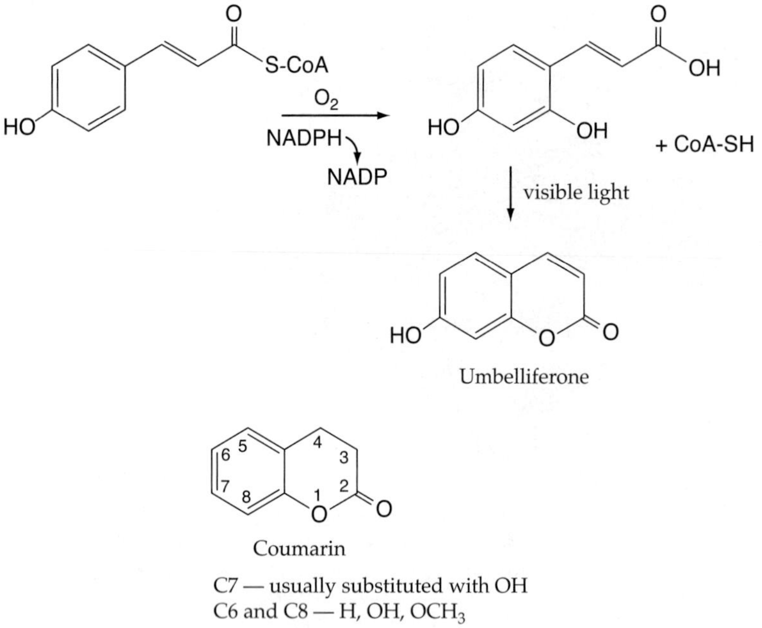

FIGURE 8.38 Synthesis of umbelliferone. Coumarins are synthesized from *p*-coumaroyl-CoA, the key phenylpropanoid pathway intermediate. A specific cytochrome P450 monooxygenase hydroxylates the ring at the *ortho* position with respect to the side chain and removes the CoA. A light-catalyzed isomerization of the double bond in the side chain is followed by spontaneous formation of the δ-lactone. Coumarins can be further modified by hydroxylation and methylation. The inset shows the numbering system and substitution patterns of coumarins.

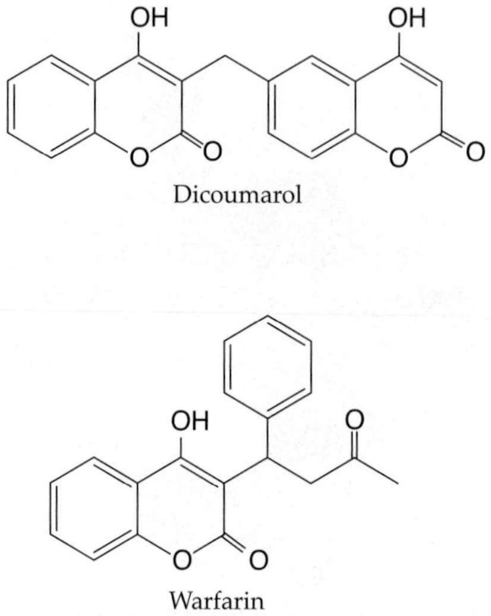

FIGURE 8.39 Structures of two coumarin derivatives. Dicoumarol is produced in damaged clover plants by dimerization of coumarol. It is a powerful anticoagulant in mammals. Warfarin is a synthetic coumarin with structural similarity to dicoumarol and is used therapeutically to prevent internal clot formation.

the double bond in the side chain catalyzed by visible light, followed by spontaneous formation of a δ-lactone. **FIGURE 8.38** illustrates the synthesis of umbelliferone and the general structural designations for coumarins.

Coumarins have a restricted distribution among angiosperms and are found in some members of the Fabaceae, Asteraceae, and Apiaceae. The sweet smell of new-mown hay is due to the release of coumarins from damaged clover (*Melilotus officinalis*). Coumarins can be dimerized to dicoumarol, which is a highly effective anticoagulant (**FIGURE 8.39**). Consumption of large amounts of sweet clover can cause internal bleeding in ruminants. Synthetic anticoagulants, such as warfarin, are based on the dicoumarol structure. Once widely used as a rat poison, warfarin is now prescribed as an anticlotting agent for the treatment of humans with atherosclerosis.

Coumarins are often modified by the addition of other chemical groups. The C-7 position is usually hydroxylated and glycosylated. A specific prenyl transferase can use dimethyl allyl diphosphate to prenylate C-6. The prenylated coumarin is then further modified by two monooxygenases to yield a linear furano derivative called a psoralen (**FIGURE 8.40**).

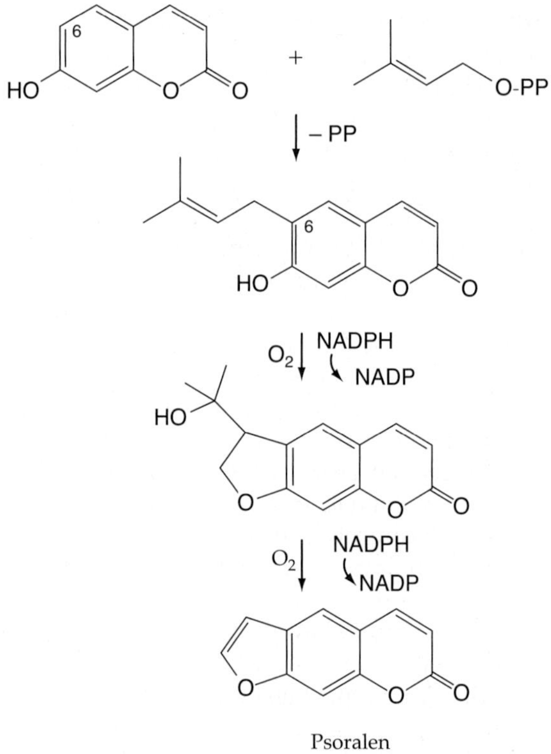

FIGURE 8.40 Synthesis of the furanocoumarin, psoralen. Umbelliferone (7-hydroxycoumarin) is prenylated at C-6 with dimethylallyldiphosphate, which can then lead to formation of a five-membered furan ring. The prenyl side chain is hydroxylated by a cytochrome P450 monooxygenase. A second P450 monooxygenase catalyzes removal of the side chain and formation of a double bond in the furan ring to form psoralen. The ring may be modified by glycosylation. The linear furanocoumarins are phytophotosensitizers that cause severe dermatitis in humans.

The furanocoumarins are produced in some crop plants such as celery in response to fungal pathogens. These coumarin derivatives act as phytophotosensitizers in humans. Workers who harvest the plants by hand can come in contact with the furanocoumarin; subsequent exposure of the skin to sunlight causes dermatitis followed by severe blistering of the skin. In the United States, cow parsnip or hogweed (*Heracleum sphondylium*) is an invasive plant that produces furanocoumarins. The plant is often found growing in open fields and along highways. Its dermatitis-causing properties present a major barrier to removal and eradication of this invasive species.

CHAPTER SUMMARY

A general outline describing the synthesis of aromatic compounds in plants is illustrated in **FIGURE 8.41**. Synthesis begins in plastids with two simple intermediates of primary carbon metabolism, erythrose-4-phosphate and phosphoenolpyruvate (PEP). These feed into the shikimic acid pathway to produce chorismic acid and the essential aromatic amino acids, tyrosine, phenylalanine, and tryptophan. Intermediates in the shikimic acid pathway can provide the building blocks for additional aromatics such as the gallotannins and the plant hormone, salicylic acid (Figure 8.1). Phenylalanine is the central product of the shikimic acid pathway; it is the substrate for phenylalanine ammonia lyase (PAL), a cytoplasmic enzyme and the gateway to phenylpropanoid synthesis. The precursor for most phenylpropanoids is coumarate or an esterified derivative, most often coumaroyl-CoA. The synthesis of primary products such as lignin draws from the pool of coumaroyl-CoA. Natural products such as flavonoids also draw from the same precursor pool and can be considered primary metabolites because of their essential functions in plant–environmental interactions. Many polyketide synthases use coumaroyl-CoA as a starting unit and produce unique natural products such as curcumin (Figure 8.34).

Additional aromatic compounds are synthesized by branches from the phenylpropanoid pathway, and only a few of these have been covered in this chapter such as the coumarins. Aromatic metabolism in plants comprises a rich assortment of chemical reactions and products. The structures and syntheses of many new compounds are still not fully resolved and others are waiting to be discovered. The aromatic amino acids are also precursors in other unique synthetic pathways, such as the glucosinolates covered in Chapter 5. The alkaloids are a rich assortment of additional natural products that are also derived from amino acids, particularly the aromatic amino acids. A small number of these are described in Chapter 9.

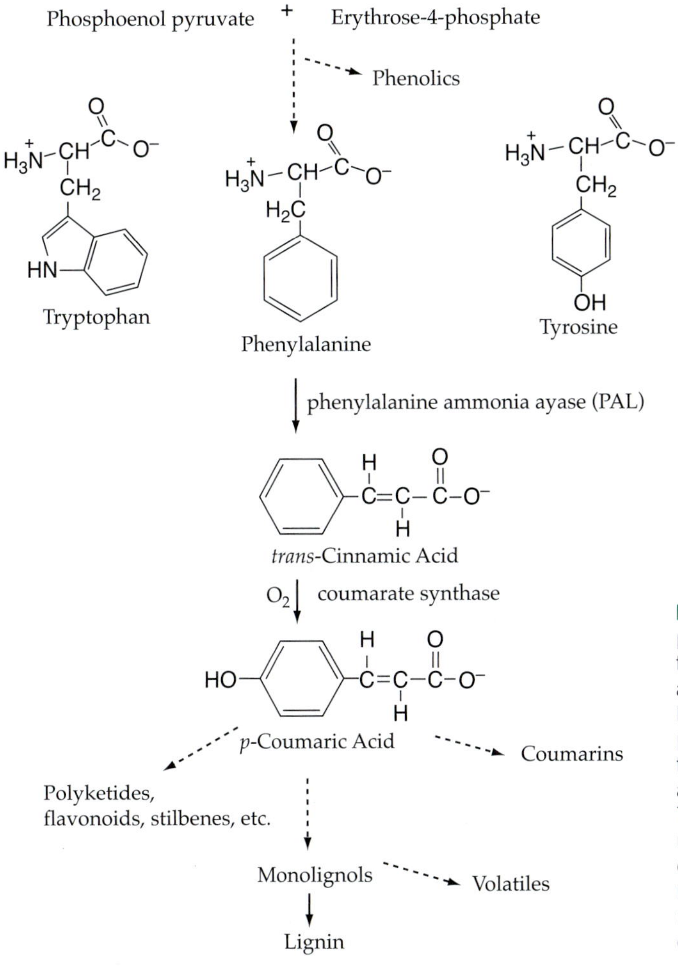

FIGURE 8.41 Summary of general aromatic metabolism in plants. The synthesis of aromatic compounds starts in plastids with the shikimic acid pathway and results in the production of the three aromatic amino acids, tyrosine, phenylalanine, and tryptophan. Phenylalanine occupies a key position as the precursor of the phenylpropanoid pathway. The cytoplasmic enzyme, PAL, removes the amino group from phenylalanine and yields *trans*-cinnamic acid, which is a substrate for a cytochrome P450 monooxygenase. The product is *p*-coumaric acid, which can be further modified, usually by esterification to coenzyme A. The resulting pool of coumaroyl-CoA in the cytoplasm is a source for the synthesis of monolignols and lignin, which are primary products, and a large number of natural products, including the vital flavonoids and their derivatives and defensive compounds such as coumarins.

REFERENCES FOR FURTHER INFORMATION

Boerjan, W., Ralph, J., & Baucher, M. (2003). Lignin biosynthesis. *Annu. Rev. Plant Biol.* **54:**519–546.

Grotewold, E. (2006). Genetics and biochemistry of floral pigments. *Annu. Rev. Plant Biol.* **57:**761–780.

Herrmann, K.M. (1995). The shikimate pathway: early steps in the biosynthesis of aromatic compounds. *Plant Cell* **7:**907–919.

Humphreys, J.M. & Chapple, C. (2002). Rewriting the lignin road map. *Curr. Opin. Plant Biol.* **5:**224–229.

Koes, R., Verweij, W., & Quattrocchio, F. (2005). Flavonoids: a colorful model for the regulation of evolution of biochemical pathways. *Trends Plant Sci.* **10:**236–242.

Tanaka, Y., Sasaki, N., & Ohmiya, A. (2008). Biosynthesis of plant pigments: anthocyanins, betalains and carotenoids. *Plant J.* **54:**733–749.

PLANT BIOCHEMISTRY *on the web*

Visit the *Plant Biochemistry* student companion Web site at http://biology.jbpub.com/gleason/plantbiochemistry/ for additional online study resources on these topics.

CHAPTER 9
Alkaloids

CHAPTER OUTLINE

9.1 Overview

Alkaloids are imprecisely defined as cyclic compounds containing one or more nitrogen atoms in a negative oxidation state (i.e., they act as bases). This classic definition does not always apply because of the effects of other substituents in the molecule. Many nitrogen-containing compounds are called alkaloids but are not "alkaline." Alkaloids are a diverse group of natural products usually associated with vascular plants, but many are also found in algae, microorganisms, insects, and some other animals, particularly marine organisms. Alkaloids in plants are a heterogeneous group of approximately 12,000 known compounds and are part of a plant's constitutive defenses against herbivores and some species of pathogens. The types of alkaloids present tend to be specific to a particular plant family or even a single species.

Alkaloids also have a range of physiological activities in humans. They are the most widely used active plant metabolites found in many pharmaceutical preparations and herbal medicines. The best known examples are the opiate alkaloids, morphine, codeine, and papaverine, which humans have used for millennia to reduce pain. They are still medically important. All these opiate alkaloids are derived from the opium poppy, *Papaver somniferum*. The biosyntheses of opiates and most alkaloids are very complex, involving intermediates from several metabolic pathways. Because simple and economical chemical syntheses are not always possible, many alkaloids are still extracted from plants. The complete biosynthetic schemes and the enzymes required for some alkaloid syntheses, such as the opiate alkaloids, have recently been described and their genes have been cloned into microorganisms. Currently, attempts to genetically "engineer" plants for their overproduction of alkaloids are being explored.

Because their origin and chemical properties are so heterogeneous, there are no straightforward ways to classify the alkaloids. They can be divided into at least four major groups based on their biosynthetic precursors:

1. Most alkaloids are derived from amino acids, in particular, the three aromatic amino acids, and certain basic aliphatics such as arginine, ornithine, lysine, and histidine. This is the largest group, including the opiates and tropanes, and their synthesis and uses are covered extensively in this chapter.
2. Purine alkaloids, containing a purine ring, are derived from adenosine. This group includes the alkaloids found in coffee, tea, and cocoa, which are the most widely used alkaloids by humans.
3. Aminated terpenes are generally synthesized from a terpene and amino acid precursor.
4. Polyketides have the nitrogen moiety introduced into a polyketide skeleton.

9.2 Alkaloids Derived from Aromatic Amino Acids

Tyrosine Derivatives

Tyrosine is the precursor for the synthesis of a variety of well-known alkaloids classified as benzoisoquinolines. The first step in their synthesis is the addition of a second hydroxyl group to the aromatic ring of tyrosine to produce dihydroxyphenylalanine (DOPA), followed by its decarboxylation to dopamine (**FIGURE 9.1**). In mammals, dopamine is a neurotransmitter and the precursor to adrenaline (epinephrine), the "flight or fight" hormone produced under stress. In the American cactus, *Lophophora williamsii*, dopamine is hydroxylated, and all the hydroxyl groups are methylated to produce mescaline. Synthesis of mescaline is relatively straightforward and the compound can be made chemically. It is a hallucinogen in humans, which is not unexpected given its resemblance to mammalian neurotransmitters. The dosage required to obtain a euphoric effect in humans is rather large, 300 to 500 mg of pure compound. The cactus "buttons" are consumed in religious ceremonies by Native Americans in the Southwest and Mexico. Mescaline is the active component in the cactus, but a few other alkaloids are also present.

Morphinan Alkaloids Dopamine can be condensed with a second tyrosine derivative, *p*-hydroxyphenyl acetaldehyde (**FIGURE 9.2**), to yield a trihydroxy intermediate called norcoclaurine. Norcoclaurine is a key intermediate in the biosynthesis of the morphinan group and additional benzylisoquinoline alkaloids that are found in many angiosperms. Norcoclaurine can then be oxidized and both O- and *N*-methylated to another key intermediate called reticuline. In *Papaver somniferum*, this compound is the biosynthetic precursor of the morphinan alkaloids, morphine and codeine, found in poppies (**FIGURE 9.3**).

Morphinan synthesis in *P. somniferum* has been reported to involve several cell types and transfer of intermediates. The enzymes of benzylisoquinoline alkaloid synthesis are located in the cytoplasm of companion cells, and the products are transported to the sieve elements

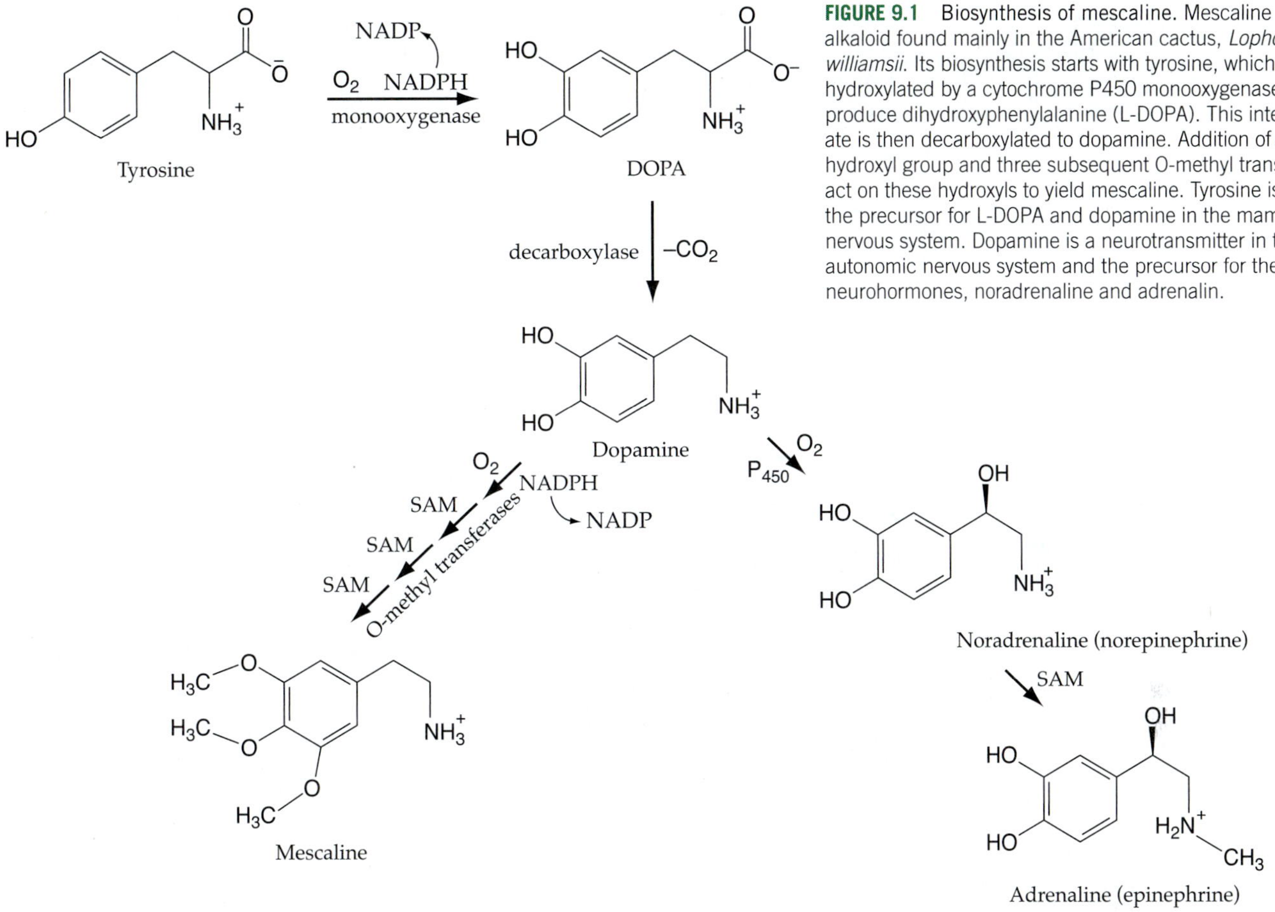

FIGURE 9.1 Biosynthesis of mescaline. Mescaline is an alkaloid found mainly in the American cactus, *Lophophora williamsii.* Its biosynthesis starts with tyrosine, which is hydroxylated by a cytochrome P450 monooxygenase to produce dihydroxyphenylalanine (L-DOPA). This intermediate is then decarboxylated to dopamine. Addition of another hydroxyl group and three subsequent O-methyl transferases act on these hydroxyls to yield mescaline. Tyrosine is also the precursor for L-DOPA and dopamine in the mammalian nervous system. Dopamine is a neurotransmitter in the autonomic nervous system and the precursor for the neurohormones, noradrenaline and adrenalin.

of the phloem where final synthesis occurs. The morphinan alkaloids are found in the phloem fluid but are mainly stored within the seed capsule in cells called laticifers, which also produce a sticky latex. The alkaloids and latex in the Papaveraceae are most likely produced to deter insect herbivores. Humans typically make small cuts in the seed capsules and allow the latex with its dissolved alkaloids to ooze out and harden. The hardened exudate, raw opium, is easily collected and dissolved in alcohol to yield laudanum, which is used as an analgesic, sleep inducer, and narcotic.

Morphine and codeine can be extracted from the raw opium. Morphine is used medically to relieve severe pain. It also causes euphoria and a state of mental detachment, thus relieving the anxiety associated with injury. Although it can be addictive, studies show that moderate use for pain relief, especially under traumatic conditions, does not usually lead to addiction. The alkaloid acts by binding to opiate receptors in the mammalian brain. The natural receptors in the brain are small peptides called endorphins that are produced in the pituitary gland. Endorphins and similar peptides act to reduce pain and promote a sense of well-being.

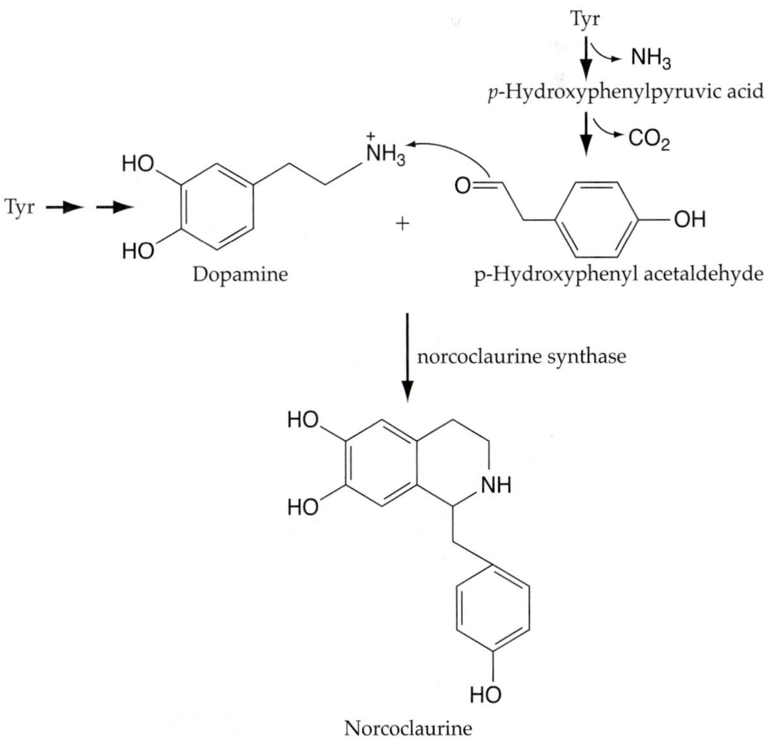

FIGURE 9.2 Norcoclaurine synthesis. Tyrosine is hydroxylated and decarboxylated to dopamine as in Figure 9.1. Tyrosine can also be deaminated to a keto derivative and decarboxylated to yield *p*-hydroxyphenyl acetaldehyde. Condensation of these two tyrosine derivatives yields norcoclaurine, a key intermediate in the biosynthesis of benzoisoquinoline class of alkaloids, including the morphinans.

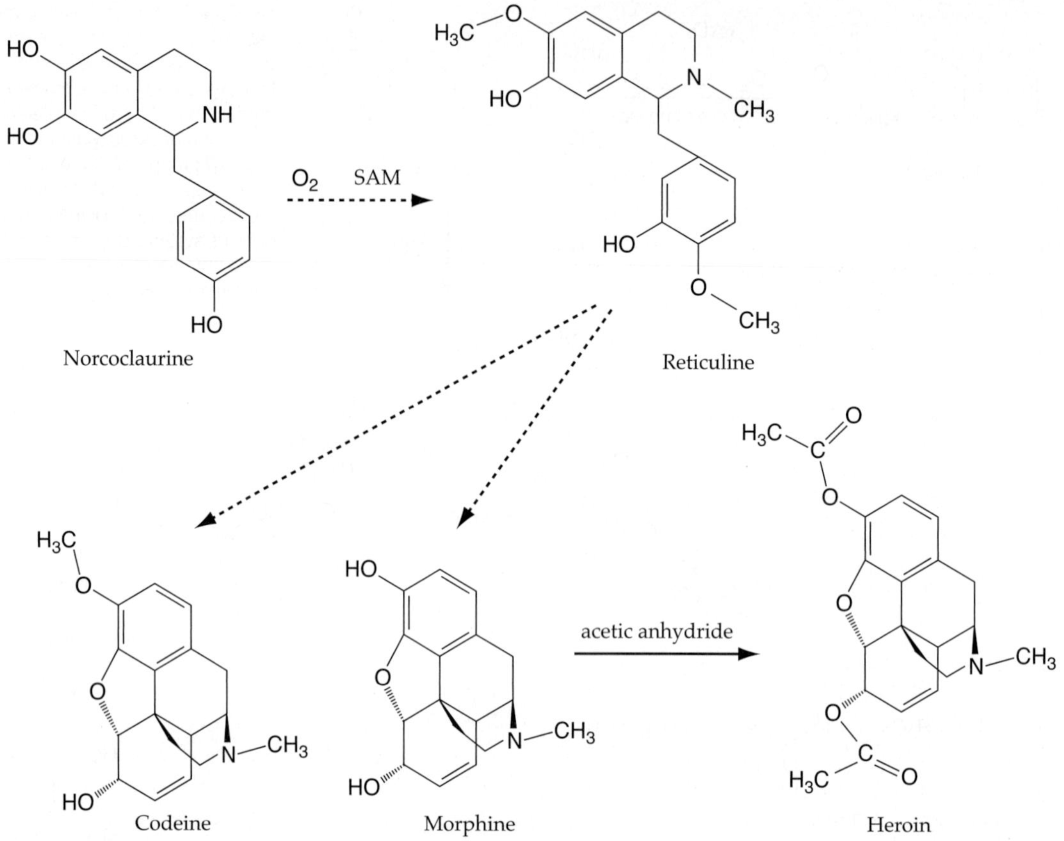

FIGURE 9.3 Synthesis of reticuline and the morphinan alkaloids. Norcoclaurine is derived from the condensation of two tyrosine derivatives as shown in Figure 9.2. A second hydroxyl group is added to the benzyl moiety of norcoclaurine by a cytochrome P450 monooxygenase. Two O-methyl transferases specifically catalyze the addition of methyl groups to one of the benzyl hydroxyls and one of the isoquinoline ring hydroxyls. An *N*-methyl transferase adds a third methyl group to produce reticuline. Reticuline is an intermediate found in many plants and is further modified to produce many species-specific alkaloid products. In the opium poppy, *Papaver somniferum*, reticuline is the precursor of morphinan alkaloids; the most abundant are codeine and morphine. Heroin is a synthetic morphinan produced by acetylation of morphine.

Although there is little chemical similarity between these endogenous peptides and morphine, a number of opiate alkaloids and synthetic opiates bind to the opioid receptors at very low concentrations and produce both the analgesic and euphoric effects. Heroin, also shown in Figure 9.3, is a product of chemical semisynthesis. It was initially produced by the Bayer Chemical Company by acetylation of morphine and sold in cough remedies. Acetylation makes heroin more hydrophobic than morphine and, therefore, more readily transported across cell membranes. Although it is an effective cough suppressant, it is highly addictive and lethal at a much lower concentration than morphine. Overdoses of opiates cause depression of the respiratory centers in the brain and result in death by asphyxiation. Codeine, another morphinan, is a moderate analgesic and relieves coughing. It is generally nonaddictive, but chronic use causes constipation.

Reticuline, the key intermediate in morphinan synthesis, is also an intermediate in the synthesis of berberine alkaloids. Reticuline is modified by formation of a C–C bond between the *N*-methyl group of the isoquinoline ring and the benzyl ring (**FIGURE 9.4**). This reaction is catalyzed by an unusual enzyme called the berberine bridge enzyme (BBE). This enzyme contains a flavin adenine dinucleotide cofactor covalently bound to histidine and cysteine residues. Molecular oxygen is required for enzymatic activity, and hydrogen peroxide is generated in this reaction. The structure and mechanism of action of the enzyme from the California poppy (*Eschscholzia californica*) have been determined. Several plant genomes contain many genes for homologous enzymes. The model plant, *Arabidopsis thaliana*, contains 12 BBE-like genes of unknown function, suggesting that the BBE-type reaction is widespread in angiosperms and is important in nonalkaloid pathways. The product of the BBE reaction in *E. californica* is scoulerine, a metabolite found in several angiosperms that is further modified into a number of different alkaloids collectively referred to as berberines (Figure 9.4).

Berberine and other benzylisoquinoline alkaloids have been characterized in many members of the Berberidaceae (barberry family) and Ranunculaceae (buttercup family). These alkaloids are effective against bacterial pathogens. Humans have used these plants for their antibiotic and other physiological activities. For example, ipecac, from *Cephaelis ipecacuanha*, is still used to induce vomiting after poison ingestion, and tubocurarine, the active component in arrow poisons, is given to patients as a muscle relaxant before undergoing surgery. Isolated berberine has been shown to have moderate antiamebic, antibacterial, and anti-inflammatory properties. Many plants containing berberine are used in traditional herbal medicines. This alkaloid is not a particularly effective antibiotic in pure form because most bacteria are able to rapidly adapt to the alkaloid and pump it out of their cells. Whole leaf extracts of *Berberis fremontii*, however, were found to be effective in killing *Staphylococcus aureus*. Assay-guided fractionation of *B. fremontii* homogenates led to the isolation of a flavonoid, 5′-methoxyhydnocarpin, that inhibits bacterial efflux pumps. Thus, these two components in this plant's leaves are capable of defending the plant against bacterial pathogens. This alkaloid-flavonoid antibiosis is an exceptionally clear example of synergistic action found in whole plants in which neither isolated component is especially active alone. It also illustrates one of the difficulties in investigating the effects of traditional herbal medicines by attempting to isolate a single active component.

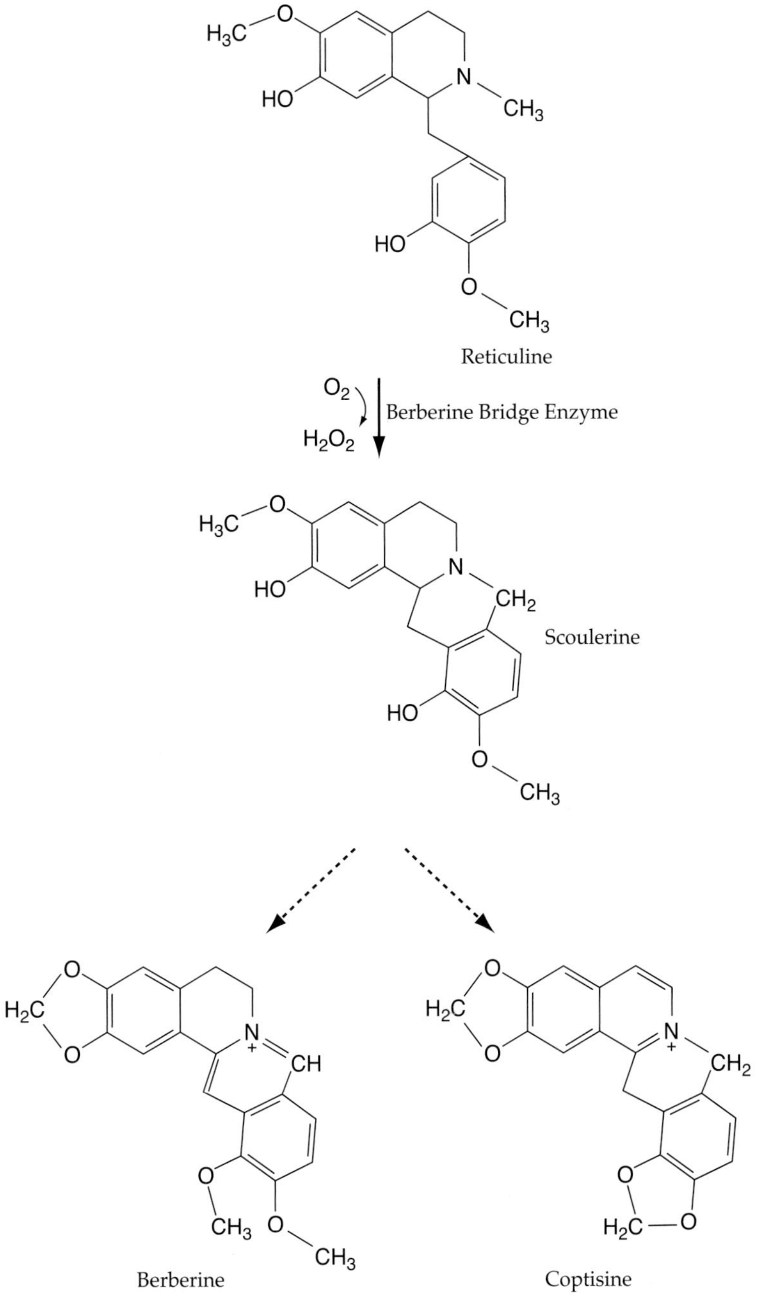

FIGURE 9.4 Modification of reticuline by the berberine bridge enzyme (BBE) and formation of benzoisoquinoline alkaloids. The BBE catalyzes formation of a C–C bond by linking the *N*-methyl group of the isoquinoline to the benzyl ring. The intermediate, scoulerine, is the precursor of several berberine-type alkaloids found in a number of different angiosperms. Berberine and coptisine are widespread in the family, Berberidaceae and Ranunculaceae, respectively. They have demonstrated antibacterial activity.

Betalains Betalains are the only known colored alkaloids and are produced in only a few plant genera that do not synthesize anthocyanins. These include red beets (*Beta vulgaris*), the prickly pear cactus (*Opuntia* sp.), *Bougainvillea* sp., and a few additional plants in the order, Caryophyllales (carnation class). Feeding radioactive metabolites to portulaca (moss roses) showed that the biosynthetic precursor is L-DOPA derived from tyrosine (Figure 9.1 and **FIGURE 9.5**). All the steps and enzymes in this pathway, however, are not known.

Betalains are usually found in the flower petals, but in some plants, such as beets, they are present in the leaves and roots as well. They are usually glycosylated and stored in vacuoles. Betanidine, shown in Figure 9.5, has an absorption maximum at approximately 520 nm and gives the flowers and other plant parts a red color. In some plants the key intermediate, betalamic acid, is condensed with derivatives of other amino acids, and the resulting betalains have yellow to orange colors. The function of these pigments in plants is not certain; like anthocyanins, they may attract pollinators or deter specific insect herbivores. The reason for their unusual distribution

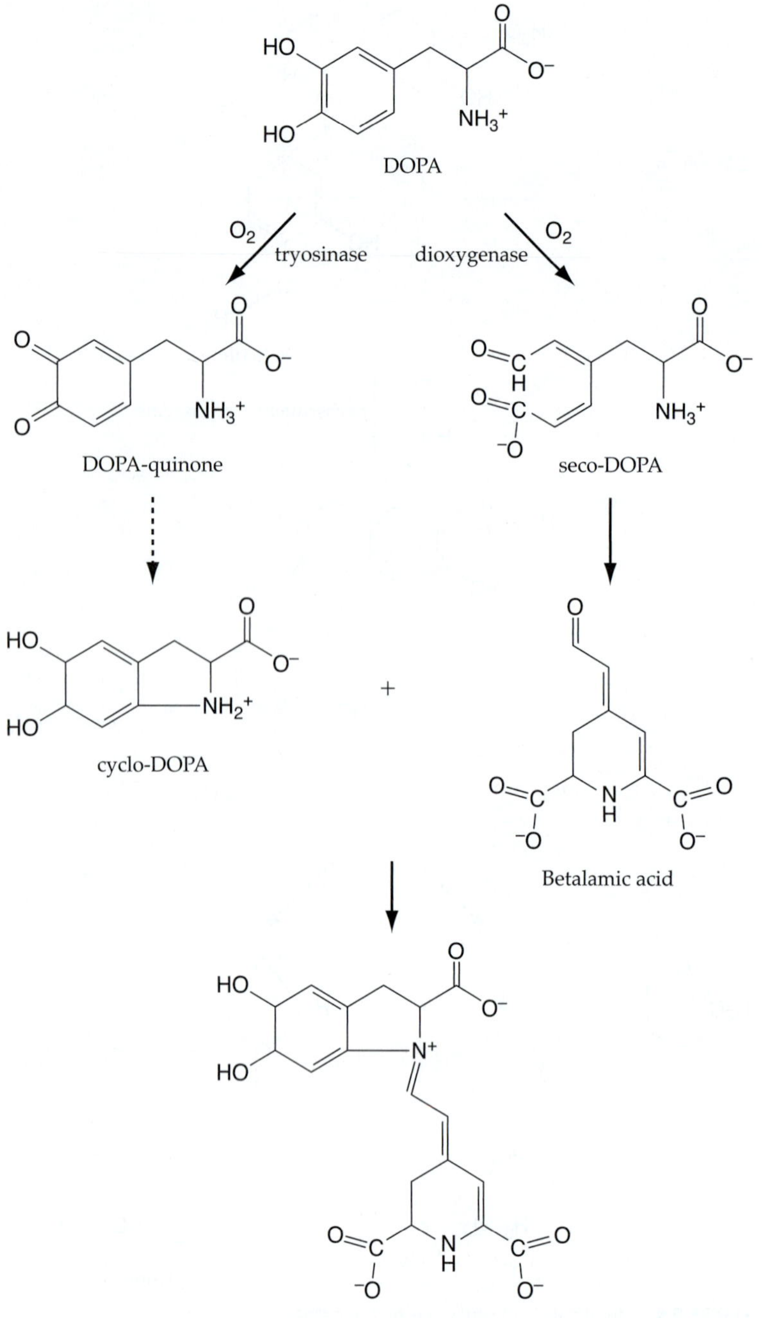

FIGURE 9.5 Synthesis of betalains. L-DOPA (see Figure 9.1) is the precursor for the biosynthesis of these colored compounds that are found in a small number of angiosperms. Tyrosinase, a copper-containing phenol oxidase, catalyzes the oxidation of DOPA to the quinone *in vitro*. Tyrosinase is found in many plants and oxidizes a number of different phenolic substrates. It is the enzyme responsible for the browning that occurs when slicing fruits and tubers. DOPA is also oxidized by a dioxygenase to seco-DOPA. Ring closure converts seco-DOPA to betalamic acid, a common precursor for the synthesis of various betalains. The red pigment found in beets, betanidine, is produced from a condensation of betalamic acid with cyclo-DOPA.

among plants and the fact that they do not occur with anthocyanin pigments is unknown because the flavonoid pathway for anthocyanin precursor synthesis exists in all betalain-producing plants.

Phenylalanine Derivatives

In contrast to tyrosine and tryptophan, phenylalanine has not been found to be the direct precursor for many alkaloids, although it is the major source of intermediates for other natural products such as flavonoids (see Chapter 8). One well-known phenylalanine-derived alkaloid is ephedrine from the gnetophyte, *Ephedra intermedia*. Gnetophytes are a group of vines and shrubs related to gymnosperms. *Ephedra* species have been used in traditional medicines by humans for centuries; a widely used Chinese herbal drug for bronchial asthma, MaHuang, contains parts of this plant. Ephedrine alkaloids are effective nasal decongestants and bronchial dilators. As illustrated in **FIGURE 9.6**, phenylalanine is the precursor for benzoyl-CoA in *Ephedra* sp. This intermediate is condensed with pyruvic acid and leads to the formation of cathinone. Cathinone is the branch point for the syntheses of a variety of ephedra alkaloids found in plant leaves and stems. Pseudoephedrine, shown in Figure 9.6, is the active component in the over-the-counter drug, Sudafed, which is used for relief of cold and allergy symptoms. The modern commercial product is produced by chemical synthesis rather than extraction from the plant. Ephedra alkaloids act on adrenergic receptors mimicking the neurotransmitter, adrenaline. Small doses provide relief from the runny nose and coughing associated with allergies. Pseudoephedrine is readily converted to methamphetamine, which has more pronounced and longer lasting adrenaline-like effects on the mammalian nervous system. The pathway intermediate, cathinone, is also the active ingredient in khat (qat). The leaves of the plant, *Catha edulis*, are rich in cathinone and are chewed for their stimulant effects by people in the Middle East at social functions.

Tryptophan Derivatives

Alkaloids derived from tryptophan are called indole alkaloids. The first step in the biosynthesis of two of these (**FIGURE 9.7**) is the decarboxylation of tryptophan to tryptamine. Two subsequent hydroxylations of the aromatic ring and *N*-methylation produce psilocin and its phosphorylated derivative psilocybin. These compounds are produced by fungi in the genus *Psilocybe* sp. Tropical plants, for example, *Peganum harmala* and *Banisteriopsis caapi*, produce similar indole alkaloids from tryptamine such as harmine. All these compounds are hallucinogenic in animals because of their resemblance to the mammalian neurotransmitter, serotonin, which

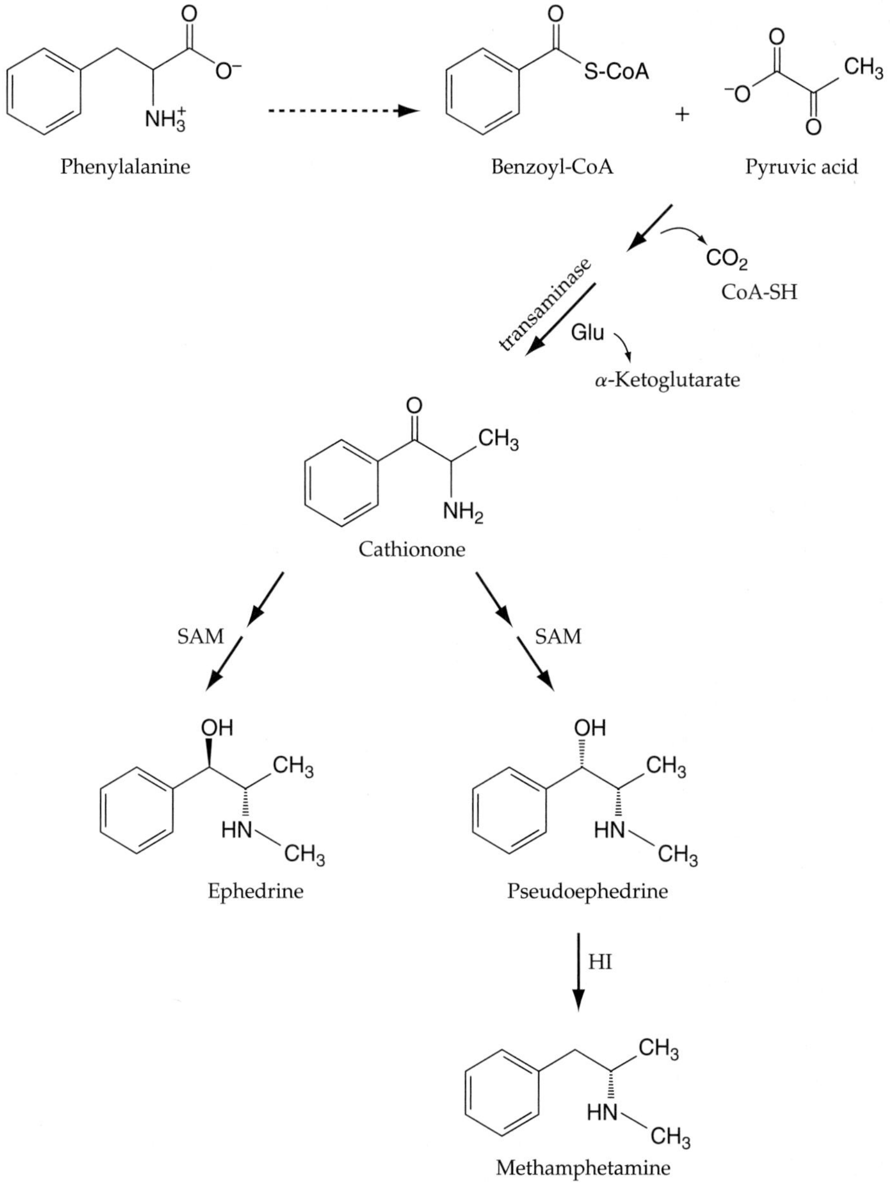

FIGURE 9.6 Synthesis of *Ephedra* alkaloids. Phenylalanine is the precursor of benzoyl-CoA in *Ephedra intermedia*. Benzoyl-CoA is condensed with pyruvic acid, and the resulting keto intermediate is the substrate for a specific transaminase that produces cathinone. Cathinone is found in several plants. In *E. intermedia*, the keto group of cathinone is reduced to a hydroxyl group and the amino group is methylated to yield several ephedra alkaloids, which differ in their configuration at the two chiral centers shown. Ephedrine has an *R*–OH, *S*–NH_2–CH_3 configuration, whereas pseudoephedrine has the *S*–OH and *S*–NH_2–CH_3 configuration. Both compounds resemble the mammalian neurotransmitter, adrenaline (see Figure 9.1). Methamphetamine can be synthesized from pseudoephedrine by removal of the OH group with a strong acid such as hydrogen iodide (HI).

is also synthesized from tryptophan (Figure 9.7). Serotonin is a monoamine neurotransmitter in the mammalian central nervous system. It mediates both central and peripheral neural functions. Melatonin is produced in the pineal gland of the brain and is a hormone that regulates circadian and other biological rhythms. The fungal and plant indole alkaloids inhibit reuptake of these neurotransmitters in the mammalian brain, leading to enhanced neural stimulation and often resulting in euphoria and hallucinations. None of these alkaloids, however, is known to be toxic to humans. Their effects on insect neurosystems will be more severe, and, most likely, their primary purpose in the plant is to deter these herbivores.

Terpenoid Indole Alkaloids Tryptamine can be condensed with a terpene derivative, usually from the 10-carbon geranyl diphosphate, to yield a large group of terpenoid-indole alkaloids. These compounds are found mainly in the

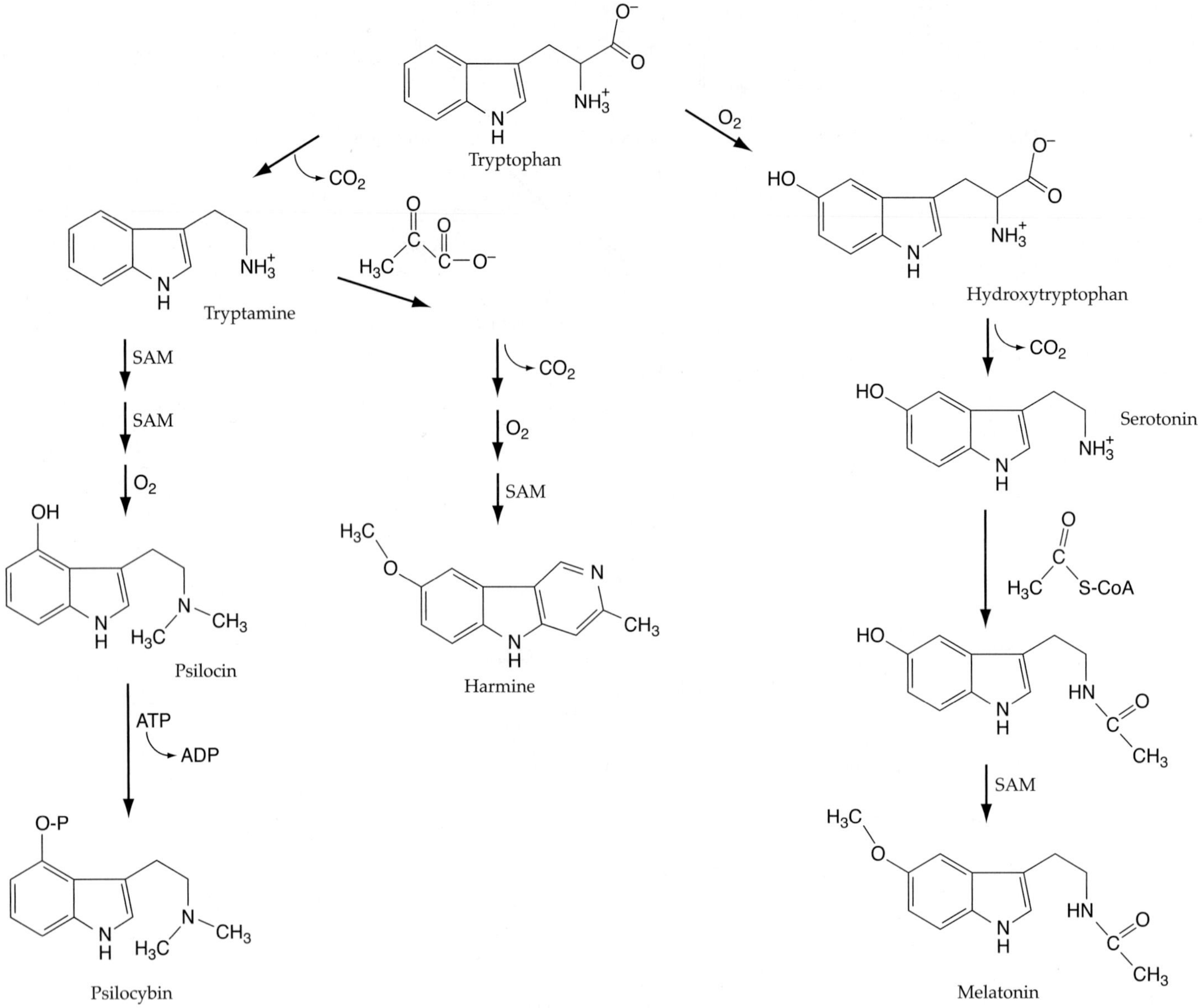

FIGURE 9.7 Synthesis of several indole alkaloids. Tryptophan is the precursor for the synthesis of indole alkaloids that have been found in plants, fungi, and animals. The initial step in plant and fungal alkaloid biosynthesis is a decarboxylation to tryptamine. Two subsequent *N*-methylations of the indole ring, followed by a hydroxylation, lead to psilocin and its phosphorylated derivative, psilocybin, produced in some fungi. Harmine, found in the angiosperm, *Peganum harmala,* is the result of condensation of tryptamine with pyruvate, followed by modification of the side chain, and ring closure. In mammals, tryptophan is hydroxylated by a specific cytochrome P450 monooxygenase to hydroxytryptophan. The resulting hydroxytryptophan is decarboxylated to the alkaloid neurotransmitter, serotonin. Serotonin is further modified by acetylation and methylation to produce the hormone, melatonin.

plant families, Apocynaceae (milkweed family), Loganiaceae, and Rubiaceae (coffee family). A key intermediate is the indole derivative, strictosidine. Strictosidine is produced by condensation of tryptamine with a terpene derivative called secologanin (**FIGURE 9.8**).

Strictosidine synthase is a key enzyme in the synthesis of these complex terpenoid alkaloids. The gene for this enzyme from *Rauwolfia serpentina,* a plant that is native to Asia, has been cloned. *R. serpentina* produces the terpenoid alkaloids, ajmaline and reserpine. The gene for strictosidine synthase can be expressed in *Escherichia coli,* although the enzyme is not glycosylated in the bacterium as it would be *in planta*. This synthase is a protein of 344 amino acids, and the x-ray structure has been solved. It has a novel six-bladed β-propeller fold with each blade composed of four β-strands (**FIGURE 9.9A**). The active site is formed by residues from all six "blades," reminiscent of the configuration of α/β-barrel proteins such as the Rubisco large subunit (see Chapter 2). Figure 9.9A shows a terpene analog bound between the blades but

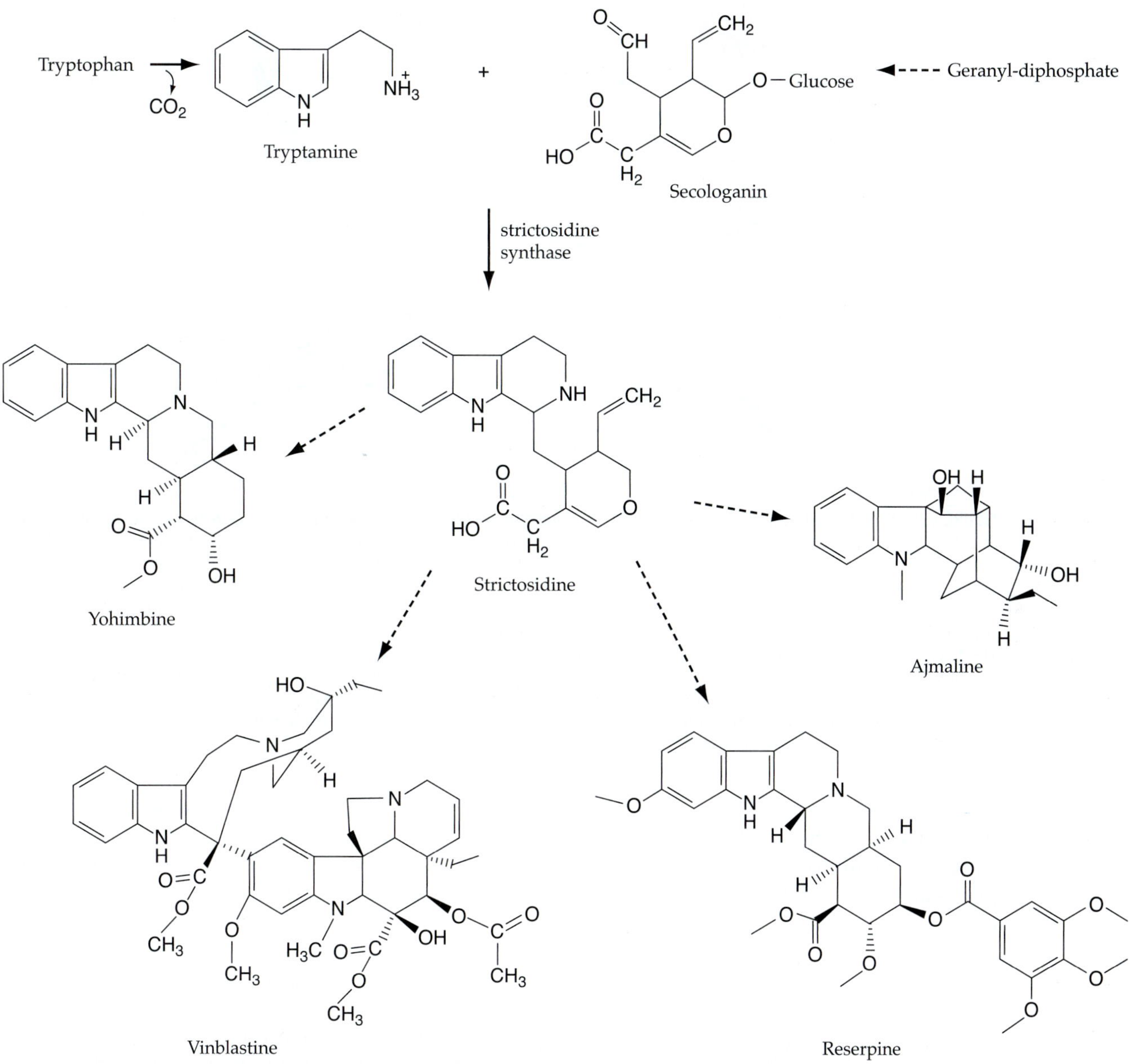

FIGURE 9.8 Strictosidine and synthesis of complex indole terpenoid alkaloids. Tryptamine (see Figure 9.7) from the decarboxylation of tryptophan is condensed with secologanin, an O-glycosylated terpenoid (see Chapter 6) to produce strictosidine. This key intermediate is produced in a variety of different plant species and is subsequently modified to produce a number of complex indole-terpenoid alkaloids. Ajmaline and reserpine are both found in *Rauwolfia serpentina*. Yohimbine is common in the bark of some tropical species. Two terpenoid alkaloids, vindoline and catheranthine, are found in *Catharanthus roseus*. These are condensed to form a di-terpenoid alkaloid, vinblastine.

close to the surface of the enzyme. In contrast, tryptamine binds deeper into the active site cavity as seen in Figure 9.9B. The active site in strictosidine synthase is mainly composed of hydrophobic amino acids with one major exception, a glutamic acid residue that is highly conserved in all strictosidine synthases (**FIGURE 9.9B**). The γ-carboxyl group of this glutamic acid residue is located between the tryptamine and the terpenoid binding sites. The γ-carboxyl is positioned to remove a proton from the amino group of tryptamine. The deprotonated *N*-moiety then attacks the carbonyl carbon of the terpenoid substrate, secologanin (**FIGURE 9.10**). Dehydration facilitates ring closure and release of the product, strictosidine.

Strictosidine synthase from *R. serpentina* is homologous to other alkaloid synthases, including those in plants that produce some of the most complex of the terpenoid indole alkaloids. Site-directed mutagenesis of the cloned gene has produced modified catalysts capable of synthesizing analogs of these complex and medically important alkaloids. Ajmaline, shown in

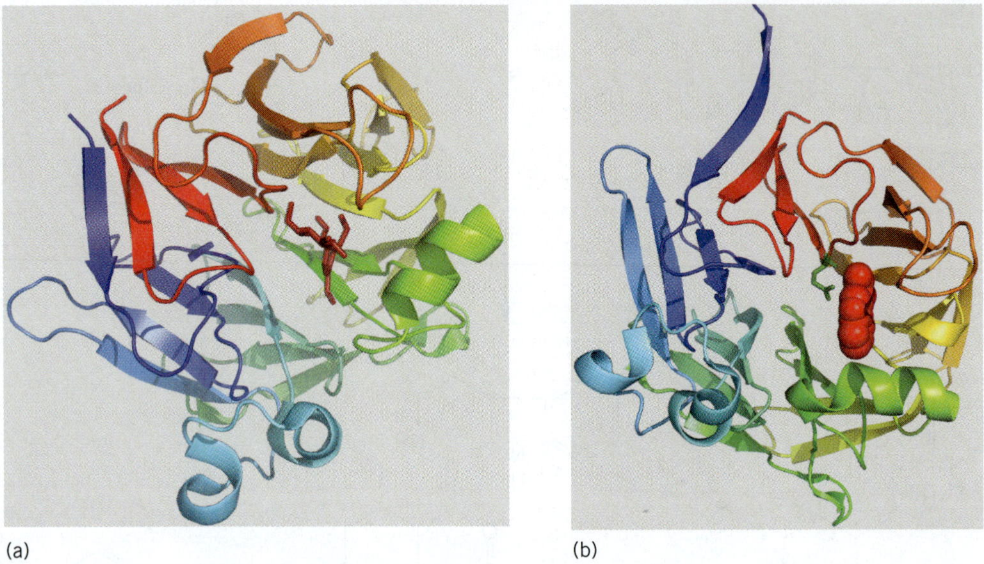

FIGURE 9.9 Structure of strictosidine synthase from *Rauwolfia serpentina*. The recombinant protein was expressed in *E. coli*, and the structure was solved with bound substrates to 2.0 to 3.0 Å resolution. (a) The enzyme with a bound secologanin in the active site is shown. The protein is composed mainly of β-strands organized into six "blades," each consisting of four β-strands. These blades are shown in orange, yellow, green, teal, cyan, blue, and red. The active site is formed by residues in the loops that connect the blades. The glycosylated terpene, secologanin, is shown as a red stick figure. It binds near the surface of the active site with the glucose moiety exposed to the solvent. (b) This model shows the enzyme with a bound tryptamine (red spheres). The tryptamine is bound in a hydrophobic environment and deeper into the active site cavity than the terpene shown in part a. The one conserved, hydrophilic amino acid, glu-309, is shown as green stick figure protruding into the active site cavity just to the left of the substrate. The carboxyl group of this glutamic acid removes a proton from tryptamine to initiate the condensation reaction with secologanin shown in Figure 9.10. (Structure a from Protein Data Bank 2fpc. S. Panjikar, *Plant Cell* 18 [2006]: 907–920. Structure b from Protein Data Bank 2fpb. S. Panjikar, *Plant Cell* 18 [2006]: 907–920.)

Figure 9.8, from *R. serpentina* is effective in treating cardiac arrhythmias. Reserpine is produced by several plants in the genus, *Rauwolfia*. These plants are also used in a variety of herbal remedies. Reserpine has been used to lower blood pressure. In addition, it has a tranquilizing effect. Yohimbine is found in the bark of *Paucinystalia yohimbe*, a tropical plant in the Rubiaceae family (which also includes the coffee tree), and *Aspidosperma* sp., a tropical woody species that also produces cardiac glycosides. Yohimbine is used in traditional herbal preparations and is supposed to act as an aphrodisiac. It can dilate blood vessels and lower blood pressure in mammals. It is a monoamine oxidase inhibitor, and humans should not take it with other antidepressant medications because the combination can result in an overdose of dopamine and possible cardiac arrest. Other well-known alkaloids in this class are strychnine, a deadly poison from *Strychnos* sp., and quinine from the bark of *Cinchona* species, which was the only treatment for malaria in the eighteenth and nineteenth centuries and is still used today.

The vinca alkaloids are a subgroup of the dimeric terpenoid indole alkaloids found in the leaves of the Madagascar periwinkle, *Catharanthus roseus* (previously, *Vinca roseus*). Vinblastine (Figure 9.8) is used in the treatment of Hodgkin's disease, a cancer affecting the lymph glands, spleen, and liver. A related alkaloid, vincristine, is effective in treating childhood leukemia. Other vinca alkaloids are currently being tested in clinical trials for

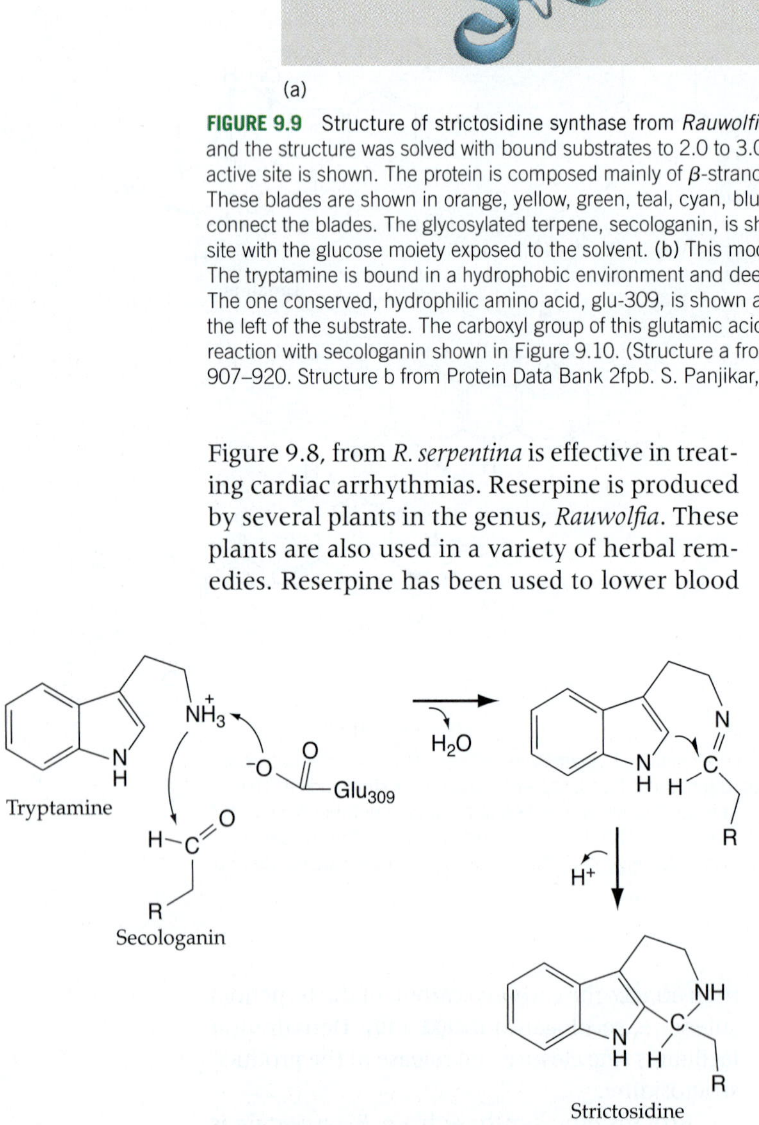

FIGURE 9.10 Proposed mechanism of action of strictosidine synthase. Based on the crystal structures shown in Figure 9.9, the carboxyl group of residue 309, a glutamic acid, is positioned between the two substrates, tryptamine and secologanin. The glutamate removes a proton from the amino group of tryptamine that then attacks the carbonyl carbon of the terpene. The position of the intermediate in the active site facilitates a C–C bond and formation of the third ring in the product, strictosidine.

treatment of certain forms of lung cancer. The vinca alkaloids bind to β-tubulin and inhibit cell division by blocking the function of the spindle apparatus. These alkaloids bind tightly to a specific site, but their precise mechanism of binding to tubulin is unknown, as is their specificity for targeting a certain cell type. Given the chemical complexity of these molecules, they are still obtained from the periwinkle. Moreover, several cell types are involved in the biosynthesis and storage of these compounds. Intense efforts are now being made to produce these compounds in plant tissue cultures. Plant cells in culture can be infected with the bacterium, *Agrobacterium rhizogenes,* to produce only highly branched roots. These hairy root cultures of periwinkle have been transformed with genes for wild-type strictosidine synthase and various mutant synthases. These cultures produce some monomeric vinca alkaloids and analogs and release them into the culture medium. To date, hairy root cultures produce only enough vinca alkaloids for clinical research but could be scaled-up in the future. In addition, a better understanding of structure/activity basis of vinca alkaloids and enzymes will provide a rationale for production of additional synthetic or semisynthetic anticancer agents.

Ergot Alkaloids The ergot alkaloids are also synthesized from tryptophan and a single isoprene unit (**FIGURE 9.11**). These are made mainly by fungi such as *Claviceps* sp. and some plants in the morning glory family (Convovulaceae) such as *Ipomoea* and *Rivea* sp. The fungus *Claviceps* infects grains such as wheat and rye and produces fruiting bodies containing these toxic alkaloids. Ingestion of ergot alkaloids in flour made from infected grains causes severe stomach pains, vasoconstriction in the limbs leading to gangrene, and neurological symptoms such as headache and hallucinations. In contrast, some of the ergot alkaloids, such as ergometrine, have been used medically in the past to control bleeding during and after childbirth. Semisynthetic analogs, such as dihydroergotamine, are currently used to treat migraines, and bromocriptine suppresses lactation and is used to treat breast tumors.

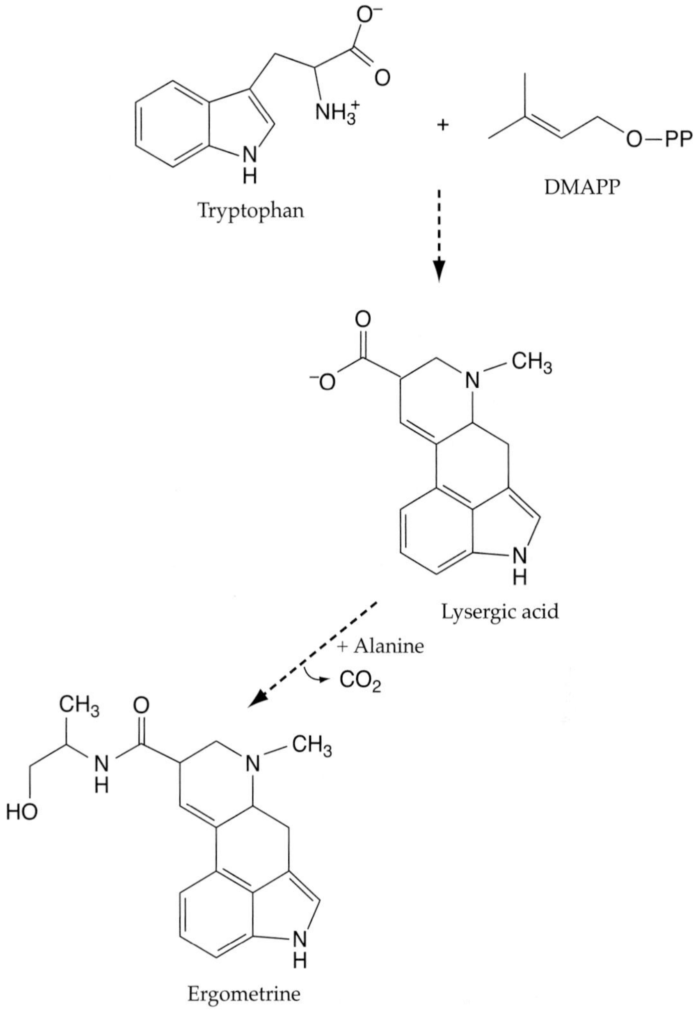

FIGURE 9.11 Synthesis of ergot alkaloids. Tryptophan is the precursor for the ergot alkaloids and is condensed with a five-carbon terpene, usually dimethylallyldiphosphate. Lysergic acid is a key intermediate and can be modified by addition of amino acids. These reactions are catalyzed by nonribosomal peptide synthetases. Ergometrine is the result of adding alanine to lysergic acid followed by a decarboxylation of the alanine side chain.

Albert Hofmann, a chemist at the Sandoz Pharmaceutical Company in the 1930s, was working on the chemical synthesis of ergot alkaloids for these types of medical uses. He synthesized several lysergic acid analogs, one of which, LSD-25, seemed to have effects in addition to vasoconstriction. He experienced strange visual anomalies when working with LSD-25, so he used himself as an experimental animal to investigate. Hofmann took what he thought was a very small dose of 0.25 mg of LSD-25. He felt very agitated and bicycled home from his lab. It was an amazing trip because this small dose of pure compound was actually a relatively large dose of the hallucinogen. Amazingly, he remained sufficiently lucid to write about his experience soon after it occurred. LSD (lysergic acid diethylamide) mimics serotonin and binds to neural receptors in the mammalian brain at micromolar concentrations. Hofmann and other researchers explored the possible use of LSD to treat psychiatric disorders but also unwittingly ushered in the "Age of Aquarius."

9.3 Alkaloids Derived from Basic Amino Acids

Many alkaloids are synthesized from ornithine, a nonprotein amino acid. Ornithine can be derived from arginine. An arginase catalyzes the hydrolysis of the terminal carbon and its two amino groups from the side chain of arginine to generate ornithine and urea (**FIGURE 9.12**). This reaction is part of the urea cycle in mammals. In plants, ornithine is more likely synthesized from glutamine. Ornithine is generally decarboxylated to putrescine, which is a precursor for spermidine and spermine. These polyamines are associated with DNA in all eukaryotic cells and act as regulators of gene activity, especially for genes active in growth and cell differentiation. In plants, these polyamines are also the precursors for the pyrrolizidine alkaloids. Putrescine in plants and some microorganisms can be subsequently *N*-methylated and oxidatively deaminated to produce a five-membered ring, the *N*-methylpyrrolinium cation (Figure 9.12). This cation is a key intermediate in the synthesis of the tropane alkaloids and nicotine.

Tropane Alkaloids The *N*-methylpyrrolinium cation is modified by condensation with acetoacetyl-CoA to generate tropinone, which is the major intermediate with a tropane ring structure (**FIGURE 9.13**). The keto-group at position 3 is reduced to an OH group, which can then form an ester bond with a variety of organic acids. Two common tropane alkaloids found in many Solanaceae are hyoscyamine and scopolamine. Their biosynthesis is shown in **FIGURE 9.14**. Tropane alkaloids occur in many angiosperm families such as Erythroxylaceae, Proteaceae, and Convolvulaceae but are especially common in the Solanaceae, which includes many edible species such as potatoes, tomatoes, and eggplant. The alkaloids can be found in the leaves, flowers, fruits, and roots of these plants, although their biosynthesis occurs only in root tissues. The enzymes required have been found in several cell types in the root and xylem tissue. They are often produced constitutively, presumably to deter herbivores because of their bitter taste and neuroinhibitory activity.

The tropane alkaloids have anticholinergic activity in mammals; that is, they compete with

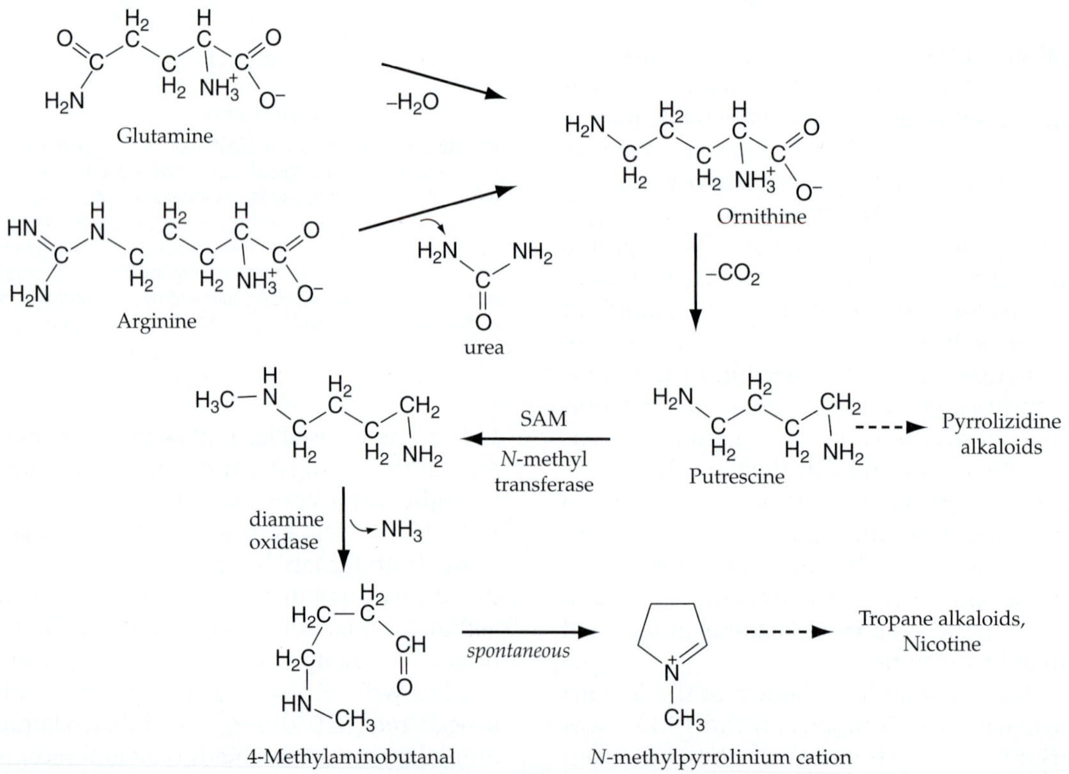

FIGURE 9.12 Synthesis of the pyrrolinium cation. The *N*-methylpyrrolinium cation is a key intermediate in the biosynthesis of several different groups of alkaloids. The precursor is the nonprotein amino acid, ornithine. In plants, this is usually derived from glutamine, whereas arginine is the precursor in animals. Ornithine is decarboxylated to putrescine, which is another important intermediate in the synthesis of pyrrolizidine alkaloids and in the polyamines, spermine and spermidine. One of the amino groups of putrescine is *N*-methylated. The other is removed to yield a transient aldehyde that spontaneously cyclizes to the *N*-methylpyrrolinium cation.

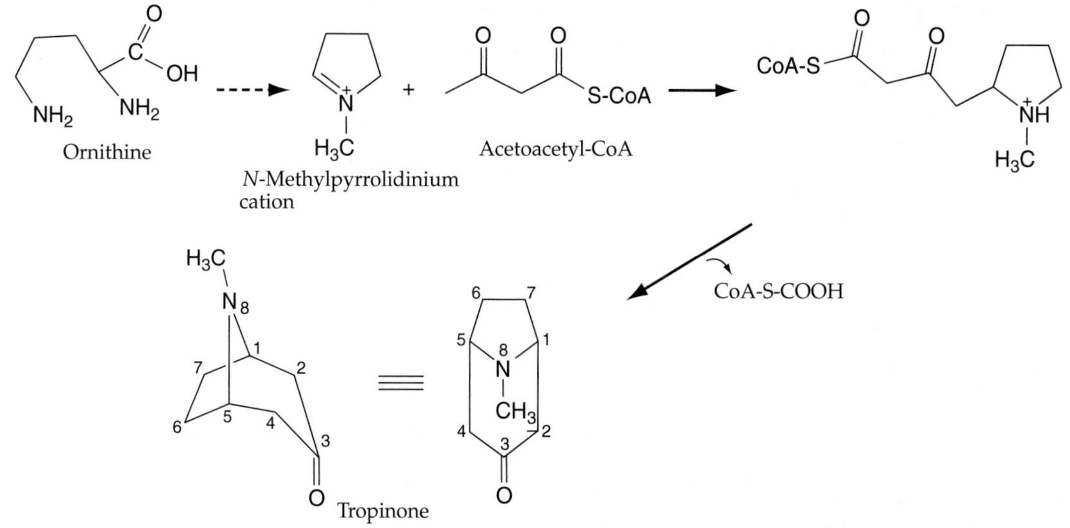

FIGURE 9.13 Synthesis of tropinone. Tropinone is the common intermediate in the synthesis of the tropane alkaloids. The *N*-methylpyrrolinium cation is produced from ornithine as shown in Figure 9.12. The cation is condensed with acetoacetyl-CoA by an unknown enzyme to yield hygrine. Decarboxylation of hygrine leads to ring closure and formation of the tropane ring. Alternate representations and numbering of the ring are shown.

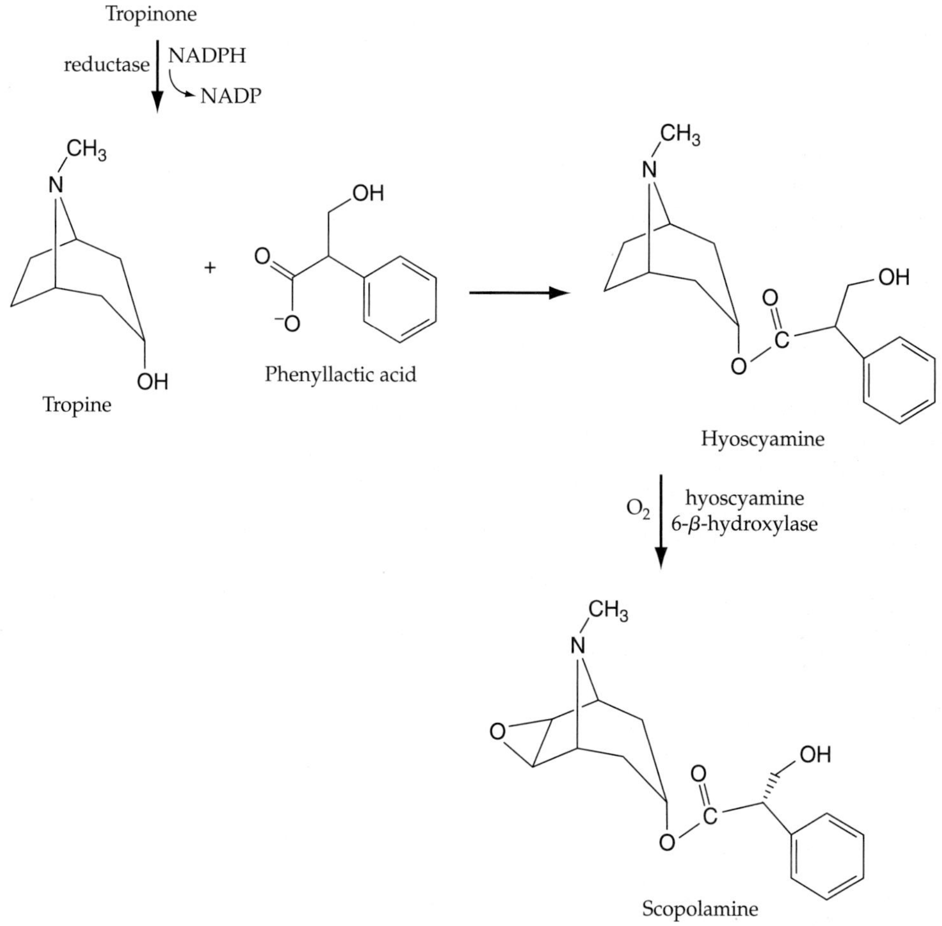

FIGURE 9.14 Synthesis of two common tropane alkaloids, hyoscyamine and scopolamine. A specific reductase reduces the tropinone keto group (see Figure 9.13) to the alcohol of the tropine ring. This alcohol can then be esterified to a number of acids to produce tropane alkaloids. Hyoscyamine, found in many Solanaceae, is the ester of tropine and phenyl lactic acid. In some plants, such as nightshade (*Atropa* sp.), a specific 2-oxoglutarate–dependent dioxygenase oxidizes hyoscyamine to scopolamine. Two isomeric forms of scopolamine are found in some plants, such as *Atropa belladonna.*

the neurotransmitter acetylcholine for the same receptors in the parasympathetic nervous system that mediates involuntary actions and muscle contraction. Tropanes can cause an increase in heart rate and a relaxation of the smooth muscles in the gastrointestinal tract. These alkaloids can also bind to receptors in the central nervous system and cause drowsiness, amnesia, delirium, hallucinations, and death from respiratory failure. Toxic doses in humans range from 3 to 10 mg kg^{-1} but are much lower in children.

Hyoscyamine and scopolamine are present in leaves and fruit of the thorn apple (*Datura stramonium*), which was once cultivated as a source of tropane alkaloids for use as sedatives. Scopolamine has sedative activity and can be used as a preanesthetic and to treat some symptoms of Parkinson's disease. A scopolamine skin patch is often prescribed to prevent motion sickness during air travel. Today, the thorn apple is often used as an ornamental plant or may be cultivated for its narcotic properties. The fruits are highly toxic.

Atropine is the name used for a racemic mixture of the (+) and (−) isomers of hyoscyamine. This tropane alkaloid is produced by nightshade plants (*Atropa belladonna* and other species). It was once used by ophthalmologists for dilating the pupils of the eyes. It has since been replaced by synthetic compounds based on the tropane structure, although the natural product is still used in some Asian countries. Atropine is currently prescribed as an antidote for poisoning due to organophosphorus pesticides and nerve gas. These toxins act by inhibiting the esterase that hydrolyses acetylcholine after neural transmission. By blocking the acetylcholine receptors, atropine relieves the convulsions caused by excess acetylcholine.

Cocaine is synthesized by condensation of a modified tropane ring with phenylalanine-derived benzoyl-CoA (**FIGURE 9.15**). It is the principal alkaloid found in the leaves of *Erythroxylum coca*, a plant native to South America. The leaves have been chewed by native Americans for thousands of years to increase alertness and endurance. Like other tropanes, cocaine affects the central nervous system of mammals. It blocks the reuptake of dopamine and noradrenaline (Figure 9.1). This blockage results in vasoconstriction and elevated blood pressure, increased heart rate, hyperthermia, decreased fatigue, intellectual stimulation, and euphoria. This cocaine "rush" is short-lived, however; both its ester bonds are readily cleaved by esterases. The "high" is followed by physical exhaustion and depression. Addiction to the purified alkaloid is more psychological than physical; abruptly quitting cocaine use does not cause severe withdrawal symptoms. Cocaine affects insect herbivores by binding to octopamine nerve receptors. The resulting neural overstimulation is lethal to insects.

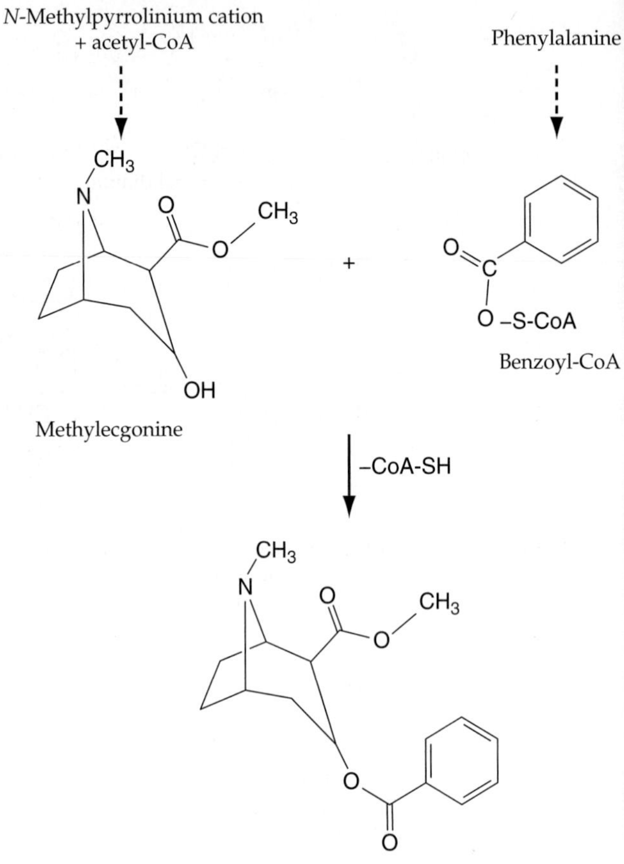

FIGURE 9.15 Synthesis of cocaine. This tropane alkaloid is produced only in species of *Erythroxylum*. In these plants the tropinone keto group (see Figure 9.13) is reduced and esterified with acetyl-CoA to yield the intermediate, methylecgonine. This is then esterified with benzoyl-CoA, derived from phenylalanine, to produce the diester, cocaine.

Cocaine also blocks the transmission of nerve impulses from the peripheral nervous system by blocking sodium ion channels. This makes it a good local anesthetic, and it is still used for this purpose in some countries. In the United States, cocaine has been replaced by synthetic compounds such as novocaine and lidocaine, which are based on the cocaine chemical structure.

Nicotine Nicotine is synthesized by a condensation of nicotinic acid with the key intermediate in the tropane pathway, the *N*-methylpyrrolinium cation. Nicotinic acid, in turn, is derived from dihydroxyacetone phosphate and aspartic acid

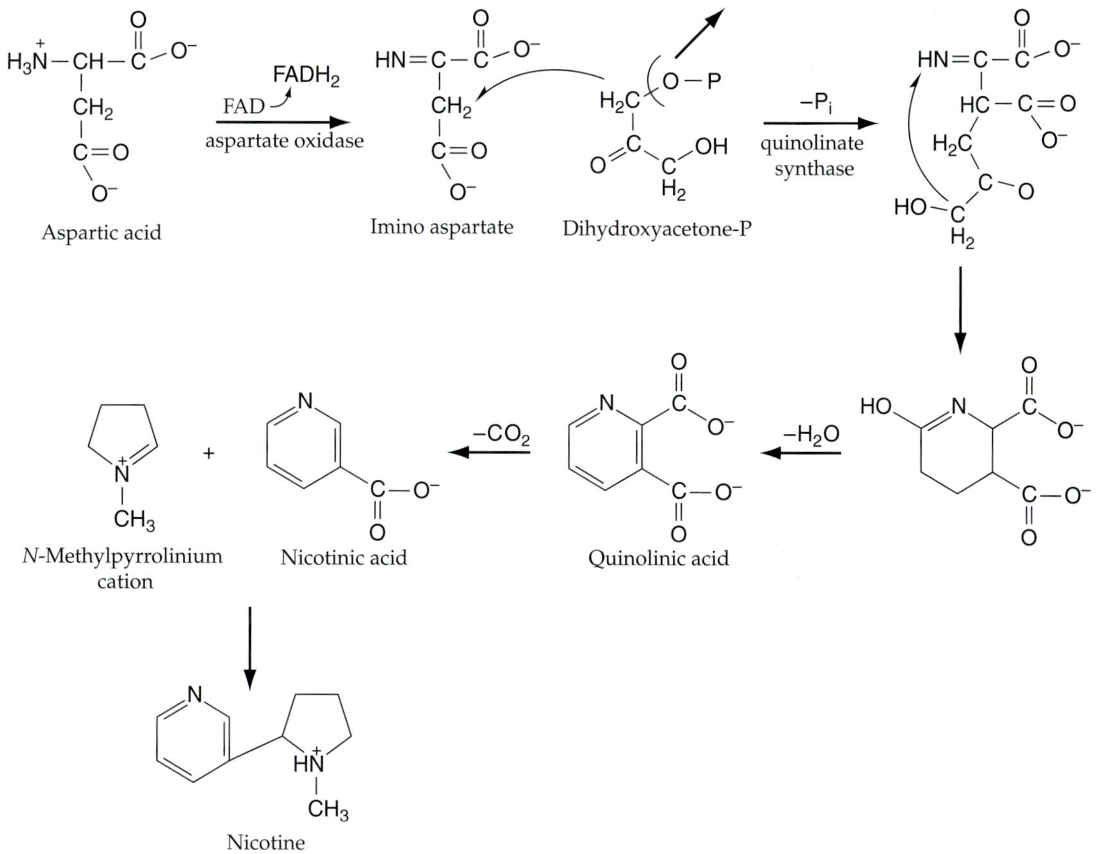

FIGURE 9.16 Nicotinic acid and nicotine biosyntheses. Nicotinic acid (vitamin B3, niacin) is synthesized from aspartic acid and dihydroxyacetone phosphate (DHAP) in plants. The amino group of aspartate is oxidized by a flavin-containing oxidase. The resulting imino acid is condensed with DHAP followed by ring closure to yield quinolinic acid. Decarboxylation then produces nicotinic acid, the precursor for redox cofactors, NAD and NADP. In *Nicotiana* sp., nicotinic acid reacts with the *N*-methylpyrrolinium cation to yield the alkaloid, nicotine.

(**FIGURE 9.16**). Nicotinic acid is produced in all plants and is the precursor for synthesis of the essential redox cofactors, NAD and NADP. In contrast, nicotine and similar alkaloids are made only in species of *Nicotiana* that are native to the Americas. Nicotine is synthesized in the plant roots and transported in the xylem. It accumulates in leaves but is also found in stems and flowers. Damage to leaves, either by insects or mechanical means, induces genes for nicotine synthesis, but only in roots. Jasmonic acid, produced as a result of leaf injury, is most likely the signal for gene induction in roots.

Nicotine is highly toxic to insects; one of the major commercial uses of nicotine is for agricultural insecticides. Nicotine is also toxic to humans; a lethal dose for the average adult is approximately 60 mg. When taken in small amounts, however, it is highly addictive. The nicotine and similar alkaloids in tobacco leaves are rapidly absorbed either through the lungs from the smoke of cigarettes or through the oral mucosa from cigars and chewing tobacco. Like the tropane alkaloids, nicotine binds to specific acetylcholine receptors, acting as a neural stimulant. It is rapidly metabolized so that repeated doses are required to maintain this excitation. Withdrawal can be physically debilitating, causing nausea, headache, and intense gastrointestinal upset. Chronic tobacco use has been shown to cause various diseases in experimental animals and humans, including cancers and heart and pulmonary diseases. In addition to nicotine, however, tobacco smoke also contains a variety of other compounds such as polycyclic aromatic hydrocarbons (PAHs) that are responsible for many of these deleterious effects.

9.4 Alkaloids Derived from Purines

The purine alkaloids are found in relatively few plant species, but they are probably the most widely used alkaloids by humans. The three major products, theophylline, theobromine, and caffeine, are found in tea, chocolate, and coffee. Xanthosine is the major precursor for the purine alkaloids (**FIGURE 9.17**). Xanthosine can be derived from the purine biosynthetic pathway

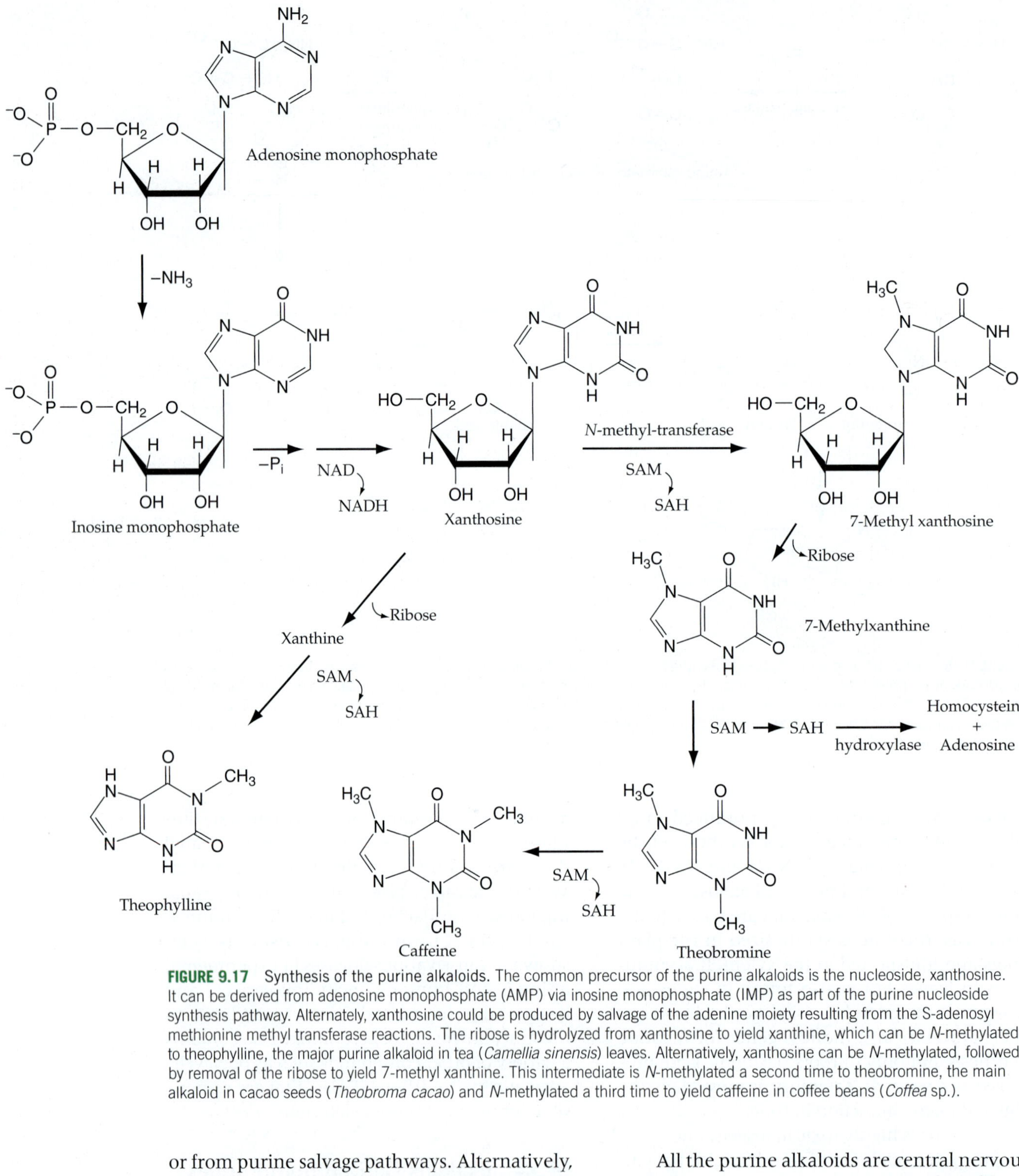

FIGURE 9.17 Synthesis of the purine alkaloids. The common precursor of the purine alkaloids is the nucleoside, xanthosine. It can be derived from adenosine monophosphate (AMP) via inosine monophosphate (IMP) as part of the purine nucleoside synthesis pathway. Alternately, xanthosine could be produced by salvage of the adenine moiety resulting from the S-adenosyl methionine methyl transferase reactions. The ribose is hydrolyzed from xanthosine to yield xanthine, which can be *N*-methylated to theophylline, the major purine alkaloid in tea (*Camellia sinensis*) leaves. Alternatively, xanthosine can be *N*-methylated, followed by removal of the ribose to yield 7-methyl xanthine. This intermediate is *N*-methylated a second time to theobromine, the main alkaloid in cacao seeds (*Theobroma cacao*) and *N*-methylated a third time to yield caffeine in coffee beans (*Coffea* sp.).

or from purine salvage pathways. Alternatively, the xanthosine used in purine alkaloid synthesis could be supplied by the recycling of adenosine from S-adenosyl methionine used in the *N*-methylation steps of this biosynthetic pathway. The purine alkaloids accumulate in young leaves and fruits. They are effective in deterring some insect herbivores and also have allelopathic effects, inhibiting the growth of plant competitors.

All the purine alkaloids are central nervous system stimulants in humans. They block adenosine receptors, which normally have a depressant action on the brain, heart, and kidneys, and their ingestion results in a general physical stimulation. Theophylline is the most potent stimulant and is a major alkaloid in *Camellia sinensis* leaves used to make tea. It is also prescribed medically for relief of bronchial spasms and is found in some asthma inhalers.

Caffeine is the main alkaloid in leaves of *Coffea arabica* and *C. robusta* but also occurs in *Camellia sinensis* and *Cola nitida* (coffee, tea, and colas, respectively). It is also abundant in *Ilex paraguariensis* leaves that are used to make maté or Paraguay tea and in *Paullinia cupana* seeds that are used to make guarana, a soft drink sold in Brazil. Theobromine is the major alkaloid in *Theobroma cacao* and the active ingredient in chocolate. In addition to being consumed in beverages, caffeine is also added to many nonsteroidal anti-inflammatory drugs, such as aspirin and ibuprofen. Caffeine in these combinations with over-the-counter drugs has a synergistic effect and helps to promote the analgesic action. Research in animals indicates that caffeine itself may have anti-inflammatory effects.

None of the purine alkaloids is especially toxic to humans. The estimated lethal dose for caffeine is about 10 g for an "average" person. There are no confirmed cases of human poisoning. The purine alkaloids are mildly addictive in humans, and withdrawal symptoms include headache, lethargy, and irritability. Although many epidemiological studies suggest that consumption of caffeine may have chronic adverse effects on humans, none has been confirmed in controlled clinical trials.

9.5 Aminated Terpenes: The Steroidal Alkaloids

Although a number of different alkaloids can be synthesized by incorporation of amino groups into terpenes (see Chapter 7), the steroidal alkaloids (glycoalkaloids), derived from sterols, are the group most likely to have an impact on humans. They often occur in the Solanaceae including the tubers of potato (*Solanum tuberosum*). The more common glycoalkaloids are produced from a condensation of cholesterol with arginine (**FIGURE 9.18**). α-Solanine is common in potatoes, whereas a related alkaloid, α-tomatine, is found in tomatoes. The glycoalkaloids are present in high concentrations in leaves and flowers but also accumulate in fruits and in potato tubers, particularly after wounding or exposure to light. These alkaloids act as saponins. They disrupt cell membranes and have hemolytic activity. Thus, they provide protection against herbivores and microbial pathogens. They can cause gastrointestinal upset in humans but are also carcinogenic after chronic ingestion. The U.S. Department of Agriculture routinely monitors commercial potato crops for high concentrations of glycoalkaloids. The tubers are still metabolically active, however, and the alkaloids can accumulate in potatoes if stored improperly after being purchased.

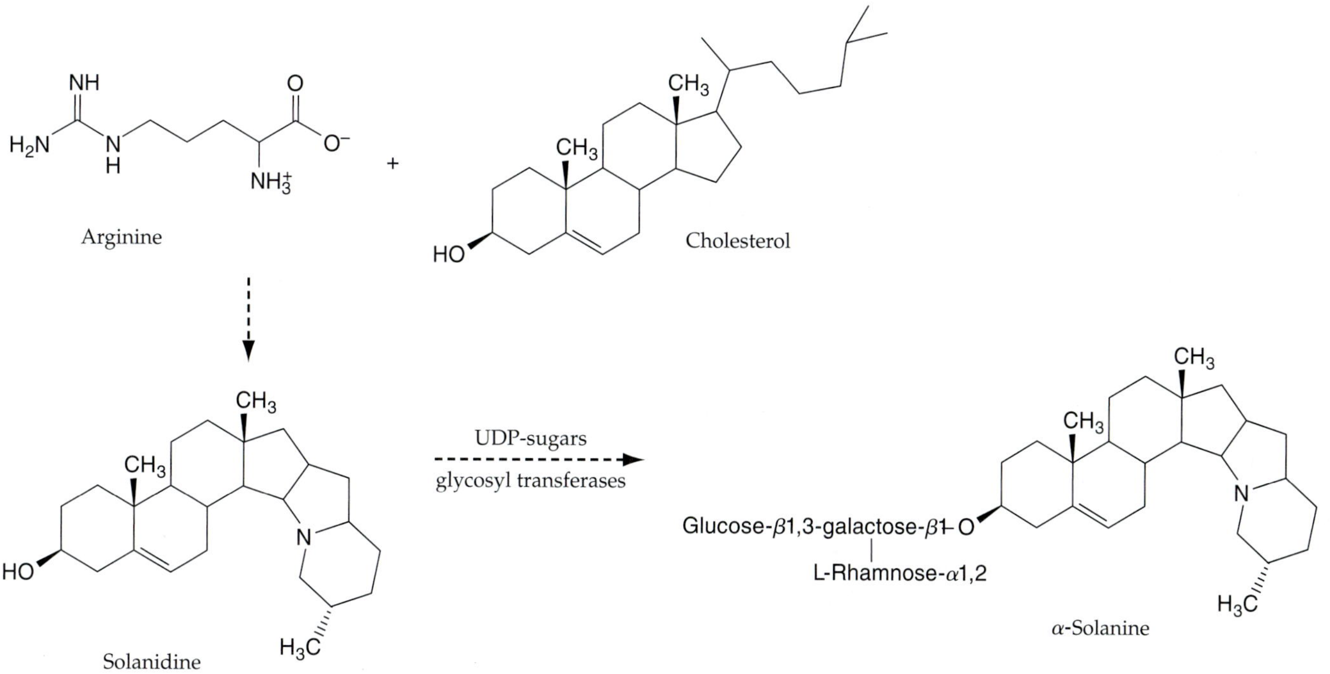

FIGURE 9.18 Synthesis of glycoalkaloids. Glycoalkaloids are typically produced from an amino acid and a sterol. The condensation of arginine with cholesterol to form solanidine, a glycoalkaloid found in leaves, flowers, and tubers in the Solanaceae is shown. These compounds are an example of the aminated terpene alkaloids. They are typically O-glycosylated with one or more sugar residues and therefore referred to as glycoalkaloids. α-Solanine is the most common glycoalkaloid produced in potato tubers when wounded or exposed to pathogens.

CHAPTER SUMMARY

The above description of alkaloids covers only a small fraction of these heterogenous compounds found in plants and other organisms. From a chemical standpoint, they have relatively little in common other than a nitrogen atom within a heterocyclic ring. Their biosyntheses also involve precursors derived from basic metabolic pathways but subsequently modified by enzymes and intermediates that are unique to a relatively small number of plants. The occurrence of alkaloids in the plant kingdom is sporadic; in some families, such as the Solanaceae, they are common, whereas other families apparently do not produce any alkaloids. In most cases alkaloids are produced constitutively, although their biosynthesis can be up-regulated during stress. Their major function seems to be defense against herbivores and pathogens. This ill-defined category of biological compounds, however, has had a major impact on humans, and alkaloids have been used for centuries in both traditional herbal medicines and modern drugs. Many of the alkaloids described in this chapter resemble human neural transmitters and bind to mammalian receptors, causing behavioral changes from stimulation (caffeine) to depression (morphine). Others, such as the vinca alkaloids, bind to tubulin and inhibit cell division, making them valuable as anticancer agents in their own right and as model compounds for production of additional drugs to cure cancers. Finally, some alkaloids that were important for historical reasons (e.g., atropine and ergot) have found new uses in the modern world.

Some of the alkaloid biosynthetic pathways are known, but in contrast to other natural products, such as flavonoids, the enzymes and their mechanism of action remain largely uncharacterized. Genome analyses in some medically important plants are now being used to further explore these pathways and to "engineer" over-production of medically useful alkaloids or provide precursors for semisynthetic analogs. Although plant biochemists have had some success with plant cells in culture, total alkaloid synthesis in plants often takes place in several different cell types. The use of tissue-specific gene promoters may be required to obtain high yields of enzymes and the finished product. The rationale for alkaloid synthesis in most plants is still unknown, but recent research shows that some of these compounds, such as nicotine, provide significant protection against insect herbivores in contrast to nonalkaloid-producing cultivars. Their widespread value to humans, however, continues to increase, and investigations combining biochemical and genetic methods are leading to rapid advances in the alkaloid field.

REFERENCES FOR FURTHER INFORMATION

Ashihara, H. & Crozier, A. (2001). Caffeine: A well known but little mentioned compound in plant science. *Trends Plant Sci.* **6**:407–412.

Facchini, P.J. & DeLuca, V. (2008). Opium poppy and Madagascar periwinkle: Model non-model systems to investigate alkaloid biosynthesis in plants. *Plant J.* **54**:763–784.

Kutchan, T.M., Frick, S., & Weld, M. (2008). Engineering plant alkaloid biosynthetic pathways: Progress and prospects. In: *Advances in plant biochemistry and molecular biology*, Vol. 1. Amsterdam: Elsevier, pp. 283–310.

Usera, A.R. & O'Connor, S.E. (2009). Mechanistic advances in plant natural product enzymes. *Curr. Opin. Chem. Biol.* **13**:485–491.

Zulak, K.G., Liscombe, D.K., Ashihara, H., & Facchini, P.J. (2006). Alkaloids. In: Crozier, A., Clifford, M.N., & Ashihara, H. (Eds.), *Plant secondary metabolites: Occurrence, structure, and role in the human diet.* Malden, MA: Blackwell, pp. 102–133.

PLANT BIOCHEMISTRY ***on the web***

Visit the *Plant Biochemistry* student companion Web site at http://biology.jbpub.com/gleason/plantbiochemistry/ for additional online study resources on these topics.

CHAPTER 10

Plant Peptides and Proteins

CHAPTER OUTLINE

10.1 Overview

As described in previous chapters, plants synthesize a large number of proteins, including the usual enzymes of primary metabolism such as ribulose-1,5-biphosphate carboxylase/oxygenase, or rubisco, and many unusual enzymes involved in natural product synthesis. In addition, structural proteins such as tubulin are common constituents of the plant cytoskeleton. Plants also synthesize a variety of unique proteins that function in amino acid storage, defense, and intercellular signal transduction. Many of these proteins are found only in a small number of species, and their precise function is often unknown.

Historically, the seed storage proteins are the most thoroughly characterized. The reason is obvious: they can be isolated from abundant and readily available sources. In addition, most are very stable, which facilitates biochemical analysis. Many of these storage proteins are not known to have enzymatic or other activities in the plant. Their sole function is presumably to provide a source of amino acids, fixed nitrogen, and sulfur for the growing embryo. The genes for some of these storage proteins, however, have recently been shown to be expressed in tissues other than seeds, and the proteins have various biological activities. Some of the thionin proteins are a good example; many classes of these proteins may function as storage molecules, but others serve to defend seeds against pathogens and herbivores. Proteins synthesized in germinating seeds, the germins, are detected in tissues other than seeds and probably function in cell wall remodeling rather than as amino acid reserves. Lectins found in cell walls may function in cell–cell recognition, and the notorious ribosome-inactivating proteins (RIPs) can also deactivate RNA viruses. Other classes of proteins are extensively processed in the endoplasmic reticulum (ER) and secreted from the cell to produce small proteins and peptides that are active as signaling molecules (i.e., they are peptide hormones). The structures and functions of a small number of these unique proteins are described here with the caveat that much of this information will need to be extensively updated in the near future.

10.2 Germins

The term *germins* was first used to describe a class of homologous proteins found in germinating wheat seeds. Germins have since been found in many plant tissues other than seeds. Proteins with similar sequences also occur in gymnosperms and even nonvascular plants such as mosses. A subclass of the germins that has limited sequence similarity but has similar tertiary structures and functions has been described. They are referred to as germin-like proteins. Most of the germins and germin-like proteins are glycoproteins with molecular masses of 20 to 30 kDa and are associated with the cell wall matrix. Some germins contain metal ions, usually manganese, and have been shown to have oxalic acid oxidase or superoxide dismutase (SOD) activity (**FIGURE 10.1**). Both activities yield hydrogen peroxide, which can function in defense against microorganisms. H_2O_2 can be directly toxic to bacteria or can be used by the plant for restructuring and strengthening the cell wall. Hydrogen peroxide can promote the formation of cross-links between wall matrix carbohydrates and proteins and serve as a radical initiator in lignin synthesis (see Chapter 8). Catabolism of oxalate releases Ca^{2+} ions, which can serve as an intracellular cell signal and promote expression of defense genes.

Approximately 30 genes for germins and germin-like proteins have been annotated in the *Arabidopsis thaliana* genome. These proteins are abundant in developing tissues such as embryos and flowers, and gene expression can be enhanced by stress, such as exposure to pathogens. All the germin genes contain a signal sequence that directs ribosomal synthesis to the ER. The newly synthesized proteins are sequestered in the ER where the ER-signal sequence is removed, the protein is glycosylated, and disulfide bridges are formed. The germins are

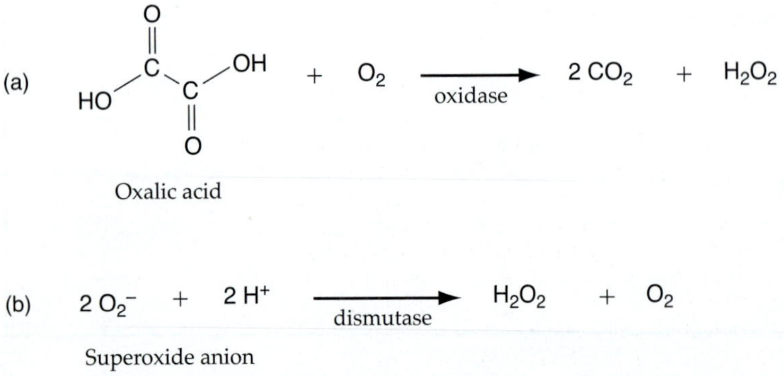

FIGURE 10.1 Enzymatic activities of germins. Some germins have been shown to oxidize oxalic acid to carbon dioxide and hydrogen peroxide (a). These same proteins may also exhibit superoxide dismutase activity (b). Both germin-catalyzed reactions occur in the cell wall matrix and can provide hydrogen peroxide for cell wall cross-linking reactions and lignin synthesis. Oxalic acid is a product of the glyoxylate cycle. Superoxide anion can be produced in the wall matrix by NADPH oxidases located in the cell membrane.

then loaded into Golgi vesicles and transported to the cell wall matrix.

Germins and germin-like proteins belong to a protein superfamily called cupins. This superfamily encompasses proteins in many organisms in addition to plants. All cupins have a highly conserved sequence of 10 amino acids, P-**H**-I-**H**-P-R-A-T-**E**-I. The two histidine residues (H) and the glutamic acid residue (E) in this conserved sequence constitute a metal binding site. Cupins can bind a number of different metals but most often bind manganese, as in the germins. Cupins also have a highly conserved tertiary structure, called a β-jelly roll, also referred to as a β-barrel. The name cupin is derived from the Latin *cupa,* meaning small barrel, and indicates this common tertiary structure in this family. The x-ray crystal structure of germin from wheat has been solved to 1.6-Å resolution and has a classic β-barrel structure (**FIGURE 10.2A**). The central β-barrel domain is flanked by an α-helical domain. A Mn^{2+} ion is liganded to three histidine residues and a glutamic acid in wheat germin. This is also the binding site for oxalate.

Germins are usually found as homodimers, but three dimers readily associate to form highly stable hexamers. Interactions of the neighboring α-helical domains stabilize this hexameric quaternary structure. The hexameric structure of the germins is very similar to that of phaseolin (Figure 10.2B), a seed storage protein found in French bean (*Phaseolus vulgaris*). These seed storage proteins lack the metal-binding site of the cupins and are not known to have enzymatic activity. It is probable that some of the seed storage proteins evolved from cupin-like ancestral proteins.

10.3 Seed Storage Proteins

As mentioned above, the seed storage proteins have been characterized extensively because of their ready availability and also their importance in human and animal nutrition. They were initially divided into various categories based on their solubility in water. More recent protein and gene sequence analyses have shown that many of the proteins placed in a particular category bear little relationship to each other. Because much of the original terminology is still in use, the solubility classification is presented here with subdivision into more related classes.

Globulins

The globulins are soluble in dilute salt solution (0.5–1.0 M NaCl). They are found in most dicot and some monocot seeds. There is little sequence similarity among the proteins from different plants, but all belong to the cupin superfamily based on their tertiary structure. The hexameric structure of the globulin, phaseolin,

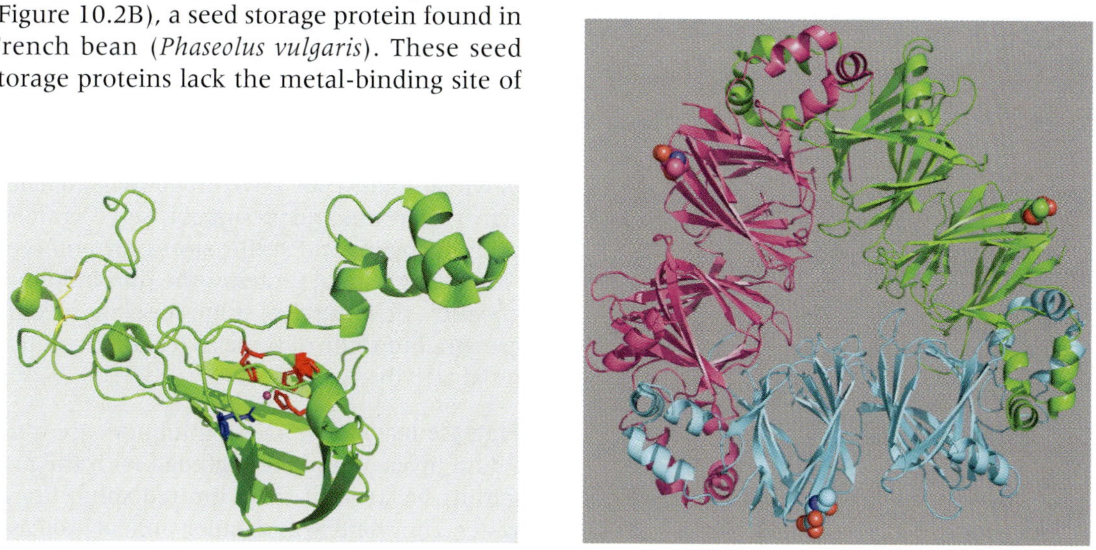

(a) (b)

FIGURE 10.2 X-ray crystal structures of wheat germin monomer (a) and the phaseolin hexamer (b). (a) The wheat germin monomer exhibits the characteristic cupin β-barrel structure with a Mn^{2+} ion (purple sphere) in the center of the barrel. The metal is coordinated by His 88 (red), His 90 (red), and Glu 95 (blue) residues in the conserved 10 residue cupin sequence. In wheat germin, a third His at position 137 (red) also coordinates the Mn^{2+} ion. A single disulfide bond (yellow) is present, connecting Cys 10 to Cys 26. (b) The phaseolin monomer readily forms dimers shown in blue, magenta, and green. The dimers then form a trimer of dimers to produce the stable hexameric structures found *in vivo.* One *N*-acetyl glucosamine residue is attached to the outer surface of each dimer, shown as space-filling spheres. The phaseolin monomer belongs to the cupin protein superfamily. It has the characteristic β-barrel structure comparable to wheat germin shown in part a. Unlike germin, phaseolin does not have a metal binding site or any known enzymatic activity. (Structure a from Protein Data Bank 1fi2. E. J. Woo, J. M. Dunwell, P. W. Goodenough, A. C. Marvier, and R. W. Pickersgill, *Nat. Struct. Biol.* 7 [2000]: 1036–1040. Structure b from Protein Data Bank 2phl. M. C. Lawrence, T. Izard, M. Beuchat, R. J. Blagrove, and P. M. Colman, *J. Mol. Biol.* 238 [1994]: 748–776.)

is shown in Figure 10.2B. It is a trimer of dimers where each monomer has the β-barrel fold that is characteristic of cupins. The globulins apparently serve only as an amino acid and nitrogen store for developing embryos, and therefore there is little evolutionary pressure to conserve their primary structures. The hexameric cupin structure, however, is very stable and resistant to chemical or biological degradation, which is a desirable trait in a seed protein.

The globulins have been classified as either 7S or 11S proteins based on their rate of sedimentation in the ultracentrifuge. These S-values correspond to masses of 45 to 50 kDa and 60 to 65 kDa, respectively. Like the germins, the genes for globulins encode precursor proteins whose synthesis occurs on the ER. The synthesis of the globulins has been characterized in several legumes. In the case of legumin from soybean (*Glycine max*), a preprolegumin undergoes cleavage of its N-terminal signal sequence in the ER, glycosylation, and formation of a disulfide bond between the N- and C-terminal domains of the protein to yield prolegumin. A specific chaperone assists trimerization of prolegumin, and the trimers are packed into Golgi vesicles that fuse with a vacuole. The vacuoles then form large inclusions in cotyledons called protein bodies or protein storage vacuoles (**FIGURE 10.3**).

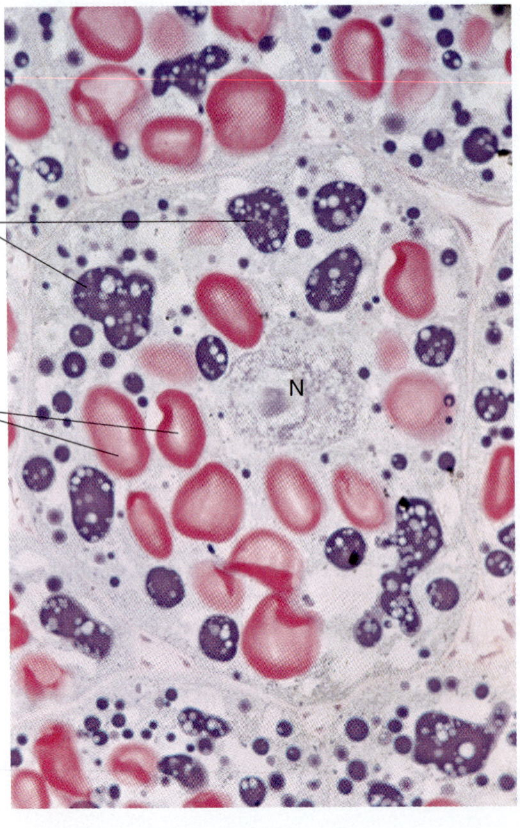

FIGURE 10.3 Protein bodies in seed cells of *Lathyrus* sp. (sweet pea). Seed proteins are made in the ER and packaged into Golgi vesicles. In some cases these vesicles are targeted to vacuoles that are specialized for protein storage. These vacuoles form dense membrane-bound organelles called protein bodies (PB), or protein storage vesicles. Also visible in this micrograph are the nucleus (N) and starch grains (S). Magnification is approximately 660×.

A specific cysteine protease in the storage vacuole cleaves a peptide bond in prolegumin, producing two peptide chains connected by a disulfide bond. This cleavage alters the conformation of the mature legume protein and facilitates the formation of the highly stable hexamer, which is the final state of legumin. Seed germination triggers the activation of specific proteases, called legumains in legumes, that catalyze the breakdown of these storage proteins to nourish the growing embryo.

Prolamins

These storage proteins are found mainly in the monocot endosperm and are soluble in dilute ethanol. They are named for their high content of proline-glutamine residues. Prolamins are subdivided into a low-molecular-weight fraction, with proteins having masses of about 30 kDa and no disulfide bonds, and a high-molecular-weight fraction, with masses of 90 kDa and containing variable numbers of cysteine residues in their C-terminal domain that can form disulfide bonds. The high-molecular-weight prolamins dimerize and form even more complex oligomers. The prolamins of wheat seeds have been studied extensively because of their economic and nutritional importance. The wheat proteins are known collectively as glutens and have a high content of cysteine residues that form disulfide bonds. The alcohol soluble glutens, the gliadins, are mainly monomers that provide viscosity to wheat flour. The alcohol-insoluble glutenins form disulfide bonded polymers and determine the "risibility" of wheat flour. They have no known biological activity in plants and presumably serve as amino acid, nitrogen, and sulfur stores for embryos. A related group of proteins, the puroindolines, are also a rich source of sulfur for plant seedling growth, but some of these also exhibit antibacterial activity.

Protease Inhibitors These inhibitors are often found in seeds and are included with the prolamins because of their limited solubility in water. They bind to and inhibit animal proteases and, therefore, probably function in defense.

A well-known example is the soybean trypsin inhibitor that limits herbivory of soy seeds. This inhibitor is a protein of 21 kDa with several disulfide bonds. It binds to mammalian trypsin and other serine proteases with a dissociation constant of approximately 5×10^9 mol^{-1}. Trypsin cleaves a peptide bond in the inhibitor between residues arginine-63 and isoleucine-64. Given the high binding constant, however, the fragments remain bound to the protease. Because it is cleaved by its protease target, soy trypsin inhibitor is considered a suicide inhibitor. Similar protease inhibitors are found in other seeds such as pea (*Pisum sativum*) and chick pea (*Cicer arietinum*). Protease inhibitors may also be found in leaves and other plant parts as reported in species of poplar and potato.

Proteolytic Enzymes Proteases are produced by a limited number of plants to protect seeds from insect herbivores. They have some structural resemblance to the protease inhibitors. The best-known plant proteolytic enzymes are papain from papaya (*Carica papaya*), bromelain from pineapple (*Anana comosus*), and ficin from figs (*Ficus* sp.). These plant enzymes are cysteine proteases in which the cysteine sulfhydryl group takes the place of the serine hydroxyl group present in the usual protease catalytic triad of aspartate, histidine, and serine.

Papain from papaya is characteristic of the plant proteases. It is synthesized in the leaf mesophyll and transported to the pericarp of developing fruit via laticiferous ducts. The concentration of this protease is highest in the latex of unripe fruits. Papain has a molecular mass of 23 kDa and is stabilized by four disulfide bridges. The enzyme is activated on reduction of these disulfide bonds. Cysteine-25 is the active site residue that initiates catalysis and forms a sulfoester bond with the carboxyl group of the cleaved peptide. Papain is resistant to heat denaturation and has a pH optimum of approximately 5.5. It has broad specificity but generally cleaves a protein at a site one residue beyond a phenylalanine residue. Papain is used commercially as a digestive aid, as a meat tenderizer, for treating necrotic tissue around wounds, and for cleaning contact lenses. Bromelain from pineapple fruits has a mass of 33 kDa and is a glycoprotein. Its activity is similar to papain.

Sweet Proteins The sweet proteins are found in a small number of plant species. They resemble the protease inhibitors having similar molecular weights and several disulfide bonds. They, however, have no known inhibitory or enzymatic activity. The thaumatins are found in the fruits of the African arrow root (*Thaumotococcus danielli*). These proteins are approximately 1,000 times sweeter than sucrose on a weight basis. Monellins are present in fruits of *Dioscoreophyllum cumminsii*, a tropical vine related to the yam. They resemble thaumatins but form heterodimers. These proteins are 3,000 times sweeter than sucrose. Presumably, the proteins are stored in fruits to attract seed dispersal agents. They are also used as sweeteners by humans as a substitute for sucrose. The thaumatins are the only sweet proteins in commercial use.

A similar group of proteins modify the human taste receptors to give a sweet taste to normally sour compounds such as ascorbate or citrate. The proteins themselves are tasteless. Miraculin, a protein from an African shrub, *Richadella dulcifica*, is a glycosylated protein of 220 residues. It modifies human taste receptors after chewing on the leaves to provide a sweet flavor several thousand times more intense than sucrose. The genes for these taste-modifying proteins have been identified, and attempts to clone these genes into other plants have been tried with varying success.

Albumins The seeds of many angiosperms, especially dicots, contain small proteins called 2S albumins that have a high content of cysteine and methionine. Similar to most of the seed storage proteins, they are made from large precursor proteins that are processed in the ER and packaged into protein storage vacuoles. They have a higher content of the sulfur-containing amino acids than most prolamins and are believed to function as a sulfur store for the germinating embryos.

Lectins (*Agglutinins*)

The term lectin was originally coined to describe seed proteins that bind to sugars and cause red blood cells to aggregate. These proteins are glycosylated and form cross-links to carbohydrates on the surfaces of other cells such as mammalian erythrocytes. They are abundant in some seeds but have also been found in other plant tissues. Similar proteins with lectin properties have also been isolated from microorganisms and vertebrate animals.

Plant lectins range in molecular masses from 12 to 62 kDa and have relatively little sequence similarity. They generally contain several disulfide bonds and can form multimers.

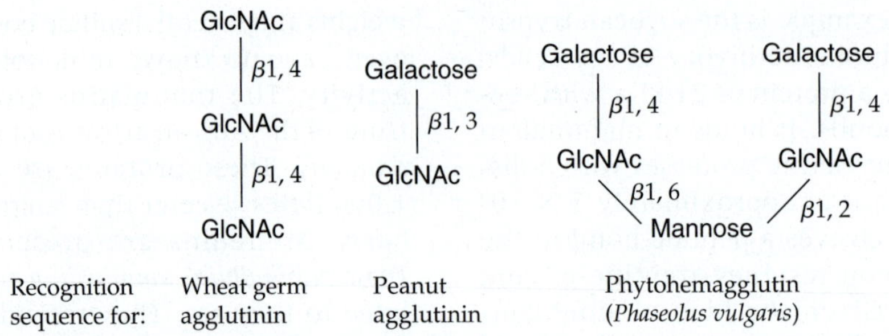

FIGURE 10.4 Sugar-binding specificity of three major plant lectins. Lectins bind to and cause agglutination of cells with specific carbohydrate sequences protruding from their cell membranes. Wheat seed lectin binds to cells with a β-1,4-*N*-acetyl glucosamine (GlcNAc) trimer. Peanut lectin recognizes galactose with a β-1,3 bond to *N*-acetyl glucosamine. Phytohemagglutin from French bean binds to a pentasaccharide consisting of galactose with a β-1,4 link to *N*-acetyl glucosamine and a β-1,6 linkage from the glucosamine to a mannose residue. The mannose has a β-1,2 link to another GlcNAc-galactose unit. Their specificity enables lectins to recognize and cross-link specific cell types.

Lectins are very specific in their binding properties. For example, wheat germ lectins bind to a series of three β-1,4-linked *N*-acetyl glucosamine residues on the surface of cell membranes. Similar specificities have been described for other lectins (**FIGURE 10.4**). The function of lectins in plants is not known; they may simply be a form of storage protein with fortuitous cell-agglutination properties. Their ability to recognize specific carbohydrate groups suggests, however, a possible role in intercellular communication. Their specific cell binding properties have made them a valuable tool in cell biology research.

Ribosome-Inactivating Proteins Ribosome-inactivating proteins (RIPs) are enzymes that cleave a specific *N*-glycosidic bond in the RNA of the 28S ribosomal RNA of the 60S subunit resulting in ribosome that can no longer function in protein synthesis. These proteins are found mainly in the seeds of some plants but can also occur at lower concentrations in roots, leaves, and bark. The RIPs are divided into two classes. Class 1 proteins are composed of a single polypeptide with a molecular mass of approximately 30 kDa and a basic isoelectric point. The best-known member of this class is the antiviral protein from pokeweed (*Phytolacca americana*). Class 1 RIPs are also found in a number of monocots. They have a low toxicity in animals because of their limited ability to penetrate cell membranes; LD_{50} of 1 to 50 mg per kg in mice when injected.

Class 2 RIPs are not as common as class 1 proteins. They consist of a heterodimer in which the monomers are linked by a disulfide bond. The best-characterized class 2 RIP is ricin from castor bean (*Ricinus communis*). Similar proteins are viscumin in mistletoe (*Viscum album*) and abrin in the tropical legume *Abrus precatorius*. Unlike class 1 proteins, class 2 RIPs can cross cell membranes and are extremely toxic. The toxicity in mice is 0.5 to 3 μg when injected intraperitoneally. The tertiary structure of ricin, shown in **FIGURE 10.5**, has been determined by x-ray crystallography.

The A chain of the ricin heterodimer has a mass of approximately 30 kDa and has *N*-glycosidase activity. It is homologous to class

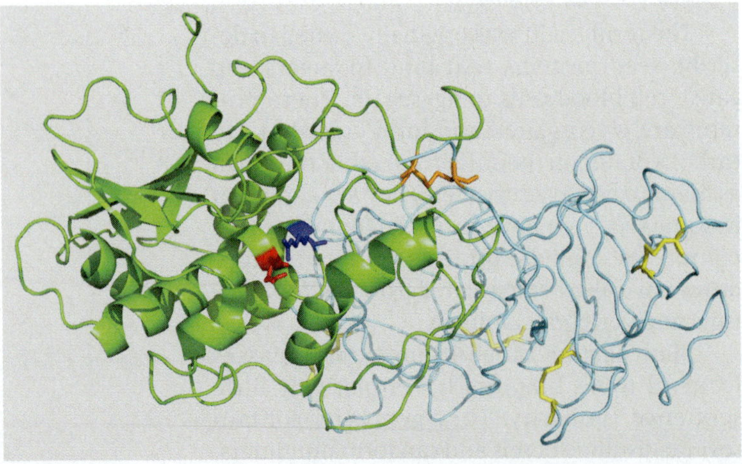

FIGURE 10.5 X-ray crystal structure of ricin. This ribosome-inactivating protein from castor bean is a class 2 RIP consisting of two chains. The A chain, shown in green, has *N*-glycosidase activity. The B chain, shown as the C-α backbone in blue, is a lectin. The two chains are connected by a disulfide bond shown in orange. The B chain has four additional intramolecular disulfide bonds shown in yellow. Two residues in the active site of the A chain, Glu-177 (red) and Arg-180 (blue), are found in the long α-helix that runs through the center of this protein. (Structure from Protein Data Bank 2aai. E. Rutenber, B. J. Katzin, W. Montfort, J. E. Villafranca, S. R. Ernst, E. J.Collins, D. Mlsna, A. F. Monzingo, M. P. Ready, and J. D. Robertus, *Proteins* 10 [1991]: 240–250.)

1 RIP peptide. It is connected to the B chain by a single disulfide bond (Figure 10.5). The B chain is similar in size to the A chain and has lectin activity. It binds to galactose and *N*-acetyl glucosamine residues found on the exterior of many animal cells. Binding promotes uptake of this RIP by endocytosis. The protein is then sequestered internally in an endosome that can then fuse with a lysosome, and the protein is degraded. Some endosomes, however, back transfer and fuse with the Golgi apparatus, transferring their sequestered proteins to the lumen of the ER. In the ER, the disulfide bond connecting the A and B chains of ricin is cleaved. It seems likely that the A chain is then recognized as an aberrant protein by chaperones in the ER and is moved across the membrane to the cytoplasm for degradation by the proteasome. When the A chain enters the cytoplasm, it folds into its enzymatically active conformation and can act on ribosomes. The A chain has been shown to cleave the *N*-glycosidic bond between ribose and adenine at position 4324 in the rat 28S rRNA. This adenosine residue is in a stem-loop structure that recognizes and binds elongation factor complexes. Loss of the purine base deactivates the ribosome. It is estimated that one molecule of the A chain can kill a cell by this process.

The A and B chains of ricin are coded for by one gene and synthesized in the ER as a single precursor protein. The two proteins are joined by a 12-residue linker peptide. In the ER, the N-terminal signal peptide is removed and the protein is glycosylated. Four disulfide bonds are made in the B chain and a fifth disulfide bond is produced that links the A and B domains (Figure 10.5). The linker peptide serves as a targeting sequence, and the precursor RIP is transported to protein storage vacuoles in the seed endosperm. In the acidic environment of the vacuole, an endoprotease removes the 12-residue linker, generating the disulfide-bonded dimer. The RIP does not encounter ribosomes *in planta* and is degraded when the seeds germinate. Its precise function in plants is unknown. The protein is not toxic when ingested because the disulfide bond between the two subunits is readily broken in the digestive tract and the liberated A chain cannot readily penetrate into the interior of cells. Other functions have been proposed for RIPs. They can deglycosylate other nucleic acids as well as ribosomal RNA and may be a defense against RNA viruses. Alternatively, they may mediate apoptosis in cells that are invaded by pathogens and act as a suicide defense mechanism.

The RIPs, particularly ricin, have recently gained a notorious reputation as possible agents of biological warfare. However, because they must be injected to be effective, they are unlikely to be used for such purposes. Their use as antitumor agents is also being investigated. The lectin subunits can be engineered to bind to specific cells, such as cancerous tumor cells, and destroy these cells leaving normal cells unharmed.

10.4 Antimicrobial Peptides

A number of small proteins, 30 to 50 amino acid residues in length, that can kill bacteria and fungi are found in almost all eukaryotes, including plants. These peptides are believed to be part of an organism's innate immune system, that is, they are constitutively expressed and relatively general in their antimicrobial activity. They have some features in common, such as their small size and an abundance of arginine and lysine residues. As expected from their amino acid composition, the antimicrobial peptides have an overall positive charge and an isoelectric point in the alkaline range. Although they have diverse tertiary structures, the antimicrobial peptides are generally amphipathic; part of the structure is hydrophilic, whereas another is hydrophobic. Antimicrobial peptides act by binding to and damaging the cell membranes of bacteria and fungi. The minimum inhibitory concentrations of these peptides *in vitro* are approximately 1 to 20 μM.

Defensins

Defensins are plant antimicrobial peptides that were previously described as γ-thionins. Thionins are a large class of seed proteins that are rich in the sulfur-containing amino acids, cysteine and methionine. Most thionins are relatively large proteins, 45 to 50 kDa, but several subclasses have been created to distinguish different types. In general, thionins of any class were believed to function as a sulfur store for developing embryos. Defensins are all small γ-thionins with approximately 50 residues, including eight cysteines that form four disulfide bonds in the mature peptide. They have a unique three-dimensional structure consisting of three β-strands and a single β-helix (**FIGURE 10.6**). The helix is tethered to the central β-strand by two disulfide bonds and is referred to as a cysteine-stabilized helix. This tertiary structure is common to antimicrobial peptides from both plant and nonplant sources.

FIGURE 10.6 Nuclear magnetic resonance structure of a defensin from *Pisum sativum*. Defensins from many different organisms exhibit a conserved tertiary structure despite very little sequence homology. The peptide consists of three β-strands and a single α-helix. Four disulfide bonds (yellow) stabilize the tertiary structure. Two of these disulfide bonds anchor the α-helix to the central β-strand, generating a cysteine-stabilized helix. These disulfide bonds provide rigidity to the molecule and maintain the helix in a fixed orientation. (Structure from Protein Data Bank 1jkz. M. S. Almeida, K. M. S. Cabral, E. Kurtenbach, F. C. L. Almeida, and A. P. Valente, *J. Mol. Biol.* 315 [2002]: 749–757.)

Their amphipathic properties allow defensins to bind to the anionic head groups of the lipid bilayers in membranes. They bind especially well to membranes of bacteria and fungi where the anionic groups are exposed on the surface. Animal cell membranes are highly glycosylated and thus protected from defensin binding. The defensins may penetrate membranes but generally stay in the lipid bilayer and alter membrane permeability. They cause leakage of ions through the cell membrane, which appears to be the basis for their antimicrobial activity.

Defensins from seeds have been extensively characterized, but the genes encoding these peptides are also expressed in epidermal tissues of plants and in cells surrounding the vascular bundles. Analysis of the genome of the model legume, *Medicago truncatula*, shows a large family of defensin genes. These genes are expressed in all plant tissues, although some defensins are tissue specific and others are inducible rather than constitutive. There are 13 putative defensin genes in *Arabidopsis thaliana*. Some exist in tandem copies and are obviously the products of gene duplication. Some defensin genes encode proteins with a secretion signal, which suggests that the products can be found in the cell wall matrix. Although they are believed to be part of a plant's innate immunity, some of these peptides may have other functions. For example, a recent report describes the secretion of two defensin-like proteins by the synergids in the ovary of *Torenia fournieri*. These peptides contain 62 and 70 residues with three disulfide bonds. Homology modeling suggests that they have the characteristic cysteine-stabilized helix. These particular peptides, however, are secreted from the synergids and serve as a signal that guides the growth of the pollen tube through the style to the ovary. Similarly, a defensin-like protein in some Brassicaceae is expressed only in the pistil of flowers and promotes the self-incompatibility response. It has been suggested that many of the defensins evolved to provide protection against pathogens but have since been co-opted for other functions such as cell-signaling.

Cyclotides

Cyclotides are small cyclic peptides that serve to defend plants against microorganisms and insect herbivores. They generally consist of about 30 residues and contain three disulfide bonds forming an unusual knotted structure (**FIGURE 10.7**). Cyclotides have a patch of hydrophobic residues exposed on the surface of the molecule that are locked into position by the placement of the disulfide bonds.

Cyclotides apparently occur only in certain plant families, the Rubiaceae and Violaceae. The best characterized in the kalata B1 protein from the leaves of *Oldenlandia affinis* (Rubiaceae) shown in Figure 10.7. The leaves of this plant have been used in traditional herbal remedies by women in the Congo to accelerate childbirth. When the isolated peptides were tested *in vitro*, one low-molecular-weight fraction had antibacterial, antiviral, and cytotoxic properties. If leaves that overexpress the peptide are fed to insects under laboratory conditions, growth and development are arrested and the insects die. Some cyclotides have been shown to inhibit proteases and α-amylases in insect guts, causing starvation.

As is the case with many of these plant peptides, the genes for cyclotides encode large precursor proteins that are synthesized and processed in the ER. A simple diagram of a cyclotide precursor protein is illustrated in **FIGURE 10.8.** This diagram shows a precursor protein of approximately 11 kDa with one cyclotide domain. Larger precursor proteins with up to three cyclotide domains have been described. A cyclotide fraction has recently been found in a member of the Cucurbitaceae (squash family), and it is highly probable that

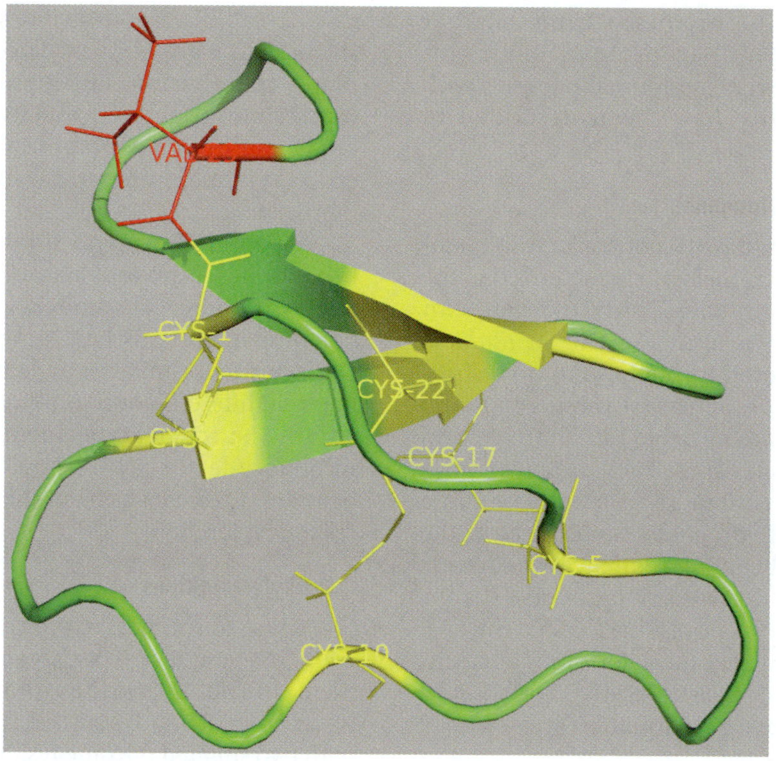

FIGURE 10.7 Nuclear magnetic resonance structure of a cyclotide from *Oldenlandia affinis*. The cyclic kalata B1 peptide from *O. affinis* consists of 29 residues with three disulfide bonds. The N-terminal cysteine residue (yellow) forms a peptide bond with the C-terminal valine (red). In addition, cysteine 1 is bonded to cysteine 15 producing a loop on one side of the molecule. Two additional disulfide bonds are shown in yellow. The disulfide bond between cysteine 5 and cysteine 17 is on the opposite side of the 1-15 disulfide bridge, whereas the cysteine 10 and cysteine 22 bond runs through the center of the molecule producing the cystine knot motif. (Structure from Protein Data Bank 1jt4. S. R. Brych, S. I. Blaber, T. M. Logan, and M. Blaber, *Protein Sci.* 10 [2001]: 2587–2599.)

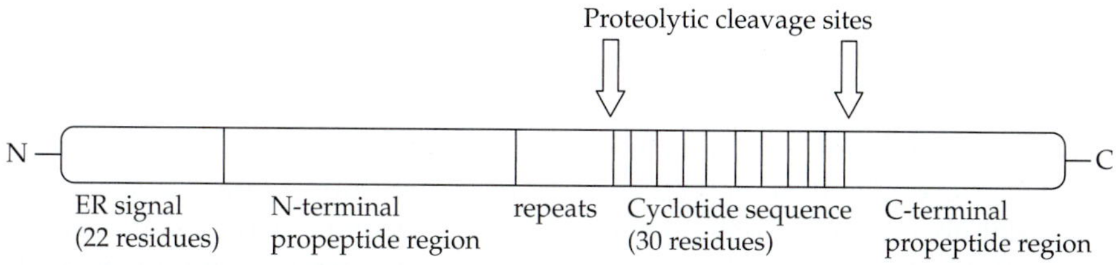

FIGURE 10.8 Diagram of a cyclotide precursor protein. Cyclotides, like many other plant peptides, are synthesized as large precursor proteins that are processed in the ER. The cyclotide precursor protein typically has an N-terminal signal sequence of approximately 20 amino acids that targets the ribosomes during protein synthesis to the external face of the ER. The nascent protein then enters the ER lumen. The precursor protein has an N-terminal propeptide region (46–68 amino acids) that is removed by proteases in the ER. These enzymes hydrolyze peptide bonds in the precursor at a site after a conserved region of repeated residues. In some cyclotide precursor proteins, there is more than one cyclotide sequence preceded by the same repeat region, and each one is cut independently from the precursor protein. A C-terminal propeptide domain is also removed. Oxidation of the cysteines to produce disulfide bonds is catalyzed by a protein disulfide isomerase. The formation of a peptide bond between the N and C terminal of the processed cyclotide (see Figure 10.7) must also occur in the ER, but enzymes that catalyze this reaction have not been described.

these cyclic peptides are widespread in the plant kingdom.

10.5 Peptide Hormones

Peptide hormones are common in animals. For example, glucagon is a 29 amino acid peptide made from a larger precursor protein. It is excreted by pancreatic cells and regulates glucose metabolism. Plant hormones are mainly small organic molecules, and, until recently, no peptide hormones were described. In 1991, however, an 18 amino acid peptide was found in tomato leaves that induced protease inhibitor gene expression as part of the plant wound response. This peptide was named systemin and shown to be produced from a larger prosystemin precursor protein. Since this initial report, several additional peptides with hormone activity have been described in angiosperms.

■ CLE Peptide Hormones

The clavata3 (CLV3)/endosperm (CLE) family in *Arabidopsis* is the best characterized of the plant peptide hormones. Thirty-two genes for CLE peptides are found in the *A. thaliana* genome distributed throughout all five chromosomes. The sequences of these genes and their products are variable but share certain common characteristics. They all have a molecular mass less than 15 kDa and contain a short stretch of hydrophobic amino acids at the N-terminus. All contain sequences that target them to the cell secretory pathway. A conserved sequence of 12 residues, called the CLE domain, is found near the C-terminus and consists of R-T-V-P-S-G-P-D-P-L-H-H (**FIGURE 10.9**). Various CLEs have been found expressed in all the plant tissues.

The CLV3 peptide is a member of this CLE family that is expressed in the apical meristem of *Arabidopsis*, specifically in cells near the center of the apex. As expected for proteins with a secretory signal, the precursor is secreted to the cell wall matrix (apoplast). Once outside the cytoplasm, it is cleaved to yield an active, 18-residue CLV3 peptide. CLV3 binds to a receptor in adjacent cell membranes called CLV1. This receptor is a large integral membrane protein with latent serine/threonine protein kinase activity. It contains 21 leucine-rich repeats in its extracellular domain and belongs to the superfamily of leucine-rich repeat-receptor kinases (LRR-RK). Another integral membrane protein, CLV2, is similar to CLV1 but lacks a kinase domain. It forms a heterodimer with CLV1. Binding of CLV3 to the CLV1–CLV2 heterodimer induces the latent kinase activity of CLV1. This triggers an intracellular phosphorylation cascade. The final product of this cascade is a transcription factor called WUS-TF. The unphosphorylated WUS-TF stimulates cell division, but the phosphorylated transcription factor is inactive. Thus, CLV3 binds and initiates the phosphorylation cascade to ultimately limit the size of the apical meristem.

A peptide hormone similar to CLV3 has been found in the root apical meristem. Other members of the CLE family regulate cell differentiation in other tissues. Although not as well characterized, similar peptide hormones regulate cell differentiation in other plants. For example, the **tracheary element differentiation inhibitory factor** (TDIF) homolog of the CLE peptides in *Zinnia* has been shown to limit vascular development.

■ Phytosulfokines

These hormones, phytosulfokines (PSKs), were first described in asparagus but have since been found in various plants, including *Arabidopsis*. They consist of five amino acids with two sulfated tyrosine residues (**FIGURE 10.10**). They have been shown to bind to a membrane-localized receptor in the leucine-rich repeat-receptor kinase family. The phytosulfokines promote cell proliferation and organogenesis and are active at nanomolar concentrations.

PSY1 is an 18 amino acid peptide with one sulfated tyrosine residue also characterized in *Arabidopsis*. It is usually glycosylated and has been shown to promote cell division. The phytosulfokines are encoded by large precursor proteins that are processed in the ER. A specific tyrosyl protein sulfotransferase has been found associated with ER fractions in *Arabidopsis*. It recognizes and adds sulfate groups to specific sequences of the hormone precursor protein. The sulfate donor for this reaction is 3′-phosphoadenosine-5′-phosphosulfate (PAPS).

Analogs of the phytosulfokine precursor proteins are found in the genomes of many plants, including the moss, *Physcomitrella* and the green

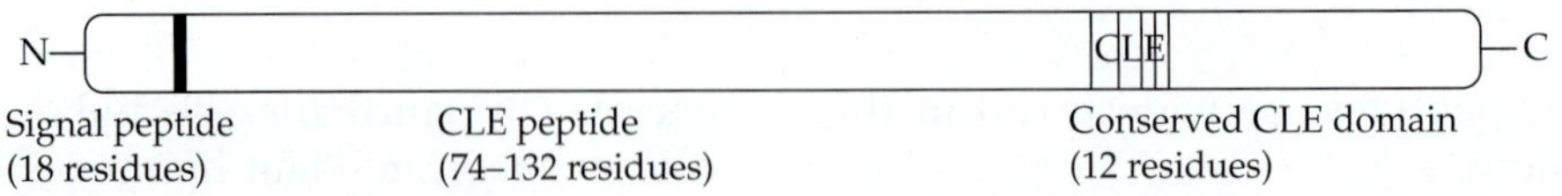

FIGURE 10.9 Diagram of a precursor protein in the CLE family. Peptide hormones, such as CLV3 in *Arabidopsis*, are produced in the ER. A signal peptide directs synthesis to the lumen of the ER where a protease removes this N-terminal peptide and other modifications, such as disulfide bond formation, occur. Because it lacks any additional signaling sequence, this protein will be packaged into Golgi vesicles and exported to the cell wall matrix (apoplast). Integral cell membrane receptor proteins in the CLV family can recognize sequences in the peptide and bind the processed peptide hormone, triggering an intracellular phosphorylation cascade.

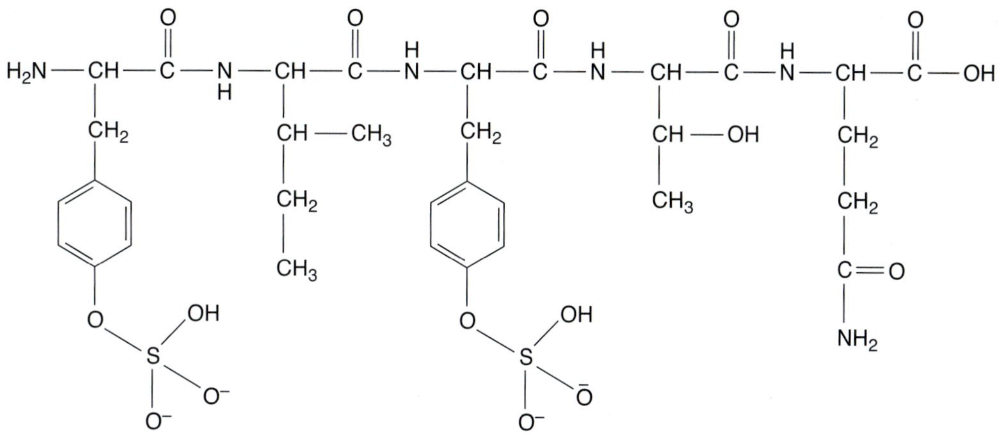

FIGURE 10.10 Structure of a phytosulfokine. The peptide hormone, PSK, is composed of five residues with two O-sulfated tyrosines. The hormone is cleaved from a larger precursor protein that is processed in the ER. A specific 3′-phosphoadenosine-5′-phosphosulfate–dependent tyrosyl protein sulfotransferase in the ER adds the sulfate groups to specific sites in the precursor protein. Like other peptide hormones, phytosulfokines are exported to the exterior of the cell and bind to specific receptors.

alga, *Chlamydomonas*. The gene products and functions in these organisms have not been characterized. Genome sequencing, however, has made it obvious that peptide hormones are ubiquitous in the plant kingdom.

10.6 Cyanogenic Glycosides

These glycosides are not peptides but are small molecules synthesized from an amino acid precursor that liberate hydrogen cyanide (HCN) on hydrolysis of their sugar moiety. The most common precursor amino acids are phenylalanine and tyrosine or the aliphatics, leucine, isoleucine, and valine. The glycosyl moiety is usually glucose. Cyanogenic glycosides are produced in a wide variety of plants and can be found in any plant tissues but most often occur in seeds. The biosynthesis of these compounds has been characterized in several plants. The pathway for the synthesis of dhurrin, found in *Sorghum bicolor*, is shown in **FIGURE 10.11**. The amino acid precursor for dhurrin is tyrosine. A specific family of cytochrome P450 monooxygenases catalyzes the hydroxylation of the amino group to yield a nitro intermediate. A decarboxylase and reductase act on the nitro compound to yield an oxime intermediate. This series of reactions and intermediates is similar to those seen in the glucosinolate synthesis pathway in the Brassicaceae (see Chapter 5). Dehydration of the oxime then leads to formation of a nitrile. Synthesis of the cyanogenic glycosides has been shown to occur in the cytoplasm catalyzed by enzymes associated with the outer surface of the vacuole membrane, suggesting that some of the enzymes are membrane bound. The resulting nitriles are transported into the vacuole and further modified and O-glycosylated for storage.

The sugar can be hydrolyzed from the cyanogenic glycosides by a specific β-glycosidase. A lyase then liberates HCN from the resulting aglycone (**FIGURE 10.12**). Under certain physiological conditions, such as low pH, cyanide release can occur spontaneously without enzymatic catalysis. In plant tissues the glycosidase is physically separated from the cyanogenic glycosides. The glycosides and degrading enzymes mix only when the cells are damaged.

The cyanogenic glycosides are believed to deter insect herbivory. They are not especially toxic to humans or large herbivores in small amounts. HCN is lethal to mammals in the range of 0.5 to 3.5 mg kg^{-1} and acts by binding to cytochrome oxidase and halting electron transport in the mitochondria. Most mammals have an enzyme in their livers called rhodanese (thiosulfate sulfur transferase) that produces thiocyanates from HCN:

$$CN^- + S_2O_3^- \text{ (thiosulfate)} \rightarrow SCN^- \text{ (thiocyanate, rhodanide)} + SO_3^-$$

Thiocyanates are less reactive than cyanide, are water soluble, and can be excreted. Animals that graze on sorghum or certain varieties of cyanogenic clover, however, can suffer neurological damage from chronic exposure to small amounts of cyanide. Cyanogenic glycosides are most effective against insect herbivores. Some insect specialists, however, have adapted and

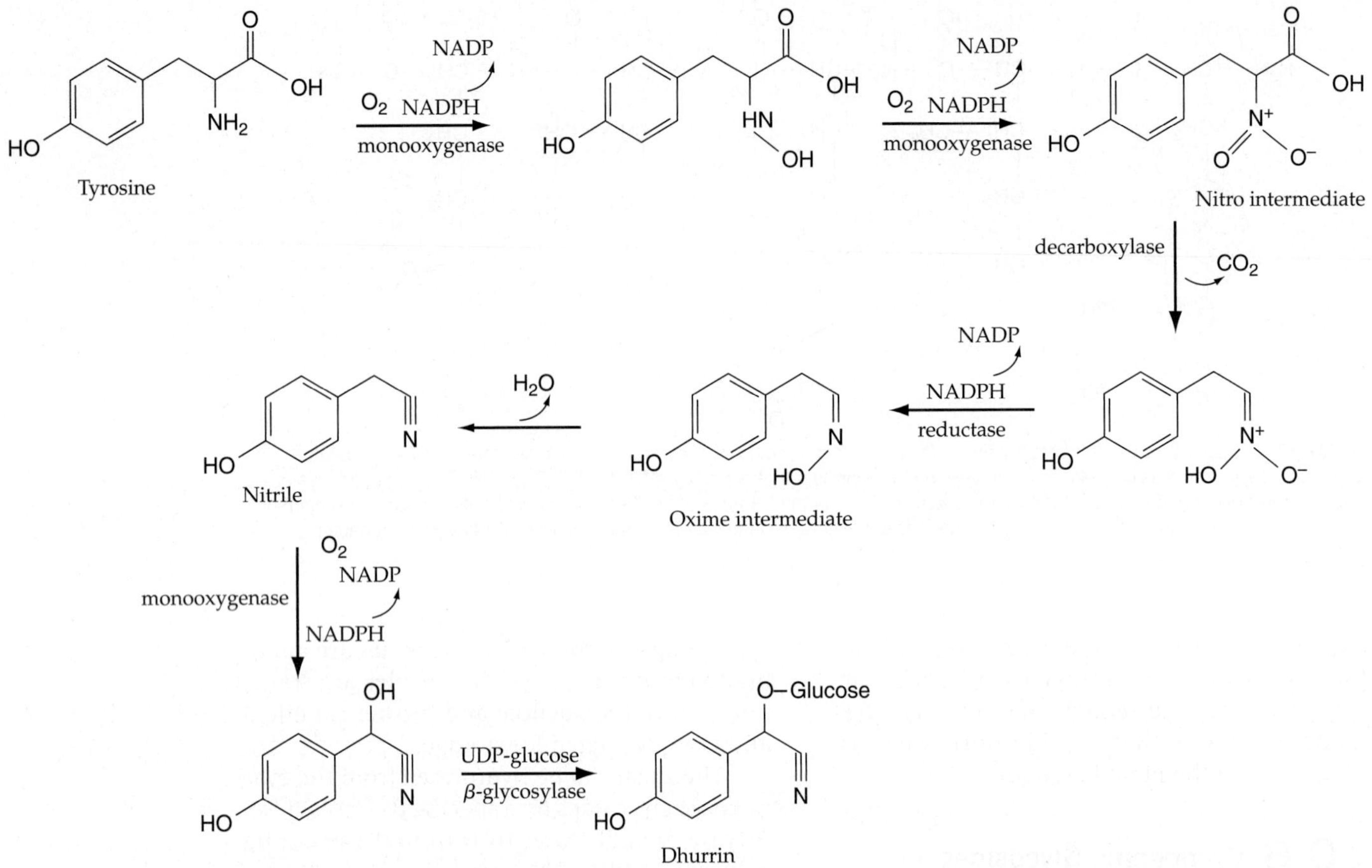

FIGURE 10.11 Synthesis of dhurrin, a cyanogenic glycoside found in *Sorghum*. Tyrosine is the precursor for dhurrin synthesis. In the first step, a specific cytochrome P450 monooxygenase hydroxylates the amino group of tyrosine. A second P450 monooxygenase adds a second oxygen to generate a nitro-intermediate. Subsequent decarboxylation and reduction yield an oxime. Dehydration of the oxime produces a nitrile that is the substrate for another specific monooxygenase that hydroxylates the carbon of the side chain. Glucose is added to this hydroxyl group with the β-configuration.

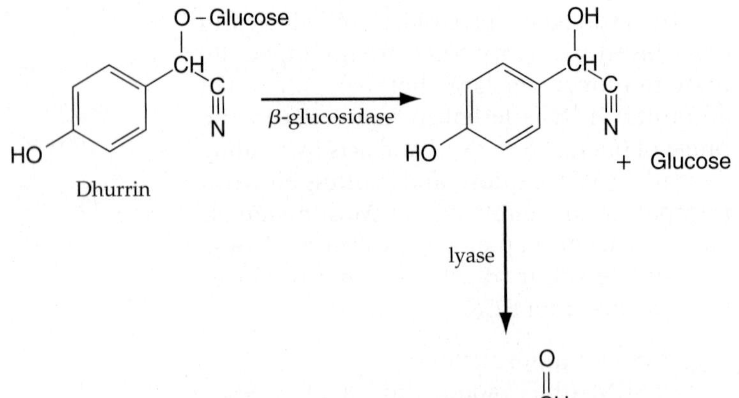

FIGURE 10.12 Hydrolysis of a cyanogenic glycoside. While in the plant, the cyanogenic glycosides are stored in vacuoles or in special cells. When the plant tissues are damaged, the glucose residues are removed by a β-glucosidase. HCN can then be released by decomposition of the nitrile either catalyzed by a lyase or spontaneously in an acidic environment.

can detoxify these compounds or store them in their tissues for their own defense.

The structures of four common cyanogenic glycosides are shown in **FIGURE 10.13**. Prunasin and amygdalin are synthesized from phenylalanine. They are found in the seeds and occasionally in stems and leaves of many members of the Rosaceae family. Amygdalin is the compound responsible for the cyanide released from bitter almonds. Amygdalin from apricot pits is the active component in the hypothetical anticancer drug, Laetrile, which is banned in the United States because of its toxicity. Linamarin is synthesized from valine and lotaustralin, from isoleucine. Both are widespread in seeds, such as flax, and in the roots of the cassava plant (*Manihot esculenta*). Cassava originated in South America but is now cultivated in many tropical countries where it is the main source of starch in the human diet. The cyanogenic compounds are eliminated by scraping the roots and steeping them extensively in water before use. Small

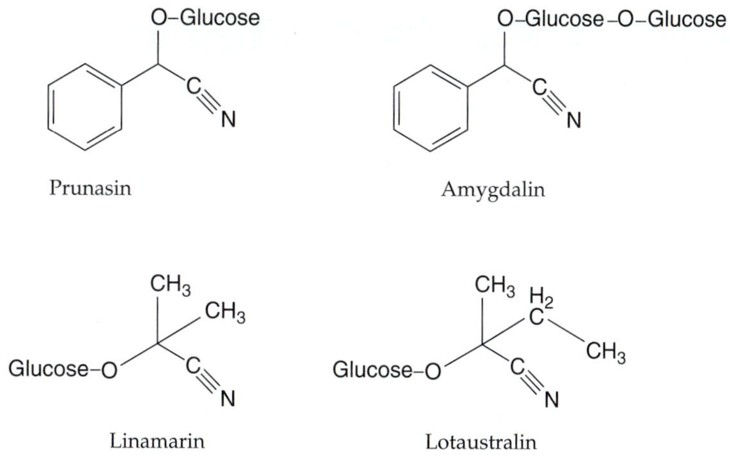

FIGURE 10.13 Example of four common cyanogenic glycosides. Prunasin and amygdalin are both synthesized from phenylalanine and are common in the seeds of the Rosaceae family such as apricots and almonds. Linamarin and lotaustralin are synthesized from aliphatic amino acids. They are also common in seeds. Lotaustralin is also found in the roots of the cassava plant but can be removed by shredding the tissues and steeping in water.

amounts of cyanide may still remain in the resulting flour and other products made from cassava. Some neurological problems in humans have been traced to chronic ingestion of small amounts of cyanide from cassava root. Because it is a staple of the diet in some countries, there have been attempts to breed "sweet" varieties of cassava that contain a much lower concentration of cyanogenic glycosides. None of these varieties is completely free of cyanide and must still undergo processing before consumption. Unlike the bitter cassava, the sweet varieties have the disadvantage of being more prone to insect and fungal attack.

CHAPTER SUMMARY

In the mid-twentieth century, several plant proteins from seeds were isolated and grouped according to their solubility in water. These included the albumins, globulins, prolamins, and lectins. Their main function was presumed to be the storage of amino acids to nourish the germinating embryo. More recent research has shown that many of these proteins have a variety of functions in addition to storage. The germins, from germinating embryos, are actually found in all plant tissues. Some of them have oxalic acid oxidase and superoxide dismutase activities and play a role in cell wall remodeling. The proteases and protease inhibitors from seeds are important in defense against herbivores but have also been exploited for commercial and research purposes. The RIPs in the lectin class appear to be an enigma as far as their function in plants but may act as antiviral agents by degrading viral nucleic acids.

Many antimicrobial peptides have also been described in plants, some of which are unique, but others seem to be homologs of bacterial and animal defense-related peptides. The defensins are widespread in the plant kingdom and provide protection against pathogens. In the past few years, some of the many defensin peptides have been shown to function in cell signaling. The cyclotides are the latest member of the antimicrobial peptides to be described. They are notable for their unusual knotted tertiary structure but may also have functions other than defense.

The relatively new field of plant peptide hormones was opened in the 1990s with the discovery of systemin. Research on the model plant *Arabidopsis* has shown that genes for a large number of peptide hormones and their receptors are present, and a few, such as the CLE family, have been shown to function in regulating cell proliferation. Further research on the unknown peptides revealed by genome sequencing will greatly expand the number of these peptide hormones and other peptides important in defense and signaling functions.

The cyanogenic glycosides are synthesized from single amino acid precursors and are most often found in seeds. Hydrolysis of the sugar moiety yields an unstable intermediate that produces hydrogen cyanide. The cyanogenic glycosides are effective against insect herbivores but can cause neurological damage in larger animals, including humans, if they are a common constituent of the diet. Plants such as cassava have been bred to produce less of the cyanogenic compounds, but they are then subject to greater insect damage.

REFERENCES FOR FURTHER INFORMATION

Bernier, F. & Berna, A. (2001). Germins and germin-like proteins: Plant do-all proteins. But what do they do exactly? *Plant Physiol. Biochem.* **39:**545–554.

Farrokhi, N., Whitelegge, J.P., & Brusslan, J.A. (2008). Plant peptides and peptidomics. *Plant Biotech. J.* **6:**105–134.

Robertus, J.D. & Monzingo, A.F. (2004). The structure of ribosome inactivating proteins. *Mini-Rev. Med. Chem.* **41:**477–486.

Thomma, B.P.H.J., Cammune, B.P.A., & Thevissen, K. (2002). Plant defensins. *Planta* **216:**193–202.

Vetter, J. (2000). Plant cyanogenic glycosides. *Toxicon* **38:**11–36.

PLANT BIOCHEMISTRY *on the web*

Visit the *Plant Biochemistry* student companion Web site at http://biology.jbpub.com/gleason/plantbiochemistry/ for additional online study resources on these topics.

APPENDIX 1

Structure and Properties of the 20 α-Amino Acids Commonly Found in Proteins

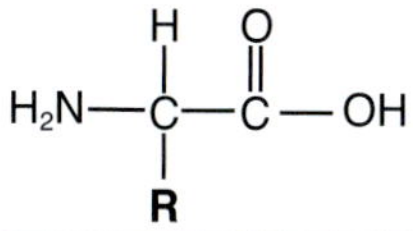

Table A.1 Properties of α-Amino Acids

Amino Acid	Structure of R Group	Abbreviations	R Class	pKa Values COOH	pKa Values NH_2	pKa Values Side-chain
Alanine	$-CH_3$	Ala, A	Alkyl, non-polar	2.34	9.69	
Arginine	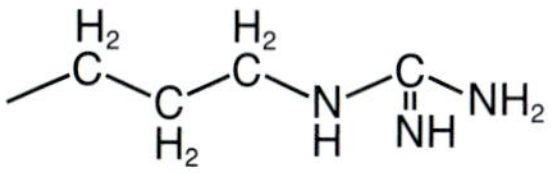	Arg, R	Guanidine, polar, basic	2.17	9.04	12.48
Asparagine	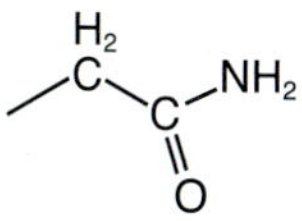	Asn, N	Amide, polar	2.01	8.8	
Aspartic Acid	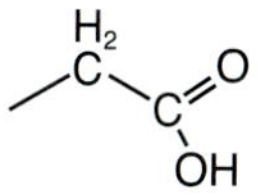	Asp, D	Carboxylic acid, polar	1.89	9.60	3.65
Cysteine	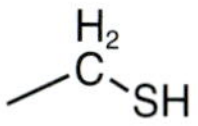	Cys, C	Thiol, polar	1.96	8.18	10.29
Glutamic acid	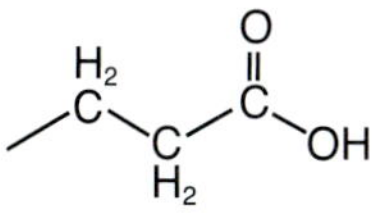	Glu, E	Carboxylic acid, polar	2.19	9.67	4.25
Glutamine	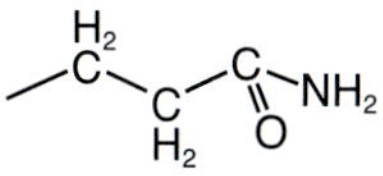	Gln, Q	Amide, polar	2.17	9.13	
Glycine	$-H$	Gly, G	Hydrogen, non-polar	2.34	9.6	
Histidine	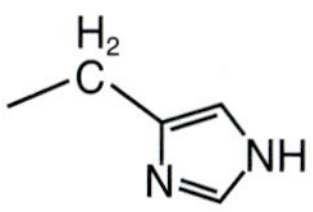	His, H	Imidazole, polar, basic	3.8	9.17	6.00
Isoleucine	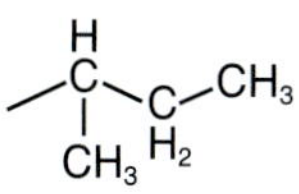	Ile, I	Alkyl, non-polar	2.35	9.68	

(*continues*)

Table A.1 (continued)

Amino Acid	Structure of R Group	Abbreviations	R Class	pKa Values COOH	pKa Values NH_2	pKa Values Side-chain
Leucine	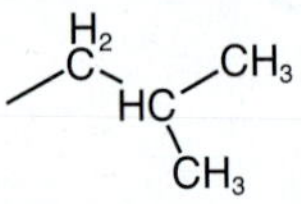	Leu, L	Alkyl, pon-polar	2.36	9.60	
Lysine	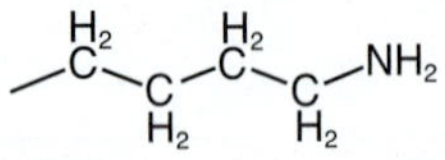	Lys, K	Amine, polar, basic	2.18	8.95	10.52
Methionine	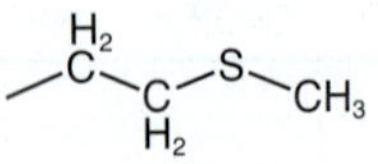	Met, M	Thioether, pon-polar	2.38	9.2	
Phenylalanine	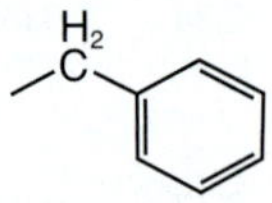	Phe, F	Aromatic, pon-polar	2.21	9.12	
Proline	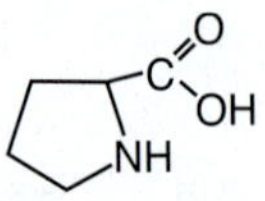	Pro, P	Cyclic, pon-polar	1.99	10.96	
Serine	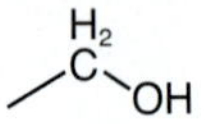	Ser, S	Alcohol, polar	2.21	9.95	13.6
Threonine	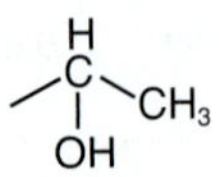	Thr, T	Alcohol, polar	2.11	9.62	13.6
Tryptophan	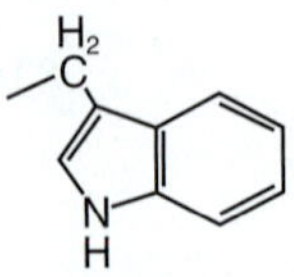	Trp, W	Aromatic, pon-polar	2.38	9.39	
Tyrosine	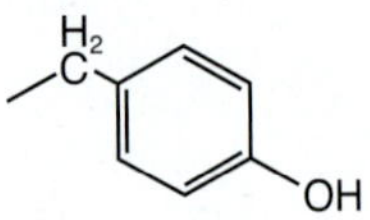	Tyr, Y	Phenol, polar	2.2	9.11	10.06
Valine	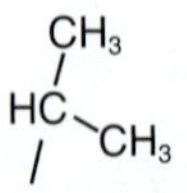	Val, V	Alkyl, pon-polar	2.32	9.61	

The structures of the R groups for the amino acids are shown. Column 3, Abbreviations, displays both the three-letter and one-letter abbreviations. The R class indicates the type of side chain and its polarity. pK_a values are those determined for the free amino acids.

APPENDIX 2

Oxidation–Reduction (Redox)

A2.1 Definitions

Oxidation = loss of electrons
Reduction = gain of electrons

The two always go together!

Oxidizing agent = a compound that can accept electrons and is reduced in the course of the reaction

Reducing agent = a compound that can donate electrons and is oxidized in the course of the reaction

A2.2 Example

$$A_{oxidized} + B_{reduced} \leftrightarrows A_{reduced} + B_{oxidized}$$

$$\text{e.g., } Fe^{3+} + Cu^{+} \leftrightarrows Fe^{2+} + Cu^{2+}$$

Reduction half-reaction:

$$A_{oxidized} + e^{-} \rightarrow A_{reduced}$$
$$Fe^{3+} + e^{-} \rightarrow Fe^{2+}$$

The standard redox potentials (E_0') in the table below are written as reduction half-reactions. They were measured at pH 7, which is indicated by the prime ('). They were determined relative to the hydrogen half-reaction, which is set to 0.0 volts at 1 M H^+, 1 atm, 298 K, and calculated to be −0.414 V at pH 7.

The values in the table can be used to estimate the extent of a redox reaction.

For example, O_2 is an excellent oxidizing agent with an E_0' of 0.816 V for its reduction to water. It can oxidize any other reactant in the table with a more negative E_0'. Conversely, H_2O is a very poor reducing agent.

Reduction Half-Reaction	E_0' in Volts
$2H^+ + 2e^- \rightarrow H_2$	−0.414
$Fe^{3+} + e^- \rightarrow Fe^{2+}$	0.771
$Cu^{2+} + e^- \rightarrow Cu^+$	0.150
$O_2 + 2H^+ + 4e^- \rightarrow 2H_2O$	0.816
$O_2 + 2H^+ + 2e^- \rightarrow H_2O_2$	0.295
$P680^+ + e^- \rightarrow P680$	1.10
$P700^+ + e^- \rightarrow P700$	0.490
Ferredoxin-Fe^{3+} + e^- → Ferredoxin-Fe^{2+}	−0.432
Cytochrome *c*-Fe^{3+} + e^- → Cytochrome *c*-Fe^{2+}	0.250
Ubiquinone + $2H^+ + 2e^-$ → Ubiquinol	0.045
DCIP + $2H^+ + 2e^-$ → DCIP reduced	0.220
Methylviologen + $2H^+ + 2e^-$ → Methylviologen reduced	−0.550
$NAD(P^+) + H^+ + 2e^- \rightarrow NAD(P)H$	−0.320
FAD + $2H^+$ + 2 e^- → $FADH_2$	−0.180
$S^0 + 2H^+ + 2e^- \rightarrow H_2S$	−0.230
$SO_4^{2-} + 2H^+ + 2e^- \rightarrow SO_3^{2-} + H_2O$	0.480
Cystine-S–S + $2H^+ + 2e^-$ → 2 Cysteine-SH	−0.220
GSSG + $2H^+ + 2e^-$ → 2GSH	−0.230
E. coli Thioredoxin 1-S-S + $2H^+ + 2e^-$ → EcT1-SH	−0.265
Oxaloacetate + $2H^+ + 2e^-$ → Malate	−0.175
Glycerate-1,3-bisP + $2H^+ + 2e^-$ → Glyceraldehyde-3-P + Pi	−0.290
Glycerate-3-P + $2H^+ + 2e^-$ → Glyceraldehyde-3-P + H_2O	−0.450
$NO_3^- + 2H^+ + 2e^- \rightarrow NO_2^- + H_2O$	0.420
α-Ketoglutarate + NH_3 + $2H^+ + 2e^-$ → glutamate + H_2O	−0.140

The values of the redox potentials of the half-reactions can be used to calculate the change in standard free energy (ΔG) for a redox reaction.

$$\Delta E_0' = E_0' \text{ half-reaction of oxidizing agent } - E_0' \text{ half-reaction of reducing agent}$$

$$\Delta G' = -n\mathcal{F}\Delta E_0'$$

Where

n is the number of electrons transferred and $\mathcal{F}$ is the Faraday constant = 96,485 J V^{-1} mol^{-1}.

To determine the reduction potential of a half-reaction in which the concentrations of components are present in nonstandard (i.e., not 1 M) concentrations, the Nernst equation can be used.

$$E = E_0' + 2.3\ RT/n\mathcal{F}\ (\log\ [\text{ox form}]/[\text{red form}])$$

R = gas constant = 8.3145 JK^{-1} mol^{-1}

T = degrees K (273.15K = 0°C)

APPENDIX 3

Steady-State Enzyme Kinetics

An enzyme-catalyzed reaction can be described by Equation 1. The free enzyme (E) binds the substrate (S) and forms a transient enzyme–substrate complex (ES). This decomposes to product (P) and free enzyme.

$$E + S \underset{k_2}{\overset{k_1}{\leftrightarrows}} ES \underset{k_4}{\overset{k_3}{\leftrightarrows}} E + P \quad (1)$$

Assume that the back reaction, $P \leftarrow ES$, is negligible and k_4 can be ignored. This is true under initial conditions. Theoretically, all reactions are reversible.

Most enzyme reactions are studied by using steady-state kinetics. In the steady state, $[ES]$ is constant, $[S]$ is decreasing, and $[P]$ is increasing.

$$k_1[E][S] = (k_2 + k_3)[ES] \quad (2)$$

Rearrange (2) to

$$[ES] = [E][S]/(k_2 + k_3)/k_1 \quad (3)$$

$$K_m = k_2 + k_3/k_1 \quad (4)$$

where K_m is the Michaelis constant. It approximates the substrate–enzyme binding constant under steady-state conditions. The true binding constant, K_D, equals k_2/k_1 and can be determined by measuring the rate in the initial milliseconds of the reaction.

$$[ES] = [E][S]/K_m \quad (5)$$

$$[E] = [E_{Total}] - [ES] \quad (6)$$

where $[E]$ is free enzyme.

$$[ES] = [E_{Total}] - [E] \quad (7)$$

Substitute from Equations (3) and (4) and solve for $[ES]$:

$$[ES] = [E_{Total}][S]/[S] + K_m \quad (8)$$

v, the initial rate of the reaction, is proportional to $[ES]$ when $[S]$ is small.
V_{max}, the maximal rate, is proportional to $[E_{Total}]$ when $[S]$ is saturating.

The Michaelis–Menten equation can be rewritten in terms of reaction rate:

$$v = V_{max}[S]/[S] + K_m \quad (9)$$

To determine K_m, measure the reaction rate at various $[S]$, while $[E_{Total}]$ is constant.

At low $[S]$, v is proportional to $[S]$ (**FIGURE A3.1**).

$$K_m = \text{the } [S] \text{ at which } v = V_{max}/2$$

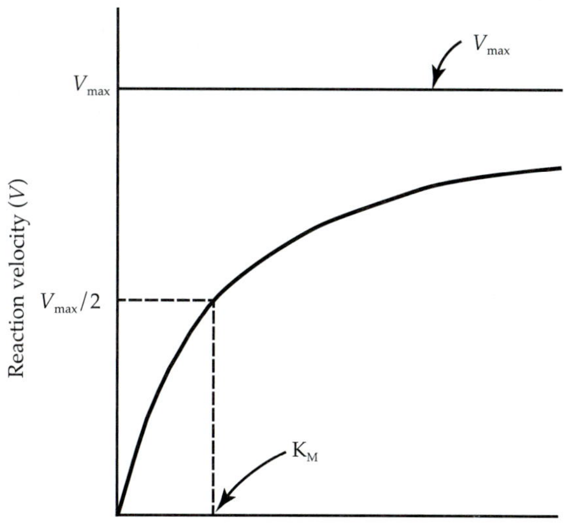

FIGURE A3.1 Adapted from: Segel, I.H., Blackwell calculations, 2nd ed. Wiley-Blackwell, 1976. pp. 214–272.

It is often easier to determine kinetic parameters by plotting the reciprocals of the Michaelis–Menten Equation (9); this is called a Lineweaver–Burk plot (**FIGURE A3.2**).

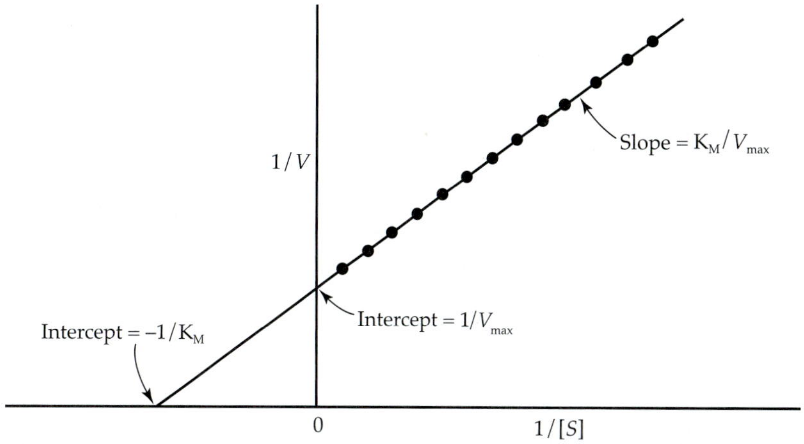

FIGURE A3.2 Adapted from: Segel, I.H., Blackwell calculations, 2nd ed. Wiley-Blackwell, 1976. pp. 214–272.

$$1/v = 1/V_{max} + K_m/V_{max}\ 1/[S] \qquad (10)$$

At V_{max}, $[ES]$ is approximately equal to E_{Total} and the rate of the reaction is proportional to k_3. k_3 is referred to as k_{cat} or the turnover number. It can be determined if $[E_{Total}]$ is known.

$$k_3 = V_{max}/[E_{Total}] \qquad (11)$$

$$k_{cat}/K_m = \text{catalytic efficiency} \qquad (12)$$

There are a number of other ways to analyze kinetic data, such as a Hanes–Woolf plot. Alternatives may be useful in cases where a reaction is inhibited at high substrate concentrations or cannot be measured accurately at low substrate concentrations. Rearranging the Michaelis–Menten equation:

$$[S]/v = 1/V_{max}\ [S] + K_m/V_{max} \qquad (13)$$

A plot of $[S]/v$ (y axis) versus $[S]$ (x axis) (Hanes–Woolf plot) should give a straight line. The slope of the line equals $1/V_{max}$. The intercept on the y axis equals K_m/V_{max}; the intercept on the x axis equals $-K_m$.

A3.1 Competitive Inhibition

In this case the enzyme inhibitor (I) binds to the same site as the substrate or otherwise inhibits substrate access to the enzyme's active site.

$$\begin{array}{c} E + S \leftrightarrows ES \rightarrow E + P \\ + \\ I \leftrightarrows EI \end{array} \qquad (14)$$

The apparent K_m is increased in the presence of an inhibitor, but the V_{max} is approximately the same as the uninhibited reaction (**FIGURE A3.3**). In the presence of inhibitor, both constants are referred to as "apparent" K_{mapp} and V_{maxapp} using subscripts or some other symbol to distinguish them from the constants of the uninhibited reaction.

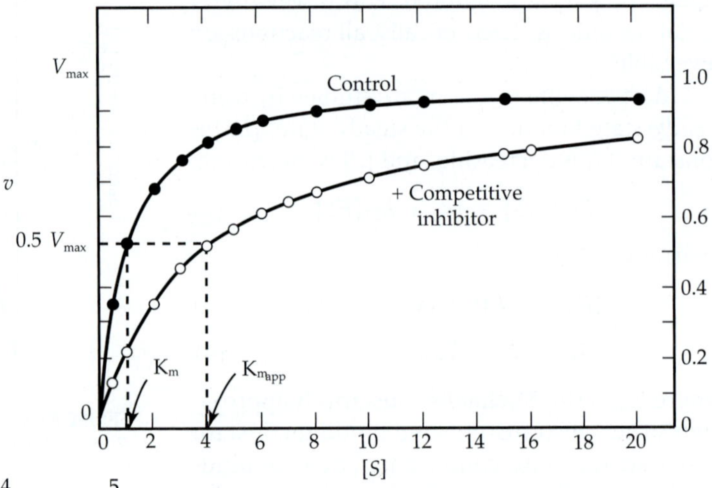

FIGURE A3.3 Adapted from: Segel, I.H., Blackwell calculations, 2nd ed. Wiley-Blackwell, 1976. pp. 214–272.

In a Lineweaver–Burk analysis, the slope of the line changes in the presence of the inhibitor. Both inhibited and uninhibited lines cross the y axis at the same point (**FIGURE A3.4**). A K_i can be calculated from the modified Michaelis–Menten equation or by plotting the slopes of the lines from a reciprocal plot versus $[I]$ (**FIGURE A3.5**).

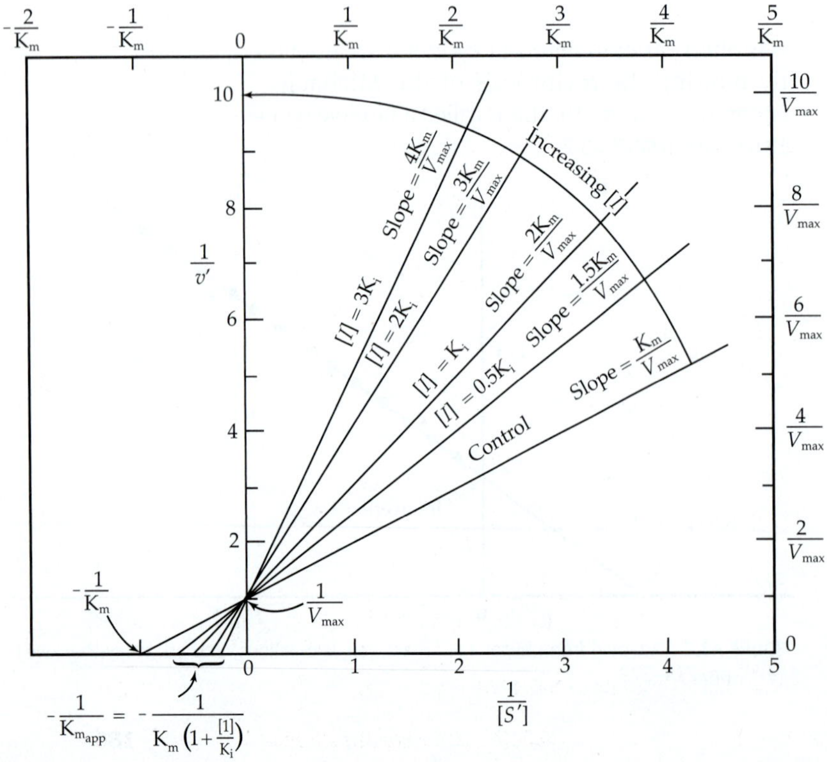

FIGURE A3.4 Adapted from: Segel, I.H., Blackwell calculations, 2nd ed. Wiley-Blackwell, 1976. pp. 214–272.

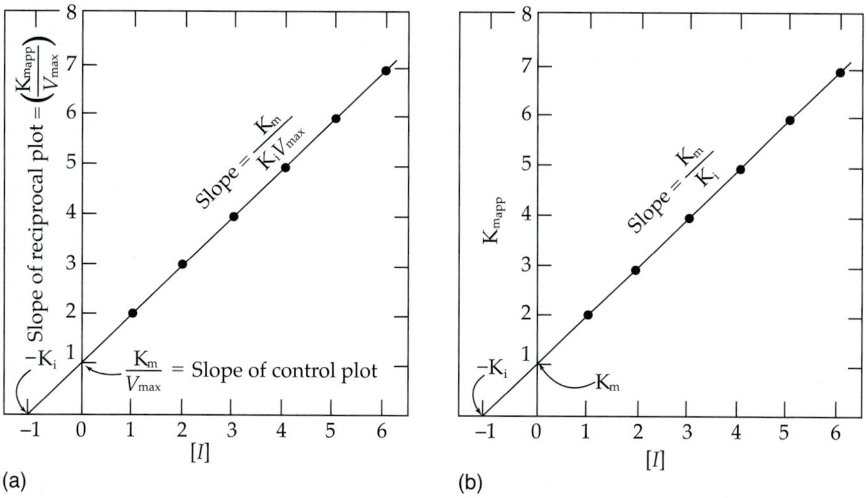

FIGURE A3.5 Adapted from: Segel, I.H., Blackwell calculations, 2nd ed. Wiley-Blackwell, 1976. pp. 214–272.

A3.2 Noncompetitive Inhibition

In this case the inhibitor binds to a site other than the enzyme active site to form an inactive complex. The inhibitor can bind to either the free enzyme (E) or the ES complex.

$$\begin{array}{lll} E + S \leftrightarrows ES \rightarrow E + P & & \\ + \quad\quad\quad \updownarrow + I & & \\ I \quad\quad\quad\quad ESI & & (15) \\ \updownarrow \quad \uparrow & & \\ EI + S & & \end{array}$$

The K_m is approximately the same in the presence of a noncompetitive inhibitor. The V_{max} is lower. Adding a noncompetitive inhibitor is kinetically similar to using a lower enzyme concentration in the reaction (**FIGURE A3.6**).

In a Lineweaver–Burk analysis, the slope of the lines is changed. The lines of inhibited and uninhibited reactions converge on the x axis at $-1/K_m$. A K_i can be calculated from a modified form of the Michaelis–Menten equation or by plotting $1/v$ versus $[I]$ (**FIGURE A3.7**).

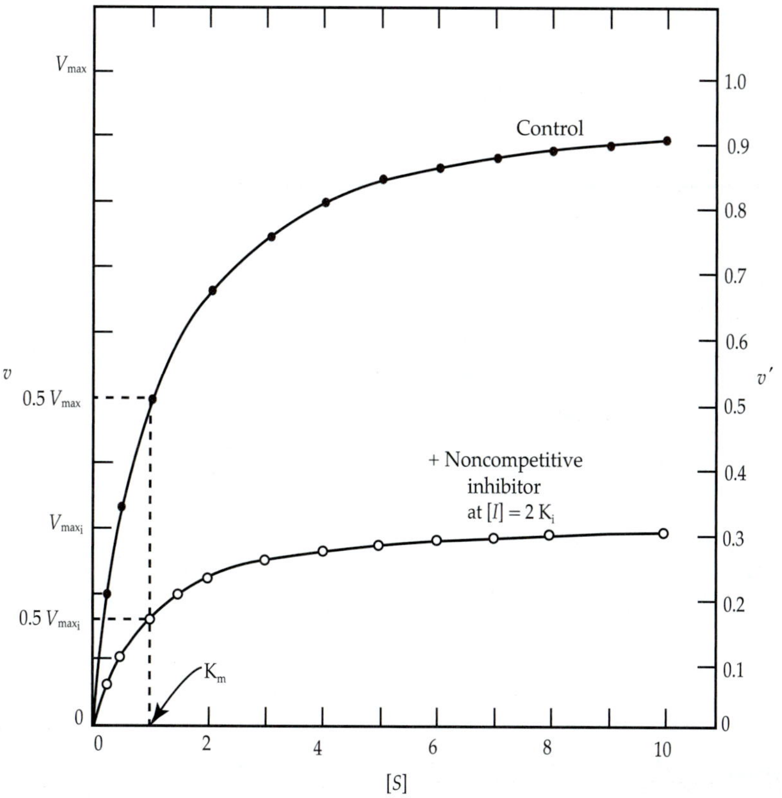

FIGURE A3.6 Adapted from: Segel, I.H., Blackwell calculations, 2nd ed. Wiley-Blackwell, 1976. pp. 214–272.

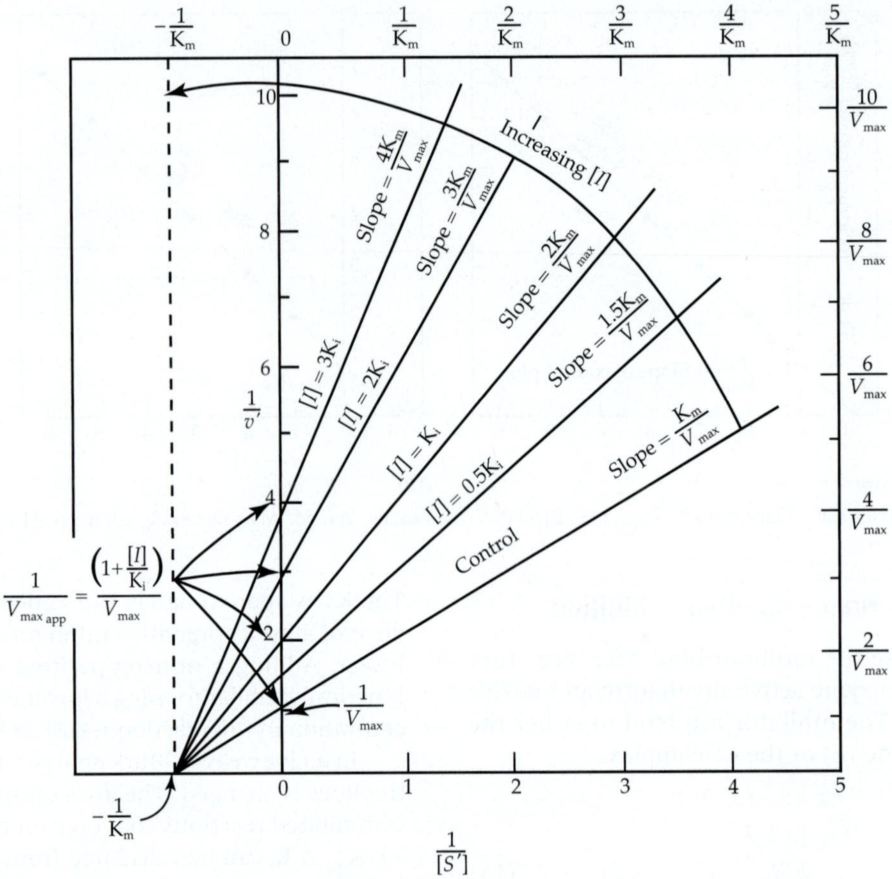

FIGURE A3.7 Adapted from: Segel, I.H., Blackwell calculations, 2nd ed. Wiley-Blackwell, 1976. pp. 214–272.

A3.3 Uncompetitive Inhibition

In this case the inhibitor binds to the ES complex but not to the free enzyme. Such inhibitors are often transition state analogs.

$$E + S \leftrightarrows ES \rightarrow E + P$$
$$+ I \leftrightarrows ESI \qquad (16)$$

In the presence of inhibitor, both K_m and V_{max} decrease (**FIGURE A3.8**).

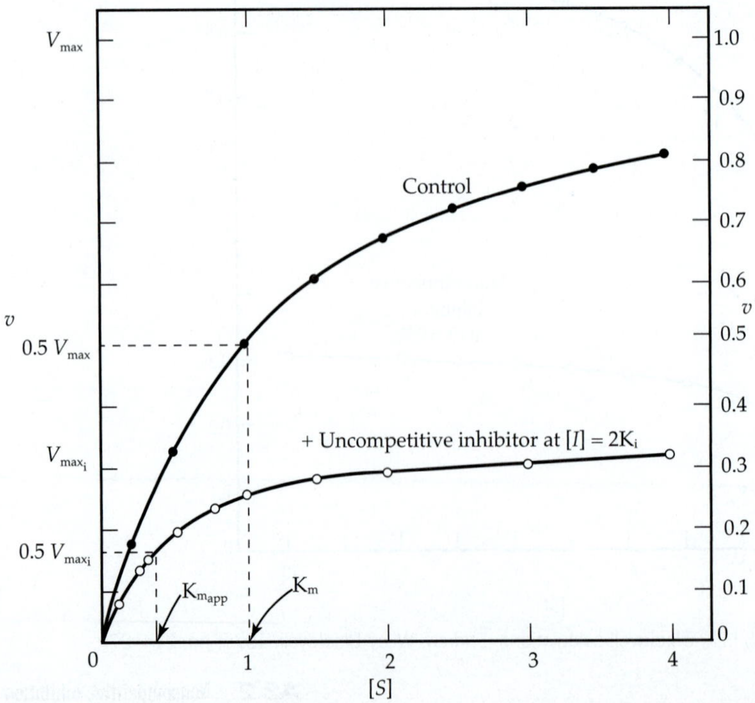

FIGURE A3.8 Adapted from: Segel, I.H., Blackwell calculations, 2nd ed. Wiley-Blackwell, 1976. pp. 214–272.

In a Lineweaver–Burk analysis, the slope of the lines for inhibited and uninhibited reactions are similar and the lines are parallel to each other (**FIGURE A3.9**). A K_i can be calculated from a modified form of the Michaelis–Menten equation.

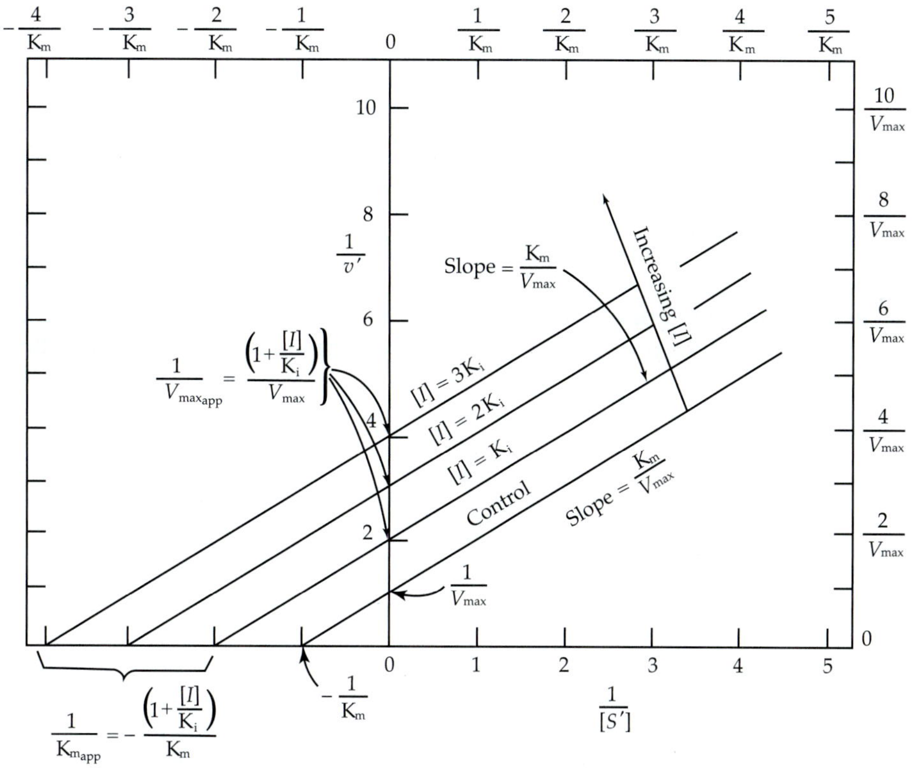

FIGURE A3.9 Adapted from: Segel, I.H., Blackwell calculations, 2nd ed. Wiley-Blackwell, 1976. pp. 214–272.

REFERENCES FOR FURTHER INFORMATION

Berg, J.M., Tymoczko, J.L., & Stryer, L. (2002). *Biochemistry*, 5th ed. New York: WH Freeman, pp. 200–206, 221–222.

Segel, I.H. (1976). *Biochemical calculations*, 2nd ed. New York: John Wiley & Sons, pp. 214–272.

APPENDIX 4

Protein Three-Dimensional Structure Determination

There are two principal methods to determine the complete three-dimensional structures of proteins: x-ray crystallography and nuclear magnetic resonance (NMR) spectroscopy. Both methods require several milligrams (10–100) of highly purified (>90% homogeneity) protein. Protein structure determination by x-ray diffraction was developed in the middle of the twentieth century. The first structures determined were of readily available, water-soluble proteins such as myoglobin (Kendrew et al., 1958). The first attempts to use NMR methods to solve protein structures in solution were made in the 1980s on small proteins such as bovine pancreatic trypsin inhibitor (Wuthrich et al., 1982).

Improvements in instrumentation combined with powerful computational methods have greatly expanded the ability of researchers to determine protein structures. Current molecular techniques in gene isolation, cloning, and protein expression in heterologous hosts, especially *Escherichia coli*, have made it possible to provide large amounts of previously rare proteins for analysis. Recent advances in protein purification procedures, including high pressure and affinity chromatography, have also facilitated production of relatively large amounts of highly purified proteins.

A4.1 X-Ray Crystallography

The first step is to obtain relatively well-ordered, stable crystals of the protein to be analyzed. This is often difficult, and there is no guaranteed method. The most common technique is the "hanging drop" method. A small drop of concentrated protein solution is placed on a glass slide, and the slide is inverted and suspended over a solution of a precipitant such as ammonium sulfate; the edges of the slide are sealed to provide an air-tight chamber. The water in the protein solution slowly evaporates into the precipitant and provides a super-saturated protein droplet. Crystalline protein then forms in the droplets which may be suitable for x-ray analysis. This procedure works best with water-soluble proteins and can be modified for some membrane-associated proteins. Attempts have been made to speed up and automate this process by using 96-well plates with different precipitants.

The next step is to place the protein crystal in an x-ray beam and collect diffraction data. The wavelengths of the x-rays must be in a range that is compatible with the chemical bond distances in proteins, such as C–C or C–N bonds. This range is typically 0.3 to 2.5 Å (1 Å = 10^{-10} m = 0.1 nm). Single-source x-rays are generated by heating a metal such as copper. Crystallographers also make use of synchrotron facilities that provide multiple-wavelength x-rays. In the latter case, it is often necessary to cool the crystals to liquid nitrogen temperatures (about −200°) to prevent beam damage to the crystal. In either case, the x-rays are scattered or diffracted by the electrons in the atoms of the protein. Because of the highly ordered, periodic arrangement of protein molecules in the crystal, the diffracted x-rays interact by either adding or canceling each other. Diffracted x-rays that interact in an additive manner produce spots on a photographic film placed on the opposite side of the crystal. Film may be used to detect diffracted x-rays in a preliminary analysis, but most data are now collected by charge-coupled devices (CCDs) and stored directly in computer files (**FIGURE A4.1**).

The diffracted x-rays have different intensities that are referred to as reflections. These reflections define the atomic structure of the protein molecules in the crystal. The "spots on the film" or CCD are comparable with a two-dimensional photograph of a three-dimensional object. It is possible, using computational methods, to reconstruct the three-dimensional structure of the molecule from these reflections. Two parameters must be known: the amplitude of the reflections, which is proportional to the intensity of the diffracted x-rays, and the phase of the diffracted x-rays. The intensity and phase provide the structure

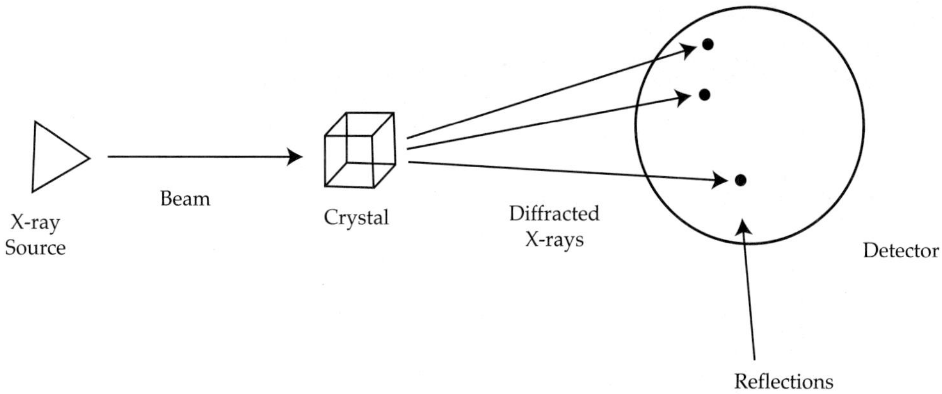

FIGURE A4.1 Basic experimental procedure for x-ray diffraction data collection. X-rays with wavelengths in the range of 0.3 to 2.5 Å are generated either by a conventional x-ray tube to obtain monochromatic radiation or by a synchrotron to obtain multiple-wavelength radiation. The x-ray beam is focused onto a crystallized protein sample that is usually cooled to minimize radiation damage. The positions and intensities (reflections) of the diffracted x-rays are recorded on film or a charge-coupled device for further analysis.

factors that are used to calculate an electron density map.

Although the intensities are easily determined, it is often difficult to obtain the phases of the reflections. The general procedure is to determine the phase of one particular atom in the structure and then calculate the rest of the phases from this single reference point. Protein crystallographers can use three different methods to obtain phases. One method, multiple anomalous dispersion (MAD), requires that the protein contain a heavy metal, such as iron, that is bound in a known position. The diffraction of x-rays is then measured at several wavelengths near the energy of known transition states of the metal and the phase can be calculated for the metal. The metal position and phase then defines the phases of the reflections of the other atoms in the molecule. If the protein does not naturally contain a metal, one can be introduced in a process called multiple isomorphous replacement (MIR). The most common technique is to incorporate seleno-methionine into the protein in place of the normal sulfur-containing amino acid. The diffraction patterns of the protein with and without the introduced metal are compared and the phase of the metal diffraction can then be defined. The easiest way to obtain phases is to use molecular replacement (MR). In this method, the diffraction pattern of the protein is compared with that of a known homologous structure obtained from the protein structure database. This method obviously depends on the primary structural similarity of the two proteins and both proteins must crystallize in the same way (space group).

The known intensities and phases of the reflections, called structure factors, are then analyzed by a Fourier transform method, and a three-dimensional model of the protein structure called an electron-density map is constructed. From this map a model of the protein structure can be made. It is then necessary to calculate how well the model structure agrees with the experimentally obtained reflections. The process of structure refinement consists of producing and modifying the models and comparing them with the data. The process is greatly aided by computer programs, but almost always requires expert human input. The final result is a set of atomic coordinates. These coordinates define the location of each atom in the molecule in x, y, z dimensions and its mobility, called a B or temperature factor. These coordinates are deposited in the Protein Data Bank jointly administered by Rutgers University and Brookhaven National Laboratory. They can be obtained from www.rcsb.org/pdb.

The electron-density map and its interpretation depend on the resolution of the x-ray analysis that is determined by the number of intensities (**FIGURE A4.2**). Data with greater resolution allow a more accurate placement of atoms in the three-dimensional structure. Disorder and large B factors are often due to alternate conformations in parts of the protein structure or to the presence of unbound solvent, usually water molecules. The bulk water in most crystals is approximately 50% of the total mass and does not diffract x-rays in a regular fashion. This decreases the resolution. Bound water molecules, on the other hand,

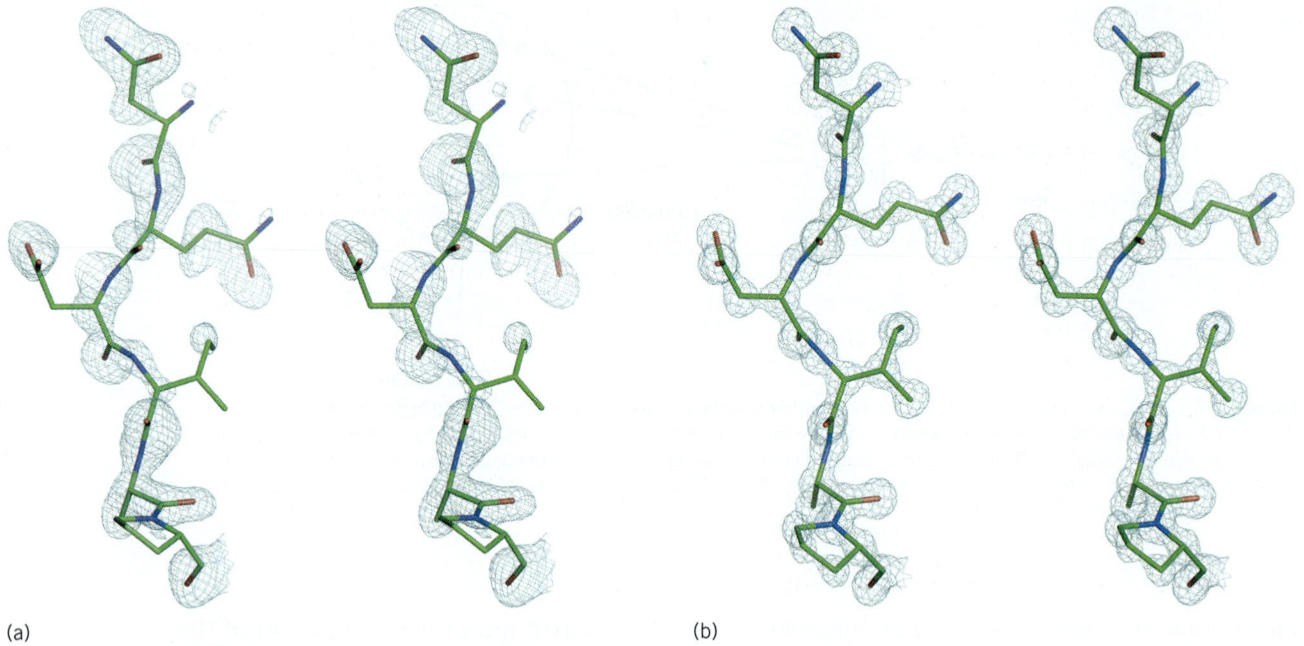

FIGURE A4.2 Stereoviews of two electron-density maps and models of a fragment of the protein, invasin. A model of the C-α backbone and the residue side chains of the protein are superimposed on the maps. (a) At 1.75 Å resolution the general outline of the protein backbone and the positions of the residues are visible, but placement of some of the side chains must be modeled from known structures of the amino acids. (b) At 1.0 Å resolution it is possible to place a more accurate model of the residue side chains on the electron-density map. (*Reproduced from*: Wlodawer A, Minor W, Dauter Z, and Jaskolski M., 2008. "Protein crystallography for non-crystallographers, or how to get the best (but not more) from published macromolecular structures," *FEBS Journal* 275[1]:1–21. Courtesy of Alexander Wlodawer.)

increase resolution and can be fit into the models. The resolution of a structure is reported in angstrom units and in most cases depends on the quality of the protein crystal and how well it diffracts x-rays. A protein structure at 3 Å or greater resolution does not show detailed residue structure. Regular secondary structures such as helices are discernable as well as some residue side chains such as bulky aromatic rings. Most protein structures are reported at medium resolution (2.0–2.7 Å). Residue side chains and bound water molecules are clearly defined. A high resolution structure is at 1.5 Å or lower, in the range of the usual C–C single bond distance. Hydrogen atoms, with only one electron in a bonding orbital, cannot be directly placed in most protein structures. Their presence and the presence of hydrogen bonds are inferred from known atomic parameters.

Other parameters that are important in judging the quality of the final protein structure are the B factors. These are a measure of the mobility of atoms that is usual in loop regions or in disordered domains in the crystallized protein. They are reported in $Å^2$ and are normally in the range of 2 to 100 $Å^2$. The R factor is a measure of how well the model structure fits the data. R factors in reliable protein structures should be less than 20%.

A4.2 NMR Spectroscopy

NMR spectroscopy detects the transitions between spin states of magnetically active atomic nuclei in a magnetic field. This includes all nuclei with an odd number of protons (½ spin states); the relevant nuclei for biological molecules are ^{1}H, ^{13}C, ^{15}N, and ^{31}P. In an applied magnetic field, the spin states of these nuclei align either with the field or against it. The applied energy needed to cause these transitions is the nuclear resonance. The effective resonances of various nuclei are further influenced by their close neighboring nuclei (two to three bonds), which allows a plausible chemical structure to be inferred from an NMR spectrum.

NMR spectroscopy does not require that the proteins crystallize in a well-ordered fashion. However, the protein must be soluble in aqueous solution, usually at a concentration of 1 to 2 mM in a final volume of 0.3 to 0.5 ml. The protein in solution must not aggregate or interact at these high concentrations. In addition, the protein must be stable at room temperature and remain in solution for several hours or days that it may take to collect sufficient data for structure determination. Currently, NMR spectroscopic methods can be used to determine structures of proteins with a mass of approximately 30 kDa or less. Cloning and over-expression of a protein-encoding gene in *E. coli* make it possible to produce large amounts of protein and incorporate additional stable isotopes into the protein. For example, the bacteria can be grown in the presence of uniformly labeled ^{13}C-glucose or ^{15}N- NH_3Cl to yield highly labeled protein.

A one-dimensional proton NMR spectrum consists of resonance peaks at characteristic energies in a magnetic field. The reference point in a spectrum is the ^{1}H chemical shift from the methyl group of 2,2-dimethylsilapentane-5-sulfonic acid (DSS) that is set at 0 ppm (parts per million). The frequencies of the sample resonances are set in relation to this reference standard; these are the chemical shifts (**FIGURE A4.3**). The interpretation of the proton-NMR spectra of even small proteins presents a considerable challenge. Many different techniques are available to provide information in addition to the usual proton spectra. Nuclear Overhauser effect (NOE) spectroscopy depends on through-space dipolar interactions between nuclei. Nuclear Overhauser effect spectra can provide information on the interactions of nuclei that are less than 5 Å apart in the three-dimensional structure. Correlation spectroscopy (COSY) yields spectra with peaks between nuclei that are covalently connected through one or two other atoms (e.g., between the proton on an alpha carbon and an adjacent amide proton). Multidimensional NMR spectra, using resonances from protons, ^{13}C-, and ^{15}N-nuclei, yield further data on interactions between nuclei in the three-dimensional structure.

Other NMR variations have been devised to generate structural information. Most of these involve perturbing the standard resonances with magnetic pulses of different frequencies. All these techniques combine to enable the assignment of resonances to the amino acids in the sequence of the protein and to their positions with respect to each other in the three-dimensional structure. Solving an NMR structure relies heavily on computational methods and databases of known secondary elements

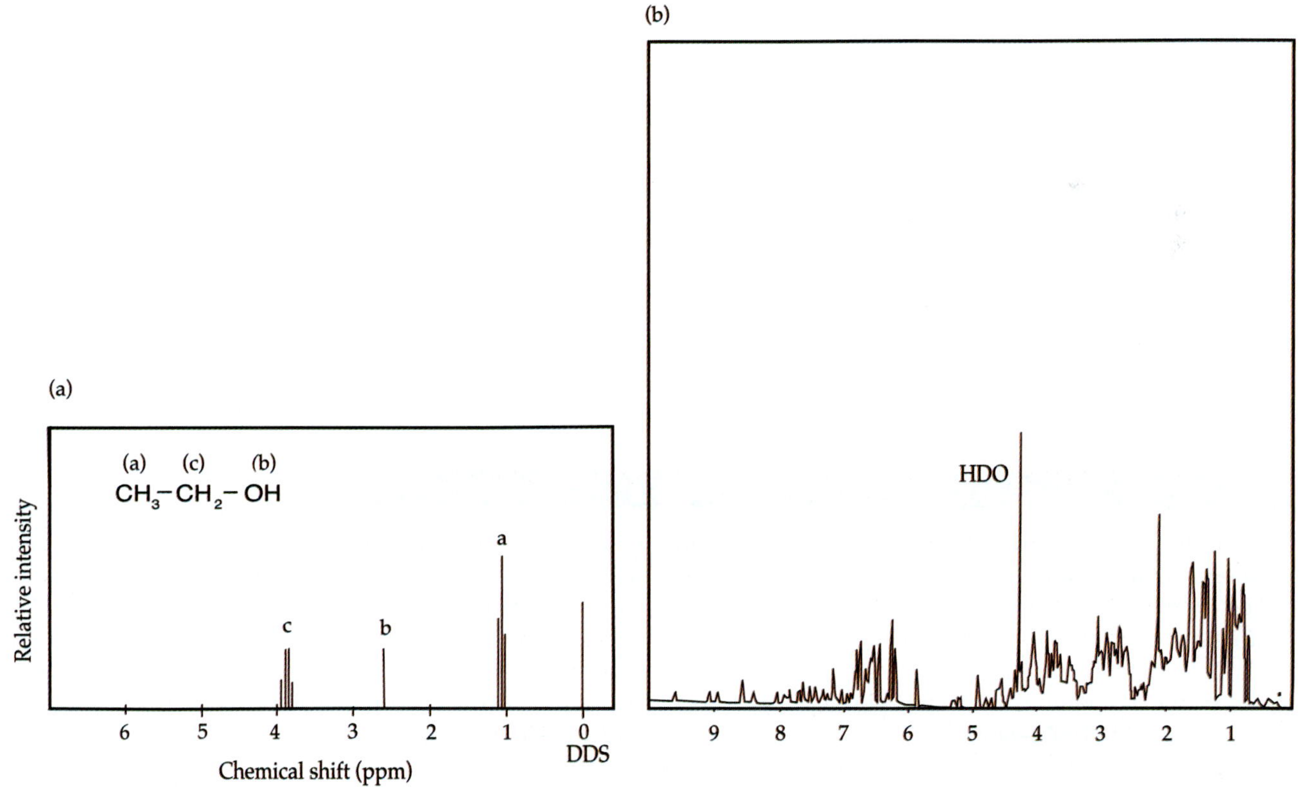

FIGURE A4.3 One-dimensional proton NMR spectra of ethanol and the C-terminal peptide of the enzyme, cellulase. (a) The resonance intensity of the reference compound, 2,2-dimethylsilapentane-5-sulfonic acid (DSS), is set at 0 ppm. Resonances of the different protons in ethanol are shifted downfield from DSS. Each type of proton, that is, hydroxyl (b), methylene (c), and methyl (a), has a characteristic resonance frequency. The peaks in each case are split by the proximity to neighboring protons. (b) The resonance intensities of the peptide fragment of cellulase, which is 36 amino acid residues, show considerable overlap. Additional spectral techniques, including resonances from other nuclei such as ^{13}C, are required to obtain a three-dimensional structure. (Part b reproduced from: K. Wüthrich, G. Wider, G. Wagner and W. Braun, *Journal of Molecular Biology* 155, No. 3 [1982]: 311–319. Reprinted with permission from Elsevier. Courtesy of Dr. Kurt Wüthrich, Ph.D.)

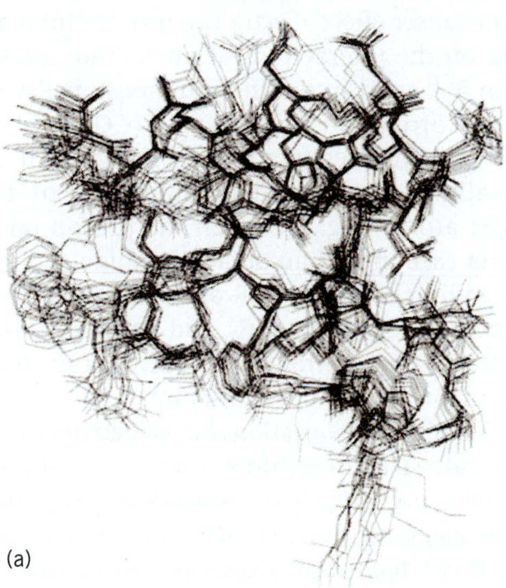

(a)

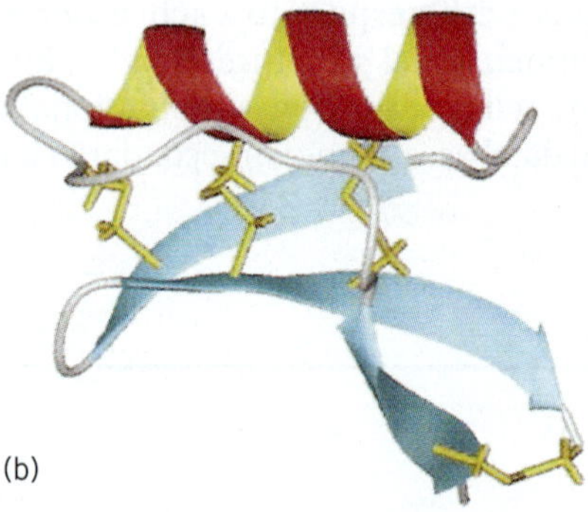

(b)

FIGURE A4.4 NMR structure of the *Pisum sativum* protein, defensin 1. (a) The set of 20 structures that satisfy the NMR parameters are shown superimposed. (b) One representative structure is shown as a cartoon. The α-helix is in red and the β-strands are in blue. Four disulfide bonds are shown as sticks. (Structure from Protein Data Bank 1JKZ. M. S. Almeida, K. M. S. Cabral, E. Kurtenbach, F. C. Almeida, and A. P. Valente, *J. Mol. Biol.* 315 [2002]: 749–757.)

such as helices. The final result is a set of possible structures that match the data (**FIGURE A4.4**). The multiple sets of structures shown for some parts of the protein, particularly loops, do not necessarily indicate that the protein has different conformations but only that any of these conformations fit the data. It is usual to select one conformer as representative of the tertiary structure.

There is no R factor for accuracy in NMR spectroscopy comparable with that in x-ray crystallography. NMR structures are usually reported with the lowest root mean square deviation (RMSD) of atomic coordinates achieved through computational methods and structure adjustments by the experimenter. In general, the resolution of an NMR structure is comparable with that of a medium resolution (2.5 Å) x-ray structure. Atomic coordinates for published structures are also deposited in the Protein Data Bank. NMR spectroscopy is also useful to study dynamic processes such as protein folding, pK_a values of ionizable groups, and ligand binding in solution.

[Standard abbreviations such as RMSD should be included in this appendix. These are used in papers in the scientific literature but may not be obvious to students.]

REFERENCES FOR FURTHER INFORMATION

Branden, C., & Tooze, J. (1999). Determination of protein structures. In: *Introduction to protein structure*, 2nd ed. New York: Garland, pp. 373–392.

Kendrew, J.C., Bodo, G., Dintzis, H.M., Parrish, R.G., Wyckoff, H., & Phillips, D.C. (1958). A three-dimensional model of the myoglobin molecule obtained by x-ray analysis. *Nature* **181:**662–666.

Wlodawer, A., Minor, W., Dauter, Z., & Jaskolski, M. (2008). Protein crystallography for non-crystallographers, or how to get the best (but not more) from published macromolecular structures. *FEBS J.* **275:**1–21.

Wuthrich, K., Wider, G., Wagner, G., & Braun, W. (1982). Sequential resonance assignments as a basis for determination of spatial protein structure by high resolution proton nuclear magnetic resonance. *J. Mol. Biol.* **155:**311–319.

APPENDIX 5

Reactive Oxygen Species

Reactive oxygen species (ROS) in biological systems refer to derivatives of molecular oxygen (O_2) and include singlet oxygen ($^1O_2$), superoxide anion ($O_2^{\bullet -}$ or O_2^-), hydroxyl radical ($\bullet OH$), and hydrogen peroxide (H_2O_2) and associated perhydroxyl radical ($\bullet OOH$). These molecules are highly reactive radicals or oxidants. Molecular oxygen itself is very reactive and a good oxidizing agent. It is a diradical with two unpaired electrons (**FIGURE A5.1**). Because the unpaired electrons have parallel spins, O_2 generally accepts one electron at a time from reactants, such as two hydrogen atoms to form water. In contrast, ROS such as superoxide anion and hydroxyl radical have one unpaired electron and are much stronger oxidizing agents as shown by their redox potentials listed in **TABLE A5.1**.

A5.1 Generation of ROS

Singlet oxygen ($^1O_2$) is produced from molecular oxygen by absorption of energy that boosts the two unpaired electrons into a singlet state with opposite spins (Figure A5.1). Singlet oxygen is not a radical because it has no unpaired electrons, but it is a more efficient oxidizing agent than O_2. Superoxide anion (O_2^-) is produced by adding one electron to O_2. It is a radical with one unpaired electron and is a somewhat better oxidizing agent than molecular oxygen (Table A5.1). Hydrogen peroxide is not a radical but is unstable. In the presence of ultraviolet light, it undergoes homolytic cleavage to generate two hydroxyl anions. H_2O_2 can also serve as a radical initiator when reacting with organic

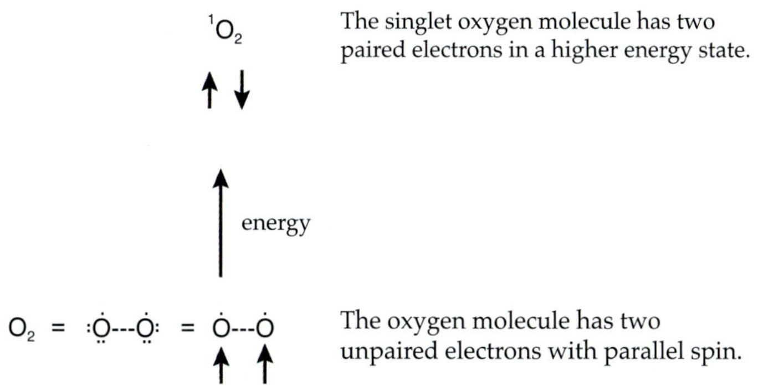

FIGURE A5.1 Formation of singlet oxygen. An oxygen atom has six electrons in its outer (valence) shell. In molecular oxygen, one electron from each atom is shared in the molecular bond. There are two unpaired electrons with parallel spins in the O_2 molecule (triplet state). On absorption of energy, these two electrons are boosted to a higher orbital (singlet state) and with opposite spins. Singlet oxygen ($^1O_2$) reacts more readily than molecular oxygen but has a very short half-life of approximately 1 μs.

Table A5.1 Redox potentials of oxygen species

Reduction half-reaction	E_0' at pH 7.0 (V)
$O_2 + 4H^+ + 4e^- \rightarrow 2H_2O$	+ 0.816
$O_2 + 2H^+ + 2e^- \rightarrow H_2O_2$	+ 0.295
$O_2^- + 2H^+ + 1e^- \rightarrow H_2O_2$	+ 0.940
$\bullet OH + H^+ + 1e^- \rightarrow H_2O$	+ 2.31
$R\text{-}O\text{-}O\bullet + H^+ + 1e^- \rightarrow R\text{-}O\text{-}O\text{-}H$	0.77–1.44 (depending on identity of R)

compounds. Hydroxyl radical (•OH) is produced mainly from the decomposition of hydrogen peroxide. It is the strongest oxidizing agent of all the ROS.

ROS can be generated in almost any organelle in plant cells and in the apoplast. The major sites include chloroplasts, the peroxisomes and glyoxysomes, and mitochondria. The water-oxidizing reaction in photosystem II (PSII; see Chapter 1) could be a source of ROS, but this is unlikely in healthy chloroplasts. PSII can generate singlet oxygen under high light intensity. Excess energy that cannot be used in the PSII reaction center is absorbed by molecular oxygen. Under similar conditions, photosystem I (PSI) can donate an electron to molecular oxygen instead of ferredoxin, producing superoxide anion. This is known as the Mehler reaction. O_2^- is a substrate for a chloroplast superoxide dismutase (SOD), which converts it to hydrogen peroxide.

$$PSI + O_2 \xrightarrow{\text{light}} PSI^+ + O_2^-$$

$$2O_2^- + 2H^+ \xrightarrow{\text{SOD}} H_2O_2 + O_2$$

In the presence of free metals, such as Fe^{2+}, hydrogen peroxide can give rise to hydroxyl radicals *via* the Fenton reaction:

$$H_2O_2 + Fe^{2+} \rightarrow \bullet OH + OH^- + Fe^{3+}$$

The concentration of free iron or other metals is very low in cells, but many proteins contain iron sulfur clusters. In chloroplasts, these clusters are found in the PSI reaction center and ferredoxin (see Chapter 1). Reaction of these proteins with superoxide anion releases iron from the clusters and drives the Fenton reaction.

$$[4Fe - 4S]^{2+} + O_2^- + 2H^+ \rightarrow H_2O_2 + [3Fe - 4S]^{3+} + Fe^{2+}$$

Glyoxysomes can be considered a specialized form of organelles called peroxisomes. In photosynthetic tissues, glyoxysomes serve to process glycolic acid, which is a product of the oxygenase reaction catalyzed by Rubisco (see Chapter 2):

$$\text{Glycolate} + O_2 \xrightarrow{\text{glycolate oxidase}} \text{glyoxylate} + H_2O_2$$

The hydrogen peroxide that is generated can serve as a reactant in the Fenton reaction to further generate hydroxyl radicals. Peroxisomes, however, contain catalase that rapidly converts hydrogen peroxide to water and oxygen, thus minimizing the formation of ROS.

$$2H_2O_2 \xrightarrow{\text{catalase}} 2H_2O + O_2$$

In mitochondria, the cytochrome oxidase complex (Complex IV) catalyzes what is essentially a reversal of the water-oxidizing reaction in the chloroplasts. Four electrons from reduced cytochrome *c* proteins are transferred to molecular oxygen to produce two molecules of water (see Chapter 3). Like the water-oxidizing complex of PSII, cytochrome oxidase is seldom the source of ROS. Complexes I and II in the mitochondrial electron transport system are more apt to give rise to superoxide anion when the ubiquinone pool is saturated with electrons:

$$\text{Complex } I_{red} + O_2 \rightarrow \text{Complex } I_{ox} + O_2^-$$

A5.2 Cellular Damage Caused by ROS

ROS can damage a number of different cellular components. A good example is the D_1 protein in the PSII reaction center (see Chapter 1). This protein is known to have a very high turnover rate because its residues are readily oxidized by singlet oxygen generated in this reaction center. The two sulfur-containing amino acids, cysteine and methionine, are the most susceptible to oxidation and give rise to various sulfated derivatives (**FIGURE A5.2**) or disulfide bonds. Singlet oxygen also oxidizes tryptophan, tyrosine, and histidine residues (**FIGURE A5.3**). Polyunsaturated fatty acids, such as 18:2 and 18:3, which are common in chloroplast membrane lipids, are oxidized by $^1O_2$ to produce various short-chain aldehydes called oxylipins (see Chapter 6). Singlet oxygen also damages DNA by reacting with guanine bases (**FIGURE A5.4**). Organellar DNA is especially susceptible to oxidation by ROS because it has relatively few associated proteins compared with nuclear DNA. Likewise, organelles do not contain the extensive DNA repair systems found in the nucleus.

Both singlet oxygen and superoxide anion have relatively short half-lives (~1 μs) and do not diffuse far from the site of generation. Superoxide anion reacts mainly with the iron-sulfur clusters in redox proteins as discussed above to release free iron and generate hydroxyl radicals. These ROS have a very short half-life of approximately 1 ns and do not readily diffuse. Hydroxyl radicals rapidly react with amino acid residues in proteins, unsaturated fatty acids, and DNA bases producing products similar to singlet oxygen. In addition, •OH can oxidize free sugars and release formic acid (**FIGURE A5.5**).

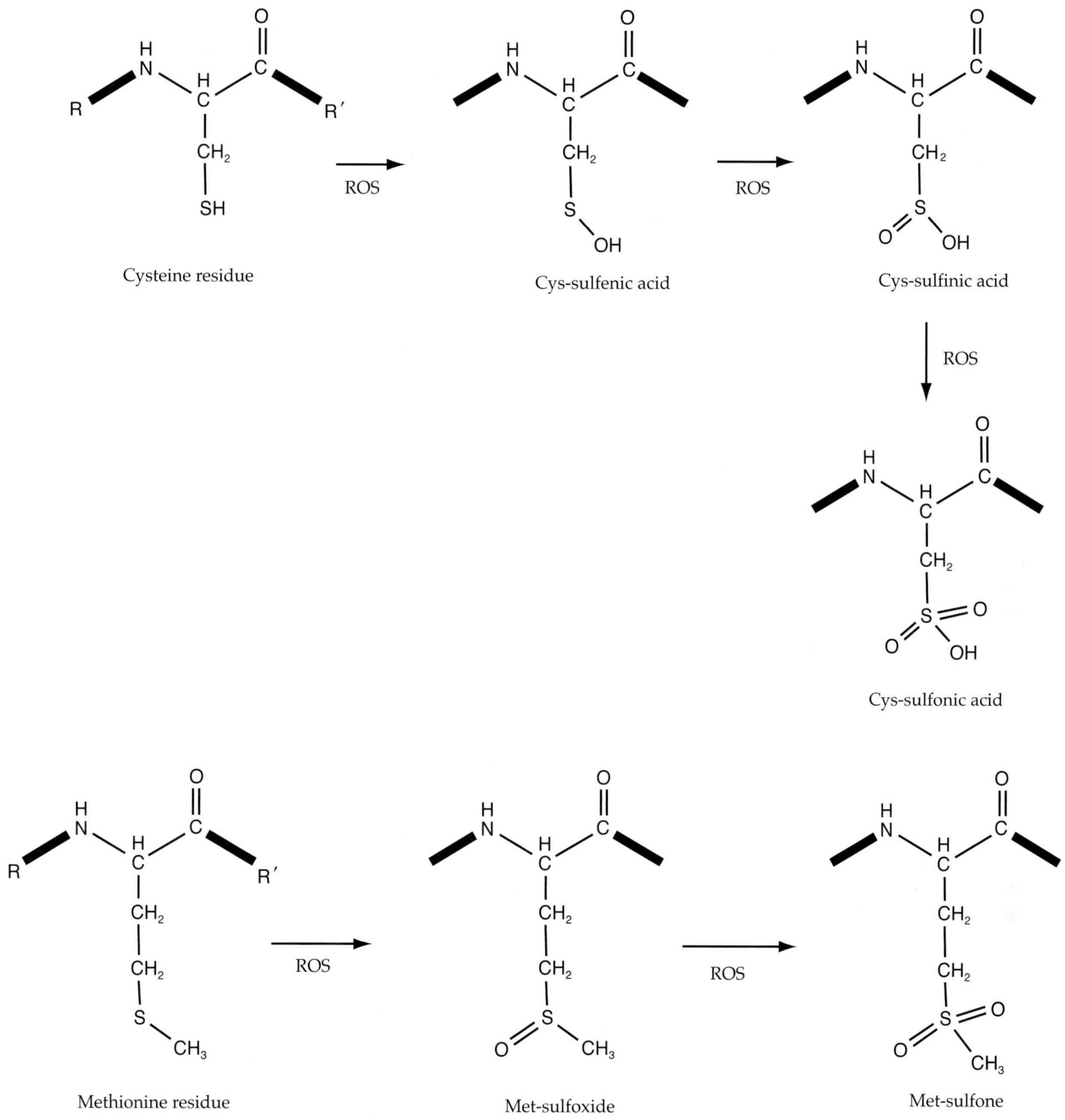

FIGURE A5.2 Oxidation of sulfur-containing amino acids. Cysteine and methionine residues in proteins are the most likely residues to react with ROS. The thiol group in cysteine can be oxidized by singlet oxygen to several derivatives or form a disulfide bond with nearby cysteine residues. Oxidation of methionine residues to the sulfoxide can be repaired by methionine sulfoxide reductase, but the sulfone cannot be reduced *in vivo*.

Hydrogen peroxide has a half-life of approximately 1 ms, longer than most ROS. Because it has no charge, it can diffuse through membranes. It reacts with cysteine residues in proteins as shown in Figure A5.2. Although H_2O_2 is not a radical, it can initiate radical reactions. It will decompose to form a hydroxyl radical and a hydroxyl anion by the Fenton reaction. Peroxidases can initiate loss of a hydrogen atom from H_2O_2 and generate a perhydroxyl radical that reacts with other molecules, especially phenolic compounds:

$$H_2O_2 \rightarrow \text{H-O-O}\bullet + \text{HR} \quad \text{or} \quad \text{ROH} \rightarrow \text{H-O-O-H} + \text{R}\bullet \quad \text{or} \quad \text{R-O}\bullet + H_2O$$

In the apoplast, these radical-generating reactions of hydrogen peroxide initiate lignin synthesis (see Chapter 8).

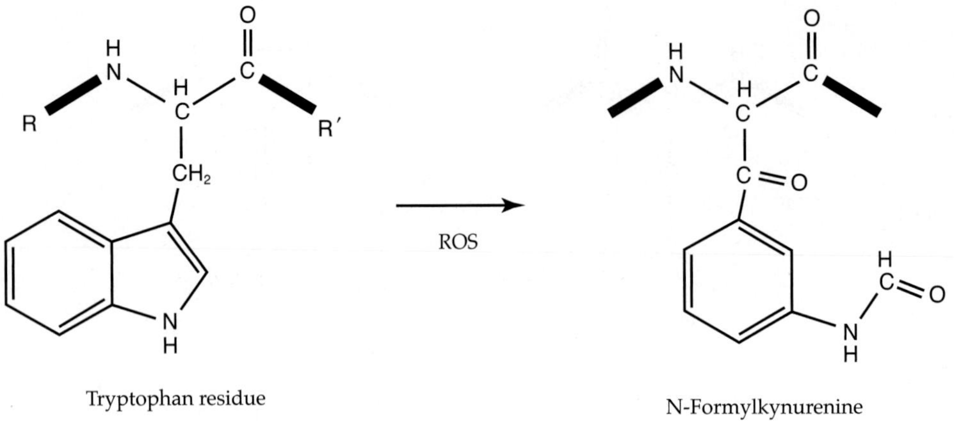

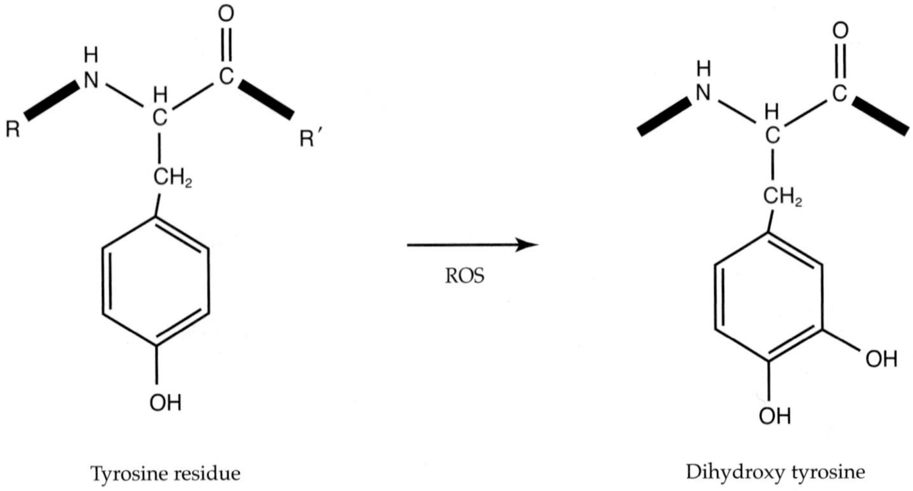

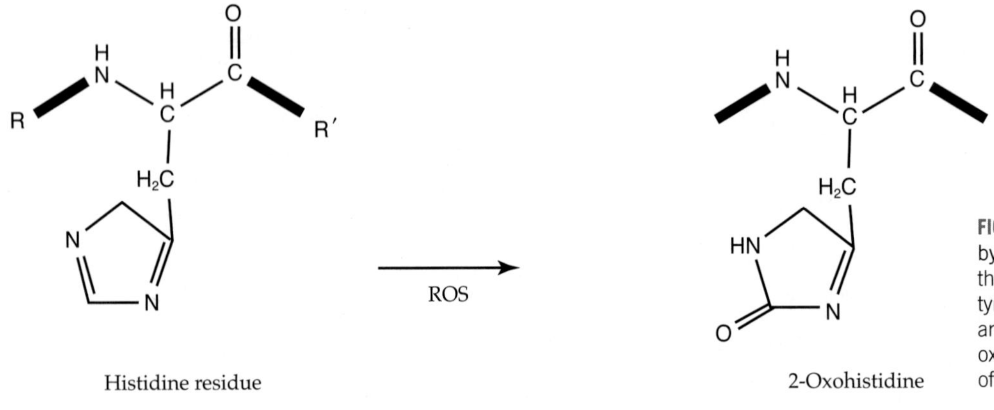

FIGURE A5.3 Oxidation of aromatic residues by ROS. Although sulfur-containing residues are the most likely to react with ROS, tryptophan, tyrosine, and histidine residues in proteins are also rapidly oxidized. These amino acid oxidation products cannot be repaired and often result in an inactive protein.

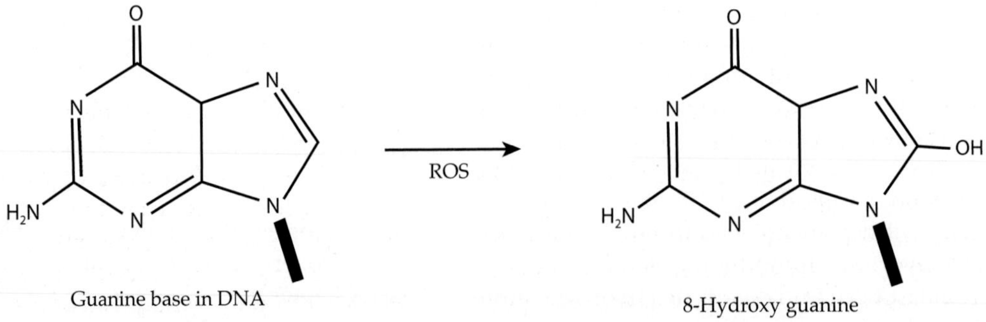

FIGURE A5.4 Reaction of a guanine base in DNA with ROS. The guanine bases in nucleic acids are the most susceptible to oxidation by ROS. Oxidized bases may not pair properly when DNA is replicated, leading to mutations.

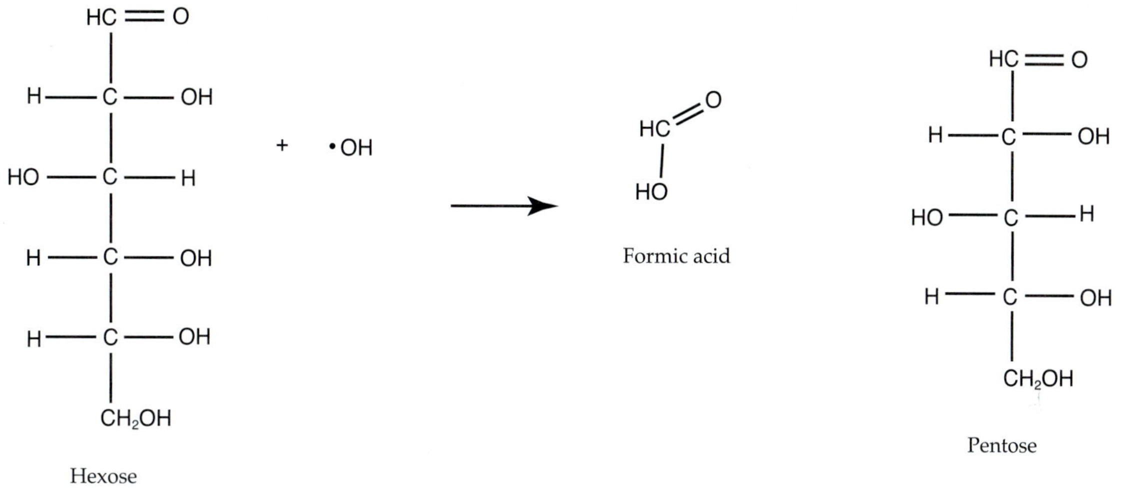

FIGURE A5.5 Reaction of hydroxyl radical with glucose. The hydroxyl radical (•OH) is the most reactive of all the ROS and is the only species that is known to react with aldoses. The radical attacks the carbonyl carbon of glucose in this figure and breaks the C–C bond to yield formic acid and the pentose sugar, xylose.

A5.3 Protection from ROS

Plants have several mechanisms to quench ROS and repair the damage caused by these reactive species. Singlet oxygen generated in PSII reacts readily with the carotenoids that are present in the reaction center and in the surrounding light-harvesting complexes. Lycopene is particularly good at accepting an extra electron from an ROS and forming a stable radical that is quenched by reaction with other antioxidants. The energy that drives the formation of singlet oxygen can be dissipated by nonphotochemical quenching in the xanthophyll cycle (see Chapter 1). In addition, state transitions can direct excess light energy away from PSII by moving the light-harvesting complexes away from the PSII reaction center.

The superoxide anion generated in PSI is a substrate for superoxide dismutase, as noted previously. The resulting hydrogen peroxide can diffuse out of the chloroplast and be reduced to water and molecular oxygen by catalases in the cytoplasm and peroxisomes. In mitochondria, the alternate oxidase (AOX) associated with the inner membrane can prevent the formation of superoxide anion by transferring electrons from ubiquinol to molecular oxygen (see Chapter 3). This redirection of electrons does not contribute to the proton gradient in the intermembrane space but may serve as a "safety valve" to prevent formation of O_2^- by Complexes I and II.

Two important small molecules that can dissipate ROS in plants are ascorbic acid (vitamin C) and α-tocopherol (vitamin E). Ascorbate can donate an H atom to radicals such as •OH to form an ascorbate radical (**FIGURE A5.6**). Delocalization of the unpaired electron in ascorbate makes it unlikely to transfer the radical to other molecules. Superoxide anion can also be neutralized by a similar mechanism using both ascorbate and α-tocopherol. Reaction of the tocopherol radical with an ascorbate radical regenerates α-tocopherol and forms oxidized ascorbic acid. The acid can then be reduced by glutathione (GSH) and a glutathione peroxidase (see Chapter 5). Oxidized ascorbate also reacts with hydrogen peroxide catalyzed by ascorbate peroxidase:

$$H_2O_2 + 2\ \text{ascorbate}_{ox} \xrightarrow{\text{peroxidase}} 2H_2O + 2\ \text{mono-dehydroascorbate radical}$$

The ascorbate radical can then be reduced by GSH, which in turn is reduced by NADPH and glutathione reductase, or by reduced ferredoxin in the chloroplast.

Flavonoids and stilbenes are very effective at quenching radicals *in vitro* (**FIGURE A5.7**). The B-ring, especially a derivative with two OH substituents, can accept one electron to become a semiquinone that is stabilized by electron delocalization in the aromatic ring. By accepting a second electron, the semiquinone is reduced to a flavonoid quinone. These molecules are often stored in vacuoles or polymerized into proanthocyanidins (see Chapter 8). To be an effective antioxidant *in vivo*, however, the flavonoids must be recycled, as in the example of ascorbate and tocopherol. Flavonoid quinones can be reduced by ascorbate, but this reaction apparently does not occur in plant tissues. Flavonoid synthesis is

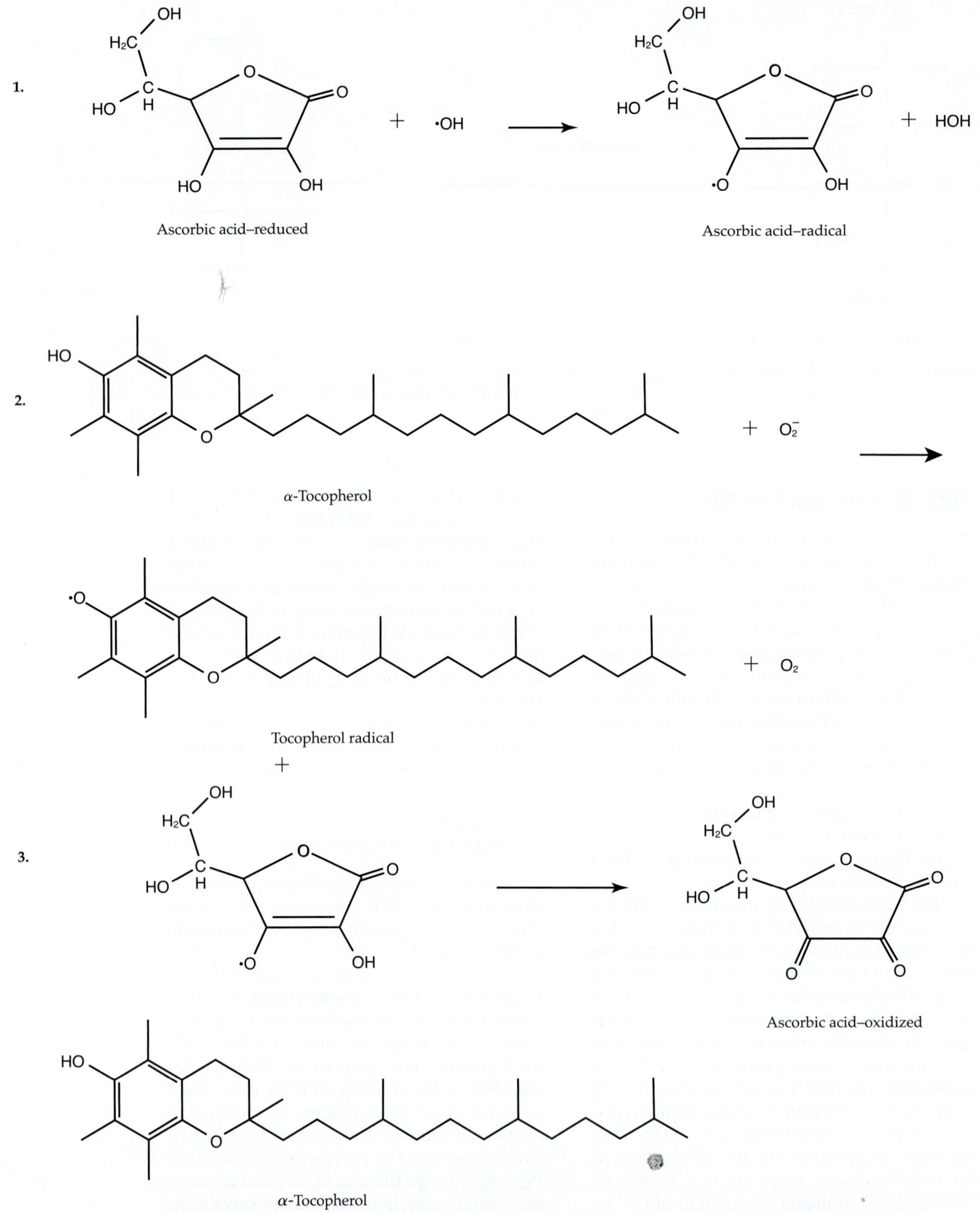

FIGURE A5.6 Ascorbic acid and α-tocopherol dissipate radical ROS. In reaction 1 reduced ascorbic acid donates a hydrogen atom to a hydroxyl radical to generate water and an ascorbate radical. This organic radical is resonance stabilized and relatively unreactive. In reaction 2 α-tocopherol is reduced by accepting one electron from a superoxide anion. The resulting tocopherol radical is resonance stabilized. As shown in reaction 3, ascorbate radical donates one electron to a tocopherol radical to regenerate α-tocopherol. The resulting oxidized ascorbic acid can be subsequently reduced by an enzymatic reaction with glutathione (GSH).

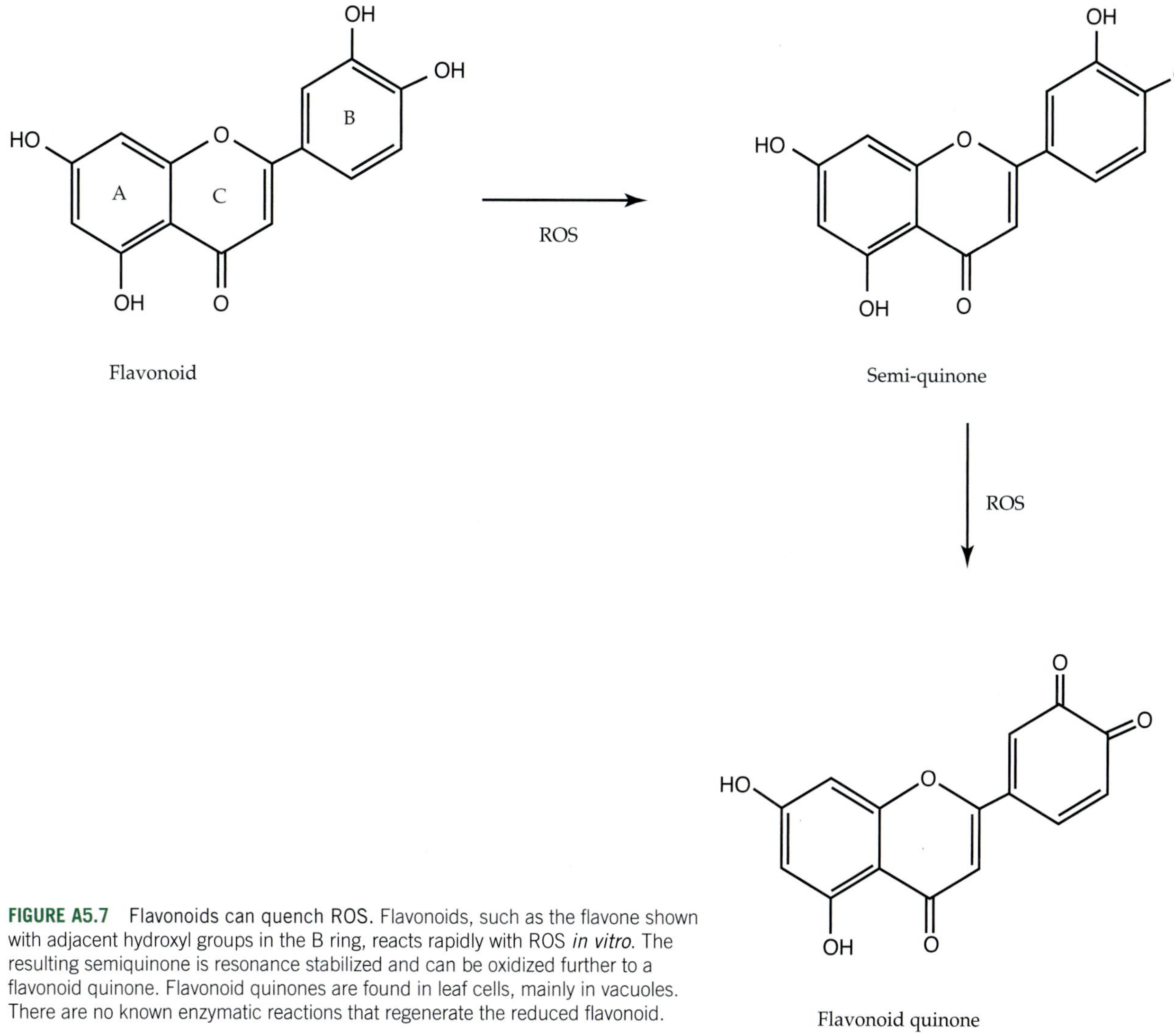

FIGURE A5.7 Flavonoids can quench ROS. Flavonoids, such as the flavone shown with adjacent hydroxyl groups in the B ring, reacts rapidly with ROS *in vitro*. The resulting semiquinone is resonance stabilized and can be oxidized further to a flavonoid quinone. Flavonoid quinones are found in leaf cells, mainly in vacuoles. There are no known enzymatic reactions that regenerate the reduced flavonoid.

up-regulated by stress, but plants produce these compounds mainly as sunscreens and their use as antioxidants *in planta* is equivocal.

The glutathione–glutathione reductase system is an established mechanism for reducing radicals *in vivo*. In addition, plants contain peroxiredoxins that react with ROS, such as hydrogen peroxide. Peroxiredoxins are small proteins containing one or two cysteine residues in their active site. Reaction with H_2O_2 oxidizes the active site cysteine to a sulfenic acid (Figure A5.2) or a cystine disulfide in the case of two-cysteine peroxiredoxins. The peroxiredoxins are recycled by reduced thioredoxin that in turn is reduced by NADPH and a thioredoxin reductase (see Chapter 2). Thioredoxin also serves as the reducing agent for methionine sulfoxide reductase, which repairs oxidized methionine residues in proteins.

A5.4 ROS May Be Active in Innate Immunity

Although the generation of ROS is generally deleterious to cellular structures, these reactive molecules can serve some useful functions. The ROS are a part of the plant's innate immune defenses because of their ability to kill microbial invaders on contact. The ROS are believed to act in concert with reactive nitrogen species (RNS), mainly nitric oxide (NO). Plant cell membranes contain integral NADPH oxidases that use NADPH produced in the cytoplasm to generate H_2O_2 in the apoplast. These enzymes are often referred to as "respiratory burst oxidases," and their primary function appears to be production of ROS as a defense against microbial pathogens. The oxylipins that result from ROS-mediated fatty acid oxidation include jasmonic acid derivatives, which are plant hormones that stimulate the synthesis of defensive compounds such as phytoalexins (see Chapter 6). Oxidation of cysteine residues in some transcription factors has been shown to activate these proteins and result in gene transcription for proteins active in defense. Some oxidized transcription factors are also important in the "hypersensitive response," i.e., a series of reactions that leads to apoptosis [cell death] when cells are invaded by pathogens. Although the production of oxygen in photosynthesis and its utilization in respiration include the inevitable yield of deleterious ROS, it seems that living cells have been able to harness some of these reactive molecules to enhance cell survival.

REFERENCES FOR FURTHER INFORMATION

Asada, K. (2006). Production and scavenging of reactive oxygen species in chloroplasts and their functions. *Plant Physiol.* **141:**391–396.

Halliwell, B. (2006). Reactive species and antioxidants. Redox biology is a fundamental theme of aerobic life. *Plant Physiol.* **141:**312–322.

Jaspers, P., & Kangasjärvi, J. (2010). Reactive oxygen species in abiotic stress signaling. *Physiol. Plant.* **138:**405–413.

Møller, I.M., Jensen, P.E., & Hansson, A. (2007). Oxidative modifications to cellular components in plants. *Annu. Rev. Plant Biol.* **58:**459–481.

Winterbourn, C.C. (2008). Reconciling the chemistry and biology of reactive oxygen species. *Nat. Chem. Biol.* **4:**278–286.

APPENDIX 6

Amino Acid Biosynthesis

Amino acid production is a central part of primary metabolism in plants. Amino acids provide the building blocks of proteins and are also precursors of other metabolites. For example, monolignols derived from phenylalanine are not only the building blocks of lignin, an essential component of secondary cell walls, but are also used in the synthesis of natural products such as lignans and volatile benzoids (see Chapter 8). Plant hormones such as indole-3-acetic acid and systemin are produced from amino acids. Most plants and many prokaryotes must synthesize all 20 of the major amino acids found in proteins and the nonprotein amino acids, such as ornithine, which are used in secondary metabolite production (see Chapter 9). Most of the early research on amino acid biosynthetic pathways was done in bacteria. The elucidation of these prokaryotic pathways provided a guide for the research done in plants. More recent advances in plant molecular genetics have allowed researchers to confirm and further refine the steps in amino acid biosynthesis and their regulation.

Glutamic acid and the corresponding amide, glutamine, play a central role in the processing of nitrate and ammonia as described in Chapter 5. Glutamate is also a major participant in the majority of transaminase (amino transferase) reactions required to produce amino acids from oxo-acid precursors, such as oxoglutarate. The sulfur-containing amino acids, cysteine and methionine, play important roles in sulfur and methyl transfer reactions; their biosynthesis is also covered in Chapter 5. The aromatic amino acids, phenylalanine, tyrosine, and tryptophan, are derived from branches of the shikimic acid pathway. The aromatic amino acids are vital for both primary and secondary metabolism, and details of their biosyntheses are described in Chapter 8. A condensed outline of the biosynthesis of the remaining 13 major amino acids is described here. For a more complete description of the reactions and pathway regulation, the references at the end of this appendix are recommended.

The carbon skeletons that are the building blocks for amino acid syntheses are derived from intermediates in primary metabolism, mainly the C-3 cycle of CO_2 fixation, glycolysis, and the citric acid cycle. An overview of the origin of 20 major amino acids and their precursors is shown in **FIGURE A6.1**. Most of the enzymes of these pathways are located in plastids, but amino acid precursors can be derived from other cellular compartments such as glyoxysomes and mitochondria. Some of the biosynthetic pathways share common intermediates. A good example is the aspartate family of amino acids, which include lysine, threonine, methionine, and isoleucine. Several levels of regulation must exist in such an interconnected scheme to ensure that sufficient pools of each amino acid are available for protein synthesis and other metabolic demands. Many of the pathway enzymes are subject to feedback inhibition by the final amino acid products. In addition, several isoforms of the enzymes have been described with different kinetic properties and sensitivity to regulators. These isoforms are often specific to a particular plant tissue or developmental stage. Finally, there is regulation at the transcriptional level in almost all these pathways that allows adaptation to environmental changes. Because of their importance in animal and human nutrition, there is a major interest in manipulation of the amino acid content of plants, in particular in seeds of wheat, rice, and maize. Molecular genetics has already provided information on the complexity and interdependence of these pathways and will undoubtedly add to our knowledge of amino acid synthesis in the future.

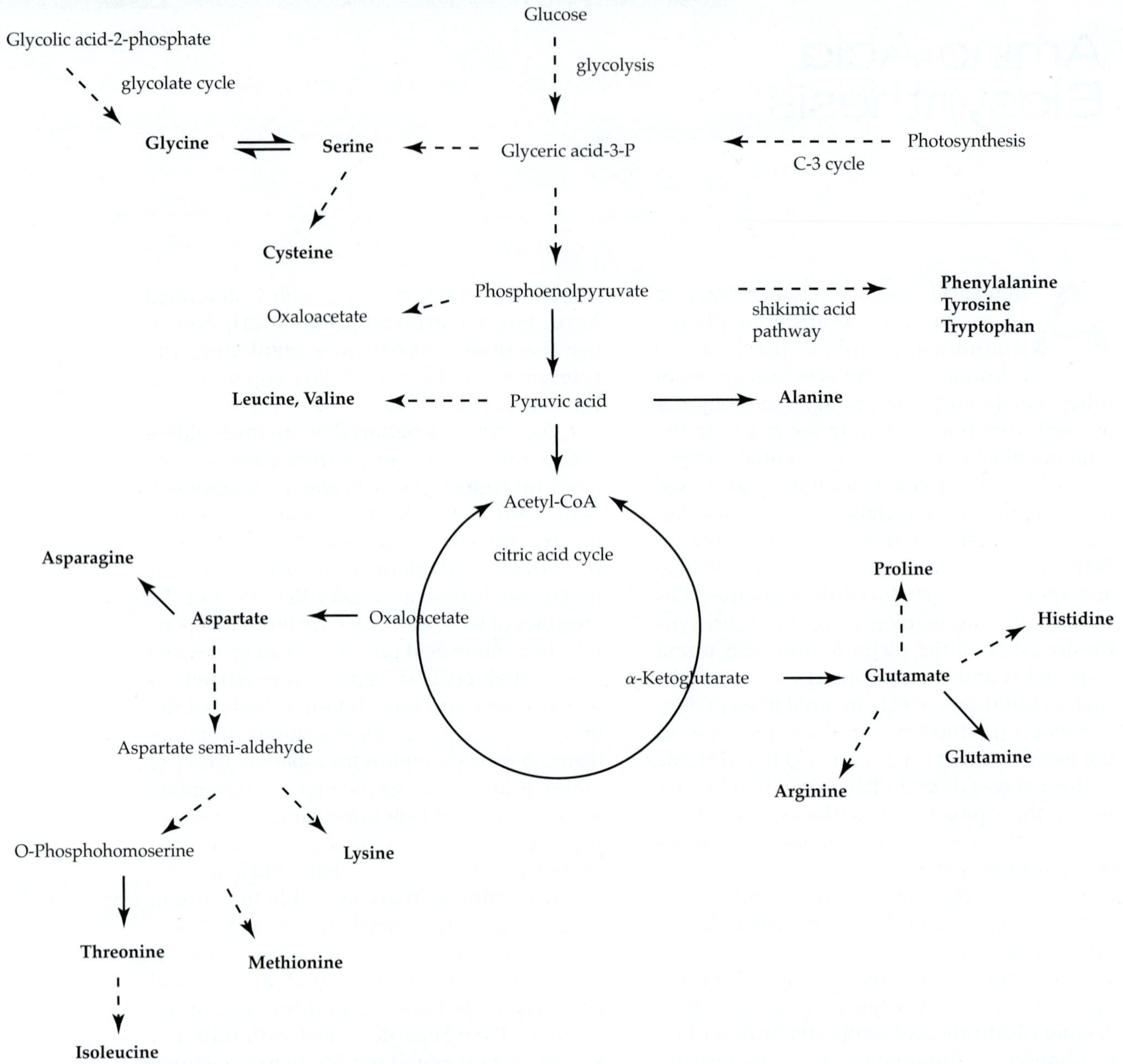

FIGURE A6.1 Overview of the biosynthesis of the 20 amino acids in proteins. The precursors that provide the carbon skeletons for amino acid synthesis are outlined together with sources of these metabolites in primary carbon metabolism. The product amino acids are shown in bold letters. Glutamate is the amino donor for most of the amino acid biosyntheses.

A6.1 Amino Acids Derived from Glyceric Acid-3-P

Glyceric acid-3-phosphate (3-phosphoglycerate) is the initial product of the C-3 cycle in chloroplasts (see Chapter 2). It is also an intermediate in glycolysis (see Chapter 3). Phosphoglycerate is the precursor for the synthesis of serine as shown in Figure A6.1 and **FIGURE A6.2**. Serine, in turn, undergoes dehydroxy demethylation to form glycine. Serine can be acetylated by acetyl-CoA to form O-acetylserine, which reacts with bisulfate anion (HS^-) to produce cysteine (see Chapter 5, Figure 5.13). Glycine and serine are also key intermediates in the glycolate cycle that occurs in leaf peroxisomes. This pathway recycles most of the glycolic acid that results from fixation of oxygen by Rubisco (photorespiration, see Chapter 2).

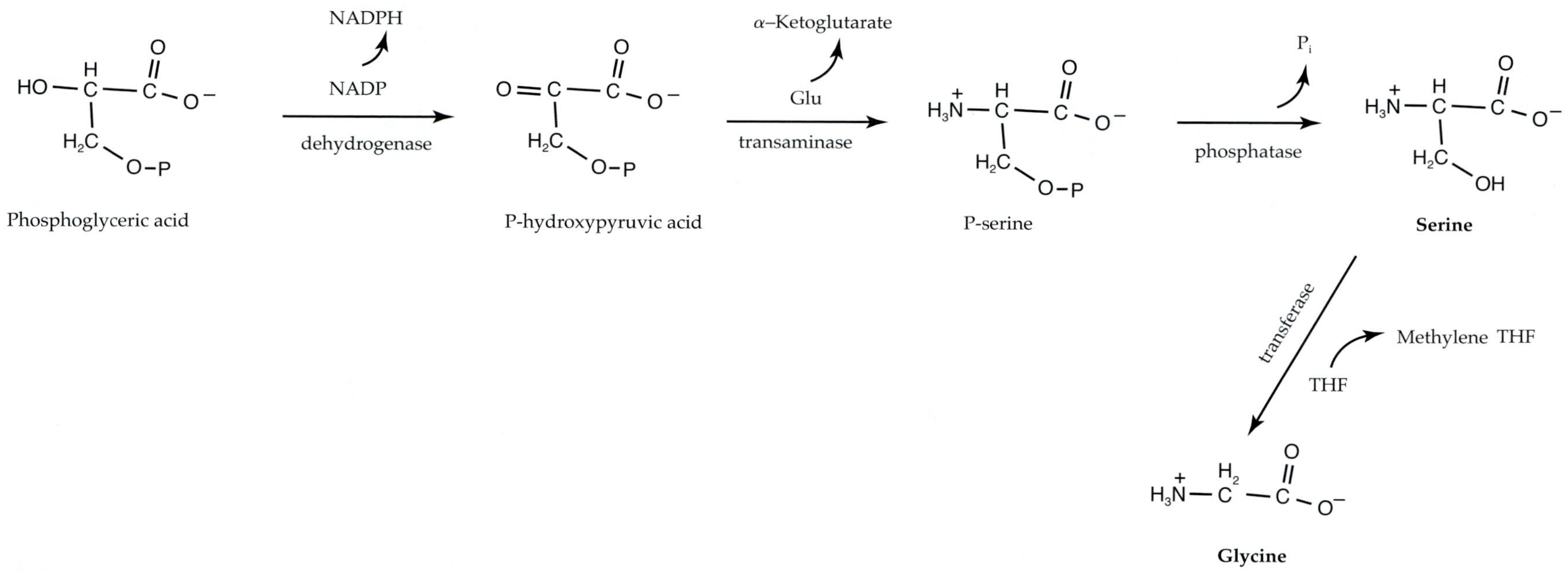

FIGURE A6.2 Biosynthesis of serine and glycine. Phosphoglyceric acid (glyceric acid-3-phosphate), derived from the C-3 cycle or glycolysis, is a precursor for serine synthesis. Glycine is synthesized from serine by transfer of a methyl group to tetrahydrofolate (THF). These two amino acids are also key intermediates of the photorespiratory glycolate cycle described in Chapter 2.

A6.2 Amino Acids Derived from Phosphoenolpyruvate

Phosphoenolpyruvate (PEP) and erythrose-4-phosphate are the initial substrates in the shikimic acid pathway described in Chapter 8. The products are the aromatic amino acids, phenylalanine, tyrosine, and tryptophan. PEP is also a substrate for PEP carboxylase, which is the enzyme that fixes CO_2 into the four-carbon acid, oxaloacetic acid, in C-4 metabolism (see Chapter 2). This dicarboxylic acid is also an intermediate in the citric acid cycle (see Chapter 3). In both C-3 and C-4 plants, oxaloacetate from both sources is the precursor for the aspartate family of amino acids. Their synthesis is described below under Amino Acids Derived from Oxaloacetate.

A6.3 Amino Acids Derived from Pyruvate

Pyruvic acid is the immediate precursor of alanine (**FIGURE A6.3**). Glutamic acid is the amino donor in this reaction catalyzed by a specific amino transferase (transaminase). Transaminases are a large family of enzymes that catalyze the transfer of an amino group from a specific amino acid, usually glutamate, to a specific oxo-acid to yield a new amino acid, which in the case of glutamate is α-ketoglutaric acid (2-oxoglutarate). Amino transferases use a pyridoxal-5-phosphate cofactor that is bound to the ϵ-amino group of a lysine residue in the transferase active site. Binding of an amino acid such as glutamate displaces the lysine, allowing the cofactor to bind to the amino group of the substrate, break the C–N bond, and release the corresponding oxo-acid. The pyridoxamine cofactor can now transfer the amino group to a different oxo-acid substrate such as pyruvic acid in Figure A6.3 to form a new amino acid. All the transaminases show specificity for a specific oxo-acid substrate. The amino donor must be an acidic amino acid and is usually glutamate or, in some cases, aspartate.

Pyruvate is also the precursor for valine and leucine biosynthesis and participates in the synthesis of isoleucine. These three amino acids are referred to as the branched-chain amino acids. They have similar hydrophobic properties, and their biosynthetic pathways use some of the same enzymes. The synthesis of valine and leucine is outlined in **FIGURE A6.4**. Two molecules of pyruvate are condensed to form 2-acetolactate. The enzyme, acetolactate synthase (acetohydroxy acid synthase), also catalyzes the condensation of pyruvate and α-ketobutyrate in isoleucine biosynthesis (**FIGURE A6.5**). This enzyme is the target for several major commercial herbicides, the sulfonylureas and imidazolinone (**FIGURE A6.6**). All these inhibitors are highly specific for acetolactate synthase; treatment leads to decreased pools of the branched-chain amino acids. These inhibitors do not bind directly to the active site residues but instead block access of the substrate, pyruvate, to the active site cavity. Sulfonylurea-resistant mutant plants often have one or two amino acid residue changes near the active site that allow substrate access in the presence of the herbicide.

The second enzyme, a keto-acid reductoisomerase, is also common to both valine/leucine and isoleucine pathways. The enzyme catalyzes both a group migration in the 2-acetolactate or 2-acetohydroxy butyrate substrate and a reduction using NADPH to yield a keto (oxo) acid intermediate. The third enzyme, a dehydratase, is also found in both pathways. α-Keto isovalerate (2-oxo-isovalerate) is the immediate precursor of valine (Figure A6.4). The transaminase is, again, a common branched-chain amino transferase that uses glutamate as the amino donor. The four common enzymes in the synthesis of the branched-chain amino acids are subject to several allosteric controls by the end products to regulate the pools of these amino acids.

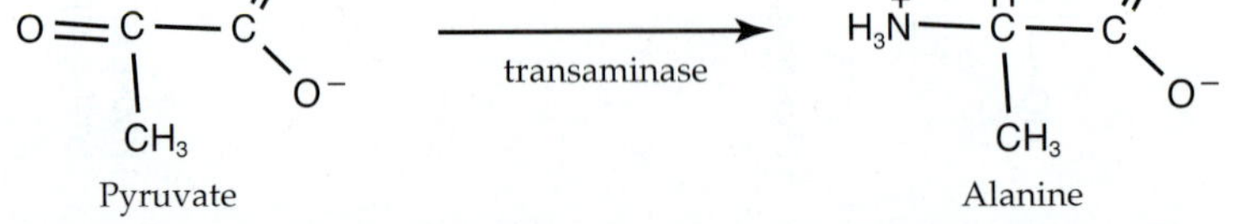

FIGURE A6.3 Alanine is a product of the transamination of pyruvic acid.

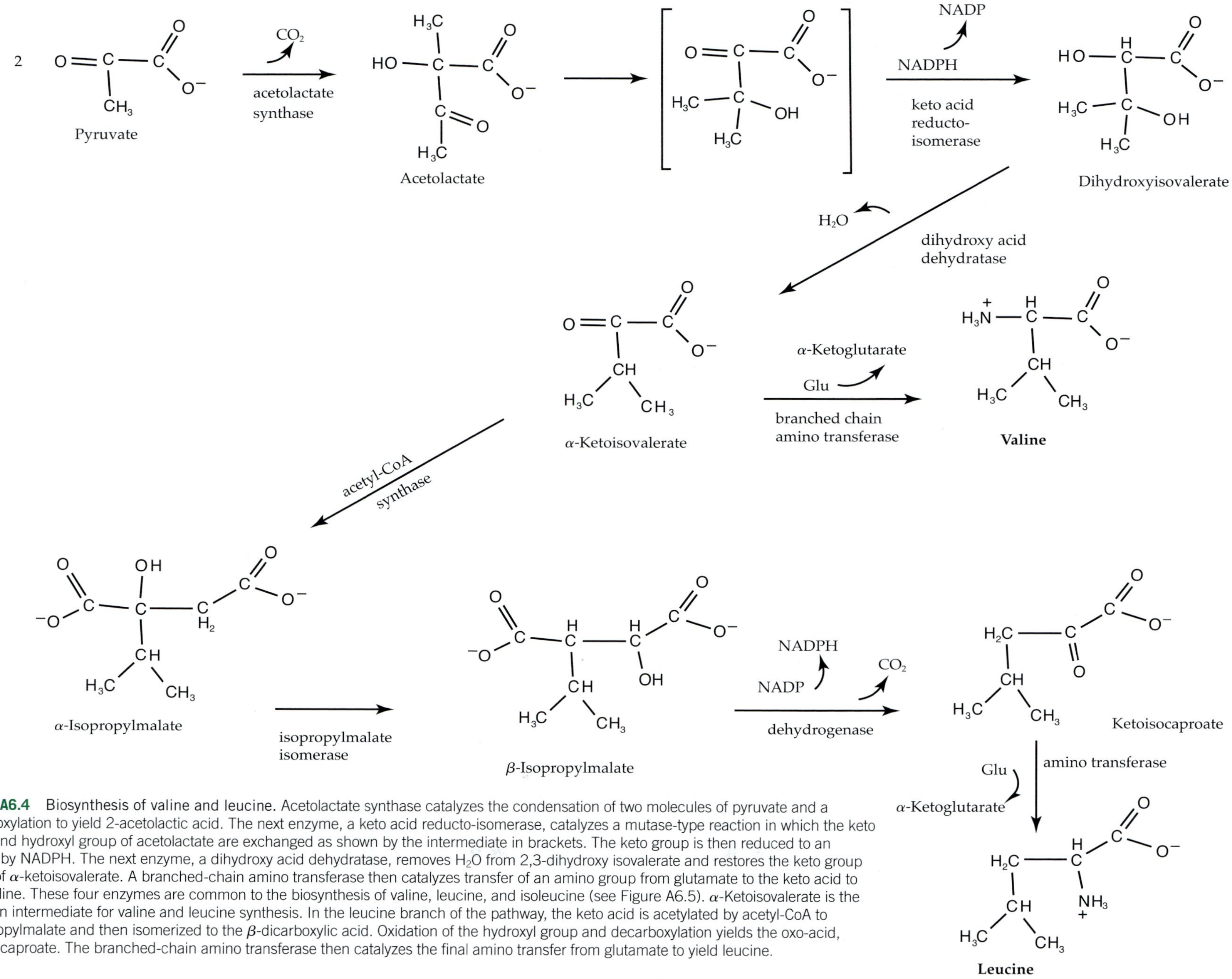

FIGURE A6.4 Biosynthesis of valine and leucine. Acetolactate synthase catalyzes the condensation of two molecules of pyruvate and a decarboxylation to yield 2-acetolactic acid. The next enzyme, a keto acid reducto-isomerase, catalyzes a mutase-type reaction in which the keto group and hydroxyl group of acetolactate are exchanged as shown by the intermediate in brackets. The keto group is then reduced to an alcohol by NADPH. The next enzyme, a dihydroxy acid dehydratase, removes H_2O from 2,3-dihydroxy isovalerate and restores the keto group at C-2 of α-ketoisovalerate. A branched-chain amino transferase then catalyzes transfer of an amino group from glutamate to the keto acid to yield valine. These four enzymes are common to the biosynthesis of valine, leucine, and isoleucine (see Figure A6.5). α-Ketoisovalerate is the common intermediate for valine and leucine synthesis. In the leucine branch of the pathway, the keto acid is acetylated by acetyl-CoA to α-isopropylmalate and then isomerized to the β-dicarboxylic acid. Oxidation of the hydroxyl group and decarboxylation yields the oxo-acid, keto-isocaproate. The branched-chain amino transferase then catalyzes the final amino transfer from glutamate to yield leucine.

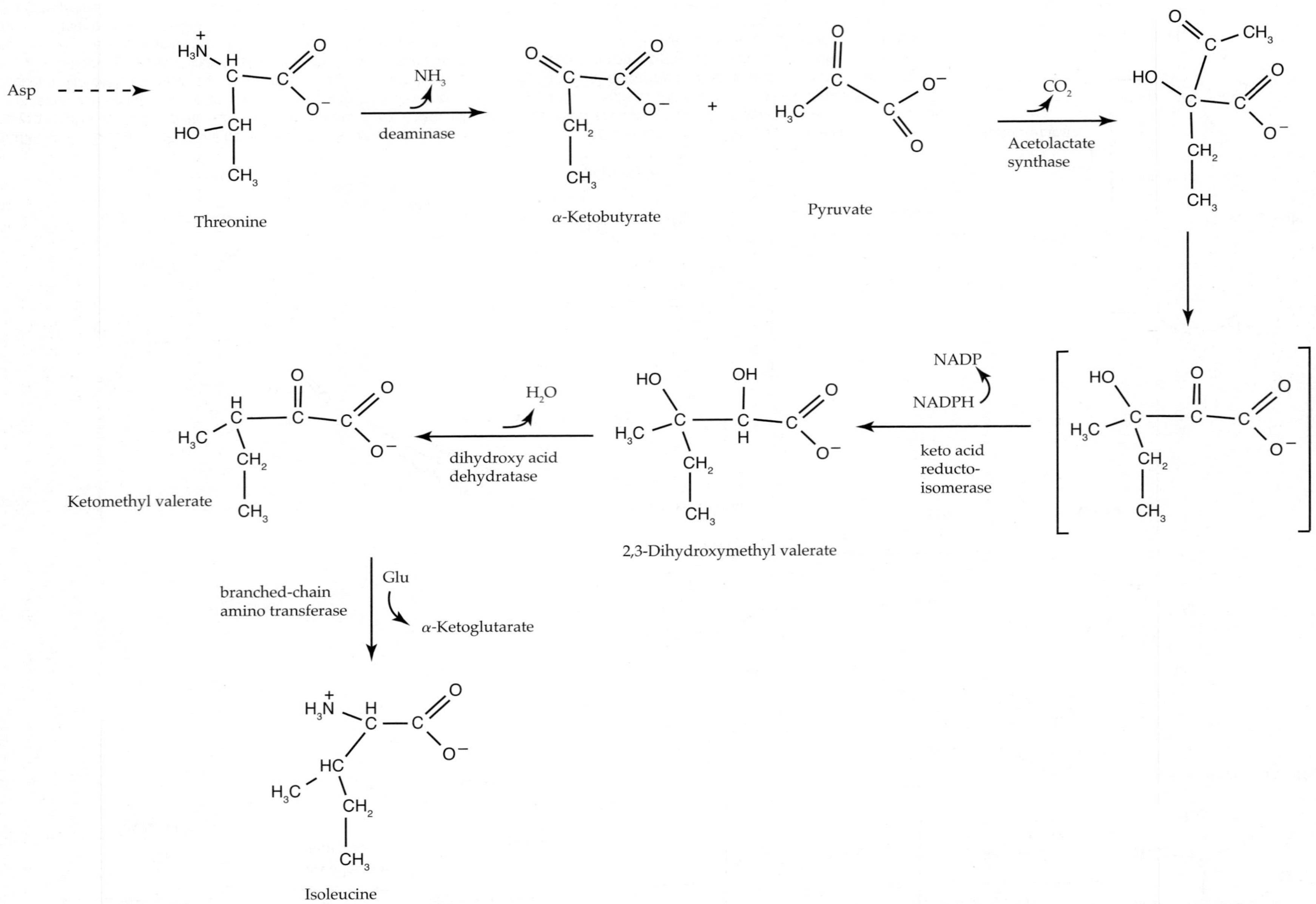

FIGURE A6.5 Biosynthesis of isoleucine. The initial substrate for isoleucine synthesis is the amino acid threonine, which is derived from aspartate. Threonine is deaminated to α-ketobutyrate, which then condenses with pyruvic acid. This condensation is catalyzed by the same acetolactate synthase that initiates valine and leucine synthesis (see Figure A6.4). Likewise, the next three enzymes and reactions in this pathway are also common to valine and leucine synthesis, although the substrates are different.

Sulfonyl ureas

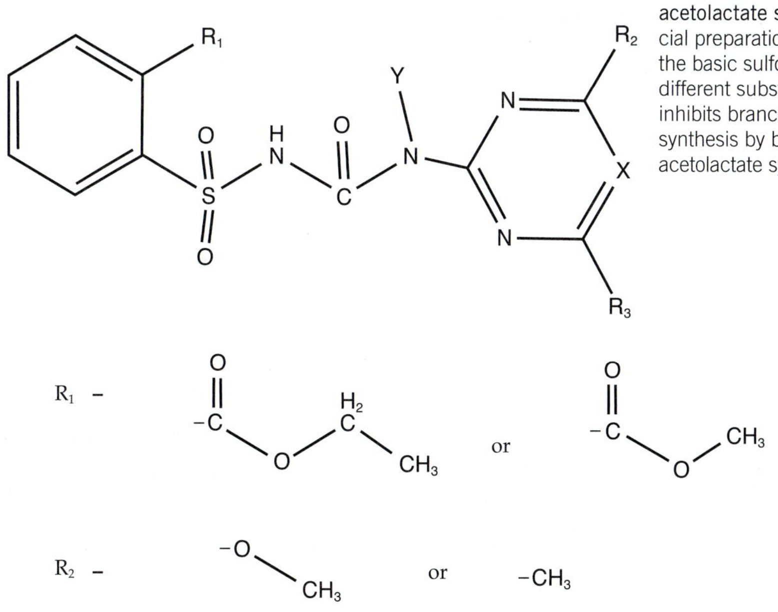

R3 - $-CH_3$ or $-Cl$

X - N or CH

Y - H or CH_3

Imidazolinone

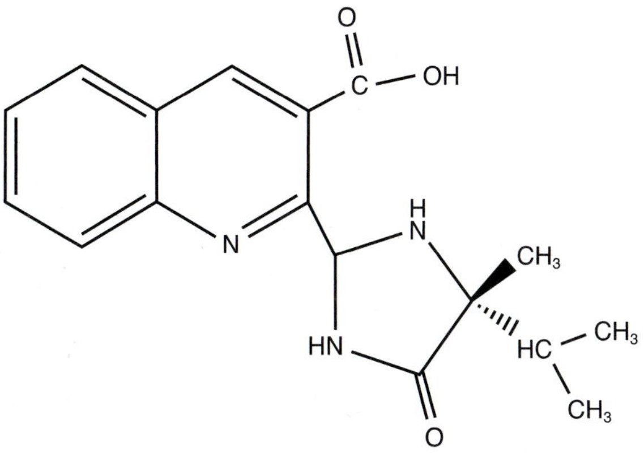

FIGURE A6.6 Herbicides that inhibit acetolactate synthase. Several commercial preparations are available that have the basic sulfonyl urea structure with different substituents. Imidazolinone also inhibits branched-chain amino acid synthesis by blocking the active site of acetolactate synthase.

A6.4 Amino Acids Derived from Oxaloacetate

Oxaloacetic acid is a product of the PEP carboxylase reaction and an intermediate in the citric acid cycle (Figure A6.1). It can be produced in several compartments in the cell, but synthesis of the aspartate family of amino acids from this precursor presumably occurs in plastids. Oxaloacetate is aminated by glutamate and a specific amino transferase to yield aspartate (**FIGURE A6.7**).

Aspartate is a substrate for a specific synthetase that produces asparagine. This Asn synthetase utilizes glutamine as the amino donor to form the amide and hydrolysis of ATP to AMP and pyrophosphate to provide energy. Unlike the transaminase reactions, the asparagine synthetase catalyzes an irreversible reaction. Asparagine is a major transport form of fixed nitrogen in some plants. It is also important as a nitrogen storage compound.

Aspartic acid is also phosphorylated and then reduced to aspartate semialdehyde, the common precursor for lysine, threonine, and methionine synthesis. Reduction of the semialdehyde yields homoserine, which is then phosphorylated to O-phospho homoserine, the immediate precursor of threonine. This same phosphorylated precursor is used in the synthesis of methionine (see Chapter 5). Threonine is also the starting compound for isoleucine synthesis. Threonine deamination yields α-ketobutyrate, which then condenses with pyruvate catalyzed by the acetolactate synthase enzyme that is common to the branched-chain amino acid synthetic pathway (Figure A6.5).

Lysine is the final member of the aspartate family. The aspartate semialdehyde intermediate is condensed with pyruvate and dehydrated to yield dihydrodipicolinic acid as shown in Figure A6.7. After a second reduction to tetrahydrodipicolinate, succinyl-CoA, an intermediate in the citric acid cycle, is added (**FIGURE A6.8**). A second amino group is transferred from glutamate, and the succinate moiety is removed to yield 2,6-diaminopimelic acid. A final decarboxylation reaction produces lysine.

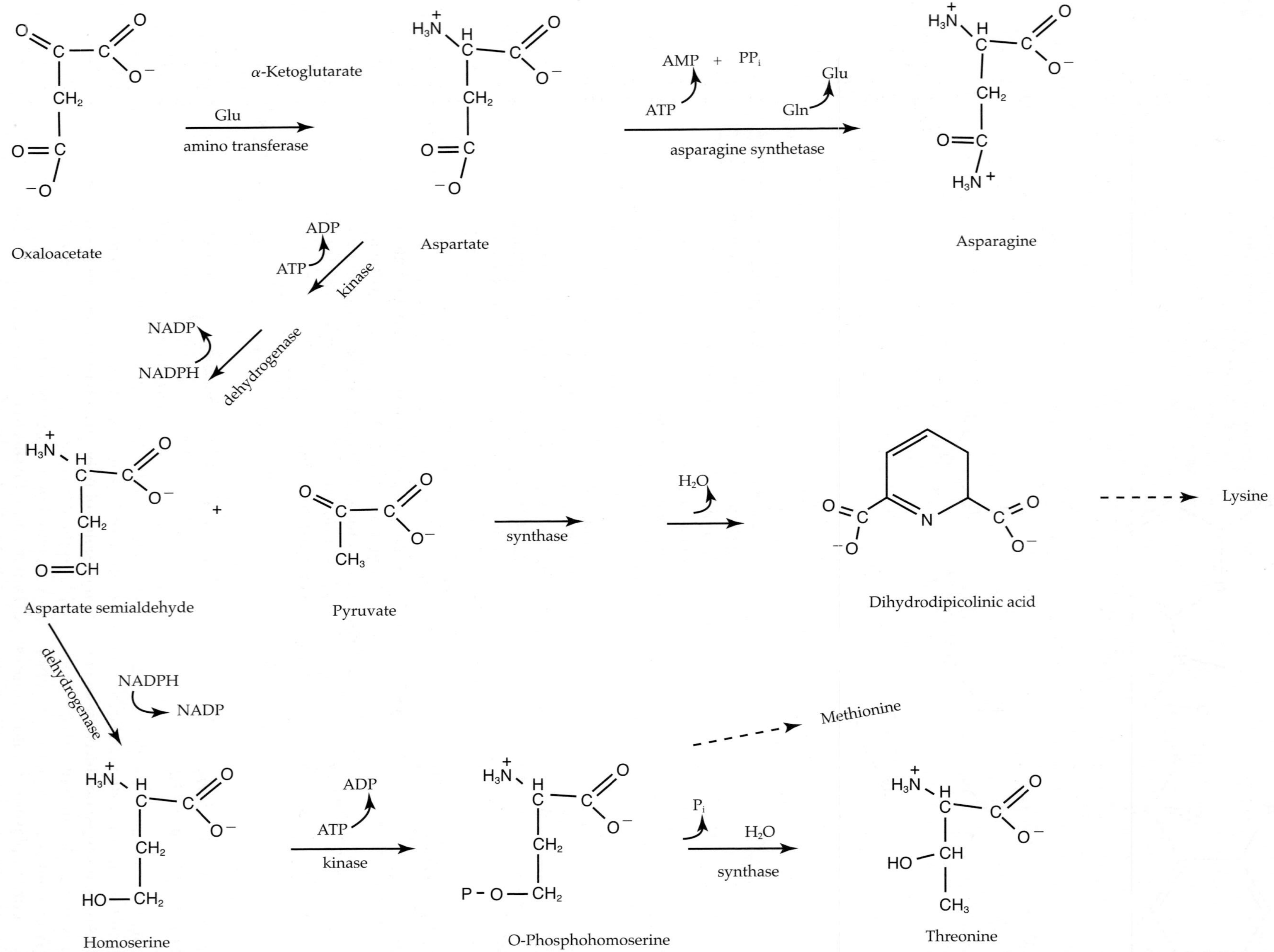

FIGURE A6.7 Synthesis of amino acids from oxaloacetate. Oxaloacetic acid is the precursor for the aspartate family of amino acids. Transamination of oxaloacetate yields aspartic acid. A specific synthetase utilizes glutamine as an amino donor to add an amino group to the γ-carbon of aspartate to produce the amide, asparagine. Aspartate is also phosphorylated and subsequently reduced to aspartate semialdehyde, the common intermediate in lysine, threonine, and methionine synthesis. Reduction of the aldehyde group and phosphorylation yield O-phosphohomoserine, the common intermediate in methionine and threonine synthesis. Removal of the phosphate group produces threonine. The steps from phosphohomoserine to methionine are described in Chapter 5. The aspartate semialdehyde can also be condensed with pyruvate and dehydrated to dihydrodipicolinic acid leading to lysine synthesis as shown in Figure A6.8.

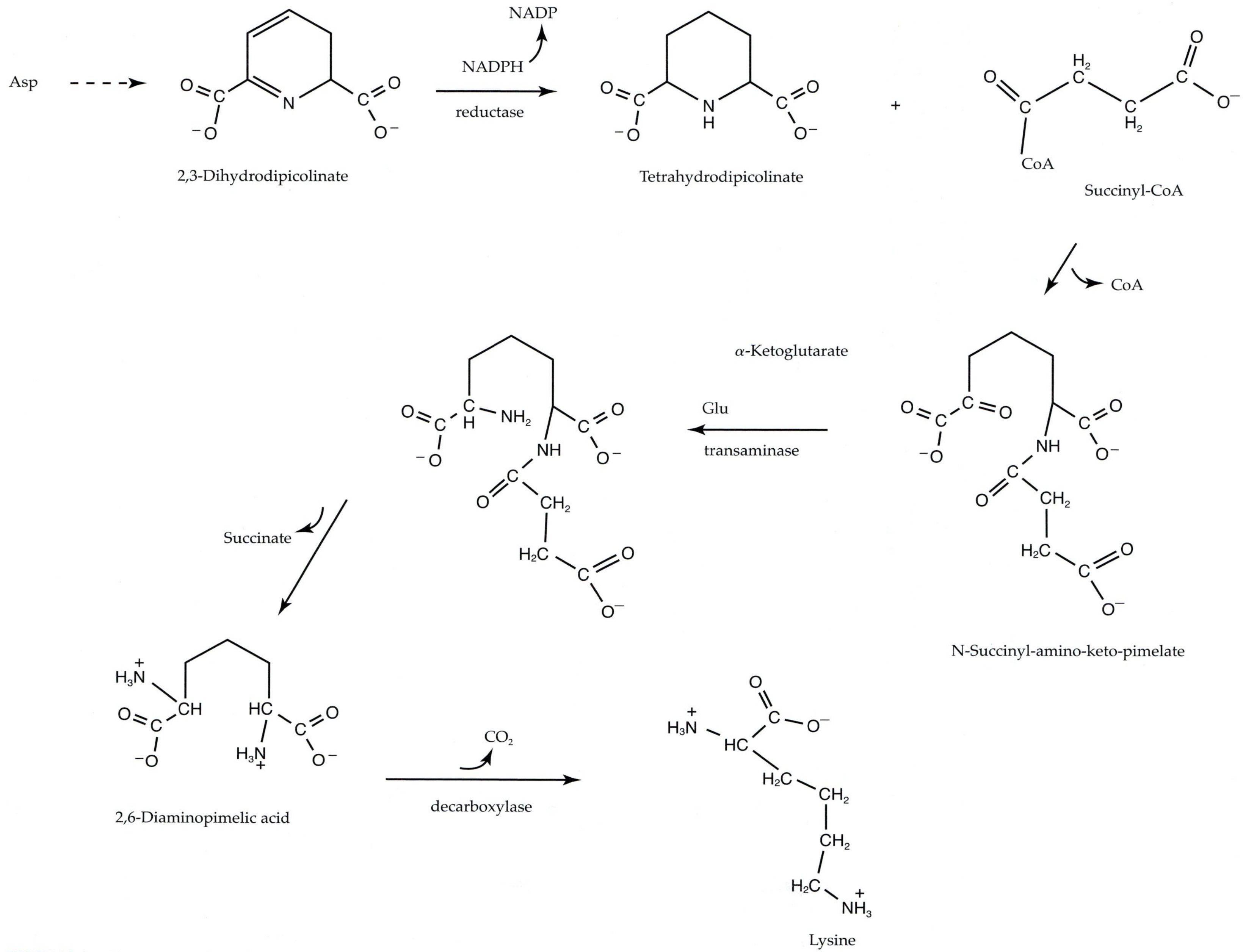

FIGURE A6.8 **Biosynthesis of lysine.** The synthesis of lysine is a branch of the aspartate pathway from the aspartate semialdehyde intermediate shown in Figure A6.7. Tetrahydrodipicolinic acid is N-succinylated to an 11-carbon intermediate and a second amino group is added by a specific transaminase. The succinate moiety is then removed to yield 2,3-diaminopimelic acid. Decarboxylation of the acid yields the basic amino acid, lysine.

A6.5 Amino Acids Derived from α-Ketoglutarate (Oxoglutarate)

The cyclic reactions of 2-oxoglutarate (α-ketoglutarate) and glutamine to form two equivalents of glutamate and the subsequent reformation of glutamine are central to NH_4^+ assimilation in plants and have been described in Chapter 5. **FIGURE A6.9** reviews this interaction, referred to as the glutamine synthetase-GOGAT cycle. The enzyme, glutamine-oxoglutarate-amino transferase (GOGAT), catalyzes the transfer of the δ-amino group of glutamine to α-ketoglutarate and requires a reducing agent. This reaction occurs in chloroplasts or plastids, and reduced ferredoxin or NADH is the most common reducing agent. Glutamine synthetase is the enzyme involved in ammonium assimilation. The reaction uses ATP hydrolysis for energy and is irreversible. Most plants also have a glutamate dehydrogenase that may contribute to the pool of glutamic acid:

$$\alpha\text{-ketoglutarate} + NH_4^+ + NADH \leftrightarrow \text{glutamate} + NAD + H_2O.$$

The above reaction is readily reversible and may be more important in degradation and turnover of glutamate rather than its synthesis.

As noted in the reactions described above, glutamate contributes its amino group in most of the amino acid syntheses. Glutamate also provides the carbon skeleton for the biosynthesis of proline and arginine. Proline synthesis is outlined in **FIGURE A6.10**. Glutamate is phosphorylated and subsequently reduced with loss of the phosphate to generate glutamate semialdehyde. This reaction sequence is similar to that in the initial steps of the synthesis of the aspartate family of amino acids. Glutamate semialdehyde is the common precursor for the synthesis of proline, arginine, and ornithine. The aldehyde spontaneously cyclizes to pyrroline 5′-carboxylic acid, which is then reduced to proline. Alternatively, the aldehyde is aminated a second time to form the amino acid ornithine, which is an intermediate in the synthesis of some alkaloids such as the tropanes (see Chapter 9).

Another amino group is added to ornithine by reaction with carbamyl phosphate (**FIGURE A6.11**) to yield citrulline. A third amino group is added to citrulline with aspartate as the amino donor producing arginine. In addition to its role in protein synthesis, this N-rich amino acid can also serve as a nitrogen storage compound in some plants.

The synthesis of histidine is unusual in that its carbon skeleton is not derived from the common oxo-acids used in the synthesis of other amino acids. The precursor is the sugar-phosphate, 5-phosphoribosyl-1-diphosphate, which is also the initial substrate for the synthesis of purine and pyrimidine bases. The two nitrogens in the imidazole ring are derived from the purine ring of adenine. The final steps of histidine synthesis from imidazole acetol-phosphate include amination of the keto group by glutamate, which is the contribution of glutamate to this pathway (**FIGURE A6.12**). The characterization of the steps and enzymes involved in the histidine biosynthesis pathway in plants are still under investigation.

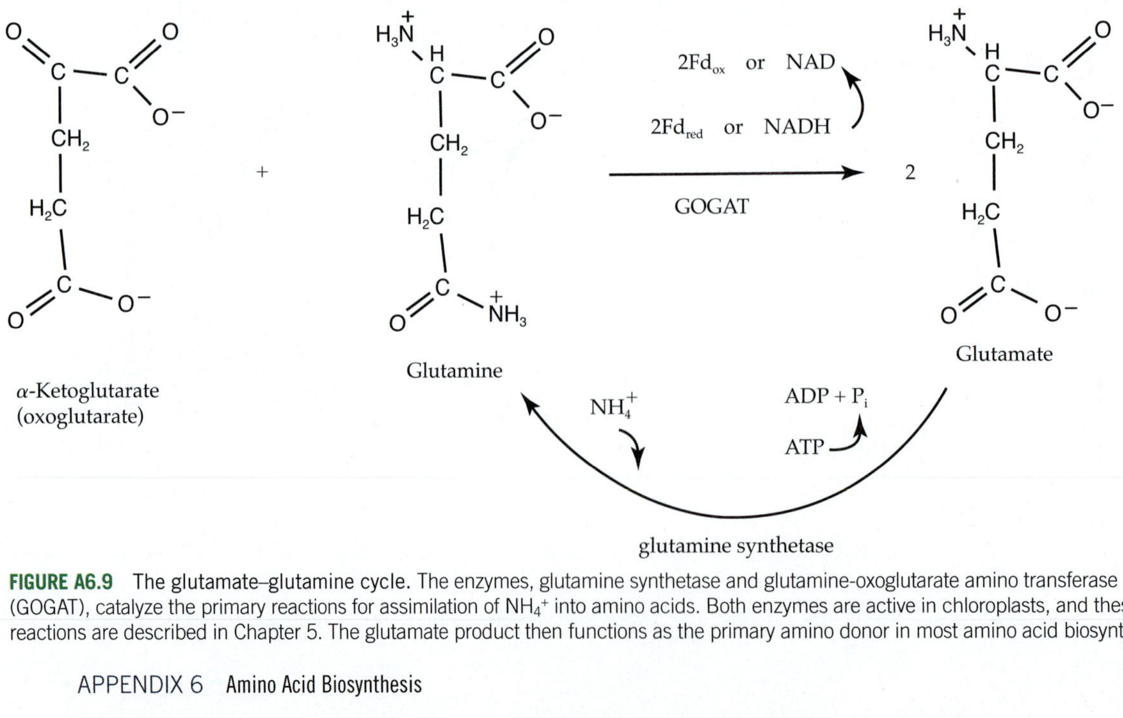

FIGURE A6.9 The glutamate–glutamine cycle. The enzymes, glutamine synthetase and glutamine-oxoglutarate amino transferase (GOGAT), catalyze the primary reactions for assimilation of NH_4^+ into amino acids. Both enzymes are active in chloroplasts, and these reactions are described in Chapter 5. The glutamate product then functions as the primary amino donor in most amino acid biosyntheses.

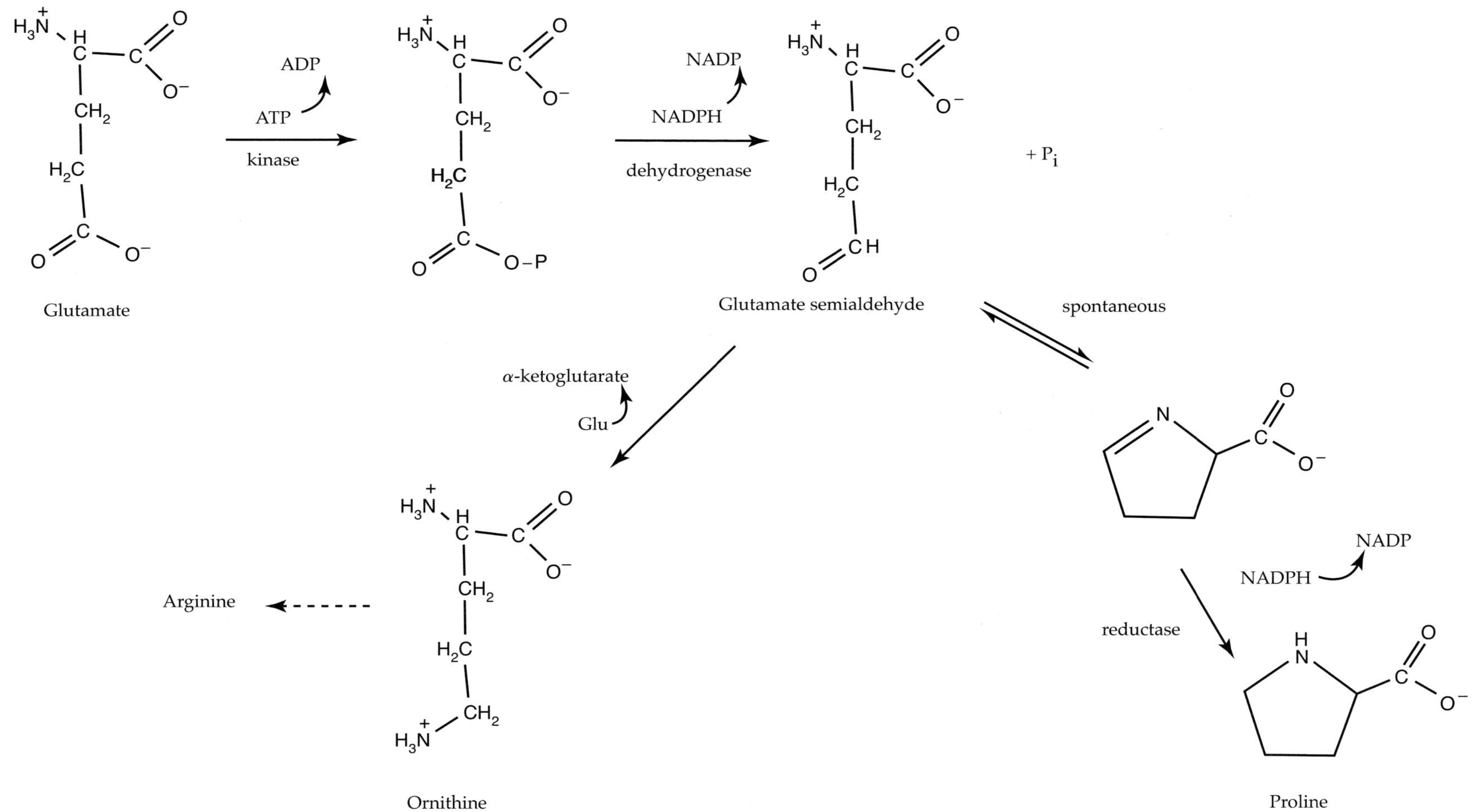

FIGURE A6.10 Biosynthesis of proline. Glutamate contributes both its carbon skeleton and amino group for the synthesis of proline. Glutamate is phosphorylated and subsequently reduced to produce a semialdehyde similar to the initial reactions for aspartate-derived amino acids shown in Figure A6.7. This semialdehyde is the common intermediate in the synthesis of proline and ornithine. Ornithine is a nonprotein amino acid that is important in the biosynthesis of alkaloids. It is also the precursor for arginine.

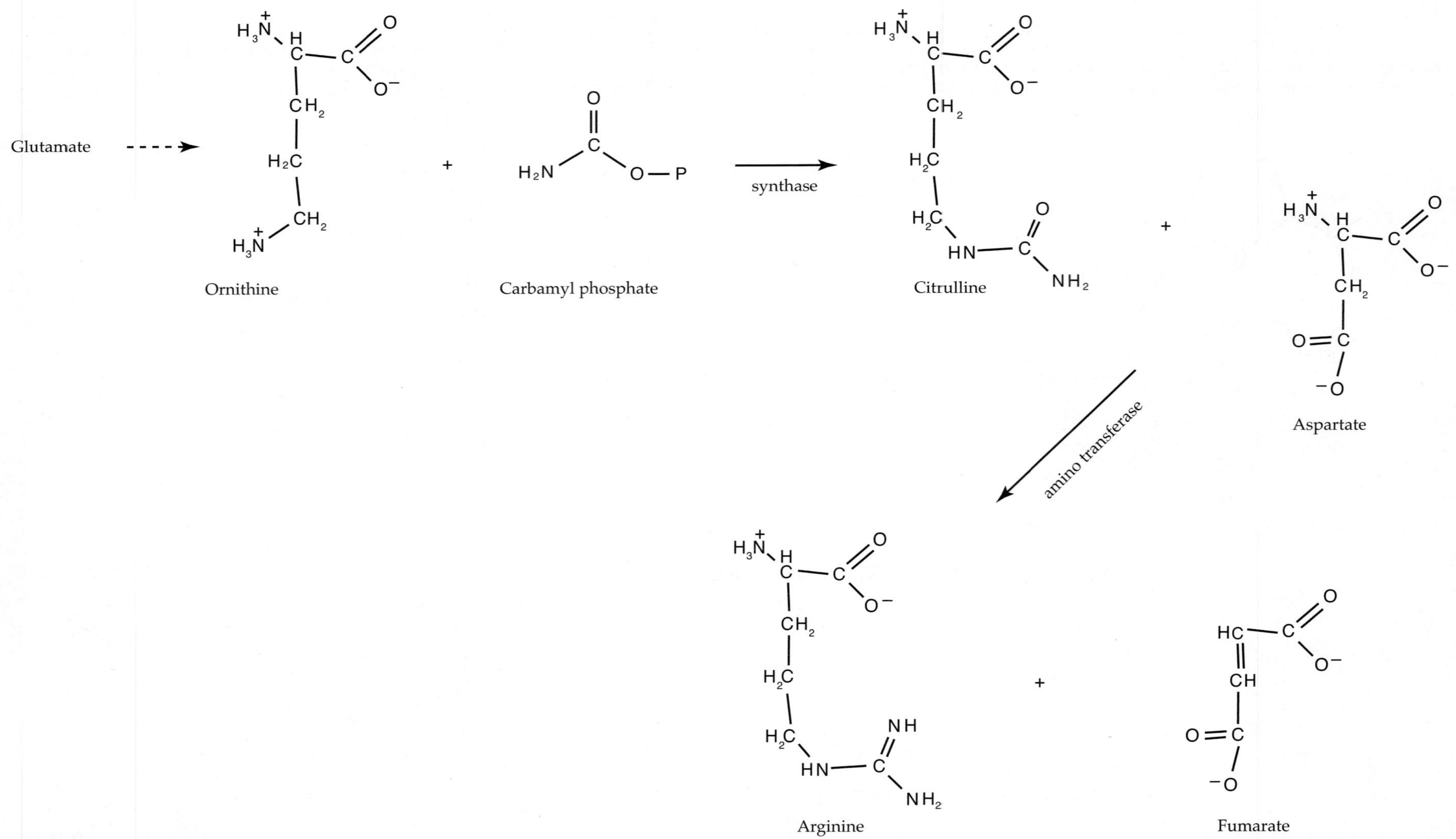

FIGURE A6.11 Biosynthesis of arginine. Continuing from the branch point shown in Figure A6.10, ornithine is condensed with carbamoyl phosphate to add a third amino group and produce citrulline. The final amino group in the arginine side chain is derived by a transamination from aspartate.

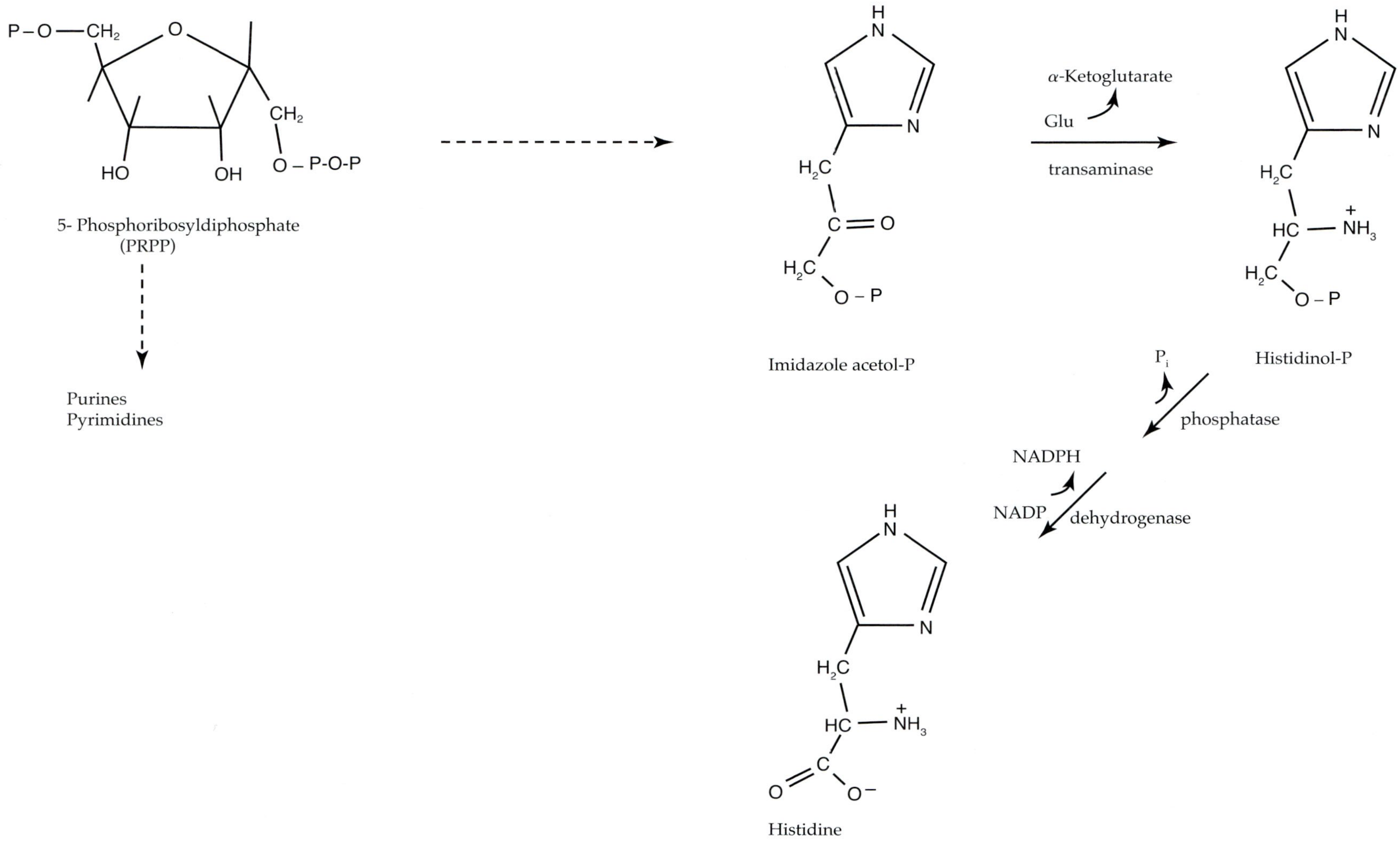

FIGURE A6.12 Partial pathway for histidine biosynthesis. A sugar-phosphate, 5-phospho-ribosyl-1- diphosphate (PRPP), is the common precursor for the synthesis of purines, pyrimidines, and histidine. This same sugar-phosphate is a substrate for a phosphoribosyl transferase reaction in the synthesis of tryptophan from anthranilic acid described in Chapter 8. The alpha-amino group of histidine is derived from glutamate. To date, the complete pathway in plants has not been characterized.

REFERENCES FOR FURTHER INFORMATION

Binder, S., Knill, T., & Schuster, J. (2007). Branched-chain amino acid metabolism in higher plants. *Physiol. Plant.* **129:**68–78.

Galili, S., Amir, R., & Galili, G. (2008). Genetic engineering of amino acid metabolism in plants. In: *Advances in plant biochemistry and molecular biology*, Vol. 1 (H.J. Bohnert, H.H. Nguyen, and N.G. Lewis, eds.). Amsterdam: Elsevier, pp. 51–80.

Singh, B.K. (ed.). (1999). *Plant amino acids: Biochemistry and biotechnology*. New York: Marcel Dekker.

APPENDIX 7

Light Properties and Analysis

Light is a form of electromagnetic radiation composed of subatomic particles called photons. These particles are pure energy and have no rest mass. They are best described as waves; thus, light has a dual particle-wave nature. Light waves have a discrete wavelength that is the distance between peaks of the wave function (**FIGURE A7.1**). Wavelength is indicated by the Greek letter λ (lambda). A light wave also has a defined frequency that is the number of peaks that pass an observer in a given time interval. The frequency is indicated by the Greek letter ν (nu) and is usually expressed as the number of peaks per second. Photons travel with a velocity or speed that depends on the medium. The speed of light is indicated by the letter c and is equal to 3×10^{10} cm s^{-1} in a vacuum. The velocity of any particular wavelength of light is given by $c = \lambda\nu$.

The light produced by the Sun consists of photons with a broad range of energies (**FIGURE A7.2**). The atmosphere of the Earth prevents radiation with wavelengths shorter than approximately 350 nm from reaching the surface of the planet. Visible light consists of wavelengths perceived by the human visual pigments in a narrow range from approximately 400 to 700 nm. This is the same range of wavelengths absorbed by plant pigments such as chlorophyll *a* (see Chapter 1, Figure 1.6) and is referred to as photosynthetically active radiation (PAR).

Early in the twentieth century Max Planck proposed his theory to describe radiant energy based on experimental observations of heated objects. Planck's theory states that energy (E) can only be transferred from one molecule to another in discrete packets called quanta. His theory is formalized in the equation

$$E = h\nu = hc/\lambda$$

where h is Planck's constant equal to 6.63×10^{-34} Joule s. For example, light with a wavelength of 650 nm would have the following energy:

$$E = (6.63 \times 10^{-34}\text{ J s})\,(3 \times 10^{8}\text{ m s}^{-1})/650 \times 10^{-9}\text{ m} = 3.06 \times 10^{-19}\text{ J}$$

In some cases it is more convenient to express light energy in terms of electron volts (eV). An electron volt is the energy acquired by an electron as it falls through a potential of 1 volt. One electron volt is equal to 1.60×10^{-19} J. The energy at 650 nm as calculated above would be 1.91 eV. (In older literature, energies are often reported in kcal. 1 kcal = 4.184 kJ.) Note that the

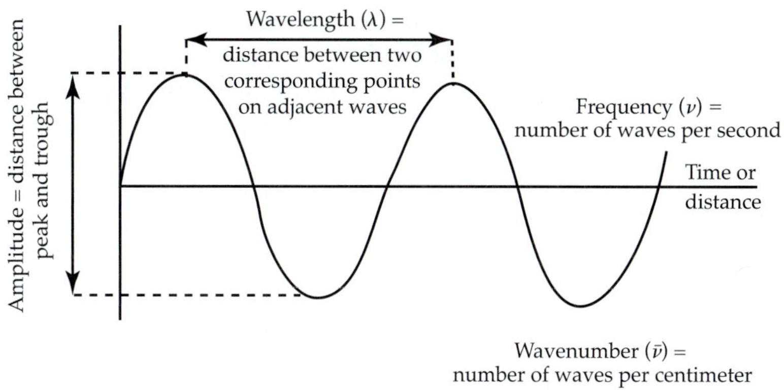

FIGURE A7.1 Light as a wave phenomenon. Light has the properties of a wave. The wavelength (λ) is the distance between two identical points on successive waves. The frequency is the number of waves (ν) that passes a given point per unit of time.

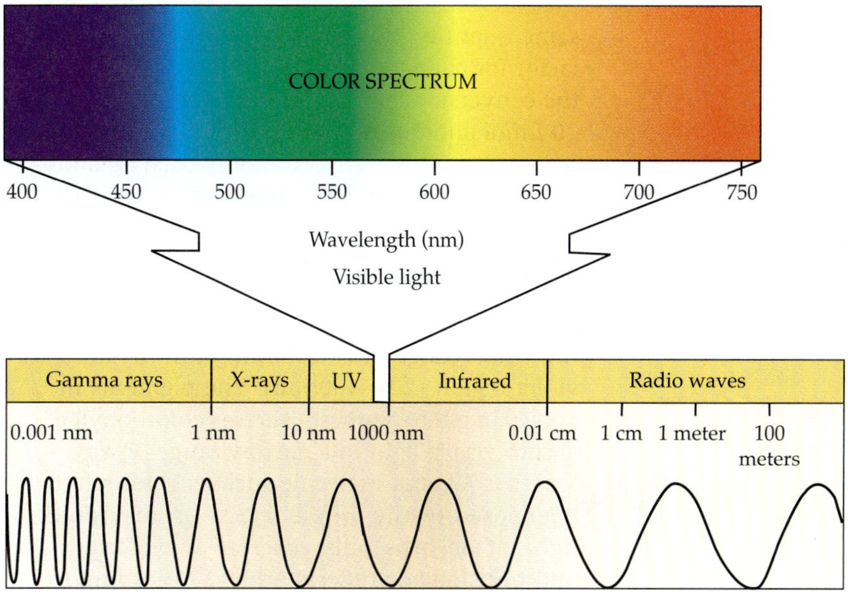

FIGURE A7.2 Spectrum of radiation from the Sun. Light, depicted as waves, is radiated from the Sun in a broad range of wavelengths. Very short energetic radiation with wavelengths from 0.001 to approximately 350 nm is prevented from striking the Earth's surface by its atmosphere. Visible light that drives oxygen-evolving photosynthesis lies in a narrow range from 400 to 700 nm.

energy in a photon is inversely proportional to its wavelength, that is, shorter wavelengths of light are more energetic than longer wavelengths.

Light intensity is a measure of the quantity of light that reaches a surface. Historically, light intensity was measured in foot candles, lumens, or lux. A lux meter measures the photon flux per unit surface at a unit distance from a standard candle. The units are lumens ft^{-2} or lux, which is lumens m^{-2}. Modern instruments measure the irradiance or radiant flux density, which is the amount of energy incident on a defined surface and is given in watts per square meter (W m^{-2}). In the process of photosynthesis, the relevant measure is the photon flux density, that is, the number of quanta that reach a defined surface per unit time. The photosynthetic photon flux density (PPFD) or photosynthetically active radiation (PAR) is the μmoles of photons per square meter per second:

$$\text{PPFD} = \text{PAR} = \mu\text{mol m}^{-2}\text{ s}^{-1}$$

(A mole of photons is Avogadro's number, 6.02×10^{23}.)

A spectroradiometer or PAR meter counts the number of photons delivered by a light source over the spectrum of wavelengths that are relevant to green plant photosynthesis. Full sunlight has a PAR value of approximately 2,000 μmol m^{-2} s^{-1}. This is equivalent to approximately 1,000 W m^{-2} or 100 klux. Because the different instruments are not measuring the same quantity, there is no precise conversion factor for changing lux into PAR. For sunlight, the conversion factor is approximately1 klux ~ 20 μmol photons m^{-2} s^{-1}. For cool-white fluorescent lamps that are commonly used in laboratory experiments, the conversion factor is 12.

A7.1 Spectrophotometry

Many biological molecules absorb light in either the visible range (400–700 nm) or in the ultraviolet region (200–400 nm). The pigments involved in photosynthesis, such as chlorophyll *a*, absorb visible light in a narrow range of wavelengths. When a molecule such as chlorophyll is exposed to the whole spectrum of visible light, it absorbs only discrete wavelengths and transmits or reflects others. According to quantum theory, a molecule can absorb only that amount of energy that excites one of its valence electrons to a higher-energy orbital. Theoretically, this should correspond to a single wavelength. Complex molecules in solution at room temperature, however, absorb over a range of wavelengths centered on an absorbance maximum. This is illustrated in Figure 1.6 in Chapter 1. Chlorophyll *a* absorbs energy in the blue region of the spectrum with a maximum at approximately 430 nm and in the red region with a maximum at approximately 660 nm. The absorption spectrum of a molecule can be determined by UV-visible spectroscopy. This is one of the most widely used techniques in biochemical laboratories due to readily available and relatively inexpensive instruments called spectrophotometers.

A typical UV-visible spectrophotometer consists of two lamps, a deuterium or hydrogen lamp, which emits light in the UV region from 200 to 350 nm, and a tungsten lamp, which generates light in the visible range from 350 to 800 nm. Discrete wavelengths are selected by a monochromator, which separates the range of photons generated by the lamp into discrete wavelengths. The monochromator can be a prism or, more likely, a diffraction grating. The sample is held in small vessels called cuvettes. These are made of either glass for determining the absorbance of samples in the visible range, or quartz, for scanning in the UV. The amount of light transmitted is detected by a photomultiplier or similar device. Many modern spectrophotometers use a diode array to detect the light that is transmitted through the sample. This array of detectors continuously monitors the entire UV-visible spectrum, and the resulting absorbance of a sample for any particular wavelength is generated by a computer. A diode-array spectrophotometer does not use a monochromator and is especially efficient for scanning an entire spectrum of wavelengths.

The spectrophotometer can provide data on several aspects of a molecule and its function. For example, the absorption spectrum of chlorophyll *a* is characteristic of the molecule and can be used to identify it in a mixture of pigments such as the chlorophyll *a* and carotenoids shown in Figure 1.6 in Chapter 1. Many other molecules are similarly identified by referring to published spectra. The redox state of metal-containing molecules is easily determined by measuring the spectral changes that occur during oxidation-reduction. The concentration of a compound can often be determined from its absorbance at a particular wavelength by using Beer's law:

$$A = \epsilon\, l\, c$$

where A is the absorbance measured at a particular wavelength, l is the path length of the light,

c is the concentration of absorbing compound, and ϵ is the extinction coefficient for the absorbing molecule at a defined wavelength.

In most instruments the sample is held in a cuvette with a 1-cm path length. The extinction coefficients of many compounds have been determined and can be found in various databases such as the Merck index. For example, the ϵ for NADPH at 340 nm is 6.29 mM. Beer's law assumes a linear relationship between the concentration of the compound and its absorbance. For many compounds at high concentrations, Beer's law may not apply and cannot be used to estimate the concentration of the compound.

Spectrophotometry is also very useful in determining steady-state enzyme kinetics as described in Appendix 3. If an enzymatic reaction yields a product with a distinct absorbance in the UV or visible range, its production can be monitored. For example, if reduced NAD or NADP is generated or consumed in a reaction, the concentration of this cofactor can be monitored and quantitated by changes in its absorbance at 340 nm. A modified spectrophotometer is often used to monitor the effluent from a chromatographic column and detect compounds during liquid and high-performance chromatography. Spectrophotometry can also yield information on the catalyst in a light-dependent reaction when this technique is used to generate an action spectrum such as that shown in Figure 1.7 in Chapter 1. In this example the rate of photosynthesis can be monitored at different wavelengths to produce an action spectrum that corresponds to the light-absorbing molecules that catalyze the overall process. In photosynthesis, these are chlorophyll and accessory pigments.

A7.2 Fluorescence Spectroscopy

As noted above, when a molecule absorbs a photon, an electron gains energy and is excited to a higher-energy orbital. This excited state is unstable and the electron quickly returns to its stable ground state by losing the absorbed energy, most often as heat. If the excited state has a relatively long lifetime of 10^{-6} to 10^{-9} seconds, however, the excitation energy can be lost as light (i.e., the molecule will fluoresce). Chlorophyll exhibits this property. The energy absorbed by chlorophyll at 430 nm is radiated much too quickly to exhibit fluorescence. When excited by light at 663 nm, however, chlorophyll will fluoresce at approximately 668 nm, depending on the solvent (see Figure 1.20 in Chapter 1). The light emitted by an excited molecule is generally at a longer wavelength than the absorbed light. This is called the Stokes shift. This shift is due to a small amount of the absorbed energy radiated as heat. There is no certain way to predict if a molecule will exhibit fluorescence, but several of the pigments in photosynthesis have this property. Determination of fluorescence parameters is a very useful technique in monitoring photosynthesis and investigating the rates of electron transfer.

An instrument for monitoring fluorescence is basically similar to a spectrophotometer. It consists of a light source and a monochromater, which selects the excitation wavelength for the molecule that fluoresces. The sample is placed in a cuvette with four transparent sides rather than two as in a spectrophotometer. The exciting light is directed into the sample, and light that is emitted by the compound is reflected into a second monochromater and then into a detector. The output is a fluorescence emission at a particular wavelength or an entire emission spectrum can be generated. Portable fluorimeters are commercially available that are adapted to measure the light emission from leaves. These instruments can be used in the field to monitor photosynthesis without damage to the plants.

The actual fluorescence output from chloroplasts *in vivo* is rather low. Most of the absorbed light energy is transferred to other molecules and fluorescence is quenched (see Chapter 1). Fluorescence, however, can be a useful tool to monitor the efficiency of energy transfer in isolated chloroplasts. When the organelles are maintained in the dark, the reaction centers, mainly of photosystem (PS) II, will be reset to the reduced or receptive state. If the chloroplasts are then illuminated with an intense flash of light, they will exhibit a fluorescence induction curve as shown in **FIGURE A7.3**. The emission will rise steeply in a few nanoseconds and then increase more slowly over milliseconds to a peak of maximum fluorescence. The emission will then decay. The initial very rapid rise (I phase) is mainly due to the emission of light from the light-harvesting chlorophylls of PSII reaction centers that have been excited. The slower rate of fluorescence increase (T) is due to transfer of excitation to the reaction centers in PSII followed by charge transfer to plastoquinones. As all the quinones become reduced, the fluorescence emission from the reaction center chlorophylls reaches its maximum. At this point there is no charge transfer. The following slow decline (S phase) reflects the further oxidation of the quinones as energy is

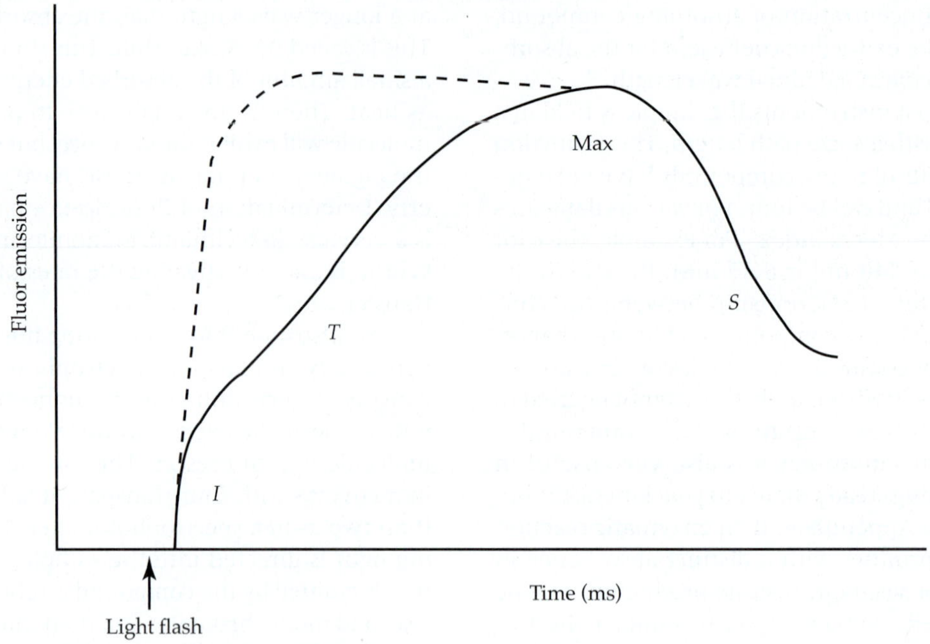

FIGURE A7.3 Fluorescence emission in isolated chloroplasts. After storage in the dark, the reaction centers in isolated organelles are reduced. After exposure to a brief, intense flash of light, a sharp rise in fluorescence emission (*I*) is seen due to energy loss from excited antenna chlorophylls in PSII. This is followed by a slower rise in emission (*T*) due to energy transfer to and loss from reaction centers, which then reduce quinones. The decline in fluorescence emission (*S*) is due to energy transfer to nonfluorescent components of the photosynthetic electron transport chain. The dashed line shows emission after addition of a PSII inhibitor such as atrazine. There is an immediate loss of energy from the reaction centers because reduction of quinones is blocked by the inhibitor.

transferred to other electron carriers in the thylakoid membrane. Addition of a chemical inhibitor of PSII electron transfer, such as the herbicide atrazine, causes a sharp rise in the fluorescence emission (dashed line in Figure A7.3) as it blocks energy transfer to the quinones and results in faster energy loss from the reaction center. This fluorescence induction in isolated chloroplasts is a valuable technique for determining the rates of energy transfer among molecules and the overall rate of energy transfer in photosynthesis. It also provides information about the site of action of inhibitors and is often used to study processes such as nonphotochemical quenching (see Chapter 1).

REFERENCES FOR FURTHER INFORMATION

Gordon, D.B. (2005). Spectroscopic techniques. In: *Principles and techniques of biochemistry and molecular biology* (6th ed.). Cambridge, UK: Cambridge University Press, pp. 551–579.

Hall, D.O., & Rao, K.K. (1999). *Photosynthesis* (6th ed.). Cambridge, UK: Cambridge University Press, pp. 4–14, 58–65, 174–177.

Glossary

Abscisic acid (ABA) A plant hormone derived from carotenoids. ABA is produced in response to water stress and regulates closing of the stomates. It also maintains dormancy in seeds and buds.

Abscission Loss of leaves, flowers, or fruits by growth or differentiation of cells in the stem that develop thick secondary walls and undergo apoptosis.

Acetylcholine A neurotransmitter that is released by an excited neuron. Acetylcholine diffuses into the synaptic cleft and triggers an action potential in a second neuron by activating a cation channel in its cell membrane.

Action spectrum A graph showing the rate of a biochemical reaction at varying wavelengths of light.

Adrenaline (epinephrine) Catecholamine produced by the adrenal medulla in animals. It is a bronchial dilator, constricts small blood vessels, and stimulates heartbeat. It is the "flight or fight" hormone produced in response to extreme stress.

Adrenergic receptors Integral membrane proteins in heart and blood vessel cells and neurons that bind catecholamines such as adrenaline.

Aldol condensation The reaction of an aldehyde or ketone with a base to generate an enolate anion. The anion attacks a second aldehyde or ketone to form a C–C bond. An aldol cleavage is the reversal of the condensation resulting in two carbonyl compounds.

Allelopathy The inhibition of growth or germination of one species by chemical compounds released by another, competing species.

Allosteric regulation Modulation of enzyme activity by small molecules that bind to sites (allosteric) on proteins other than the active site. Regulation usually involves alteration of enzyme subunit associations.

Alternate oxidase (AOX) An integral membrane protein in the mitochondrial electron transport system that transfers electrons from reduced quinones to oxygen without pumping protons into the intermembrane space.

Amino acid ammonia lyase An enzyme that catalyzes the removal of an amino group from an amino acid resulting in a derivative with a double bond and releasing ammonia.

Amphipathic An adjective used to describe a molecule having both hydrophilic and hydrophobic properties.

Analgesic A drug that relieves pain.

Angiosperm Vascular plants with flowers and seeds that develop inside a fruit (ovary).

Angstrom unit (Å) 10^{-10} cm.

Anther The upper part of the stamen (male reproductive organ in angiosperms) that includes pollen.

Antioxidant A molecule with a low redox potential that can donate electrons or hydrogen atoms; a reducing agent.

Apicoplast An organelle found in some protozoans that is apparently a nonphotosynthetic descendant of a chloroplast. The apicoplast retains some metabolic pathways present in chloroplasts, such as the methyl-erythritol phosphate (MEP) pathway for terpene synthesis.

Apocarotenoid A small molecule produced by dioxygenase-catalyzed cleavage of a carotenoid. Apocarotenoids are often volatile and impart characteristic odors to plants.

Apoenzyme The protein component of the enzyme lacking the cofactors or ligands; usually catalytically inactive.

Apoplast The area outside the cell membranes in a plant tissue, including the cell walls.

Apoptosis Programmed cell death. It is an orderly process in which the cytoplasm and organelles undergo compaction; the endoplasmic reticulum fuses with the cell membrane and forms internal membrane-bound vesicles. Cell function ceases without cell lysis.

Autotroph An organism that synthesizes all its organic constituents from carbon dioxide using light or chemical bond energy.

Auxin A plant growth hormone derived from the tryptophan biosynthetic pathway.

Beta barrel A protein tertiary structure composed of beta strands and loops in a circular orientation (barrel). The active site of β-barrel enzymes is usually located in the center of the barrel between the loops connecting the strands.

Betalain A nitrogenous plant pigment derived from tyrosine.

Biotin (vitamin H) A sulfur-containing, water soluble coenzyme that generally functions in carboxylation reactions.

Brassinolides A group of growth-promoting hormones that are derived from sterols.

Bryophytes Nonvascular plants including mosses, hornworts, and liverworts.

Bundle sheath cells Cells surrounding the vascular tissues.

Carpel Female reproductive organ of a flower that contains the ovules.

Chaperone, chaperonin A protein that binds and guides formation of tertiary and quaternary structure of nascent proteins.

Chlorotic An adjective used to describe a plant tissue or a cell that lacks chlorophyll.

Chromatophore An invagination of the cell membrane found in some photosynthetic bacteria. The chromatophore contains the pigments and proteins required for bacterial photosynthesis.

Co-evolution The selection for genetic traits in one organism in response to corresponding traits in a second organism that has evolved to out-compete it. Co-evolutionary traits may be favorable to both organisms.

Companion cell A type of cell in the phloem tissue of vascular plants that is physically connected to the anucleate sieve tube cells.

Competitive inhibitor A molecule that binds to an enzyme active site and competes with the natural substrate.

Condensed tannins Linear polymers of anthocyanidins. They are insoluble in water and generally impart a purple to brown color to cell walls such as those in grape peels.

Cotyledon A leaf-like structure in the embryos of seed plants that stores carbohydrates and lipids needed to nourish a growing embryo.

Cryptochrome A protein containing a pigment, usually a flavin, that allows plants to sense blue light.

Cultivar A variety of a plant species that has been produced by human intervention.

Cyanobacterium A gram-negative bacterium that contains chlorophyll *a* and performs oxygen-evolving photosynthesis.

Cytokinins A group of plant growth hormones consisting of purine or pyrimidine bases modified by prenyl (isoprenoid) groups.

Dehydratase An enzyme that removes water from a carbon chain usually resulting in the introduction of a double bond.

Dehydrogenase A redox enzyme that utilizes a dinucleotide cofactor. It generally transfers two electrons and two protons.

Desaturase A mono-oxygenase that uses molecular oxygen to introduce a double bond into a carbon chain; usually an integral membrane protein containing iron.

Detergent A water-soluble, organic compound that forms emulsions in water. It acts as a cleansing agent and can denature proteins.

Dicot A member of a flowering plant family having two cotyledons in the embryo.

Dictyosome Alternate term for the Golgi apparatus, the stacked system of membranes that packages molecules into Golgi vesicles for export from the cell.

Dioxygenase An enzyme that uses molecular oxygen and introduces both oxygen atoms into the substrates to form hydroxylated products.

Diterpenoid Molecule that is derived from the 20-carbon isoprenoid precursor, geranyl-geranyl diphosphate (GGPP).

Dithiothreitol (DTT), dithioerythritol (DTE, Cleland's reagent) Isomeric forms of 1,4-dimercapto-2,3-butanediol. It is used *in vitro* as a reducing agent to reduce disulfide bonds in proteins.

Diurnal cycle A periodic activity that repeats on a 24-hour (daily) basis.

Domain A section of a protein's primary structure that can fold independently of the rest of the protein into a stable, compact tertiary structure. Domains often have specific functions.

EDTA Ethylene diamine tetraacetic acid, a chelating agent that binds divalent cations such as Mg^{2+}.

Elicitor Usually a small molecule that stimulates defense responses in a plant such as the synthesis of phytoalexins.

Elongase An enzyme that adds malonyl-CoA to the carbon chain of a fatty acid to produce a fatty acid with two additional carbons.

Endodermis Sheath of cells surrounding the vascular tissues of roots.

Endorphins A group of peptides that are produced mostly in the pituitary gland. Endorphins bind to opioid receptors in animals to relieve pain.

Endosperm Polyploid tissue surrounding a plant embryo. It is usually rich in carbohydrates and lipids and provides nutrients to the developing embryo.

Endothermic (endergonic) Refers to a reaction that requires the input of energy to proceed.

Epidermis Outer layer of cells covering the plant tissues. It is usually unpigmented.

Epimerase An enzyme that changes the configuration of the substrate at a chiral center.

Essential oils Volatile monoterpenes produced in some plants that provide a characteristic odor or "essence."

Etiolated Adjective used to describe a plant part that is deficient in chlorophyll.

Excited state Absorption of a quantum of energy that boosts an electron into a higher orbital in a molecule.

Exothermic (exergonic) Refers to a reaction in which energy is released.

Flavin A cofactor consisting of an isoalloxazine ring linked to ribose.

Flavoprotein A protein containing a flavin nucleotide cofactor, either FMN (flavin mononucleotide) or FAD (flavin adenine dinucleotide). Flavoproteins are generally redox enzymes that transfer two electrons and two protons through the cofactor.

Flippase An enzyme that can change the orientation of phospholipids in a lipid bilayer.

Fluorescence Emission of light energy by an electron in an excited state as it returns to its stable ground state.

Folic acid Pteroyl glutamic acid. A cofactor that carries one-carbon units.

Fosmidomycin 3-[formyl(hydroxyl)amino] propylphosphonic acid. An antibiotic obtained from *Streptomyces lavendulae*. It inhibits the enzyme 1-deoxyxylulose-5-phosphate reductoisomerase in the methyl erythritol phosphate (MEP) pathway.

Fusion protein Fusion of two genes that result in two proteins that are expressed as one large polypeptide.

Galactolipids Diacylglycerols that contain one or more galactose moieties in the sn3 position. They are generally found in thylakoid membranes.

Gallotannins Water-soluble complexes of phenolic compounds with a sugar.

Gibberellins A family of plant hormones derived from geranyl geranyl diphosphate (GGPP). They stimulate growth, cell expansion, and seed germination.

Glucanase An enzyme that catalyzes the hydrolysis of polysaccharides containing glucose.

Gluconeogenesis The synthesis of glucose from smaller molecules such as pyruvate by a reversal of glycolysis. In plants, glucose can be made from fatty acids by gluconeogenesis.

Glycosidase An enzyme that catalyzes the hydrolysis of glycosidic bonds in polysaccharides or cleaves sugars that are attached to other molecules by O or N links.

Glyoxysome Small vesicular organelle with a single membrane. It contains enzymes for fatty acid catabolism.

Glyphosate N-phosphonomethyl glycine. It is an inhibitor of enolpyruvyl shikimate-phosphate synthase, an enzyme in the shikimic acid pathway and is a potent herbicide.

Gossypol An aromatic terpenoid formed by the condensation of two sesquiterpenes. It is the toxic compound found in cotton seeds.

Granum (pl. **grana**) Thylakoid membranes that associate in stacks.

Green leaf volatiles Short chain fatty acids and derivatives generated by the action of a lipoxygenase on membrane lipids. They produce the odor of cut leaves and grass.

Gymnosperm Vascular plant group that produces uncovered (naked) seeds.

Hemiacetal The product of a reaction of an alcohol (OH group) with an aldehyde.

Hemicellulose A short cellulose chain (β-1,4-linked glucose) with various sugar branches. They form cross-links among cellulose fibrils in cell walls.

Hemiketal The product of a reaction of an alcohol with a ketone.

Heterocyst Specialized cells in filamentous cyanobacteria that form an anaerobic compartment for nitrogen fixation.

Heterotroph An organism that obtains metabolic energy by ingesting organic (fixed) forms of carbon.

Holoenzyme Catalytically competent enzyme consisting of the protein and cofactors.

Homology Implies descent from a common ancestor.

Hydride ion H^-; a hydrogen atom with one additional electron.

Hydroxylase A type of monooxygenase that uses molecular oxygen to add an OH group to a substrate.

Hypersensitive reaction Reaction of a plant cell to microbial invasion leading to cell death.

Inclusion bodies Intracellular granules composed of denatured, insoluble proteins.

Inducible genes Genes that are normally repressed. Expression can be initiated by specific molecules called inducers.

Innate immunity Constitutive cellular defenses that are effective against a broad range of pathogens.

Invasive species A nonnative organism that becomes the dominant species in its new, invaded environment often to the detriment of the native species.

Invertase An enzyme that catalyzes the hydrolysis of the glucose-$\alpha_1 \rightarrow \beta_2$-fructose bond of sucrose to yield a mixture of glucose and fructose referred to as invert sugar.

Isozymes (isoforms) Enzymes that catalyze similar reactions and have similar structures but are coded for by different genes.

Jasmonates Various derivatives of jasmonic acid produced by the oxidation of unsaturated fatty acids. They act as plant defense hormones.

Kaurene (ent-kaurene) A tetracyclic diterpene derivative that is synthesized in chloroplasts and exported to the cytoplasm as a precursor for gibberellins.

Kinase An enzyme that transfers a phosphate group from ATP to a receptor molecule, or the reverse reaction.

Laccase A copper-containing phenol oxidase that uses molecular oxygen to generate an hydroxylated product and can generate free radicals.

Lactone A cyclic ester.

Latex The milky substance produced in the vascular tissues of some plants, mainly the Euphorbiaceae. Latex contains polyisoprenes and other natural products such as alkaloids and cardenolides. It deters insect herbivory by both mechanical means and chemical toxicity.

LD_{50} Lethal dose of a toxin for 50% of the test population.

Lectin A plant protein that recognizes and binds to specific carbohydrates on the outer surface of cell membranes.

Lipase An enzyme that catalyzes the hydrolysis of an ester bond between glycerol and a fatty acid.

Lipoxygenase (LOX) A dioxygenase that adds molecular oxygen to a fatty acid double bond to produce a hydroperoxy product.

Lodging The process of stem collapse in an herbaceous plant in a high wind.

Mesophyll Tissue that is found inside leaves and stems.

Microsomes Small vesicles formed *in vitro* from isolated endoplasmic reticulum.

Microtubules Thin, tubular intracellular structures formed from the protein, tubulin. They form part of the cytoskeleton.

Monocots Plants that have a single seed leaf, usually the grasses and related plants.

Monolignol An aromatic alcohol derived from cinnamic acid. Major monolignols are coumaroyl, coniferyl, and synapyl alcohols that are the precursors for lignin synthesis.

Monoterpene A 10-carbon isoprenoid derived from geranyl-diphosphate (GPP).

Morphinans Alkaloids found in the opium poppy (*Papaver somniferum*) that are derived from tyrosine. They are mainly morphine and codeine.

Multi-drug resistance pumps Integral membrane proteins that transport drugs and other organic compounds out of the cell or into vacuoles.

Mutase An enzyme that catalyzes an isomerization in which a chemical group is moved from one carbon to another.

Mycorrhizal fungi Soil fungi that form symbiotic relationships with plant roots. They aid in plant uptake of inorganic nutrients such as phosphate.

Myrosinase An enzyme that cleaves an S-glycosidic bond.

Nanometer 10^{-9} meters.

Noncompetitive inhibitor An enzyme inhibitor that binds to a site other than the active site. It decreases the V_{max} but does not change the K_m.

Nonvascular plants Plants lacking tissues specialized for water or sugar transport, such as bryophytes and mosses.

Oleosins Amphipathic proteins that form a membranous sheath around lipid bodies.

Operon A series of adjacent genes under control of a single promoter.

Oxoglutarate-dependent dioxygenase (ODD) A dioxygenase that uses molecular oxygen to add one oxygen atom to a substrate and transfers the other oxygen to oxoglutarate (α-keto glutarate), which is subsequently decarboxylated to yield succinate.

Oxygenase An enzyme that catalyzes the incorporation of molecular oxygen into a substrate.

Oxylipins Small molecules derived from the oxidation of fatty acids that often serve as plant signaling molecules.

Ozone Triatomic oxygen—a very strong oxidizing agent formed from molecular oxygen by high-energy sources such as UV radiation and electrical discharges.

Pathogen Refers to any microorganisms that can cause disease in a plant.

PDB (Protein Data Bank) Archive of information and data on the structure of proteins, nucleic acids, and complex assemblies. It is managed by Rutgers University and the University of California, San Diego (www.rcsb.org).

Periplasmic space Area in a bacterial cell between the cell membrane and the cell wall.

Peroxisome (microbody) An organelle with a single membrane. It contains enzymes that catalyze oxidative reactions and generate hydrogen peroxide. Peroxisomes contain catalase and often other enzymes to degrade peroxide and oxygen radicals.

Phenotype Observable characteristics of an organism such as leaf shape or type of fruit.

Pheophytin Chlorophyll molecule lacking the Mg ligand.

Phloem Cells and associated fibers that transport sugars and other organic compounds from leaves to other tissues in vascular plants.

Phosphatidic acid Lipid consisting of two fatty acyl moieties and one phosphate esterified to glycerol.

Phosphopantetheine Phosphate ester of N-(pantothenyl)-β-amino ethane thiol. It is part of coenzyme A and ligand in acyl carrier protein.

Phycobiliprotein Water-soluble pigment molecule found in cyanobacteria and some eukaryotic algae consisting of a protein with a covalently attached linear tetrapyrrole.

Phytoalexin A small molecule with cytotoxic properties. Its synthesis in plants is induced in response to pathogen attack.

Phytochrome Photochromic protein with a linear tetrapyrrole ligand. It absorbs light in the red region of the spectrum and is converted to a far-red absorbing form. It mediates growth processes in plants in response to changing light quality and quantity.

Phytoestrogen An isoflavone commonly found in legumes. It binds to mammalian estrogen receptors.

Phytol 16-carbon side chain of the chlorophyll tetrapyrrole ring. It is derived from diterpenes (2,6,10,14-tetramethylhexadec-14-en-16-ol).

pI The iosoelectric point. It is the pH at which the number of negative charges on a protein or similar macromolecule equals the number of positive charges yielding a net charge of zero.

Pistil The female reproductive organ in flowering plants consisting of the carpel or in some plants, of several fused carpels.

pKa The pH at which a weak acid is half ionized.

Polyketide A complex organic compound with several keto groups. It is synthesized from repeating units such as acetate.

Prenyl group An isoprenoid derivative consisting of five carbons or a multiple of five.

Proteasome The multisubunit protein complex in the cell cytoplasm that degrades proteins.

Protein bodies (protein storage vesicles) Single-membrane vesicles containing stored proteins such as globulins. They are generally found in seeds and cotyledons.

Pyrethrins Monoterpenes produced by chrysanthemums and a few other Asteraceae. They act on the nervous system of insects to cause paralysis.

Pyridoxal phosphate Coenzyme form of vitamin B_6 found in transaminases and decarboxylases. The cofactor often forms a Schiff base with amines.

Q-cycle Activity of a membrane-bound cytochrome complex in electron transport systems. It oxidizes quinones and transfers electrons to other carriers and transfers protons to generate a compartmentalized proton gradient.

Quantum yield In photosynthesis, the number of light quanta required to yield one oxygen molecule.

Quinone An aromatic compound with two oxygen substituents located either ortho or para to each other. It transfers two electrons and two protons.

Red drop In photosynthesis, the abrupt decline in quantum yield at wavelengths greater than 680 nm.

Reducing sugars Sugars or small oligosaccharides with a free OH group on an anomeric carbon that are characterized by their ability to reduce Cu^{2+} to Cu^{1+}.

Resolution Ability to distinguish two objects that are adjacent. In x-ray crystallography, it is the ability to distinguish separate atoms in a chemical structure.

Resonance structures (forms) Two or more representations of a molecule that differ only in the location of nonbonding electron pairs.

Resveratrol A stilbene derived from coumaroyl-CoA and malonyl-CoA *via* a polyketide synthase. It is a phytoalexin commonly found in grape peels.

Rhizome Portion of a plant stem specialized for starch storage. It may facilitate vegetative reproduction in some plants.

Saponin A glycoside of a steroid or triterpenoid that forms foamy, soap-like emulsions in water.

Schiff base Imine formed by addition of a primary amine to a carbonyl group followed by elimination of water.

Self-incompatibility The inability of pollen to initiate fertilization of eggs in the same flower.

Semi-synthesis The process of isolating natural products or their precursors from plants and subsequently modifying them by chemical procedures.

Senescence The process of deterioration of a plant organ usually involving cell death.

Sesquiterpene A 30-carbon isoprenoid derivative derived from farnesyl diphosphate (FPP).

Sieve element A type of cell found in phloem tissue of vascular plants. Several sieve elements form long tubular structures to conduct soluble sugars and other organic molecules.

Silique A fruit that splits open on two sides when dry to release seeds.

Squalene A 30-carbon isoprenoid precursor of sterols.

Stamen Male reproductive organ of a flower.

Stereoisomers Compounds with the same chemical formula having one or more chiral centers with different three-dimensional orientations of groups around the chiral centers.

Sterol Diterpene-derived alcohol with a 3β-OH and a 17β-aliphatic side chain. Sterols are an essential component of membranes that stabilize lipid bilayers and are precursors of more complex steroid hormones.

Stilbene A phenolic compound containing two benzene rings connected by a two-carbon (ethane) bridge.

Stomate (pl. **stomata**) The space between guard cells in the epidermis of vascular plants, it allows gases and water to pass into or out of the mesophyll tissues of leaves.

Style A structure that connects the top portion of the stigma to the ovary in the female reproductive organ of angiosperms.

Synergid A haploid cell that is part of the female gametophyte in angiosperms.

Tetrahydrofolic acid (THF) 6-methyl pterin bonded to para-amino benzoic acid and a variable number of glutamate residues. It is the reduced form of folic acid that is active in C-1 metabolism.

Tetrapyrrole A molecule that consists of four pyrrole rings connected by carbon bridges. A pyrrole is a five-carbon ring containing nitrogen.

Tetraterpene A 40-carbon isoprenoid derivative.

Thiamine (vitamin B_1) Generally found as a diphosphate, it is a cofactor in enzymes that transfer carbonyl groups.

Thiokinase (acyl-CoA synthetase) An enzyme that catalyzes the formation of a fatty acid-coenzyme A thioester coupled to ATP hydrolysis.

Thiolase An enzyme that acts in β-oxidation of fatty acids. It catalyzes the cleavage of the β-keto bond in β-ketoacyl-CoA to yield acetyl-CoA, and adds coenzyme A to the remainder of the molecule to yield a fatty acyl-CoA derivative that is two carbons shorter than the substrate.

Thiolate anion A negatively charged sulfur moiety, it is common in cysteine residues in enzyme-active sites.

Tonoplast The membrane surrounding a vacuole.

Topoisomerase An enzyme that catalyzes relaxation of supercoiled DNA.

Transcription factor A protein that binds to DNA upstream of the promoter region of a gene and facilitates the binding of RNA polymerase.

Transit (signal) sequence A protein sequence, 13–36 residues, usually at the N-terminal of a protein that directs the nascent protein to a specific intracellular site.

Trehalose A disaccharide consisting of glucose-$\alpha_1 \rightarrow \alpha_1$-glucose.

Triazine A 6-membered aromatic ring containing three nitrogen atoms.

Trichome Small microscopic hair that usually grows as an extension on an epidermal cell.

Triterpene A 30-carbon isoprenoid derivative.

Tuber Enlarged plant root specialized for storage.

Vacuole Organelle in plant cells that is bounded by a single membrane; usually functions in storage of organic molecules and sequestration of xenobiotics.

Vascular plants Plants that have specialized tissues to conduct water and organic nutrients between tissues.

Vasoconstrictor A chemical compound that causes small blood vessels to contract and restrict blood flow to extremities or bronchi.

Wax An ester formed from a long-chain fatty alcohol and a fatty acid. Wax is found on the external surface of leaves and stems.

Xenobiotic Any compound that is not naturally found in an organism.

Xylem Cells and fibers in vascular plants that are specialized for water transport from roots to leaves.

Index

B

C

M

Q

R

S

Photo Credits

Chapter 1
Opener, Courtesy of Florence Gleason, University of Minnesota; **1.1,** © Biophoto Associates/Photo Researchers, Inc.; **1.2 (top),** © E.H. Newcomb & W.P. Wergin/Biological Photo Service. All rights reserved.

Chapter 2
Opener, Courtesy of Florence Gleason, University of Minnesota; **2.10a,** Courtesy of Riyadh Muhaidat, Rowan Sage, and Tammy Sage, University of Toronto; **2.11,** © Dr. Kari Lounatmaa/Photo Researchers, Inc.

Chapter 3
Opener, Courtesy of Florence Gleason, University of Minnesota; **3.7,** © Biophoto Associates/Photo Researchers, Inc.; **3.16,** © Biology Pics/Photo Researchers, Inc.

Chapter 4
Opener, Courtesy of James Mauseth, University of Texas, Austin; **4.2,** Courtesy of Yuzou Sano, Hokkaido University; **4-3,** Courtesy of Dr. Thomas Giddings, University of Colorado; **4.5,** Courtesy of Hilton Mollenhauer.

Chapter 5
Opener, © Medical-on-Line/Alamy Images; **5.8,** Courtesy of James Mauseth, University of Texas, Austin; **5.11,** Courtesy of Harold Evans.

Chapter 6
Opener, © vadim kozlovsky/ShutterStock, Inc.; **6.12,** © Eye of Science/Photo Researchers, Inc.; **6.13,** *Reproduced from*: Graham, I. A., "Seed Storage Oil Mobilization," *Annu. Rev. Plant Biology* 59 (2008): 115–142, Fig. 1.

Chapter 7
Opener, Courtesy of Florence Gleason, University of Minnesota; **7.8,** Courtesy of Florence Gleason, University of Minnesota; **7.9,** © Eric Litton/ShutterStock, Inc.; **7.19a,** © Dean Evangelista/ShutterStock, Inc.; **7.19b,** © Terry Reimink/ShutterStock, Inc.

Chapter 8
Opener, © SunnyS/ShutterStock, Inc.; **8.30,** © Thomas Eisner, Cornell University.

Chapter 9
Opener, © Vladimir Pomortzeff/ShutterStock, Inc.

Chapter 10
Opener, © Piotr Skubisz/ShutterStock, Inc.; **10.3,** © B. E. S. Gunning, reproduced from "Plant Cell Biology on DVD, Information for Students and a Resource for Teachers," Educational DVD, Springer-Verlag (2009). Courtesy of Brian Gunning.